Advances in Thermodynamics

Advances in Thermodynamics

Volume 1: **C_{7+} Fraction Characterization**

Volume 2: **Fluctuation Theory of Mixtures**

Volume 3: **Nonequilibrium Theory and Extremum Principles**

Volume 4: **Finite-Time Thermodynamics and Thermoeconomics**

Volume 5: **Analytic Thermodynamics: Origins, Methods, Limits, and Validity**

Volume 6: **Flow, Diffusion, and Rate Processes**

Volume 7: **Extended Thermodynamic Systems**

ADVANCES IN THERMODYNAMICS

VOLUME 7

Extended Thermodynamic Systems

EDITED BY

Stanislaw Sieniutycz and Peter Salamon

Taylor & Francis

New York • Philadelphia • Washington, D.C. • London

USA	Publishing Office:	Taylor & Francis New York Inc. 79 Madison Ave., New York, NY 10016-7892
	Sales Office:	Taylor & Francis Inc. 1900 Frost Road, Bristol, PA 19007-1598
UK		Taylor & Francis Ltd. 4 John St., London WC1N 2ET

EXTENDED THERMODYNAMIC SYSTEMS

1 2 3 4 5 6 7 8 9 0   B R B R   9 8 7 6 5 4 3 2 1

Cover design by Berg Design.
A CIP catalog record for this book is available from the British Library.

Library of Congress Cataloging-in-Publication Data

Extended thermodynamic systems / edited by Stanislaw Sieniutycz and
Peter Salamon.
p. cm. — (Advances in thermodynamics ; v. 7)
Includes bibliographical references and index.

1. Nonequilibrium thermodynamics. 2. Gibbs' equation.
I. Sieniutycz, Stanislaw. II. Salamon, Peter, date.
III. Series.
QC318.I7E87 1992
536'.7—dc20 91-34720
ISBN 0-8448-1693-0 CIP

Contents

Contributors

J. Casas-Vázquez
Departament de Física, Universitat Autònoma de Barcelona, Bellaterra, Spain

R. Drouot
Laboratoire de Modélisation en Mécanique, Université Pierre et Marie Curie, Paris, France

B. C. Eu
Department of Chemistry, McGill University, Montreal, Canada

L. García-Colín
Departamento de Física, Universidad Autónoma Metropolitana-Iztapalapa, México

D. Jou
Departament de Física, Universitat Autònoma de Barcelona, Bellaterra, Spain

G. M. Kremer
Departamento de Física, Universidade Federal do Paraná, Curitiba, Brasil

G. Lebon
Institute of Physics, Liège University, Belgium

J. E. Llebot
Departament de Física, Universitat Autònoma de Barcelona, Bellaterra, Spain

G. A. Maugin
Laboratoire de Modélisation en Mécanique, Université Pierre et Marie Curie, Paris, France

A. Morro
DIBE, Universita Degli Studi di Genova, Italy

I. Müller
Physikalische Ingenieurwissenschaft, Technical University Berlin, Germany

R. E. Nettleton
Department of Physics, University of Witwatersrand, Johannesburg, S. Africa

D. Pavon
Departament de Física, Universitat Autònoma de Barcelona, Bellaterra, Spain

C. Pérez-García
Departmento de Física y Matématica Aplicada, Universidad de Navarra, Pamplona, Spain

P. Salamon
Department of Mathematical Sciences, San Diego State University, CA

S. Sieniutycz
Institute of Chemical Engineering, Warsaw Technical University, Poland

Introduction

This multiauthored volume sketches the applications of nonequilibrium thermodynamics to complex systems. These are characterized by an involved form of the Gibbs equation and include systems such as solutions of macromolecules, magnetic hysteresis bodies, viscoelastic fluids, polarizable media, fluids under stresses and in the presence of essential nonstationarities and high temperature gradients. As a rule, the so-called internal variables (i.e. variables which are not directly controllable through boundary conditions) and/or dissipative fluxes are essential in the thermodynamic description of such systems.

Extended irreversible thermodynamics, (EIT), the recent theory which incorporates dissipative fluxes (heat, diffusion, viscous tensor and viscous pressure) into the formalism of the Gibbs equation, is comprehensively reviewed in the context of compatibility with kinetic theory, ideal and real gases, equations of state, etc. Applications to rheology, fluids subject to shear, shear induced phase changes, etc., are discussed. Generalized reciprocity relations are shown to be useful in calculating transport coefficients in liquids and in the estimation of the nonlinear effects in heat flow and diffusion. Relations between EIT and fluctuation theory are discussed.

Magnetohydrodynamical systems and relativistic thermal inertia systems are discussed. The latter leads to EIT effects even in the limiting case of the low velocities. A novel theory of thermodynamic transformations is developed in which the extended Legendre transformations play the double role characteristic of mechanics and thermodynamics.

Thermodynamics of Complex Systems

S. Sieniutycz
Institute of Chemical Engineering
Warsaw Technical University
00-645 Warsaw, 1 Warynskiego St., Poland

P. Salamon
Department of Mathematics
San Diego State University, San Diego, CA 92182, USA

ABSTRACT
The significance of systems with Gibbs equations which incorporate additional internal variables is studied. Such systems include viscoelastic and viscoplastic materials, and fluids with high diffusive fluxes of heat, mass, and momentum. Rational and extended thermodynamic theories are seen as apropriate tools for the analysis of such systems. The relative merits of these approaches is discussed and the papers in this volume are briefly reviewed.

1.INTRODUCTION
This volume sketches the applications of nonequilibrium thermodynamics to complex systems such as solutions of macromolecules, magnetic hysteresis bodies, viscoelastic fluids, polarizable media, etc. These systems require some extra variables (so called internal variables) to be introduced into the fundamental equation of Gibbs. Extended irreversible thermodynamic (EIT) systems are also analysed. In this approach, the dissipative fluxes are the independent variables treated on an equal footing with the classical variables of thermostatics.

The concept of an internal state variable, i.e. a variable not controllable through external conditions, originated from the analysis of some models in rheology (Lemaitre and Chaboche 1988; Maugin 1990) and electromagnetic bodies (Meixner 1961, Kluitenberg 1977, 1981). The goal was a theoretical framework for studying relaxation processes such as a shock wave passing through an electrically polarizable substance. However, relaxation phenomena are not the only ones for which the concept of internal variables is useful. Theories of plasticity can also make good use of this concept (Maugin, 1990). Furthermore, it can be shown that some nontrivial analogies exist between plastic behavior and magnetic hysteresis; hence the modern developments in plasticity have implications for magnetic hysteresis systems.

It follows that hysteresis processes can be described by using the concept of the internal variables of state provided one takes into account some residual fields at vanishing load. However, one should be careful to distinguish between the relaxational recovery of equilibrium and hysteresis processes. In relaxation, time scale is an essential issue; in hysteresis the relatively slow response exhibited by a ferromagnetic sample is practically rate independent. Similarly, the low-temperature plastic effects are practically independent of the rate of strain. A detailed comparison of these effects can be found in Maugin's paper in this volume.

The continuum modelling of polarizable macromolecules in solutions is the next example showing the role of internal variables (Maugin and Drouot 1983, Maugin 1987). Their use allows one to construct unknown constitutive equations and evolution equations describing mechanical, electrical, and chemical behavior. Also, the effects that do not contribute to dissipation can be singled out from the general formalism. Flow induced and electrically induced phase transitions can be studied as well as diffusion, migration and mechanochemical effects.

Rheology of viscoelastic media (Leonov 1976, Lebon at al 1988), isotropic-nematic transitions in polymer solutions (Sluckin 1981, Giesekus 1986) and stress induced diffusion (Tirrel and Malone 1977) are other applications of internal variables. Reviews, describing the requisite nonequilibrium thermodynamic framework are available (Bampi and Morro 1984, Morro 1985, Maugin 1987). An application of the internal variable concept using rational thermodynamics to describe viscoelasticity is given in Morro' paper in this volume.

Historically, the starting point for extended thermodynamic theories was the paradox of infinite propagation rate, encountered in the classical phenomenological equations describing the transport of mass, energy, and momentum. Attempts to overcome the paradox resulted in equations of change in terms of (damped) wave equations rather than classical parabolic equations. This has been demonstrated in a number of articles. The topics ranged from theoretical developments (based on phenomenological thermodynamics or nonequilibrium statistical mechanics), to applications in fluid mechanics, chemical engineering and aerosol science. It has been recognized that taking into account local nonequilibrium and/or inertial effects (resulting from the finite mass of the diffusing particles) the constitutive equations become non-Fourier, non-Fick and non-Newtonian.

A tutorial review of the historical development of the elimination of the paradox was presented by one of us in Vol. 6 of this series (Sieniutycz 1991). The concept of relaxation time was introduced and its simplicity of evaluation for heat, mass and momentum have been shown for ideal gases with a single propagation speed. For nonideal systems, a simplest extended theory of coupled heat and mass transfer has been constructed based on nonequilibrium entropy.

In the approach discussed, the entropy source analysis yields a formula for the relaxation matrix, τ. This matrix describes the coupled transfer and corresponds to the well known values of relaxation times for pure heat transfer and for isothermal diffusion. Various simple forms of the wave equations for coupled heat and mass transfer appear. Simple extensions of the second differential of entropy and of excess entropy production can be constructed. They allow one to prove the stability of the coupled heat and mass transfer equations by the second method of Liapounov. Dissipative variational formulations can be found, leading to approximate solutions by a direct variational method. Applications of the hyperbolic equations to the description of short-time effects and high frequency behavior can be studied.

Some related ideas dealing with a variational theory of hyperbolic equations of heat can be found in Vujanovic's paper (also Vol. 6) where a class of variational principles of Hamilton's type for transient heat transfer is analyzed and valuable approximate solutions are found by direct variational methods.

Flux relaxation effects are particularly strong in the diffusion of aerosols containing particles of large mass and in polymers characterized by large molecular weight. Some experiments (Domanski 1978) show that nonclassical models of heat transfer and diffusion of mass and energy are more appropriate in the case of highly nonstationary transients, for example, laser pulses of high frequency acting on solid surfaces. Acoustic absorption at high frequency is another example (Bubnov 1976; Carrasi and Morro 1973). The phenomenological thermodynamics of diffusion links some relaxation effects with the absence of local thermodynamic equilibrium in systems with finite kinetic energy of diffusion (Truesdel 1969, Sieniutycz 1983, 1990).

2. CLASSICAL AND EXTENDED THERMAL SYSTEMS

The classical theory of nonequilibrium thermodynamics (de Groot and Mazur 1984) assumes local equilibrium. Consequently, the Gibbs equation for the entropy differential is used in the well known classical form containing the derivatives of the internal energy, volume (or density) and concentrations of chemical constituents. Application of the conservation laws for energy, mass and momentum yields the well known expression for entropy balance:

$$\rho\frac{ds}{dt} = -\ \nabla.\frac{\mathbf{q}}{T} + \mathbf{q}.\nabla T^{-1} - T^{-1}\Pi:\nabla\mathbf{u} \qquad (1)$$

The right hand side of this equation includes a source term and a divergence term which is related to the conductive entropy flow. The equation postulates phenomenological forms for these terms. The postulated forms are in agreement with the second law. We have given a brief account of the problems which arise in the classical theory in our introductory paper to Vol. 3 of this series. We do not repeat the criticisms of here; for more details the reader is referred to the literature (Meixner, 1949, 1974, Müller 1966, 1967, Lebon and Mathieu 1983). We recall two arguments which have been essential for the development of extended theories.

First; the local equilibrium postulate is not valid for continua with complex internal structure and for simple materials in the presence of high frequency phenomena and fast transients. Second; the principle of equipresence of rational thermodynamics assures that if a variable appears in one constitutive equation it has full right to appear in all the remaining equations of the theory. For example, since the mass (heat) diffusion flux is present in the first Fick's (Fourier's) law, this flux can appear also in the Gibbs equation. Indeed, the need for such terms was shown by de Groot and Mazur (1984) for diffusive mass flows in the context of the kinetic energy of diffusion, by Grad (1958) for heat flow, by Sieniutycz and Berry

(1989) for diffusive entropy flow and for other flows by many other researchers. The reader is referred to a recent review of these works (Jou, Casas-Vazquez and Lebon 1989).

Extended thermodynamics uses nonequilibrium (flux dependent) entropies. Kinetic theory supports the existence of such entropies which were first introduced in Grad (1958). The conservation laws retain their classical form in EIT. When they are substituted into the expression for the differential of the nonequilibrium entropy, they give modified expressions for the entropy flux and entropy source. In the simplest version of the extended thermodynamics of a one-component system, the second law inequality can be expressed in the form

$$\rho\frac{ds}{dt} = - \nabla.[\frac{\mathbf{q}}{T} + \mathbf{K}.(\Pi.\mathbf{q})] + \mathbf{q}.(\frac{\mathbf{C}^{-1}}{G}.\frac{d\mathbf{q}}{dt} + \nabla T^{-1}) - T^{-1}\Pi:(\frac{d\Pi}{2Gdt} + \nabla\mathbf{u}) \qquad (2)$$

Phenomena associated with bulk viscosity are neglected here and the tensor Π is the deviator of the viscous pressure tensor. From eq. (2), Π is related to the symmetrized velocity gradient, grad **u**, and the time derivative of the stress which leads to viscoelastic behavior. G is the shear modulus which is the product of the mass density and the square of the propagation speed, c_0^2. **C** is the entropy capacity matrix. For pure heat transfer $C= -c_pT^2$. The extra term in **K** appears in the entropy flux expression in accordance with the kinetic theory. The problem of frame invariance is ignored in eq. (2). Also any chemical reactions are ignored. Actually, the expression (2) is still approximate.

An exact extended theory has been constructed only recently on the basis of the rational formalism (Liu and Müller 1983; Müller 1985, Lebon and Mathieu 1983, Lebon and Boukary 1988, Kremer 1985, 1986) or based on some additional postulates added to kinetic theory (Eu 1982, 1988). The development of these ideas, from simplified early descriptions to the contemporary theory is presented in Müller's paper in this volume. Here only a few introductory remarks are given. Essential for the development of EIT has been a simple and basic form for the entropy inequality introduced by Müller (1966, 1967, 1985). He assumed the second law in the form

$$\rho\frac{ds}{dt} + \nabla.\mathbf{j}_s \geq 0 \qquad (3)$$

containing the entropy flux $\mathbf{j}_s$ which has to be described by a constitutive equation. The idea was improved in Müller's later approaches (Müller 1985) as well as in works by others. An advanced theory by Liu and Müller (Liu and Müller 1983) uses the method of Lagrange multipliers to take into account the entropy inequality (Liu 1972).

As a simplest example we consider a one-component rigid heat conductor. The set of the constitutive equations is assumed in the form

$$\begin{aligned} f &= f(T, T_{,i}, q_i) \\ s &= s(T, T_{,i}, q_i) \\ \dot{q}_i &= Q_i(T, T_{,i}, q_i) \\ j_s^i &= T^{-1} q_i + K(T, T_{,i}, q_i) \end{aligned} \tag{4}$$

where f is the free energy. These equations are substituted into the entropy inequality (3). The resulting inequality yields the classical thermodynamic conditions, e.g., $\partial f/\partial T = -s$, as well as information about the structure of the equations describing the fluxes of entropy and heat. Nonclassical terms appear in the expression for $\mathbf{j}_s$ and the Cattaneo equation is recovered in the linear case. The free energy formula contains the heat flux as an independent variable.

In the general case, the approach outlined above results in a general nonequilibrium theory of extended irreversible thermodynamics, with independent variables that include the heat flux **q**, diffusion fluxes, and viscous stresses (Jou, Casas-Vazquez and Lebon 1989). The classical role of **q** as a (dependent) state variable has already been abandoned in the kinetic theory by Grad (1958); the use of the heat flux **q** as an independent variable is in complete agreement with his moment analysis. In the general case, all diffusive fluxes (heat, mass, momentum) constitute extra independent variables in EIT. The present status of the extended thermodynamic theory is comprehensively reviewed in this volume.

The above results were obtained by assumming that the Gibbs equation is modified by the presence of diffusive fluxes, whereas the form of the conservation laws remains unchanged. However, it was recently shown (Sieniutycz 1988, 1989, 1990; Sieniutycz and Berry 1989), that extended Gibbs equations can be derived from a Lagrangian approach (Hamilton's stationary action) where an extended internal energy is extracted from the total energy. The extended internal energy is obtained within the time component G^{tt} of an energy-momentum tensor G^{ik}. The structure of this tensor corresponds to the classical conservation laws only when the sources of mass and entropy play a negligible role. For substantially irreversible processes, where these sources are large, the structure of G^{ik} deviates from the classical structure, and new important variables play a role in G^{ik} and in the energy E. The new variables are the so-called thermal phase and matter phase. For substantially irreversible processes, the energy formula and the conservation laws can be changed to take account of the sources of mass and entropy. These matters are discussed by two papers in this series. The first part (Vol. 3; Sieniutycz 1990) presents an extended hydrothermodynamics and the second (Sieniutycz, this volume) constitutes an unconventional form of the extended theory.

3. RECENT DEVELOPMENTS IN THE THERMODYNAMICS OF COMPLEX SYSTEMS

3A Internal variables

Before summarizing the works contained in this volume, we present some condensed information concerning recent investigations in the thermodynamics of complex systems. We use the adjective complex to refer to systems such as solutions of macromolecules, magnetic hysteresis bodies, viscoelastic fluids, polarizable media, etc. which exhibit involved thermodynamic behavior and for which the role of internal variables is essential in a thermodynamic description. The concept of internal variables has been succesfully used to analyse relaxation in strained solids by Kestin and Rice (1970); dielectric and magnetic relaxation (Kluitenberg 1977, 1981; Meixner, 1961); spin relaxation in ferromagnets (Maugin 1975), magnetoelasticity (Maugin 1979b), thermo-plastic fracture effects (Maugin1990), electromechanical and other couplings (Maugin 1990a) and magnetically nonsaturated fluids (Maugin and Drouot 1983a). As contrasted with previous approaches, magnetic hysteresis (Maugin 1988, Maugin and Sabir; 1990) has been cast in the framework of nonequilibrium thermodynamics describing this phenomenon exactly in terms of the constitutive equations, internal variables and dissipation function. An interesting analogy between solid mechanics, viscoplastic Bingham fluids and magnetism has been shown (Maugin 1990a). A clear distinction has been made between relaxation processes and hysteresis effects. A thermodynamic framework incorporating internal variables is summarised in the reviews by Coleman and Gurtin (1964) and Bampi and Morro (1984).

Another model based on nonequilibrium thermodynamics was extended so as to account for the new phenomena related to the presence of electrically polarizable macromolecules in solution (Aubert and Tiller 1980; Bird at al 1977; Carreau at al 1986; Drouot and Maugin 1983, 1985, 1987 1988a; Drouot, Maugin and Morro 1987). The role of dissipative effects is essential although certain less known phenomena are shown to be gyroscopic in a sense that they do not contribute to dissipation (Maugin, at al 1990). This is discussed in the paper by Maugin and Drouot in this volume where a relatively simple account of electromechanical couplings and mechano-chemical effects is presented. See also Maugin at al (1990a).

Viscoelastic bodies (solids and fluids) have been analyzed in the framework of the rational thermodynamics using internal variables (Morro 1985). Restrictions on the relaxation functions have been established (Day 1971; Boven and Chen 1973; Fabrizio and Morro 1988). Convexity of the governing functionals has been proved and minimum principles have been established. The relaxation functions satisfy the dissipation inequality. This makes the corresponding functional convex (Fabrizio at al 1989) and hence the thermodynamic limitations imply the convexity of the governing functional. An analogous result has recently been found for the nonequilibrium energy function of a flowing fluid in the field representation of hydrodynamics (Sieniutycz and Berry 1991). Extended thermodynamics of viscoelastic and non-Newtonian bodies has been estalished by Lebon and coworkers (Lebon at al 1986, 1988, 1990; Lebon this volume).

3B Extended irreversible thermodynamics.

Extended irreversible thermodynamics, (EIT), has grown to the status of a mature theory. Typical problems investigated include: compatibility with Grad's 13 moment method (Grad 1958), frame invariance, relation to rational thermodynamics, the role of Lagrange multipliers, and spped s of propagation (Müller, 1985; Boukary and Lebon 1986; Lebon and Boukary 1988). Three levels of description have been analysed: macroscopic (thermodynamic), mesoscopic (fluctuation theory) and microscopic (projection operators). The EIT description has been compared with the memory function approach based only on the conserved variables (Garcia-Colin at al 1984, Jou at al 1988). The superiority of the extended theory over the classical one has been shown. The correspondence of the theory with classical results for ideal and real gases as regards thermal and caloric equations of state has been established. Explicit results have been obtained for the classical ideal gas, for degenerate Bose and Fermi gases, and for molecular gases and mixtures (Müller 1985; Kremer 1985, 1986, 1987, 1989).

The theory was given kinetic theory foundations by means of the Boltzmann equation for a mixture of dilute monatomic gases and the generalized Boltzmann equation for a mixture of dense simple fluids (Eu 1980 1982, 1988). The theory has been applied to rheology, to fluids subject to shear and to shear induced phase changes (Eu 1989). Additional reciprocity conditions resulting from introduction of heat, matter and momentum fluxes as state variables have been shown to be useful in calculating transport coefficients in liquids and in the estimation of the nonlinear effects in heat flow and diffusion (Nettleton 1987). These conditions have also been obtained in the framework of the maximum entropy formalism (Nettleton and Freidkin 1989). The relation between EIT and fluctuation theory has been discussed (Jou at al 1988). It has also been shown that EIT provides a framework for the interpretation of memory functions and generalized transport coefficients (Perez Garcia and Jou 1986). Extended chemical systems have been treated (Garcia Colin at al 1986; Garcia Colin 1989; Sieniutycz 1987). Astrophysical and cosmological applications of EIT have been pursued in the framework of a relativistic theory (Pavon at al 1983).

3C Related approaches

Relativistic thermal inertia systems, (leading to EIT effects even in the limiting case of the low velocities) have been investigated in terms of Lagrangian and Hamiltonian structures. A theory of thermodynamic transformations has been constructed in which the extended Legendre transformations play the double role characteristic of mechanics and thermodynamics (Sieniutycz 1986, 1988, 1989, 1990, 1991). Similar qualitative effects have been found for the theory constructed on the basic of Grad's (1958) solution. Invariant thermodynamic temperatures have been found and a theory of thermodynamic transformations developed in the context of field thermo-hydrodynamics (Sieniutycz and Berry 1989, 1991). Valuable variational principles for irreversible wave systems were found and succesfully applied to obtain approximate solutions by direct variational methods (Vujanovic and Jones 1989).

4. DISCUSSION

We conclude below with a brief summary of each of the papers in this volume.

The paper by G. A. Maugin treats magnetic hysteresis as a thermodynamic process. As contrasted with many previous formal approaches, hysteresis is cast in the framework of nonequilibrium thermodynamics. The process is described at the phenomenological level, in terms of the constitutive equations, internal variables of state and the dissipation function which is homogeneous of degree one. The model allows one to account for the cumulative effect of the residual magnetization and it describes magnetic hardening in a natural way. Both local and global stability criteria for hysteresis loops are obtained. The model, being incremental, provides an operational way to construct hysteresis loops from a virgin state by alternate loads with increasing maximum amplitude. The interaction between the magnetic properties and stress or temperature variations are taken into account. An interesting analogy between solid mechanics (viscoplastic Bingham fluid) and magnetism is shown while a clear distinction is made between relaxation processes and hysteresis-like effects.

G. A. Maugin and R. Drouot develop an internal variable based thermodynamics of solutions of macromolecules. The relevant internal variables of state are the components of the conformation tensor. This quantity models some properties of macromolecules by treating them as "material" particles undergoing deformations. The dissipation inequality is obtained in the form of the Clausius-Duhem inequality involving the free energy. Two different types of dissipative processes are distinguished. The first type involve only the observable variables of the thermo-electro-mechanics. The second are the relaxation processes (of the type encountered in chemical kinetics) involving electric and conformational relaxation. Processes of a third type (plasticity), while essential in the previous work discussed above, do not appear. Evolution equations for conformation, electric polarization and chemical processes are given. An important effect for gyroscopic type reversible processes is developed. Applications to equilibrium conformations, flow-induced and electrically-induced conformational phase transitions, electric polarization, and mechano-chemical effects are discussed. Possible generalizations accounting for diffussion and interaction between vortex fields and conformations are given. This paper should be read by everyone who has to deal with complex fluids, such as liquid crystals and emulsions.

A. Morro presents the thermodynamics of linear viscoelasticity in the framework of rational thermodynamics and extremum principles. First, a general scheme for thermodynamics of simple materials is outlined. This scheme is applicable to materials with internal variables and those with fading memory. Rigorous definitions are given for states, processes, cycles and thermodynamic laws. His thermodynamic approach is then applied to viscoelastic solids and fluids. The approach is essentially entropy-free in the sense that the dissipative nature of the theory is assured by the negative definiteness of the half-range Fourier sine transform of a (Boltzmann) relaxation kernel. The fact that such negative definiteness yields all conditions derived so far by various procedures is a new result. In addition, such a condition has been proved to be sufficient for the validity of the second law. Necessary and sufficient conditions for the validity of the second law are derived for fluids. Evolution equations are analysed. The Navier-Stokes model of a viscous fluid follows as a limiting case when some relaxation functions are equal to delta-functions. The extremum principles are shown to be closely related to thermodynamic restrictions on the relaxation functions. This treatment is quite involved from the mathematical view point, but the reader gets to reexamine many basic features of the rational theory in a concise and precise way. In this context we refer the reader also to the work by Altenberger and Dahler in vol. 5 which discusses applications of convolution integrals to the entropyless statistical mechanical theory of multicomponent fluids.

I. Müller begins a series of reviews devoted (or closely related) to extended irreversible thermodynamics (EIT). His review starts with the analysis of imperfections in Cattaneo's treatment and then outlines a simplified extended theory based on a nonequilibrium (flux dependent) entropy (Müller 1966, 1967, 1985). Its compability with the entropy resulting from Grad's 13 moment theory (Grad 1958) is shown and the problem of material frame indifference is discussed. The role of rational thermodynamics as a starting point for constructing an improved extended theory is also shown. This is based essentially on Liu and Müller's approach (Liu and Müller 1983) which uses Liu's method of Lagrange multipliers to take into account the entropy inequality (Liu 1972). The nonrelativistic principle of relativity is used to make the fields explicit in velocity (

Ruggeri 1989). Constitutive equations, inertial frame balance laws and equilibrium conditions are obtained. A detailed comparison of the results of EIT with those of rational thermodynamics is made. The relativistic generalization of the theory is given and relativistic analogues of the phenomenological equations of Fourier and Navier-Stokes are derived. The mathematical structure of EIT is analyzed.

G. M. Kremer's review extends EIT theory to a broader class of processes. He investigates EIT of ideal and real monatomic gases, molecular gases and mixtures of ideal monatomic gases. For the ideal and real gases the formalism of 14 fields is developed where the 13 fields of mass density, pressure tensor and heat flux are supplemented by an additional scalar field associated with nonvanishing volume viscosities. Coefficients of shear and volume viscosities, thermal conductivity, three relaxation times and two thermal-viscous coeeficients are identified in the theory. The ideal gas theory results as a limiting case from the real gas approach. Explicit results are given for classical ideal gases and for degenerate Bose and Fermi gases. Then, an extended theory of molecular gases is formulated which contains 17 fields of density, velocity, pressure tensor, heat flux, intrinsic energy and intrinsic heat flux. The intrinsic energy pertains to the rotational or vibrational energy of the molecules. Coefficients of thermal and caloric equations of state as well as transport or rate coefficients are identified (shear viscosity, thermal conductivity, self diffusion, absorption and dispersion of sound waves). Finally, the EIT theory for mixtures of monatomic ideal gases is analyzed. Generalized phenomenological equations follow corresponding to the classical laws of Fick, Fourier and Navier-Stokes. Onsager reciprocity follows without statistical arguments. The unusual breadth of EIT theory and its capability to synthesize static, transport and rate properties of nonequilibrium systems in one unifying approach is shown in this paper.

B. C. Eu analyses the relation between kinetic theory and irreversible thermodynamics of multicomponent fluids. His approach is different from that presented in the works of Müller and Kramer above. It is based on an axiomatic formulation (Eu 1982) and on nonequilibrium entropy (Eu 1988), i.e. on exactly those features supported by kinetic theories of gases and fluids. His emphasis is on the nonlinear constitutive equations compatible with the statistical mechanical theories of matter in nonequilibrium. The Boltzmann equation of dilute gases and a modified moment method are used to construct an irreversible theory. Approximation methods are employed to obtain the distribution functions. The requirement that the second law be rigorously satisfied by the approximate solution of the Boltzmann equation is the tenet under which the modified moment methods is developed. The equations for nonconserved moments are the constitutive equations of the system. A consistency condition (in the form of a partial differential equation) is coupled to the evolution equations for the Gibb's variables. The Onsager-Rayleigh quadratic dissipation and classical phenomenological laws follow in the limit of the linear theory close to equilibrium. The theory is extended to dense multicomponent fluids, based on a set of the kinetic equations for a dense mixture of charged particles in a solvent. Tests of the constitutive equations are outlined in the framework of rheology (polymer viscoelasticity). The formal theory of nonequilibrium thermodynamic potentials is given with an extended Gibbs-Duhem relation. Applications for fluids under shear and shear-induced depression of the melting point are discussed.

R. E. Nettleton treats EIT problems from the viewpoint of the general reciprocity (and antireciprocity) relations encountered in nonequilibrium systems and predicted by extended theories. These reciprocity relations result from the introduction of diffusional fluxes of mass, energy and momentum, as (extra) state variables of the nonequilibrium Gibbs equation. They are shown to be useful in calculating transport coefficients in liquids. Reciprocity reduces the number of coefficients which must be estimated from molecular approaches. By using reciprocity, certain generalizations of the Cattaneo equation of heat and the Maxwell equation of viscoelasticity can be obtained. These equations govern coupled flows of heat and matter and viscoelasticity under several creep mechanisms. The use of reciprocity in the estimation of nonlinear effects associated with the flow of heat and diffusion is shown. Nonlinear reciprocity is derived from the Fokker-Planck equation for a distribution g of values of certain microscopic operators whose averages are variables such as heat and mass fluxes. The equation for g is obtained by applying Zwanzig or Grabert projection operators to the Liouville equation. The maximum entropy formalism provides an ansatz for the solution of the Liouville equation which yields a finite number of moments. From this ansatz one can derive kinetic equations exhibiting nonlinear reciprocity by taking moments of the equation for g. It is shown that a Lagrangian formulation can lead to the reciprocity and antireciprocity properties.

J. Casas-Vazquez and D. Jou review the theory of fluctuations in the framework of EIT. Mutual contributions of the two theories are analyzed. Fluctuation theory enables one to evaluate the values of the coefficients of the nonclassical terms in the entropy and in the entropy flux, from a microscopic basis. The fluctuation-dissipation relations focus on the time correlation function of fluctuations in the fluxes. Equilibrium fluctuations of dissipative fluxes are considered and the second moments of the fluctuations are obtained. Connections with the Green-Kubo transport formulae are shown. Microscopic evaluations of the coefficients of the nonequilibrium entropy are presented for ideal gases, dilute real gases, and dilute polymer solutions. An expression is given for the entropy flux in terms of fluctuations. Next, the nonequilibrium fluctuations of dissipative fluxes are analysed. The contribution of the irreversible fluxes to the entropy leads to explicit expressions for the second moments of nonequilibrium fluctuations. EIT predicts nonequilibrium contributions to the fluctuations of variables as a consequence of the nonvanishing relaxation times of the fluxes. An example is given discussing fluctuations of the heat flux in a rigid heat conductor. The second differential of the nonequilibrium entropy (with nonvanishing mixed terms) is introduced into the Einstein formula to compute the second moments of fluctuations in a nonequilibrium steady state. It is shown that the correlations of energy and heat flux and those of volume and heat flux, which vanish in equilibrium on account of the different parity of u and q and of v and q, are different from zero in nonequilibrium. This expresses the breaking of the time-reversal symmetry in nonequilibrium states. The results of the second moments from the nonequilibrium entropy and from microscopic (kinetic) models are of the same order but not coincide. This open problem points out the status of the present theory of fluctuations in EIT.

C. Perez-Garcia and D. Jou present EIT theory as a suitable macroscopic framework for the interpretation of generalized transport coefficients. The classical thermodnamic theory holds only in the hydrodynamic limit of small frequency and small wave vector (vanishing ω and k) for any disturbances. A generalized hydrodynamics (GH) is postulated which replaces the dissipative transport coefficients by general memory functions which are justified by some microscopic models. [See paper by Altenberger and Dahler in vol. 6 and that by Morro in this volume which also deal with memory effects]. A phenomenological theory involving higher order hydrodynamic fluxes is given for the description of GH processes. The additional variables playing the role of the currents of the traceless pressure tensor and viscous pressure are introduced into the extended entropy and balance equations. The entropy flux contains the most general vector that can be constructed with all the dissipative variables. As specific applications, the velocity autocorrelation functions are calculated around equilibrium, and ultrafast thermometry is analyzed. A parallelism of the macroscopic theory with the projection- operator techniques is outlined. Memory functions and continued fraction expansions for transport coefficients are interpreted in terms of the generalized thermo-hydrodynamics.

G. Lebon develops the thermodynamics of rheological materials in the context of the extended irreversible theory. He shows that EIT has capability of a general unifying formalism which is able to produce a broad spectrum of viscoelastic constitutive equations. In particular, linear viscoelasticity is easily interpreted within EIT. It is shown how the classical rheological models of Maxwell, Kelvin-Voight, Poynting-Thompson and Jeffreys are obtained as special cases of the formalism developed. A general Gibbs equation for viscoelastic bodies is given and its special cases, pertaining to the particular models specified above, are discussed. Then the author poses the question: what are the consequences of introducing a whole spectrum of relaxational modes for the pressure tensor, instead of working with one single mode. For that purpose the Rouse-Zimm relaxation models (Rouse 1953; Zimm 1956) are analysed. These models are based on a whole relaxational spectrum; such spectra are are frequently predicted by kinetic models. These models have proved useful for describing dilute polymeric solutions. It is shown that EIT is capable of coping with the relaxational spectrum of the Rouse-Zimm model. Finally, the analysis is extended to fluids described by nonlinear constitutive equations (non-Newtonian). This EIT-based approach to viscoelasticity can be compared with other thermodynamic approaches, based on other theories: classical (Meixner 1949; Kluitenberg 1962; Kestin and Bataille 1980) or rational (Rivlin-Ericksen 1955; Koh and Eringen 1963; Huilgol and Phan-Thien 1986).

J. E. Llebot presents extended irreversible thermodynamics of electrical phenomena. He studies a rigid isotropic solid conductor in the presence of flows of heat and electricity. By combining the balance equations for the internal energy and electron concentration with the nonequilibrium entropy an expression for the entropy production is obtained containing the generalized thermodynamic forces with the time derivatives of the energy and the electron flux. He next considers nonstationary electrical conduction. It is shown that only EIT can preserve the positiveness of the entropy production σ in general. For sufficiently large frequencies, the classical theory can yield an erroneous expression allowing negative σ during part of the cycle. EIT always gives the correct result impying $\sigma > 0$ at each time instant. Nonequilibrium fluctuations of the electric current are analyzed. Instability of the electric current in plasma is studied in the framework of EIT. The critical value of the electric field obtained within the EIT formalism is in good agreeement with that obtained on the basis of kinetic considerations (Fokker-Planck equation).

L. S. Garcia-Collin analyses chemical kinetics as described by the classical kinetic mass action law from the view point of the compatibility of this law with EIT. The difficulty lies, of course, in the nonlinearity of the kinetics. A single step homogeneous chemical reaction is considered with the well known exponential relation between the chemical flux and the ratio of the affinity and the temperature. Nonequilibrium entropy is postulated to depend on the conventional variables, as well as on diffusion fluxes, heat flux and chemical flux. Using the formalism of extended thermodynamics, an expression for the entropy source is obtained and extended phenomenological relations are postulated. It is shown that a correspondence between classical kinetics and the implications of EIT can be made provided that an ansatz is adopted concerning the appropriate form of the phenomenological function, P, appearing in the generalized relaxation formula. Fluctuations of the chemically reacting fluid aroud an equilibrium reference state are discussed in the framework of EIT. It is shown that their time decay obeys an exponential rather than a linear equation due to nonvanishing relaxation times in the fluctuation formulae. The open question concerning the existence of general thermodynamic criteria suitable for predicting how a chemical reaction evolves in time is discussed.

D. Pavon's paper deals with extended relativistic thermodynamics (ERT). The paper deals with applications to astrophysics and to cosmology, rather than dealing with the general theory. [The reader interested in causal relativistic theories is refered to the papers by Israel (1976) and Hiscock and Linblom (1983)]. The second moments of fluctuations are determined from the Einstein probability formula involving the second differential of the nonequilibrium entropy.

These moments are subsequently used to determine the transport coefficients of heat conductivity, bulk viscosity and shear viscosity of a radiative fluid. By exploiting correlation formulae, second order coefficients of transport are determined. Nonequilibrium fluctuation theory, based on the assumption of the validity of the Einstein formula close to equilibrium, is used to determine the nonequilibrium corrections of the bulk viscous pressure. An analysis of survival of the protogalaxies in an expanding universe completes the presentation of astrophysical applications of ERT. Cosmological applications are discussed next. It is shown that nonequilibrium effects and transport properties (such as, e.g., bulk viscosities) play a certain role in the cosmic evolution of the universe. In that context, the following issues are outlined: cosmological evolution (described by a Bianchi-type model with the bulk viscosity), production of entropy in the leptonic period, the inflationary universe and FRW cosmology. Some cosmological aspects of the second law are also discussed.

S. Sieniutycz's approach to extended thermodynamics views this theory as a macroscopic consequence of the recent "micro-thermodynamic" theory of de Broglie (de Broglie 1964). The treatment is in the context of Hamilton's action and the conservation laws for multicomponent fluids with source terms in the continuity equations. The de Broglie theory is microscopic, quantum and relativistic; however, its nonrelativistic macroscopic counterpart is pursued. This is sufficient to preserve the effect of thermal inertia, inherent in that theory. In this extended thermodynamics, the sources of entropy and matter result as the partial derivatives of a generalized energy, E, with respect to the action-type variables ("phases") under the condition of constancy of momenta and other natural variables of E. These sources can also be obtained by differentiation of generalized thermodynamic potentials (the suitable Legendre transforms of E) with respect to the phases. Internal symmetries of the theory are shown to yield the global mass conservation law and chemical stoichiometry. Vector external fields and charged systems can be considered in a generalized variational scheme. General equations of motion are obtained. They correspond to the phenomenological equations of irreversible thermodynamics under with a suitable kinetic potential or generalized energy function. Nonlocality of the irreversible transport and rate processes is a natural feature of the theory. It is shown that the conventional assumption allowing the energy-momentum tensor to be independent of the sources of the entropy and matter cannot be generally met, and hence the conservation laws of highly irreversible processes are modified by the presence of dissipation.

ACKNOWLEDGEMENT
One of the authors (S. S.) acknowledges the kind hospitality of the Department of Mathematical Sciences at San Diego State University. Both authors gratefully acknowledge partial support from the Office of Naval Research, grant number N00014-88-K-0491.

REFERENCES

Aubert, J., and M. Tirrell. 1980. Macromolecules in nonhomogeneous velocity gradient fields. J. Chem. Phys. 72: 2694-2701.

Bampi, F. and A. Morro. 1984. Non-equilibrium thermodynamics: a hidden variable approach. In Lecture Notes in Physics199: 221-232. Eds. J. Casas Vazquez, D. Jou and G. Lebon. Berlin: Springer.

Baranowski, B. 1974. Nonequilibrium Thermodynamics in Physical Chemistry. Warsaw: PWN.

Bird, R.B., W.E. Steward and E.N. Lightfoot. 1960. Transport Phenomena. New York: J. Wiley.

Bird, R.B., R.C. Amstrong, and D. Hassager. 1977. Dynamic of Polymer Liquids. New York: Wiley.

Boukary, M.S., and G. Lebon. 1986. A comparative analysis of binary fluid mixtures by extended thermodynamics and the kinetic theory. Physica 137A: 546-572.

Boven, R., and P.J. Chen. 1973. Thermodynamic restriction on the initial slope of the stress relaxation function. Arch. Rat. Mech. Anal. 51: 278-284.

Bubnov, V.A. 1976. On the nature of heat transfer in a acoustic wave. Inzh. Fiz. Zh. 31: 531-536.

Callen, H. 1988. Thermodynamics and an Introduction to Thermostatistics. New York: J. Wiley.

Carrasi, M., and A. Morro. 1973. Some remarks about dispersion and absorption of sound in monoatomic rarefied gases. Nuovo. Cim. 13 B: 231-249.

Carreau, P.J., M. Gmerla, and A. Ait-Kadi. 1986. A conformational model for polymer solutions. Commun. 2nd Conference of European Rheologists, Prague, June 17-20, 1986.

Coleman, B.D., and M. Gurtin 1964. Thermodynamics with internal variables. J. Chem. Phys. 47: 597-613.

Day, W.A. 1971. Restrictions on relaxation functions in linear viscoelasticity. Q. Jl Mech. Appl. Math. 24: 487-497.

Domanski, R. 1978. Analytical description of temperature field caused by heat pulses. Arch. Termod. i Spalania 9: 401-413.

Drouot, R., and G.A. Maugin. 1983. Phenomenological theory for polymer diffusion in non-homogeneous velocity gradient flows. Rheol. Acta. 22: 336-347.

Drouot, R., and G.A. Maugin. 1985. Continuum modelling of polyelectrolytes in solutions. Rheol. Acta 24: 474-487.

Drouot, R., and G.A. Maugin. 1987. Optical and hydrodynamical regimes in dilute solutions of polyelectrolytes. Rheol. Acta 26: 350-357.

Drouot, R., and G.A. Maugin. 1988. Polyelectrolytes in solutions: equilibrium conformations of the microstructure without and with external fields. Int. J. Engng. Sci. 26: 225-241.

Drouot, R., G.A. Maugin, and A. Morro. 1987. Anisotropic equilibrium conformations in solutions of polymers and polyelectrolytes. J. Non-Newtonian Fluid Mech. 23: 295-304.

Eu, B.C. 1980. A modified moment method and irreversible thermodynamics. J. Chem. Phys. 73: 2858.

Eu, B.C. 1982. Irreversible thermodynamics of fluids. Ann. Phys. (NY) 140:131

Eu, B.C. 1988. Entropy for ireversible processes. Chem. Phys. Lett. 143:65.

Eu, B.C. 1989. Irreversible thermodynamic theory of scalar induced melting point depression. Physica A160: 87.

Fabrizio, M., and A. Morro. 1988. Viscoelastic relaxation functions compatible thermodynamics. J. Elasticity19: 63-75.

Fabrizio, M., C. Giorgi, and A. Morro. 1989. Minimum principles, convexity and thermodynamics in viscoelasticity. Continuum Mech. Thermodyn. 1: 197-211.

Fitts, D.D.1962. Nonequilibrium Thermodynamics. New York: Mc Graw-Hill.

Garcia Colin, L.S. 1988. Comments on the kinetic mass action law revisited by thermodynamics. J. Phys. Chem. 92: 3017-3018.

Garcia Colin, L.S., S.M.T. de la Selva, and E. Pina. 1986. Consistency of the kinetic mass action law with thermodynamics. J. Phys. Chem. 90: 953-956.

Garcia Colin, L.S., M. Lopez de Haro, R.F. Rodriguez, J. Casas Vazquez, and D. Jou. 1984. On the foundations of extended irreversible thermodynamics. J. Statist. Phys. 37: 465-484.

Gelfand, J.M., and S.W. Fomin. 1970. Variational Calculus. Warsaw: PWN.

Giesekus, H. 1986. Constitutive models of polymer fluids: Toward a unified approach. In Trends in Applications of Pure Mathematics to Mechanics, ed. E. Kroner and K. Kirchgassner, 331-348. Berlin: Springer.

Glansdorff, P., and I. Prigogine. 1971. Thermodynamic Theory of Structure Stability and Fluctuations. New York: J.Wiley.
Grad, H.1958. Principles of the theory of gases. In Handbook der Physik 12, ed. S. Flugge. Berlin: Springer.
de Groot, S.R., and P. Mazur. 1984. Nonequilibrium Thermodynamics. New York: Dover.
Gyarmati, I. 1977. On the wave approach of thermodynamics and some problems of nonlinear theories. J. Non-equilib. Thermodyn. 2: 233-245.
Hiscock, W.A. and L. A. Lindblom. 1983. Stability and causality in dissipative fluids. Ann. Phys. (N.Y.)151: 466-496.
Huilgol, R. R., and N. Phan-Thien. 1986. Recent advances in the continuum mechanics of viscoelastic liquids. Int. J. Engng. Sci. 24: 161-261.
Israel, W. 1976. Nonstationary irreversible thermodynamics: a causal theory. Ann. Phys. (N.Y.)100: 310-313.
Jou, D. 1989. Fluctuation theory in extended irreversible thermodynamics. Physica A: 221-234.
Jou, D., J. Casas-Vazquez, and G. Lebon. 1989. Extended Irreversible Thermodynamics. Rep. Progr. Phys. 51: 1105-1179.
Jou, D., and J.E. Llebot. 1980. Electric current fluctuations in extended irreversible thermodynamics. J. Phys. A: Math. Gen.13: L47.
Kestin, J. 1979. Foundations of Non-Equilibrium Thermodynamics. Domingos at al, eds. London: MacMillan Press.
Kestin, J., and J.M. Rice. 1970. Paradoxes in the application of thermodynamics to strained solids. In A critical Review of Thermodynamics, eds. F.B. Stuart, B. Gal Or and A.J. Brainard, 275-298. Baltimore: Mono Book Corp.
Kestin, J., and J. Bataille. 1980. Thermodynamics of solids. In Continuum models of Discrete Systems. Univ. of Waterloo Press.
Koh, S., and C. Eringen. 1963. On the foundations of non-linear thermo-viscoelasticity. Int. J. Engng. Sci.1: 199-229.
Kluitenberg, G. 1966. Application of the theory of irreversible processes to continuum mechanics. In: Nonequilibrium Thermodynamics, Variational Techniques, and Stability. R. Donelly, R. Herman and I. Prigogine, eds. Chicago: Chicago University Press.
Kluitenberg, G.A. 1977. On dielectric and magnetic relaxation and vectorial internal degrees of freedom in thermodynamics. Physica 87a: 541-563.
Kluitenberg, G.A. 1981. On vectorial internal variables and dielectric and magnetic relaxation phenomena. Phenomena109 a:91-122.

Kremer, G.M. 1985. Erweiterte Thermodynamik Idealer und Dichter Gase. Diss. TU Berlin.
Kremer, G.M. 1986. Extended thermodynamics of ideal gases with14 fields. Ann. Inst. Henri Poincare. 45: 419-440.
Kremer, G.M. 1987. Extended thermodynamics of non-ideal gases. Physica 144A: 156-178.
Kremer, G.M. 1989. Extended thermodynamic of molecular ideal gases. Continuum Mech. Thermodyn.1: 21-45.
Lebon, G., and P. Mathieu. 1983. Comparison of the diverse theorie of nonequilibrium thermodynamics. Intern. Chem. Engng. 23: 651-662.
Lebon, G., C. Perez Garcia, and J. Casas-Vazquez. 1986. On a new thermodynamic description of viscoelastic materials. Physica 137A: 531-545.
Lebon, G., C. Perez Garcia, and J. Casas-Vazquez. 1988. On the thermodynamic foundations of viscoelasticity. J. Chem. Phys. 88: 5068-5075.
Lebon, G., and M. S. Boukary. 1988. Objectivity, kinetic theory and extended irreversible thermodynamics. Int. J. Engng. Sc. 26: 471- 483.
Lebon, G., P.C. Dauby, A. Palumbo, and G. Valenti. 1990. Rheological properties of dilute polymer solutions: an extended thermodynamic approach. Rheol. Acta (to appear).
Lemaitre, J., and J. L. Chaboche 1988. Mechanics of solid materials. Cambridge: Cambridge University Press.
Leonov, A.I. 1976. Nonequilibrium thermodynamics and rheology of viscoelastic polymer media. Rheol. Acta 21: 683-691.
Liu, I-Shih, Müller, I.1983. Extended Thermodynamics of Classical and Degenerate Gases. Arch. Rational Mech. Anal. 83
Liu, I-Shih. 1972. Method of Lagrange Multipliers for Exploitation of the Entropy Principle. Arch. Rational Mech. Anal. 46.
Maugin,G. A. 1975. On the spin relaxation in deformable ferromagnets. Physica 81a: 454-458.
Maugin, G.A. 1979. Vectorial internal variables in magnetoelasticity. J. Mecanique. 18: 541-563.
Maugin, G.A. 1987. Thermodynamique a variables internes et applications. In Seminar: Thermodynamics of Irreversible Processes. Institut Francais du Petrole, Rueil-Malmaison, France.
Maugin, G.A. 1990. Thermomechanics of plasticity and fracture. Cambridge: Cambridge University Press.
Maugin, G.A. 1990a. Nonlinear dissipative effects in electrodeformable solids. In Nonlinear Mechanical Couplings,

G.A. Maugin, B. Collet, R. Drouot and J. Pouget, Chapter 6. Manchester: Manchester University Press.

Maugin, G.A., and R. Drouot. 1983. Internal variables and the thermodynamics of macromolecules solutions. Int. J. Engng. Sci. 21: 705-724.

Maugin, G.A., and R. Drouot. 1983. Thermomagnetic behavior of magnetically nonsaturated fluids. J. Magnetism and Magnetic Materials. 39: 7-10.

Maugin, G.A., and M. Sabir. 1990. Mechanical and magnetic hardening of ferromagnetic bodies: Influence of residual stresses and applications to non-destructive testing. Int. J. of Plasticity (in press).

Meixner, J. 1949. Thermodynamik und Relaxationserscheinungen. Z. Naturforschung 4a: 594-600.

Meixner, J. 1961. Der Drehimpulssatz in der Thermodynamik der irreversiblen Prozesse. Zeit. Phys.16: 144-165.

Meixner, J. 1974. Coldness and Temperature. Arch. Rat. Mech. Anal. 57: 281-286.

Morro, A., 1985. Relaxation phenomena via hidden variable thermodynamics. In Thermodynamics and Constitutive equations, ed. G. Grioli, Berlin: Springer.

Müller, I. 1966. Zur Ausbreitungsgeschwindigkeit vor Störungen in kontinuierlichen Medien . Dissertation TH Aachen.

Müller, I. 1967. Zum Paradox der Wärmeleitungstheorie. Z. f. Physik 198.

Müller, I. 1985. Thermodynamics. London: Pitman.

Nettleton, R.E. 1987. Application of reciprocity to non-linear extended thermodynamics kinetic equations. Physica A 144: 219-234.

Nettleton R. E., and E.S. Freidkin. 1989. Nonlinear reciprocity and the maximum entropy formalism. Physica A158: 672-690.

Pavon, D., D. Jou and J. Casas-Vazquez. 1983. Equilibrium and nonequilibrium fluctuations in relativistic fluids. J. Phys. A 16: 775-782.

Perez-Garcia, C., and D. Jou. 1986. Continued fraction expansion and memory functions for transport coefficients. J. Phys. A: Math. Gen. 19: 2881.

Prigogine, I. 1949. On the domain of validity of the local equilibrium hypothesis. Physica 15:1942.

Rivlin, R.S., and J.L. Ericksen. 1955. Stress deformation relations for isotropic materials. J. Rat. Mech. Anal. 4: 325-425.

Rouse, P.E. 1953. A theory of the viscoelastic properties of dilute solutions of coiling polymers. J. Chem. Phys. 21: 1272-1280.

Ruggeri, T. 1989. Gallilean invariance and entropy principle for system of balance laws. The structure of extended thermodynamics. Cont. Mech. Thermodyn. 1

Sieniutycz, S. 1979. The wave equations for simultaneous heat and mass transfer in moving media- structure testing, space-time transformations and variational approach. Intern. J. Heat Mass Transfer 22: 585-599.

Sieniutycz, S. 1983. The inertial relaxation terms and the variational principles of least action type for nonstationary energy and mass diffusion. Intern. J. Heat Mass Transfer 26: 55-63.

Sieniutycz, S. 1986. On the relativistic temperature transformations and the related energy transport problem. Intern. J. Heat Mass Transfer 29: 651-654.

Sieniutycz, S. 1987. From a least action principle to mass action law and extended affinity. Chem. Engng Sci. 42: 2697-2711.

Sieniutycz, S. 1988. Hamiltonian energy-momentum tensor in extended thermodynamics of one-component fluid. Inz. Chem. Proc. 4: 839-861.

Sieniutycz, S. 1989. Energy-momentum tensor and conservation laws for extended multicomponent fluids with the transport of heat, mass and electricity. Research report of Inst. Chem. Engng. Warsaw: WTU.

Sieniutycz, S. 1990. Thermal momentum, heat inertia and a macroscopic extension of the de Broglie microthermodynamics I . In Advances in Thermodynamics 3, eds. S. Sieniutycz and P. Salamon. New York: Taylor and Francis.

Sieniutycz, S. 1991. Wave equations of heat and mass transfer. In Advances in Thermodynamics 6, eds. S. Sieniutycz and P. Salamon. New York: Taylor and Francis.

Sieniutycz, S., and R.S. Berry. 1989. Conservation laws from Hamilton's principle for nonlocal thermodynamic equilibrium fluids with heat flow. Phys. Rev. A. 40: 348-361.

Sieniutycz, S., and R.S. Berry. 1991. Field thermodynamic potentials and geometric thermodynamics with heat transfer. Phys. Rev. A 43: 2807-2818.

Sluckin, T.J. 1981. Influence of flow on theisotropic-nematic transition in polymer solutions: a thermodynamic approach. Macromolecules 14: 1676-1680.

Tirrell, M. and F. Malone. 1977. Stress-induced diffusion of macromolecules. J. Polym. Sci. 15: 1569-1583.

Truesdell, C. 1969. Rational Thermodynamics. New York: McGraw-Hill.

Wisniewski, S., B. Staniszewski, and R. Szymanik. 1973. Thermodynamics of Nonequilibrium Processes. Warsaw: PWN.

Wisniewski, S. 1984. Balance equations of extensive quantities for a multicomponent fluid taking into account diffusion stresses. Archiw. Termod. 5: 201- 213.

Zimm, B.H. 1956. Dynamics of polymer molecules in dilute solution: viscoelasticity, flow birefringence and dielectric loss. J. Chem. Phys. 24: 269-278.

Thermodynamics of Hysteresis

Gérard A. Maugin
Laboratoire de Modélisation en Mécanique (URA 229)
Université Pierre et Marie Curie
Tour 66 - 4 place Jussieu
75252 Paris Cedex 05, France

ABSTRACT

Magnetic hysteresis is cast in the framework of non-equilibrium thermodynamics. It serves as a paragon of all hysteresis phenomena described directly in terms of constitutive equations at the phenomenological level. This is to be contrasted with many mathematical approaches which simply aims at reproducing hysteresis loops independently of any thermodynamical considerations and underlying physical mechanisms. Here, a phenomenological model of magnetic hysteresis is constructed which is independent of time rates and does account for magnetization from a virgin state and the obviously differing responses in loading and unloading (in magnetizing field). The basic model uses the notions of dissipation function which is homogeneous degree *one* in its arguments, internal variables of thermodynamical state, and arguments of convex analysis. This model allows one to account for the cumulative effect in time of the residual magnetization and it describes magnetic hardening in a natural way. Both local and global stability criteria for hysteresis loops are obtained. The model itself, being incremental, provides an operational way to construct hysteresis loops from a virgin state by alternating loading of increasing maximum amplitude. The very conception of the thermodynamical model facilitates the formulation of interactions between the process described by hysteresis loops (e.g., magnetism) and other process (e.g., stresses or temperature variations). In addition, a semi-microscopic model (irregular motion of domain walls, Barkhausen jumps) justifies the thermodynamically admissible model. The same considerations apply to ferroelectric hysteresis in ceramics and locking materials. Whenever possible analogies between various domains of physics and mechanics are mentioned. Most of the formulation has been developed by the author and co-workers during the period 1984/1990.

eine Magnetisierungsintensität I an, welche als Funktion des Feldes dargestellt werden kann (Figur 22).

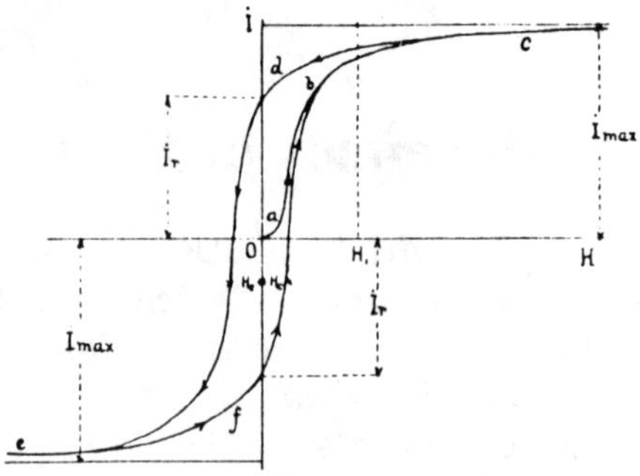

Fig. 22.

Die Intensität der Magnetisierung wächst zuerst langsam

Facsimile of Pierre Weiss' lecture notes at the Zurich Polytechnicum, Academic year 1911 (collection G.A. Maugin)

1. - INTRODUCTION

Hysteresis is generally considered as a rather well characterized response to a cyclic loading. Typically, after a while, the response itself has become periodic for a periodic loading. The peculiarity of a real hysteresis is that in spite of periodicity the response paths in loading and unloading for a complete cycle differ markedly from one another (Figure 1). This, in turn, means that *energy is dissipated* during a cycle ; this loss of energy corresponds to the area enclosed within the hysteresis *loop*. Therefore, *thermodynamics* should be *the* proper macroscopic theoretical framework for hysteresis. Hysteresis phenomena abound in Nature and they are observed in many materials and processes : to name a few, the stress-strain mechanical response (Lemaître and Chaboche, 1988), the magnetic hysteresis loop of hard ferromagnets (Hoselitz, 1952), the electric hysteresis loop of ferroelectric crystals (Maugin, 1990b), the wall properties of veins in animals, the human respitory system (Ruch and Patton, 1965), *etc*. But very few general works are devoted to this type of phenomenon, a notable exception being the book by Krasnosel'kii and Pokrovskii (1988) where hysteretic systems are approached as a class of phenomena rather than as the result of specific physical processes. Here, in fact, is the crux of the matter. Hysteresis is a *complex nonlinear* phenomenon and the question naturally arises as whether to *understand* the physical basis of this phenomenon or just to *model* it by *ad hoc* mathematical expressions. Facing such a dilemna, but interested in a macroscopic response for which a *representation* is needed in technological applications (e.g., in electrical engineering), those inclined towards mathematics and formalism reduce the problem to postulating a suitable mathematical representation which can be shown to exhibit hysteretic behavior. This line of thought goes back to V. Volterra (1956). Contemporary proponents of this line are, for example, Bouc (1971), Chua and Stromsmoe (1970, 1971), Mayergoyz (1986), Coleman and Hodgdon (1986, 1987). However, Volterra recognized the important fact that hystereis is an *hereditary* process in the sense that the response of an hysteretic system at time t depends in some way on the past history of this system, i.e., on the collection of states which the system has previously experienced. He was thus led to introducing so-called *hereditary integrals* (linear functionals). The more recent authors in this vein rather use a mathematical representation in terms of *evolution equations*, i.e., first-order differential equations, emphasizing thus the *incremental* nature of the process. None of the papers just mentioned make use of thermodynamical considerations.

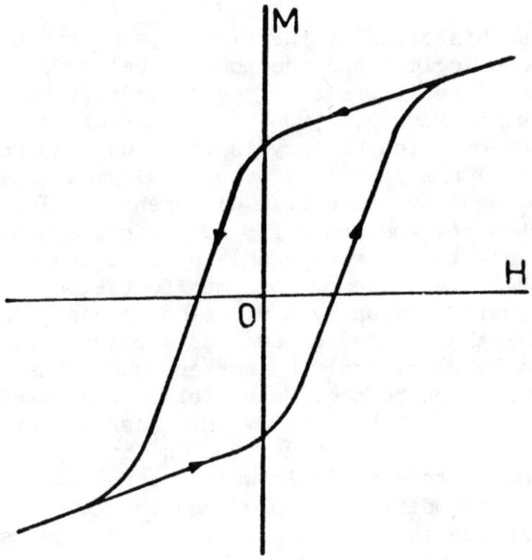

Figure 1. Typical M-H hysteresis loop for a hard ferromagnet

The most desirable approach to hysteresis phenomena is to determine the *physical laws* that govern the system but, then, one is confronted with a lack of clear understanding of the exact interactions of the physical processes involved. This can be attempted at different scales. In this respect the case of ferromagnetic hysteresis is paradigmatic. In this field Preisach (1935) was the first to devise such a model by considering a magnetic material as an assembly of small domains, each presenting an ideal, shifted, rectangular hysteresis loop, and then assuming a distribution of such domains in the *ad hoc* plane. Néel's works (1943, 1946) go further in a deeper microscopic (or *meso*scopic) understanding of magnetic hysteresis by relating ferromagnetic hysteresis to the irreversible motion of magnetic domain walls, the latter being pinned from time to time on structural defects which, themselves, are distributed at random. Having in the past been concerned with magnetic relaxation in rigid or deformable ferromagnets (Maugin, 1975, 1976, 1982) and with the coupling between defects and dynamical magnetoelastic properties in crystalline ferromagnets (Maugin, 1979a ; Maugin and Fométhé, 1982), the author was recently asked to examine, on some rational basis, the relationship and interactions between magnetic hysteresis and the mechanical (stress) state of some materials (Maugin *et al*, 1987 ; Sabir, 1988, Maugin and Sabir, 1990 ; Sabir and Maugin, 1988). This has an important practical side because of the development of nondestructive methods of stress analysis utilizing the Barkhausen effect, i.e., the magnetization by jerky motion of magnetic domain walls (see, e.g. Rudyak, 1971 ; Pashley, 1970 ; Gardner *et al*, 1971 Matzkanin and Gardner, 1976 ; Karljalainen and Moilanen, 1979 ; Rautioaho *et al*, 1987). At the phenomenological level one has to understand the influence of stresses on the shape of magnetic hysteresis loops. We had the feeling that this goal could be reached only through a better understanding of these loops and that the analysis, of necessity, should be carried out in a fully *thermo-mechanical* framework where both magnetic hysteresis and irreversible mechanical behavior, e.g., plastic flow, would be combined in a harmonious whole. Then strong analogies between the magnetic behavior and mechanical ones are revealed, but one must account for the different tensorial orders of the involved variables, axial vectors in the first case, second-order symmetric tensors in the second case.

Magnetic hysteresis was historically the first hysteretic behavior to be recognized. However, the phenomenological and thermodynamical study of irreversible stress behaviors has made much more progress in recent years (see, e.g., the modern presentation of plasticity in Maugin, 1990a) than that of macroscopic magnetic behaviors. By some face about of which history is fond, one can now apply this progress to magnetic (or electric) hysteresis by using analogies, while bearing constantly in mind limitations brought by essential differences.` Therefore, in the present contribution we present a *thermodynamically admissible* approach to magnetic hysteresis (this has been selected as a paragon) in which the full *vectorial nature* of magnetization is accepted, the *irreversible* manifestation of hysteresis is acknowledged, the *physical justification* by a microscopic analysis is outlined, and the eventual coupling with other processes such as stresses is indicated. We hold the opinion that this greatly clarifies the matter while casting it in a physical framework that was never given before. Essential in the ensuing development is the use of the fruitful notion of *thermodynamical internal* variable of state. Issued from rheological models (see, e.g., Lemaître and Chaboche, 1988 ; Maugin, 1990a), this notion of additional state variable which macroscopically accounts for irreversible processes without modifying before hand the general statements of thermodynamical laws, was introduced in electromagnetic bodies by Meixner (1961), Kluitenberg (1977, 1981), and the author (Maugin, 1979b, 1981) with a view to studying *relaxation processes* (as they may occur, for instance, after the passing of a shock wave through an electrically polarizable ceramic). Hysteretic processes can be described by using the same notion, but one should be careful to distinguish between relaxation effects (return to equilibrium) and hysteresis-like effects which imply the existence of *residual* fields when the load vanishes. We are also naturally led to distinguishing between those irreversible effects that *do* involve a *time scale* (this is the case of relaxation) and those that do not ; e.g., relatively slow magnetization of a ferromagnetic sample produces a rate-independent response in the same way as low-temperature plasticity is practically independent of the strain rate. The adequate mathematical framework for dealing with rate-independent response is *convex analysis* (e.g., Rockafellar, 1970). This is applied all along the following development. The latter also applies to ferroelectric bodies with the appropriate change of variable. Insofar as possible, mechanical analogies are always indicated. The reader is referred to other works (Bassiouny *et al*, 1988a,b ; 1989a,b ; Maugin, 1989, 199b ; Maugin and Bassiouny, 1989) for the electric analogue, in deformable ferroelectrics and ceramics, of the present development.

2. - STATE VARIABLES AND FUNDAMENTAL THERMODYNAMICAL LAWS

A. - Clausius-Duhem inequality

Traditionally in electrical engineering the magnetic constitutive equation is given in terms of scalar quantities (magnitudes). More precisely, a functional relation between the magnetic-field intensity H and the magnetic-induction intensity B, such as $B = B(H)$ is specified, $B = \mu H$ being the simplest (linear isotropic) example with $B = \mu_0 H$ in a vacuum. Obviously, this *instantaneous* relation (here H and B are specified at the *same* time t) is not sufficient to characterize magnetic hysteresis and, following Volterra (1956), we could formally write the true *functional*

$$B(t) = \mathcal{B}[H(t - s)] \qquad s \in [0,\infty) \tag{2.1}$$

where B at time t, is a functional of H on the *past* time interval. In physical reality magnetic induction and field are *axial* vectors and we prefer to characterize the response of a *magnetizable* material by that quantity which distinguishes such a material from a vacuum, the *magnetization vector* such as

$$\vec{M} = \mu_0^{-1} \vec{B} - \vec{H} , \tag{2.2}$$

in which μ_0 is the vacuum permeability. Here $\vec{M}$ is the *magnetization*. This macroscopic manifestation of microscopic magnetic dipoles and spins is a *density*. From the point of view of macroscopic thermodynamics, it is an *extensive* quantity (Maugin, 1988, Chapter 3 ; Rosensweig, 1989). In a *deformable* body magnetization always appears per unit mass (Maugin, 1988, Chapter 3 ; Rosensweig, 1989). Elementary work performed, per unit mass, by magnetization in a magnetic field $\vec{H}$ is shown to read

$$dw_m = \vec{H}.d\vec{\mu} , \tag{2.3}$$

where $\vec{\mu} = \vec{M}/\rho$, ρ being the density of mass. Per unit volume one has

$$dW_m = \rho \; dw_m = \rho\vec{H}.d\vec{\mu} . \tag{2.4}$$

For a rigid body, without loss in generality the last equation can be replaced by

$$\mathcal{P}(\vec{H}) = \frac{dW_m}{dt} = \vec{H}.\frac{d\vec{M}}{dt} = \vec{H}.\dot{\vec{M}} , \tag{2.5}$$

where $\mathcal{P}(\vec{H})$ is the power developed (per unit volume) by the magnetization in the magnetic field $\vec{H}$. The superimposed dot indicates the material time derivative. In such a formulation, $\vec{H}$ appears at the "cause" or "loading", being the *magnetizing* field, and $\vec{M}$ as the "effect" or "response" in the same way as the power developed by infinitesimal strains ε (a symmetric tensor of second order of indexed Cartesian components ε_{ij}) in a stress field σ (a symmetric tensor of second order of indexed Cartesian components σ_{ij}) is given by

$$\mathcal{P}(\sigma) = \sigma : \dot{\varepsilon} \equiv \sigma_{ij} \; \dot{\varepsilon}_{ij} = \sum_{i,j} \sigma_{ij} \; \dot{\varepsilon}_{ij} . \tag{2.6}$$

To pursue the analogy, it is assumed (but this follows from experimental evidence in a simple test) in the elastoplasticity with small strains that ε is *additively* decomposed into a reversible elastic contribution ε^e and an irreversible (plastic) contribution ε^p in such a way that $\varepsilon = \varepsilon^e + \varepsilon^p$. For finite strains a *multiplicative* decomposition of the deformation gradient replaces this additive decomposition (Maugin, 1990a, Chapter 8). Experimental evidence shows that magnetization, in total analogy with small strain, can be decomposed into a reversible part $\vec{M}^R$. While in plasticity finding the evolution equation for ε^p is the ultimate goal (ε^e is related to σ by the elasticity law of state deduced from the energy density), the *evolution equation* of $\vec{M}^R$, giving rise to magnetic hysteresis, should be the main concern of magneticians. In mechanics only the elastic strain is an *observable* thermodynamic state variable ; likewise, in magnetism only $\vec{M}^r$ is the magnetic *observable* thermodynamic state variable. Obviously, we are aware that hysteresis is a complex, thermodynamically irreversible phenomenon and it would be surprising that only $\vec{M}^r$ and $\vec{M}^R$ be sufficient to describe it. At this stage we acknowledge our *relative* ignorance of the underlying microscopic mechanicsms and we propose, for the lack of better arguments, simply to introduce new state variables, called *internal* thermodynamical state variables which we collectively denote by an n-vector $\beta = (\beta_1,\ldots,\beta_n)$. These variables should be measurable by a "gifted" experimentalist but they are not controlable by external agents. As a matter of fact we shall assume that β is composed of a traditional axial three-vector $\vec{M}^{int}$ (its components have the same physical dimension as magnetization) and an n-3 vector $\alpha = (\beta_4,\ldots,\beta_n)$. The physical significance of these variables should be found in a semi-microscopic analysis. Let Ψ be the free energy density per unit volume. To simplify the presentation we neglect here thermal effects. Then for a rigid material, on account of eqn. (25), it is usually shown that the *second law* of thermodynamics can be written in the so-called *Clausius-Duhem* form as (Maugin, 1982, 1988)

$$-\frac{d\Psi}{dt} + \vec{H}.\frac{d\vec{M}}{dt} \geqslant 0 \ . \qquad (2.7)$$

the equality sign holding at thermodynamical equilibrium. On account of above-made comments, for materials exhibiting hysteresis we will write

$$\Psi = \tilde{\Psi}(\vec{M}^r = \vec{M} - \vec{M}^R \ , \vec{M}^{int}, \alpha) \qquad (2.8)$$

Then eqn. (2.7) renders

$$\varnothing_{magn} := \vec{H}^{relax}.\frac{d\vec{M}^r}{dt} + \vec{H}.\frac{d\vec{M}^R}{dt} + \vec{H}^{int}.\frac{d\vec{M}^{int}}{dt} + A.\frac{d\alpha}{dt} \geqslant 0 \ , \qquad (2.9)$$

where we have set

$$\vec{H}^{relax} = \vec{H} - \vec{H}^r \ , \qquad (2.10)$$

and

$$\left.\begin{aligned} \vec{H}^r &= \partial\Psi/\partial\vec{M}^r \ , \\ \vec{H}^{int} &= -\ \partial\Psi/\partial\vec{M}^{int} \ , \\ A &= -\ \partial\Psi/\partial\alpha \ . \end{aligned}\right\} \qquad (2.11)$$

The last group of equations provides the so-called *laws of state*, i.e., the thermodynamical constitutive equations which are derived from the energy potential. The fields $\vec{H}^{int}$ and A are *generalized forces* according to the terminology of irreversible thermodynamics (de Groot and Mazur, 1962). The same holds true of $\vec{H}^{relax}$ and $\vec{H}$ itself. The inequality (2.9) is none other than the *dissipation inequality*. It exhibits four contributions to which one could apply the rules of irreversible thermodynamics, but the expression (2.9) is of great interest in itself.

B. - **Thermodynamical equilibrium**

Thermodynamical equilibrium (corresponding to the equality sign in eqn. (2.9)) is defined by

$$\vec{H}^{relax} = \vec{0} \Longrightarrow \vec{H} = \vec{H}^r = \partial\Psi/\partial\vec{M}^r \ , \tag{2.12}$$

$$\vec{H}^{int} = \vec{0} \ , \ A = 0 \ , \tag{2.13}$$

and

$$\frac{d\vec{M}^R}{dt} = \vec{0} \text{ or } \vec{M}^R = \vec{\mathcal{M}}_0 = \text{const.} \ . \tag{2.14}$$

Then the above energy takes on the simple form

$$\Psi_{eq} = \tilde{\Psi}\left(\vec{M}^r \ ; \ \vec{M}_0^{int}, \alpha_0\right) \ , \tag{2.15}$$

where $\vec{\mathcal{M}}_0$ is a constant (possibly zero) vector and $\vec{M}_0^{int}$ and α_0 are determined by eqns. (2.13), i.e.,

$$\vec{H}^{int} = - \frac{\partial\Psi}{\partial\vec{M}^{int}}\left(\vec{M}^r, \vec{M}_0^{int}, \alpha_0\right) = \vec{0} \ ,$$

$$A = - \frac{\partial\Psi}{\partial\alpha}\left(\vec{M}^r, \vec{M}_0^{int}, \alpha_0\right) = 0 \ . \tag{2.16}$$

In theory these n equations provide $\vec{M}_0^{int}$ and α_0 in terms of $\vec{M}^r$, so that eqn. (2.15) can be rewritten as

$$\Psi_{eq} = \hat{\Psi}\left(\vec{M} - \vec{\mathcal{M}}_0\right) = \check{\Psi}\left(\vec{M};\vec{\mathcal{M}}_0\right) \ . \tag{2.17}$$

C. - **Outside equilibrium**

Outside equilibrium (but not too far so as to respect the working hypotheses of the theory of irreversible processes), one must consider the inequality form of eqn. (2.9). This is of the same form as the inequality that one obtains in deformable solids exhibiting a dissipative mechanical behavior (see, e.g., Maugin, 1990a, Chapters 2,3) for, then, eqn. (2.9) is replaced by

$$\emptyset_{mech} := \sigma^{visc} : \dot{\varepsilon}^e + \sigma : \dot{\varepsilon}^p + \sigma^{int} : \dot{\varepsilon}^{int} + A.\dot{\alpha} \geqslant 0 \tag{2.18}$$

if the *viscous stress* is defined by

$$\sigma^{visc} = \sigma - \sigma^e \ , \tag{2.19}$$

while eqns. (2.11) are replaced by

$$\left.\begin{aligned} \sigma^{e} &= \partial\Psi/\partial\varepsilon^{e} , \\ \sigma^{int} &= -\partial\Psi/\partial\varepsilon^{int} , \\ A &= -\partial\Psi/\partial\alpha , \end{aligned}\right\} \tag{2.20}$$

if we assume a free energy with the following functional dependence

$$\Psi = \tilde{\Psi}(\varepsilon^{e} = \varepsilon - \varepsilon^{p} , \varepsilon^{int} , \alpha) . \tag{2.21}$$

We have thus the correspondence listed in Table 1. This table and a direct scrutiny of eqns. (2.9) and (2.18) reveal essential features. In particular, if $\varepsilon^{p} = \mathbf{0}$ and there are no internal variables, eqn. (2.18) reduces to $\varnothing_{mech} = \sigma^{visc} : \dot{\varepsilon} \geqslant 0$ and this, by use of standard irreversible thermodynamics, e.g., by taking σ^{visc} *linear* in $\dot{\varepsilon}$, yields the classical *Kelvin-Voigt viscoelasticity* of solids (in passing, a *not* very good representation of viscosity). Similarly, if $\vec{M}^{R} = \vec{0}$ and we discard the magnetic internal variables $\vec{M}^{int}$ and α, eqn. (2.9) reduces to

$$\varnothing_{magn} = \vec{H}^{relax} . \frac{d\vec{M}}{dt} \geqslant 0 . \tag{2.22}$$

Magnetism		**Solid mechanics in small strains**	
magnetization (axial vector)	$\vec{M}$	small strain (tensor)	ε
power	$\vec{H}.\dot{\vec{M}}$	power	$\sigma : \dot{\varepsilon}$
magnetic field (axial vector)	$\vec{H}$	stress (tensor)	σ
decomposition	$\vec{M} = \vec{M}^{r} + \vec{M}^{R}$	decomposition	$\varepsilon = \varepsilon^{e} + \varepsilon^{p}$
reversible magnetization	$\vec{M}^{r}$	elastic strain	ε^{e}
residual magnetization	$\vec{M}^{R}$	anelastic (plastic) strain	ε^{p}
dissipation	$\vec{H}.\dot{\vec{M}}^{R}$	dissipation	$\sigma : \dot{\varepsilon}^{p}$
relaxation magnetic fied	$\vec{H}^{relax}$	viscous stress	σ^{visc}
dissipation	$\vec{H}^{relax}.\dot{\vec{M}}^{r}$	dissipation	$\sigma^{visc} : \dot{\varepsilon}^{e}$
internal magnetization	$\vec{M}^{int}$	internal strain	ε^{int}
(n-3) internal variable area of domain walls oriented perpendicular to a certain direction, per unit volume	α	(n-6) internal variable densities of dislocations along different glide systems in the crystal	α

Table 1. State variables and dissipation in magnetism and solid mechanics

For the sake of example, we can apply the standard theory of irrevesible processes to this dissipation inequality for the case of isotropic rigid media. This amounts to considering $d\vec{M}/dt$ as proportional to $\vec{H}^{relax}$, so that accounting for the definition of the latter we obtain the *evolution equation*

$$\frac{d\vec{M}}{dt} = \frac{\chi}{\tau}(\vec{H} - \vec{H}^r) \ , \ \vec{H}^r = \frac{\partial \Psi}{\partial \vec{M}^r} = \chi^{-1}\vec{M} \ , \tag{2.23}$$

with $\chi \geqslant 0$, $\varphi \geqslant 0$. Equation (2.23) is a *thermodynamically admissible relaxation equation* for the magnetization. It can also be written as

$$\frac{d\vec{M}}{dt} = -\frac{1}{\tau}\left(\vec{M} - \vec{M}_{eq}\right) \ , \ \vec{M}_{eq} \equiv \chi\vec{H} \ , \tag{2.24}$$

and this is of the standard *Bloch* type. Generalizations to the case of deformable magnetically saturated ferromagnets ($|\vec{M}|$ = const.) and nonsaturated magnetic fluids where d/dt is replaced by a so-called co-rotational derivative, have been given elsewhere and allow one to produce a nice thermodynamical formulation of the Gilbert and Landau-Lifshitz types of spin-lattice relaxation in ferromagnets (Maugin, 1975) and so-called ferrofluids (Maugin and Drouot, 1983) where orientational relaxation and relaxation in magnitude are separated. Obviously, equations such as (2.23) or (2.24) involve a *characteristic time*, the relaxation time τ.

D. - **Maxwell type behaviour and standard model**

A more curious type of evolution equation can be obtained thus, albeit remaining completely thermodynamically admissible. Consider the case where **(i)** we choose $\alpha = 0$ and impose $\vec{H}^{relax} = \vec{0}$, so that $\vec{H} = \vec{H}^r$; next **(ii)** $\vec{M}$ is made of two parts, the reversible part $\vec{M}^r$ and an irreversible contribution, $\vec{M}^{int}$ *itself*, which does not have here any *residual* nature; We have thus $\vec{M}^R = \vec{0}$, while $\vec{M}^r = \vec{M} - \vec{M}^{int}$; **(iii)** suppose that Ψ depends on $\vec{M}$ and $\vec{M}^{int}$ only through $\vec{M}^r$. It follows from eqns. (2.11) that $\vec{H}^{int} = \vec{H}$, and the inequality (2.9) reduces to

$$\varnothing_{mag} = \vec{H}.\frac{d\vec{M}^{int}}{dt} \geqslant 0 \ . \tag{2.25}$$

Taking $d\vec{M}^{int}/dt$ as linear isotropic in $\vec{H}$, we have

$$\frac{d\vec{M}^{int}}{dt} = \frac{\chi}{\tau}\vec{H} \ , \ \tau > 0 \ , \ \chi \geqslant 0 \ , \tag{2.26}$$

and eliminating $\vec{M}^{int}$ between this and the simple law of state

$$\vec{H} = \vec{H}^r = \chi^{-1}\vec{M}^r \ , \tag{2.27}$$

we are led to a magnetic constitutive equation of the type

$$\vec{H} + \tau \frac{d\vec{H}}{dt} = (\tau/\chi) \frac{d\vec{M}}{dt} \tag{2.28}$$

which is the magnetic equivalent of *Maxwell's model* in the rheology of viscous materials (Maugin 1990a, Chapter 3). It can also be rewritten as a time integral representation for the magnetic field :

$$\vec{H}(t) = \chi^{-1} \int_0^t \vec{M}(t') \exp\left(- \frac{t - t'}{\tau}\right) dt' \ . \tag{2.29}$$

Again, eq. (2.28) or (2.29) contains a time scale. It is of the type of Chua's magnetic constitutive equation (Chua and Stromsmoe, 1970, 1971) as used in practical computations of electrical engineering by Saito *et al* (1987, 1988) although these authors use B and H (as scalars ; this is valid in homogeneous situations in which the direction of one of the axes of reference is taken in the direction of the field) as magnetic variables instead of $\vec{M}$ and $\vec{H}$. It is clear that by weighting adequately the role of classical magnetization relaxation and magnetic-field relaxation by an appropriate choice of internal variables it is possible to generate magnetic constitutive equations of the general *differential* type (presenting *two* time scales ; magnetic equivalent of the so-called *standard model* in rheology)

$$\vec{M} + \tau_1 \frac{d\vec{M}}{dt} = \chi \left(\vec{H} + \tau_2 \frac{d\vec{H}}{dt}\right) \ . \tag{2.30}$$

However, such laws, which can be thermodynamically admissible in the sense that they obey or are consequences of the second law of thermodynamics, will not reproduce the essential fact of hysteresis which is the differing response path in loading and unloading. Appropriate laws must, in some way, involve the *sign of the time increment in the load*, i.e., the sign of $d\vec{H}/dt$. Such a law, not based on thermodynamics, is postulated in the *scalar* case by Coleman and Hodgdon (1986, 1987), following an idea previously given by Bouc (1971). Their general "hysteretic" magnetic constitutive equation reads (using B and H as variables)

$$\dot{B} = \alpha|\dot{H}|[f(H) - B] + \dot{H}\, g(H) \ , \tag{2.31}$$

where α is a positive constant, f is an odd, monotone increasing, piecewise continuous real-valued function (with (df/dH) ($H \to \infty$) having a limit) and g is an even, piecewise continuous, real-valued function. Simple plausible examples of functions f and g in the form of linear piecewise and crenel-type functions are given in Figure 2. The function f is called the *skeleton* function since in practice it fixes the overall shape of the hysteresis loop. Equation (2.31) presents *no* time scale so that the magnetic response does not depend on the rate of magnetization or on the speed at which the specimen is being magnetized. This equation is *ad hoc* in the sense that it is postulated, is not governed by thermodynamics, and has no microscopic physical support. However, it *does* contain an essential ingredient since the absolute value of $\dot{H}$ present in the first contribution in the right-hand side makes that loading and unloading will yield a different response. Note also that since there is no intrinsic time scale, the time parameter present in the derivative can be replaced by any whatsoever convenient monotonically increasing parameter. In particular, elementary increment or any other type of "derivative" can be used instead of the superimposed dot.

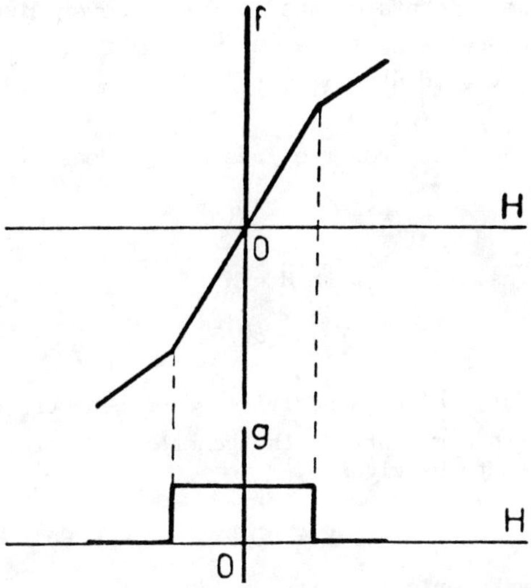

Figure 2. f and g functions for the model (2.31)

The next section has for purpose to show that a thermodynamically admissible model of hysteresis evolution equation (eventually containing *no* time scale) can be deduced by exploiting the dissipation inequality (2.9) within the framework of the classical theory of irreversible processes. For ease in presentation, we shall discard relaxation and assume that α is reduced to a single scalar α, so that enq. (2.9) is reduced to

$$\varnothing_{magn} = \vec{H}.\frac{d\vec{M}^R}{dt} + \vec{H}^{int}.\frac{d\vec{M}^{int}}{dt} + A\,\dot{\alpha} \geqslant 0 \ . \qquad (2.32)$$

3. - THERMODYNAMIC MODEL OF MAGNETIC HYSTERESIS

We base our modeling solely on energy considerations, i.e., on the data of a *free energy* density, which is usually *convex* in its arguments (the notable exception being temperature, but thermal effects are discarded) and a *dissipation potential* also assumed *convex* in its arguments. The dissipation inequality will, therefore, be automatically satisfied. However, the determination of the precise expression of the free energy is a problem in itself. A polynomial expansion, containing very few terms, often is accepted (e.g., a quadratic expression in many cases, with third order and even fourth order terms being added in exceptional cases, in particular when dealing with phase transitions ; then the convexity may be lost).

A. - **Dissipation potential and evolution equations**

At this point it is convenient to introduce a lighter, short hand, unified notation. Let X, Y and Z be elements of the following seven-dimensional spaces :

. "irreversible" state variables : $X = (\vec{M}^R, \vec{M}^{int}, \alpha) \in \mathcal{X}$,

. generalized forces : $Y = (\vec{H}, \vec{H}^{int}, A) \in \mathcal{F}$,

. thermodynamic state variables : $Z = X - (\vec{M}, \vec{0}, 0) \in \mathfrak{Z}$.

The laws of state are derived from the *free energy* density (convex in Z) :

$$\Psi = \Psi(Z) \tag{3.1}$$

through (compare to eqns. (2.11) with $\vec{H} = \vec{H}^r$)

$$Y = -\partial\Psi/\partial Z \ , \tag{3.2}$$

while, $\mathcal{D}(\dot{X})$, $\dot{X} \equiv dX/dt$, being a positively homogeneous (of degree n) *dissipation potential* convex in its arguments $\dot{X}$, the generalized forces present in the inequality (2.32) are assumed to be given by

$$Y = \partial\mathcal{D}/\partial\dot{X} \ . \tag{3.3}$$

Because of homogeneity Euler's identity yields

$$\varnothing := Y\dot{X} = \frac{\partial\mathcal{D}}{\partial\dot{X}}\dot{X} = n\mathcal{D} \geqslant 0 \ , \tag{3.4}$$

so that (2.32) is automatically satisfied. Moreover, if $\dot{X}_\alpha$ and $\dot{X}_\beta$ are any two components of $\dot{X}$, then

$$\frac{\partial^2\mathcal{D}}{\partial\dot{X}_\alpha\partial\dot{X}_\beta} = \frac{\partial^2\mathcal{D}}{\partial\dot{X}_\beta\partial\dot{X}_\alpha} \tag{3.5}$$

is a statement of *Onsager's reciprocity relations*. The evolution equations for X are obtained by eliminating Y between equations (3.2) and (3.3), yielding thus

$$\frac{\partial\mathcal{D}}{\partial\dot{X}} + \frac{\partial\Psi}{\partial Z} = 0 \ . \tag{3.6}$$

For instance, if $\mathcal{D}$ is quadratic in $\dot{X}$ (thus homogeneous of degree two), this gives symbolically

$$\dot{X} \propto \partial\Psi/\partial Z \ . \tag{3.7}$$

A more purposeful presentation is as follows. As $\mathcal{D}$ is supposed to be convex in $\dot{X}$, we can define its so-called *Legendre-Fenchel transform* (see Rockafellar, 1970, for all notions on convexity ; also in abridged form, Appendix I in Maugin, 1990a) :

$$\mathcal{D}^*(Y) = \underset{\dot{X}\in\mathcal{H}}{\text{Sup}}\,[Y\dot{X} - \mathcal{D}(\dot{X})] \tag{3.8}$$

so that

$$\dot{X} = \partial\mathcal{D}^*/\partial Y \ . \tag{3.9}$$

The transform (3.8) conserves both homogeneity and convexity properties. Explicitly, eqn. (3.9) reads

$$\frac{d\vec{M}^R}{dt} = \frac{\partial\mathcal{D}^*}{\partial\vec{H}} \ , \ \frac{d\vec{M}^{int}}{dt} = \frac{\partial\mathcal{D}^*}{\partial\vec{H}^{int}} \ , \ \frac{d\alpha}{dt} = \frac{\partial\mathcal{D}^*}{\partial A} \ , \tag{3.10}$$

while eqns. (3.3) read

$$\vec{H} = \partial\Psi/\partial\vec{M}^r \ , \ \vec{H}^{int} = -\partial\Psi/\partial\vec{M}^{int} \ , \ A = -\partial\Psi/\partial\alpha \ . \tag{3.11}$$

If $\mathcal{D}$ and $\mathcal{D}^*$ are homogeneous of degree *two*, then $\varnothing = 2\mathcal{D}$, and eqns. (3.10) exhibit a characteristic time because Y will be homogeneous of degree *one* in $\dot{X}$. For instance, the first of (3.10) will then necessarily reads

$$\frac{d\vec{M}^R}{dt} = \frac{1}{\tau_R}\vec{G}(Y) \ , \tag{3.12}$$

where τ_R is a characteristic time and dim $|\vec{G}|$ = dim $|\vec{M}^R|$. Equation (3.12) is of the relaxation type and, therefore, is *not* what we look for. Much more interesting is the case where $\mathcal{D}$ and $\mathcal{D}^*$ are homogeneous of degree *one* only. Then Y will be homogeneous of degree *zero* in $\dot{X}$, i.e., it will *not* depend on $\dot{X}$ although there is a dissipation for $\dot{X} \neq 0$! Such a curious case happens when the generalized forces Y are restrained to a convex set C in space $\mathcal{F}$, interior points of C corresponding to a thermodynamically reversible behavior ($\varnothing = 0$) and points on the boundary of C corresponding to a *possible* dissipation. Let $f(Y) = 0$ denote the boundary of C in $\mathcal{F}$-space. The above statement takes the following rigorous mathematical form (see the proof in Chapter 3 in Maugin, 1990a for the case of elastoplasticity)

$$\varnothing = \underset{Y^*\in C}{\text{Sup}}\,(Y^*\dot{X}) \geqslant 0 \tag{3.13}$$

or, equivalently,

$$(Y - Y^*)\dot{X} \geqslant 0 \ , \ \forall\, Y \in C \qquad \text{for } Y^* \in C \tag{3.14}$$

or

$$\dot{X} \propto \partial f/\partial Y \ , \tag{3.15}$$

where $\mathcal{D}^*$ is the so-called *indicator* of the convex C, i.e.,

$$\mathcal{D}^* = \text{Ind } C = \begin{cases} 0 & \text{for } f \leqslant 0 , \\ +\infty & \text{for } f > 0 . \end{cases} \tag{3.16}$$

Because of rate-independence following from the homogeneity of degree one of $\mathcal{D}$ or $\mathcal{D}^*$, the proportionality relation (3.15) can be written as

$$\dot{X} = \dot{\lambda} \frac{\partial f}{\partial Y} , \tag{3.17$_1$}$$

where $\dot{\lambda}$ is a multiplier to be determined. The notation in eqn. $(3.17)_1$ shows that this evolution equation contains *no* time scale. Obviously, to respect (3.13) or (3.14) we must have $\dot{\lambda} = 0$ for $f < 0$ and $\dot{\lambda} > 0$ for $f = 0$. As a matter of fact, the same as in rate-independent elastoplasticity (Maugin, 1990a), one can show the more precise conditions

$$\begin{aligned} &\dot{\lambda} = 0 \quad \text{if} \quad f < 0 \quad \textit{or} \quad \dot{f} < 0 , \\ &\dot{\lambda} \geqslant 0 \quad \text{if} \quad f = 0 \quad \textit{and} \quad \dot{f} = 0 . \end{aligned} \tag{3.17}$$

Equations (3.13) or (3.14) are *variational statements* (in the form of *variational inequalities*) of the normality property (3.15) or (3.17) -the latter property means that $\dot{X}$, if not zero, is directed along the local normal at $f = 0$ in $\mathcal{F}$-space. Equation (3.14) is a *maximal-dissipation* principle akin to the analogous principle by Hill and Mandel in elastoplasticity with hardening (Maugin, 1990a, Chapter 5). Before specifying f, we note a few general properties (which can be proved as in elastoplasticity so that we do not repeat the proof here) :

a) Orthogonality property : If all evolutionary quantities are time-differentiable to the right, then we have the following orthogonality between members of $\dot{\mathcal{H}}$ and $\dot{\mathcal{F}}$:

$$\dot{X}\,\dot{Y} = 0 \tag{3.18}$$

b) Local statibility : If Ψ is convex in $\vec{M}^{int}$ and α, then thanks to the properties of second-order derivatives of convex functions eqn. (3.18) yields

$$\frac{d\vec{H}}{dt} \cdot \frac{d\vec{M}^R}{dt} \geqslant 0 . \tag{3.19}$$

This is analogous to *Drucker's stability "postulate"* in elasto-plasticity with hardening (Maugin, 1990a, Chapter 5). In the scalar case eqn. (3.19) means that increments in H and M^R must have the *same* sign, or $\partial H/\partial M^R \geqslant 0$, thus guaranteeing the *local* stability of all points of the irreversible magnetic response (see Figure 3). We refer to the fact that $\partial H/\partial M^R > 0$ (strictly) as the existence of a *magnetic hardening*. The latter wording is introduced in analogy with mechanical behaviors. Greater hardening means that we need to apply a *larger* increment in the magnetizing field to obtain the *same* increment in magnetization (e.g., material 2 in Figure 3 has a larger magnetic hardening than material 1).

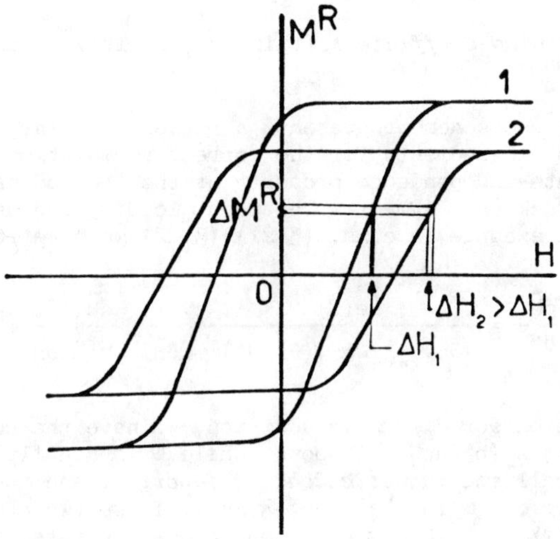

Figure 3. Magnetic hardening : Material 2 is "harder" than material 1

c) Global stability : By direct calculation and using convexity properties and eqn. (3.14) it can be shown that

$$\int_0^1 \vec{H}.d\vec{M} = \int_0^1 \vec{H}.\frac{d\vec{M}}{dt}\,dt \geqslant 0 \quad , \quad \vec{M}(t=0) = \vec{M}(t=1) \tag{3.20}$$

for any closed magnetization cycle (the time interval has been normalized to one in the absence of time scale). Equation (3.20) means that magnetic hysteresis loops are always described *counterclockwise* (in a plane where H is the abscissa and M is the ordinate -see Figure 1).

d) Evaluation of the multiplier $\dot{\lambda}$: Suppose that both Ψ and f are known. Then $\dot{\lambda}$, when it is not zero, is evaluated for $\dot{f} = 0$ [see eqn. (3.17)]. By way of example consider the case where $\vec{M}^{int} = \vec{0}$, so that Ψ reduces to $\Psi(\vec{M}^r, \alpha)$. Furthermore, to simplify the argument, we assume the separability $\Psi = \Psi_m(\vec{M}^r) + \Psi_{int}(\alpha)$. Then we obtain

$$\dot{\lambda} = \frac{He(f)}{h}\left\langle \frac{\partial f}{\partial \vec{H}}.\dot{\vec{H}} \right\rangle \tag{3.21}$$

where $\langle \ldots \rangle$ denotes the positive part of its enclosure,

$$He(f) = \begin{cases} 0 & \text{if} \quad f < 0 \\ +1 & \text{if} \quad f \geqslant 0 \end{cases} \tag{3.22}$$

is a Heaviside characteristic function, and the scalar h defined by

$$h = \left(\partial^2\Psi_{int}/\partial\alpha^2\right)(\partial f/\partial A)^2 > 0 \tag{3.23}$$

is the *magnetic hardening coefficient* (this is a positive quantity because Ψ_{int} is convex in α).

Remark - The above-given scheme represents a rather *singular* behavior. However, it already contains two ingredients for the correct formulation of magnetic hysteresis, namely, the rate-independence property of the thermodynamically irreversible response and the lack of symmetry between loading and unloading (in magnetic fields) since, in the example in eqns. (3.21)-(3.22) we have for instance

$$\frac{d\vec{M}^R}{dt} = \frac{He(f)}{\left(\partial^2\Psi_{int}/\partial\alpha^2\right)(\partial f/\partial A)^2}\left\langle\frac{\partial f}{\partial\vec{H}}\cdot\frac{d\vec{H}}{dt}\right\rangle\frac{\partial f}{\partial\vec{H}} \tag{3.24}$$

with a similar equation for $d\alpha/dt$. In addition, we have the characteristic properties mentioned at points (b) and (c) above. While Ψ is usually known, it remains to specify f, which we call the *magnetic loading function*, and the significance of internal variables (e.g.,, α in the case (3.24)). These two are intimately related, for a proposal for f(Y), which may be based on measurements, determines the significance of the internal variables through eqn. (3.17) (see below).

B. - Magnetic loading function

As a typical loading function we consider a factorizable function that involves only the magnitude (or norm) of the magnetizing field $\vec{H}$ and the scalar generalized force A if we pursue the example in eqns. (3.21)-(3.22) :

$$f(\vec{H},A) = \left(\|\vec{H}\| + A\right)^2 - H_c^2 \tag{3.25}$$

or, obviously,

$$f = f^+ . f^- \quad , \quad f^{\pm} = \|\vec{H}\| + A \mp H_c \ . \tag{3.26}$$

Applying eqn. (3.17) to $\vec{M}^R$ and α, eliminating $\dot{\lambda}$ between the resulting two equations and integrating in time, we find that

$$\alpha(t) = \int_0^t \left[\dot{\vec{M}}^R(t').\chi_f^{-1}.\dot{\vec{M}}^R(t')\right]^{1/2} dt' \equiv \hat{M}^R(t) \ , \tag{3.27}$$

if χ_f^{-1} is the inverse of the invertible matrix of constant coefficients χ_f which accounts for a possible anisotropy ; the norm $\|\vec{H}\|$ has been defined for an anisotropic body by

$$\|\vec{H}\| = \left(\vec{H}.\chi_f.\vec{H}\right)^{1/2} \ . \tag{3.28}$$

We see that for the special case (3.25) of magnetic loading function, the scalar internal variable α is found to be the ***cumulative residual magnetization*** $\hat{M}^R(t)$, a quantity which is determined by the whole integrated past history of irreversible magnetization. This is in total analogy with the introduction of the cumulative plastic strain (also called Odqvist parameter) in elastoplasticity with mechanical hardening (see Maugin, 1990a). We have thus :

$$\begin{aligned} \Psi(t) &= \hat{\Psi}\,(\vec{M}^r(t),\hat{M}^R(t)) \quad , \\ f(\vec{H}(t),A(t)) &= \left[\|\vec{H}(t)\| - (\partial\Psi/\partial\hat{M}^R)(t)\right]^2 - H_c^2 \ . \end{aligned} \tag{3.29}$$

Other cases may be envisaged. For instance, if we set $\alpha = 0$ and consider a *vectorial* internal variable $\vec{M}^{int}$ (with dual generalized force $\vec{H}^{int}$), instead of eqn. (3.25) we would take

$$f(\vec{H},\vec{H}^{int}) = \|\vec{H} + \vec{H}^{int}\|^2 - H_c^2 \ . \tag{3.30}$$

Then by the same argument as in the previous example we find that

$$\vec{M}^{int} = \vec{M}^R + \vec{m} \ , \tag{3.31}$$

where $\vec{m}$ is a constant vector. In that case the internal variable is none other than the residual magnetization itself, up to this constant vector. We have thus

$$\begin{aligned} \Psi(t) &= \tilde{\Psi}(\vec{M}^r(t),\vec{M}^R(t)) \ , \\ f(\vec{H}(t),\vec{M}^R(t)) &= \left\|\vec{H}(t) - (\partial\tilde{\Psi}/\partial\vec{M}^R)(t)\right\|^2 - H_c^2 \ . \end{aligned} \tag{3.32}$$

The examples (3.29) and (3.32) could be combined, yielding a more involved scheme. Now we may ask what is the quantity H_c and what is the convex set C in the two examples examined ?

Coercive field - At a given time t the magnetic loading function yields the equation $f = 0$, and this provides a value of H_c by

$$\begin{aligned} H_c &= \pm\ \|\vec{H}\| + A \qquad \text{(example (3.29))} \\ H_c &= \pm\ \|\vec{H} + \vec{H}^{int}\| \quad \text{(example (3.32))} \ . \end{aligned} \tag{3.33}$$

This defines specific values of the magnetic field intensity which may be called the *magnetic coercive field*. This field is *evolutive* in a *magnetically hardened* material since A or $\vec{H}^{int}$ depends on the past history of loading. At time t, the line segment $(-H_c,+H_c)$ defines a convex on the real line $\mathbb{R}$ of magnetic field intensities. The *normal* to this convex is given by the unit vectors in the positive and negative directions at the right and left end points of this segment. But usually there is only prescribed one *fixed coercive* field for a given magnetic material (when all other parameters than H are fixed). This must correspond to the largest convex segment $(-H_c,+H_c)$ that can be attained during the evolution. This is reached for *saturation* and this leads us to considering the notions of saturation and first magnetization curve.

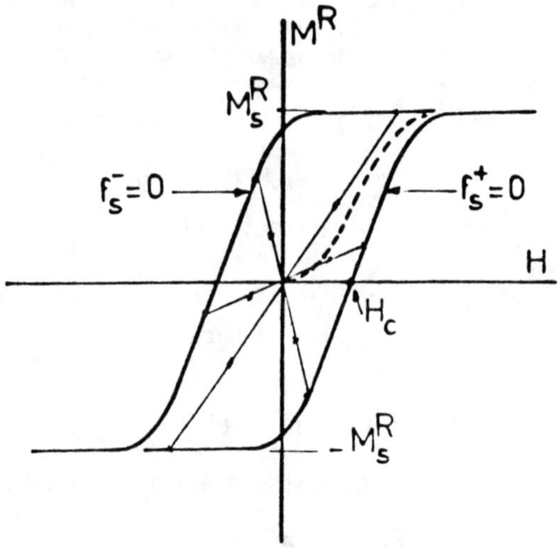

Figure 4. Magnetic loading function f = 0 and coercive field H_c

C. - **Saturation and first magnetization curve**

Magnetic hysteresis is characterized by a *saturation in magnetization*, more precisely, in the residual magnetization. Bearing in mind the mechanical analogy in Table 1, this would correspond to a *saturation in strain* in solid mechanics, i.e., to bodies with bounded deformations. This is rare but so-called *locking* materials, first considered by Prager (1957), are just that. Some biological materials behave in this way. It is only very recently that some mechanicians have recognized the similarity between the plasticity of locking materials and magnetic hysteresis (e.g., Suquet, 1985). Thus the branches $f^{\pm}$ of the magnetic loading function must be such that (see Figure 4)

$$\lim_{|\vec{H}| \to \pm\infty} |\vec{M}^R| \ (H \in f^{\pm}) = \pm M_s^R \ . \tag{3.34}$$

This, in turn, means that $\partial H/\partial M^R \longrightarrow \pm\infty$ as $|\vec{H}| \longrightarrow \pm\infty$, and thus *magnetic hardening* increases to reach infinity at saturation. It remains to determine in what way do we reach the final hysteresis loop, hence also saturation and the widest convex C defining the coercive field, by starting from a virgin state ($\vec{H} = 0$, $\vec{M}^R = 0$). This relates to the notion of *first magnetization curve*, i.e., an odd, monotone increasing, real-valued function $H = \varphi(M^R)$ such that $\partial\varphi/\partial M^R > 0$, for any H. This curve, in some way, provides the one-line *skeleton* for the final hysteresis loop. It should be given by a microscopic or mesoscopic analysis. For a defective magnetizable material this curve is usually of the type represented in Figure 5b, where each region corresponds to a different mechanism of magnetization. The best mesoscopic analysis providing a *differential equation* for that curve is, in our opinion, the one given in Atherton and Jiles (1986) and Jiles and Atherton (1983, 1984), which combines the notion of *Langevin function* $\mathcal{L}$ (that would describe the magnetization curve in the absence of structural defects in the ferromagnet) and the notion of pinning of magnetic walls on structural defects, to provide an equation of the type

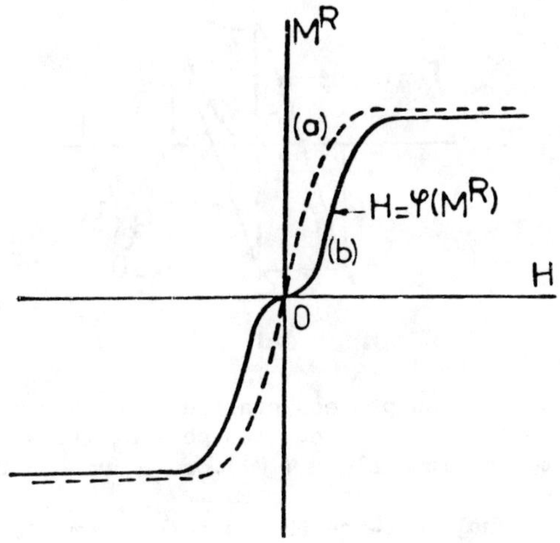

Figure 5. **(a)** Magnetization curve for a perfect ferromagnetic substance
(b) First magnetization curve for a defective ferromagnet

$$H_A = (\text{const.}) \times \mathcal{L}^{-1}\left(-\,k\,\frac{dM}{dH_e}\right) , \tag{3.35}$$

where $\mathcal{L}(x) = \coth x - x^{-1}$, $H_e = H + \alpha M$ is an effective field in the tradition of the molecular field of Pierre Weiss, k is a *parameter of pinning*, and H_A is a phenomenological (magnetic) pinning field (see Sabir, 1985, pp. 148-155). J.A. and B. Szpunar (1985) have given an expression for k in terms of the average of the change in orientation of magnetization through domain walls and the pinning energy of the walls. Thus we may assume that some theory provides a satisfactory first magnetization curve. One can then envisage a scenario for constructing the ferromagnetic hysteresis loop from a virgin state with a curve $\varphi(M^R)$ such that

$$\varphi^{-1}(H \longrightarrow \pm \infty) = \pm M_S^R . \tag{3.36}$$

D. - Magnetic loading scenario

The point at which we reverse the sign of the increment of loading along the first magnetization curve fixes the evolutionary convex C introduced above. That is, loading from a virgin state and increasing H until point A along the curve φ, defines the branch $f_A^+ = 0$, as also the branch $f_A^- = 0$ obtained by symmetry through the origin. A decrease of H from A then imposes that we employ (e.g.) eqn. (3.24) with $\dot{\vec{M}}^R = \vec{0}$ and thus we jump to f_A^- as indicated in Figure 6. Had we not changed sign in dH/dt at A, we would have continued to load along $\varphi(M^R)$ until we reach saturation and then a change in sign of $\dot{H}$ will make the representative point go along $f_S^- = 0$ until a further increase in H brings this point back on $f_S^+ = 0$ with a width

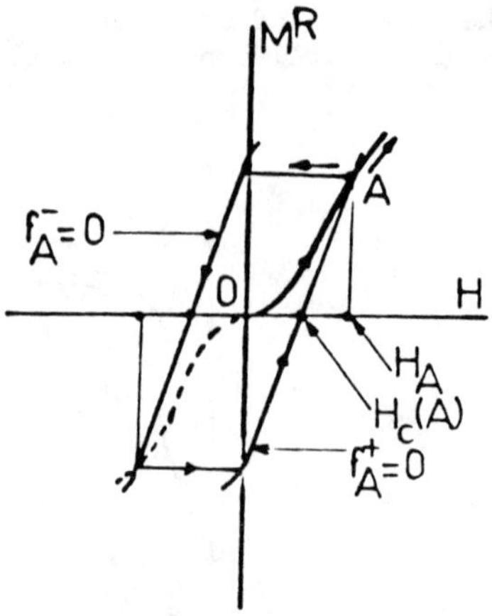

Figure 6. Magnetization process starting from a virgin state 0 :
At point A, there are two possibilities :
either to continue along $\varphi(M^R)$ or to jump on $f_A^- = 0$.

of the loop at zero M^R equal to twice the *coercive magnetic field of the material.* This last scenario is the one obtained when a sufficiently intense alternating H is applied at once. For alternating H fields growing slowly in amplitude we whill have a response of the type in Figures 7 and 8 : Figure 7 shows a schematic scenario where both φ and $f^{\pm} = 0$ curves are piecewise linear. Figure 8 gives a numerical simulation for a more plausible hysteresis loop. Figure 9 shows how, after identification of parameters, we can make the theoretical *final* hysteresis loop practically coincide with an experimentally measured one (the latter is borrowed from Jiles and Atherton, 1984).

4. - MICROSCOPIC BASIS FOR THE EVOLUTION OF M^R

According to the above scheme where f assumes a factorizable expression f^+f^- as in eqns. (3.25) and (3.30), we obtain an evolution equation for $\vec{M}^R$ in the *rate-independent form*

$$\frac{d\vec{M}^R}{dt} = \pm\, 2\, \dot{\lambda}\, \frac{H_c}{\|\vec{H}\|}\, \chi_f . \vec{H}\ , \qquad (4.1)$$

where H_c is defined by $f^{\pm} = 0$ at H(t), and the expression of $\dot{\lambda}$, containing a *positive part*, is given by (3.21) in the case where α is the only internal variable. If it is possible to give a microscopic support to (4.1), then the theory will be practically complete. This has been attempted by Sabir and MAugin (1988), in terms of the irreversible motion of a population of magnetic domain walls by exploiting ideas of Néel's (1946) pioneering work. Their essential result is as follows. For *one* domain wall of area S and unit normal $\vec{n}$, and across which the magnetic moment density $\vec{m}$ suffers a finite jump $[\![\vec{m}]\!]$, with an applied magnetic field $\vec{H}$ and a mean free path L_w between two successive stops of the wall on pinning points, we have the following *microscopic* evolution equation

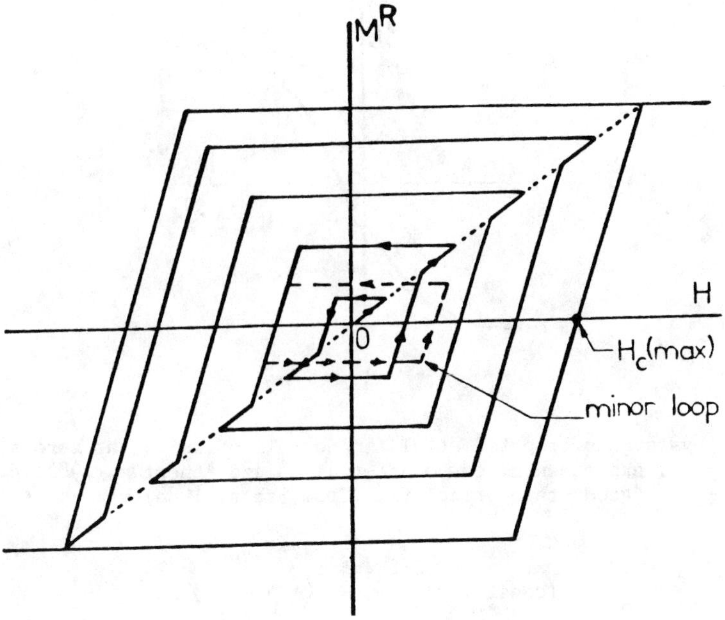

Figure 7. Alternating magnetic loading of a fictitious ferromagnetic sample (both first magnetization curve and loading function are piecewise linear).

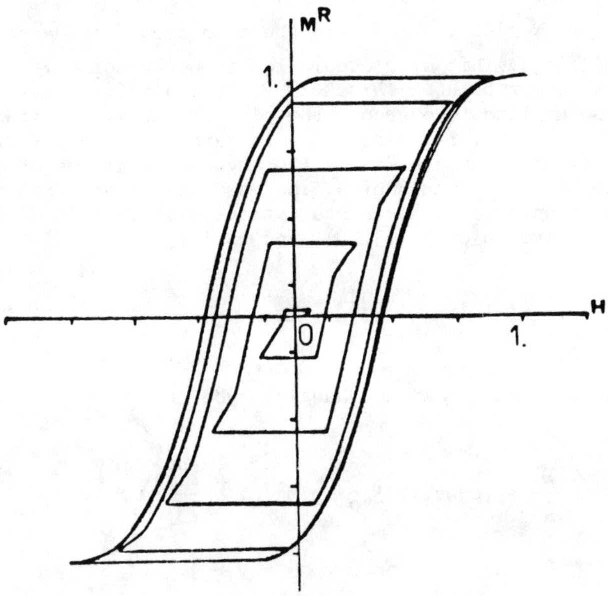

Figure 8. Realistic alternating magnetic loading from a virgin state until saturation (from Sabir, 1988).

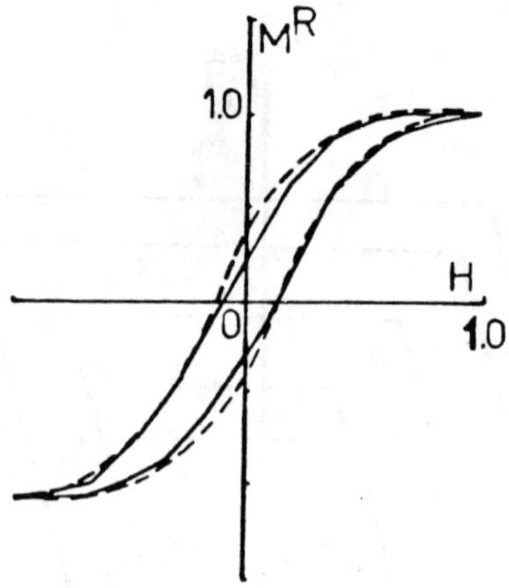

Figure 9. Hysteresis loop for steel (in normalized units) at zero stress :
___ : experimental curve after Jiles and Atherton (1984, p.1273)
--- : "good" theoretical fit (from Sabir, 1988).

$$\frac{d\vec{m}^R}{dt} = (\text{const.}) \times \frac{2SL_m \|\vec{m}\|^2}{\tau_T} \left\langle 1 - \frac{\|\vec{H}_A\|}{\|\vec{H}\|} \right\rangle \mathcal{A}.\vec{H} \ , \tag{4.2}$$

where $\mathcal{A}$ is a simple matrix whose numerical expression depends on the type of wall (e.g., so-called 180° and 90° walls), $\|\vec{H}_A\|$ is related to the pinning energy W_p of the wall by $\|\vec{H}_A\| = W_p/|[\![\vec{m}]\!]|$, and τ_T is the mean *transit time* of the wall between two successive pinnings. This microscopic law is *not* precisely of the same form as eqn. (4.1). It involves both a *threshold* (i.e., a sufficiently intense magnetic field needs to be applied to overcome the pinning energy) typical of plastic-like behaviors and a *characteristic time* which is rather typical of viscous-like behaviors. We may say that eqn. (4.2) is the magnetic analogue of a *viscoplasticity* evolution equation in solid mechanics. The essential properties of (4.2) are carried to the *macroscopic* level for a statistical distribution of domain walls. The final result is [compare to eqns. $(3.10)_1$ and (3.12)]

$$\frac{d\vec{M}^R}{dt} = \frac{\partial \mathcal{D}^*}{\partial \vec{H}} \tag{4.3}$$

where it can be shown that (Sabir and Maugin, 1988)

$$\frac{\partial \mathcal{D}^*}{\partial \vec{H}} = (\text{const.}) \times \frac{S'L_W}{2\tau_T} M_S^2 \left\langle 1 - \frac{\|\vec{H}_A\|}{\|\vec{H}\|} \right\rangle \chi_f.\vec{H} \ , \tag{4.4}$$

where S' is the total area swept out by all walls in their motion and $\vec{H}_A$ is an averaged pinning magnetic field. A complete analogy with the viscoplastic behavior of so-called *Bingham fluid* (viscous fluids, like tooth past, that flow as a rigid block for sufficiently high shear -cf. Germain, 1973) can be drawn. This is reported in Table 2 together with other analogies which are more or less straightforward.

Because of the presence of a time scale, equations such as (4.2) and (4.3) correspond to a rather smooth behavior which contrasts with the rather *singular* correspondence between $\vec{H}$ and $\dot{\vec{M}}^R$ obtained in the rate-independent modeling. The latter case can be recovered in a limit procedure where the transit time τ_T becomes zero so that wall jumps (corresponding to so-called Barkhausen jumps) become quasi-instantaneous. In this limit which is well known in viscoplasticity (see, e.g., Maugin 1990a, Chapter 3), the singular correspondence between $\vec{H}$ and $\dot{\vec{M}}^R$ is recovered while $\mathfrak{D}^*$ becomes the indicator of the convex C defined by $\|\vec{H}_A\|$ and the magnetic loading surface $f(\vec{H})$ is obtained as a *singular* sequential limit of surfaces at constant potential $\mathfrak{D}^*$ (this also occurs in plasticity considered as a limit of viscoplasticity for vanishing viscosity, Kestin and Rice, 1970).

Magnetism	Solid mechanics
Saturation in magnetization	locking
magnetic loading surface	yield surface
cumulative residual magnetization	cumulative plastic strain (Odqvist parameter)

Micromagnetism	Viscoplasticity of crystals
transition time of wall	waiting time of dislocation
Pinning pressure	Peierls-Nabarro force
Evolution equation :	Orowan's formula :
$\vec{m}^R = - [\![\vec{m}]\!] \, S \, V_{wn}$,	$\dot{\varepsilon}^p = \rho_m \, [\![u]\!] \, \bar{v}$,
$[\![\vec{m}]\!]$: jump in magnetic moment (a domain wall is a two-dimensional defect related to a discontinuity in magnetization)	$b = [\![u]\!]$: Burgers vector (a dislocation is a one-dimensional defect related to a discontinuity in elastic displacement)
S : area of domain wall	ρ_m : density of mobile dislocations
V_{wn} : mean normal speed of the wall	$\bar{v}$: mean velocity of dislocation
$(\vec{H} - \vec{H}^*).\vec{m}^R \geqslant 0$	Hill-Mandel maximal dissipation principle

Residual magnetization resulting from the motion of a distribution of domain walls	**Evolution equation for Bingham fluids (rheology)**
Magnetization rate : $\dot{\vec{M}}^R$	strain rate (tensor) : $\mathbb{D}$
transition time : τ_R	viscosity : η
pinning field : $\|\vec{H}_A\|$	stress threshold : g
magnetic field : $\vec{H}$	deviator of stress tensor : σ^d
norm of magnetic field : $\|\vec{H}\|$	(second invariant of stress)$^{1/2}$: $(\sigma_{II})^{1/2}$

Table 2. Analogy between irreversible magnetization and viscoplasticity

5. - COUPLING WITH OTHER FIELDS

Not only does the above-given approach provide a sound thermodynamical treatment while showing how a correlation with mesoscopic analyses for both the first magnetization curve and the magnetic hardening can be established, but it also points at the direction to be taken to account for the effects of various agents such as temperature, stresses, and others like irradiation, etc. Indeed these have effects not only on reversible magnetic properties (e.g., through pyromagnetism, magnetostriction, etc) but they also appreciably affect magnetic hysteresis loops. If we know how to "construct" the hysteresis loop, then we also know where external influences can be felt since the latter will mainly affect the level of saturation (hence the first magnetization curve through its limit value), the witdth of the hysteresis loop (hence the value of the coercive field) and the slope of the hysteresis loop at the coercive field. The latter is related to the Barkhausen noise measurements (Rudyak, 1971) which, therefore, can be envisaged as a means to study the effects of the external agents on the magnetic properties and possibly to "measure" these external agents in a more or less accurate manner. This is the case of *residual stresses*. From the microscopic viewpoint, it is clear that residual stresses intervene in the definition of the coercive field since both H_A in eqn. (3.35) and $\vec{H}_A$ in eqn. (4.4) are related to the energy of pinning on structural *defects*. Experimentally, it is noticed that the effect of stresses on the saturation level M_S^R is strongly *nonlinear* (see the experimental curves in Jiles and Atherton, 1984). The same holds true of the effect on the value of the coercive field. At least we know that if stresses have for effect to symmetrically widen or narrow the hysteresis loop, then they must be taken into account in the magnetic loading function exactly like a scalar internal force A. If they cause a nonsymmetric change, then they must intervene like a vectorial internal variable (these differences would disappear in a fully scalar case). Thus, Maugin (1989) and Sabir (1988) have been able to reproduce the *tendency* (but not quite exactly from a quantitative viewpoint) the change in shape of magnetic hysteresis loops of stressed steels. Influence of a *bias magnetic field* having in some cases the tendency to lock in magnetic moments in a preferred direction (thus shifting the hysteresis loop towards a definite direction must be accounted for like a vectorial internal variable [compare eqn. (3.30)]. Other possibilities can be envisaged for the effects of temperature and neutron irradiation. A similar study has been given in detail elsewhere for ferroelectric hysteresis in stressed ceramics (see Bassiouny *et al*, 1988a,b ; 1989 a,b).

6. - CONCLUSION

The modeling briefly outlined in the preceding section has, in our opinion, inherent qualities which clearly delineate it from other modelings : (**i**) it does not consist in a purely *ad hoc* set mathematical representation, as was the case of many previous models, since it satisfies non-equilibrium thermodynamics (a not surprising requirement for a modeling involving *dissipative* processes) while it looks like the limit, for short characteristic times, of a physically motivated dynamical model of microscopic magnetization process. It *does* provide hysteresis, which is not the case of simple differential models such as (2.30) which, as noticed before, also receive a thermodynamical support (Kluitenberg, 1981) ; (**ii**) it provides an operational way to *construct* the hysteresis loops : the very incremental form of the relevant equations is what is needed for that. This has no relation whatever to analog simulations (Abe, 1985). Finally, (**iii**) contrary to other recent approaches (e.g., Visintin (1984a,b ; 1986), while it involves relatively simple mathematics and mathematical formalism (obviously, *convexity* plays a predominant role but this can be dealt with in simple, albeit correct, terms), it also shows how influence of other fields on magnetic hysteresis can be accounted for in a rational way. Only the unpublished Ph.D Thesis of A.F. Ghaleb (Univ. of Moscow, Institute of Mechanics, 1979) and a work by Chernyi (1983) seem to go in the same direction.

Summary - Hysteresis, of which magnetic hysteresis is the most representative example, is a complex nonlinear irreversible behavior which results from rather complicated microscopic mechanisms (e.g., irreversible wall motion). While many works offer only a mathematical representation of this behavior, a modeling which automatically satisfies the laws of thermodynamics is presented. It uses the notions of dissipation function, internal variables of thermodynamical state, and arguments of convexity. Such notions as residual magnetization, cumulative (or past history of) residual magnetization, magnetic hardening, magnetic loading function and first magnetization curve are introduced in a natural way. This thermodynamical formulation yields local and global stabilities along hysteresis loops, allows a constructional approach to these loops, and shows how couplings with other fields (such as stresses) of interest in non-destructive testing techniques, should be accounted for in the alteration of hysteresis loops. A clear distinction between relaxation processes and hysteresis-like effects is made while strong analogies with certain mechanical behaviors of deformable solids are mentioned. Illustrations are given. This phenomenological, but thermodynamically admissible, approach is corroborated by the microscopic model of jumping magnetic domain walls in the limit of quasi-instantaneous motion of these walls otherwise pinned on structural defects. The same type of approach applies to electric hysteresis and its couplings with the strain field in ferroelectrics and ceramics.

Acknowledgements - For many enlightening discussions and help in clarifying several points we are indebted to M. Sabir, A.F. Ghaleb and E. Bassiouny. In the past we have also benefited from discussions with A. Visintin and illuminating papers kindly provided by G.A. Kluitenberg.

References

Abe, M. 1985. Analog simulation of magnetic hysteresis and its applications. *Int. J. Nonlinear Mech., 20* : 343-358.

Atherton, D.L., and D.C. Jiles. 1986. Effects of stresses on Magnetization. *NDT International 19* : 15-19.

Bassiouny, E., A.F. Ghaleb, and G.A. Maugin. 1988a. Thermodynamical formulation for coupled electromechanical hysteresis effects. I-Basic equations. *Int. J. Engng. Sci., 26* : 1279-1295.

Bassiouny, E., A.F. Ghaleb, and G.A. Maugin. 1988b. Thermodynamical formulation for coupled electromechanical hysteresis effects. II-Poling of ceramics. *Int. J. Engng. Sci., 26* : 1297-1306.

Bassiouny, E., and G.A. Maugin. 1989a. Thermodynamical formulation for coupled electromechanical hysteresis effects. III-Parameter identification. *Int. J. Engng. Sci., 27* : 975-987.

Bassiouny, E., and G.A. Maugin. 1989b. Thermodynamical formulation for coupled electromechanical hysteresis effects. IV-Complex loadings. *Int. J. Engng. Sci., 27* : 989-1000.

Bouc, R. 1971. Modèle mathématique d'hystérésis. *Acustica 24* : 16-25.

Callen, H.B. 1960. *Thermodynamics* First Edition. John Wiley and Sons, New York.

Chernyi, L.T. 1983. Models of ferromagnetic continuous media with magnetic hysteresis. In *Macroscopic theories of matter and Fields*, ed. L.I. Sedov, 116-140. Mir Publishers, Moscow.

Chua, L.O., and K.A. Stromsmoe. 1970. Lumper-circuit model for a nonlinear inductor exhibiting hysteresis loops. *I.E.E.E. Trans. Circuit Theory CT-17* : 564-574.

Chua, L.O., and K.A. Stromsmoe. 1971. Mathematical model for dynamic hyseresis loops. *Int. J. Engng. Sci., 9* : 564-574.

Coleman, B.D., and M.L. Hodgdon. 1986. A constitutive relation for rate-independent hysteresis in ferromagnetically soft materials. *Int. J. Engng. Sci., 24* : 897-919.

Coleman, B.D., and M.L. Hodgdon. 1987. On a class of constitutive relations for ferromagnetic hysteresis. *Arch. Rat. Mech. Anal., 99* : 375-396.

Gardner, C.G., G.A. Matzkanin, and D.L. Davidson. 1971. The influence of mechanical stress on magnetization process and Barkhausen jumps in ferromagnetic materials. *Nondestructive Testing 3* : 131-169.

Germain P. 1973. *Mécanique des milieux continus*. vol. I. Masson, Paris.

de Groot, S.R., and P. Mazur. 1962. *Non-equilibrium Thermodynamics*. North-Holland, Amsterdam (reprint, Dover, New York, 1986).

Hoselitz, K. 1952. *Ferromagnetic properties of metals and alloys*. Oxford University Press, Oxford, U.K.

Jiles, D.C., and D.L. Atherton. 1983. Theory of ferromagnetic hysteresis. *J. Appl. Phys., 55* : 2115-2120.

Jiles, D.C., and D.L. Atherton. 1984. Theory of magnetization process in ferromagnets and its application to the magnetomechanical effects. *J. Phys. D. Appl. Phys., 17* : 1255-1281.

Karjalainen, L.P., and M. Moilanen. 1979. Detection of plastic deformation during fatigue of mild steel by the measurement of Barkhausen noise. *NDT International, April 1979* : 51-55.

Kestin, J., and J.M. Rice. 1970. Paradoxes in the application of thermodynamics to strained solids. In *A Critical Review of Thermodynamics*, ed. F.B. Stuart, B. Gal'Or and A.J. Brainard, 275-298. Mono Book Corp., Baltimore.

Kluitenberg, G.A. 1977. On dielectric and magnetic relaxation and vectorial internal degrees of freedom in thermodynamics. *Physica, 87a* : 541-563.

Kluitenberg, G.A. 1981. On vectorial internal variables and dielectric and magnetic relaxation phenomena. *Phenomena 109A* : 91-122.

Krasnosel'skii, M.A., and A.V. Pokovskii. 1988. *Systems with hysteresis*. (translation from the 1983 Russian edition). Springer-Verlag, Berlin.

Lemaître, J. and J.L. Chaboche. 1988. *Mechanics of solid materials* (translation from the 1985 French edition). Cambridge University Press. Cambridge, U.K.

Matzkanin, G.A., and C.G. Gardner. 1976. Measurement of residual stress using Magnetic Barkhausen noise analysis. In *Proc. ARPA/AFML Review of Quantitative NDE*, Techn. Rep. AFML-TR-75-212, 791-812, AFML, California.

Maugin, G.A. 1975. On the spin relaxation in deformable ferromagnets. *Physica 81a* : 454-458.

Maugin, G.A. 1976. Micromagnetism. In *Continuum Physics*, vol. III, ed. A.C. Eringen, 213-312. Academic Press, New York.

Maugin, G.A. 1979a. Classical magnetoelasticity in ferromagnets with defects. In *Electromagnetic interactions in elastic solids*, ed. H. Parkus, 243-324, Springer-Verlag, Vienna.

Maugin, G.A. 1979b. Vectorial internal variables in magnetoelasticity. *J. Mécanique, 18* : 541-563.

Maugin, G.A. 1981. Electromagnetic internal variables in electromagnetic continua. *Arch. Mechanics, 33* : 927-935.

Maugin, G.A. 1982. A Simple thermodynamical model for rigid ferromagnets. *Phys. Rev. B25* : 7015-7025.

Maugin, G.A. 1988. *Continuum Mechanics of electromagnetic solids*. North-Holland, Amsterdam.

Maugin, G.A. 1990a. *Thermomechanics of plasticity and Fracture*. Cambridge University Press, Cambridge, U.K.

Maugin, G.A. 1989. Coupled magnetomechanical and electromechanical hysteresis effects. In *Applied Electromagnetics in Materials*, ed. K. Miya, 5-18. Pergamon Press, Oxford.

Maugin, G.A. 1990b. Nonlinear dissipative effects in electrodeformable solids. In *Nonlinear Electromechanical Couplings*, G.A. Maugin, B. Collet, R. Drouot and J. Pouget, Chapter 6, Manchester University Press, Manchester, U.K.

Maugin, G.A., and E. Bassiouny. 1989. Thermodynamics of electromechanical hysteresis in Ceramics. In *Continuum Mechanics and its Applications*, ed. G.A.C. Graham and S.K. Malik, 225-235, Hemisphere Publ., New York.

Maugin, G.A., and R. Drouot. 1983. Thermomagnetic behaviour of magnetically nonsaturated fluids. *J. Magnetism and Magnetic Materials, 39* : 7-10.

Maugin, G.A., and A. Fomethé. 1982. On the viscoplasticity of ferromagnetic crystals. *Int. J. Engng. Sci., 20* : 885-908.

Maugin, G.A. and M. Sabir. 1990. Mechanical and Magnetic hardening of ferromagnetic bodies : Influence of residual stresses and applications to non-destructive testing. *Int. J. Plasticity*.

Maugin, G.A., M. Sabir and P. Chambon. 1987. Coupled magnetomechanical hysteresis in ferromagnets : Application to non-destructive testing. In *Electromagnetomechanical Interactions in Solids and Structures*, ed. Y. Yamamoto and K. Miya, 255-264, North-Holland, Amsterdam.

Mayergoyz, I. 1986. Mathematical models of hysteresis. *I.E.E.E.Tran. Magnetics MAG-22* : 603-608.

Meixner, J. 1961. Der Drehimpulssatz in der Thermodynamik der irreversible Prozesses. *Zeit. Phys., 16* : 145-155.

Néel, L. 1943. Théorie des lois d'aimantation de Lord Rayleigh, Parts I and II. *Cahiers de Physique* (Paris) *12* : 1-20 (1942) ; *13* : 18-30.

Néel, L. 1946. Bases d'une nouvelle théorie générale du champ coercitif. *Ann. Univ. Grenoble, 22* : 299-343.

Pashley, R.L. 1970. Barkhausen effect-An indicator of stress. *Mater. Evaluation, 28* 157-161.
Prager, W. 1957. On ideal locking materials. *Trans. Soc. Rheology, 1* : 169-175.
Preisach, F. 1935. Uber die Magnetische Nachwirkung. *Zeit. Phys., 94* : 277-302.
Rautioaho, R., P. Karjalainen, and M. Moilanen. 1987. Stress response of Barkhausen noise in a tempered C-Mn steel. *J. Magnetism and Magnetic Materials, 68* : 314-320.
Rockafellar, R.T. 1970. *Convex analysis*. Princeton University Press, Princeton, New Jersey.
Rosensweig, R.E. 1989. Thermodynamics of Electromagnetism. In *Thermodynamics : An Advanced Textbook for Chemical Engineers*, ed. G. Astarita (to be published).
Ruch, T.C., and H.O. Patton. 1965. *Physiology and biophysics*. New York.
Rudyak, V.M. 1971. The Barkhausen effects. *Soviet Phys. Uspekhi, 13* : 451-479.
Sabir, M. 1988. Phenomenological Modelling of ferromagnetic noise in stressed materials (in French). *Doctoral Thesis*, Université Pierre et Marie Curie, Paris.
Sabir, M. and G.A. Maugin. 1988. Microscopic foundations of the Barkhausen effect. *Arch. Mechanics, 40* : 829-841.
Saito, Y., K. Fukushima, S. Hayano, and N. Tsuya. 1987. Application of a Chua type model to the loss and skin effect calculation. *I.E.E.E. Trans. Magnetics MAG-23* : 2227-2229.
Saito, Y., Y. Kishino, K. Fukushina, S. Hayano, and N. Tsuya. 1988. Modelling of magnetization characteristics and faster magnetodynamic field computation. *J. Appl. Phys., 63* : 3174-3178.
Suquet, P.M. 1985. Locking materials and hysteresis phenomena. In *Unilateral Problems in Structural Mechanics*, ed. G. del Pietro and F. Maceri, 339-373, Springer-Verlag, Vienna.
Szpunar, J.A., and B. Szpunar. 1985. Modelling of the ferromagnetic hysteresis in textured and strained steel. *J. Appl. Phys., 57* : 4232-4234.
Visintin, A. 1984a. On the Preisach model of hysteresis. *Nonlinear Anal. Theory Meth. and Appl., 8* : 977-996.
Visintin, A. 1984b. On the evolution of ferromagnetic media. *Mathematical Modelling, 5* : 43-51.
Visintin, A. 1986. On some microscopic models of ferromagnetism. *Rend. Sem. Nat. Univ. Politecn. Torino, 43* : 371-388.
Volterra, V. 1956. *Theory of functionals and integro and integrodifferential equations*, Reprint, Dover, New York.

Nonequilibrium Thermodynamics of Solutions of Macromolecules

G. A. Maugin and R. Drouot
Laboratoire de Modélisation en Mécanique (URA 229)
Université Pierre et Marie Curie
Tour 66 - 4 place Jussieu
75252 Paris Cedex 05, France

ABSTRACT

This contribution sets forth in one place the basic elements and most fruitful applications of the phenomenological thermodynamics of dilute solutions of macromolecules. The continuum modelling presented is based on a strict use of non-equilibrium thermodynamics extended so as to account for the new phenomena related to the presence of electrically polarizable macromolecules in the solution. That is, macromolecules alter rheological and optical properties of the solution while the surrounding flow and applied electric fields influence the conformation of macromolecules, or macroions in the case of polyelectrolytes. The conformation in fact is the relevant new thermodynamical variable of state, here considered as *internal*, in the sense that it is not directly controllable through boundary conditions. The application of the theory of irreversible processes allows a rational construction, not only of the missing constitutive equations, but also of evolution equations for conformation, electric polarization, and chemical processes. Some of these phenomena couple in such a way that they do not, in fact, contribute to the dissipation, being very similar to gyroscopic phenomena. Direct applications of the thermodynamically admissible formulation of these and of true dissipative processes include equilibrium conformations in the presence or not of external forcing, flow-induced conformational phase transitions and electric polarization, electrically-induced conformational phase transitions, and the mechanico-chemical effect. Diffusion and migration of macromolecules and coupling between local vorticity and conformation can be included in the thermodynamical formulation to the price of some complications. Liquid crystals and other complex fluids can be described in the like manner. Finally, the whole approach can be shown to be given a Hamiltonian metriplectic structure proper to many systems of evolutions in spite of their obvious dissipation. The contribution reports works by the authors and co-workers in the last ten years.

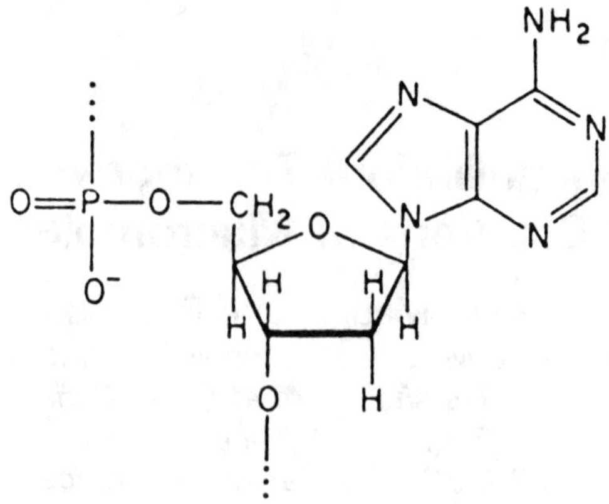

Example of chemical structure of polyelectrolyte : Deoxyribo-nucleic acid neutralized (after Oosawa, 1971).

1. - INTRODUCTION

In this contribution we set forth in one place and in a somewhat unified manner the basic elements of the phenomenological thermodynamics of dilute solutions of poly-electrolytes and briefly review the consequences of this type of thermodynamical approach insofar as the discussion of the evolution, equilibrium figures and stability of the conformation of macromolecules in a fluid carrier is concerned. As a matter of fact, starting with a work of Lhuillier and Ouibrahim (1980) and the review paper of the authors (Maugin and Drouot, 1983) which somewhat organized and systematized the subject matter (at least for dilute polymeric solutions), a phenomenological thermodynamics of solutions of polymers exhibiting or not electrical properties and electromechanical couplings has been developed step by step in a series of papers (Drouot, 1987 ; Drouot and Maugin, 1985, 1987, 1988a,b ; Drouot Maugin and Morro, 1987 ; Maugin and Drouot, 1983, 1988 a,b ; Morro *et-al*, 1985, 1987, 1990), which combines as main ingredients simple elements of the theory of continua with electromagnetomechanical interactions (Maugin, 1980, 1988) and the thermodynamical formalism of so-called *internal variables* (Bampi and Morro, 1984, Maugin, 1987). The latter formalism, which is the slightest deviation from the standard theory of irreversible processes (de Groot and Mazur, 1962)-, assumes that at the macroscopic phenomenological scale, in addition to usual *observable* state variables, there exist other state variables, called *internal* variables, that we introduce to mask our (relative) ignorance of underlying microscopic processes, and of which we know that they can possibly be measured by some gifted experimentalist but they cannot be controlled. In addition, these variables, finite in number, have associated with them characteristic (relaxation) times which are much shorther than usual time scales of macroscopic thermodynamical processes. When they are frozen, a certain equilibrium state follows. Then the true thermo*dynamics* of continua with internal variables can be envisioned as a sequence of such accompanying local state equilibria (Kestin and Bataille, 1975). Internal variables therefore belong in the same class as chemical reaction progress rates in thermo-chemistry since evolution equations must be given for them. The modern formalism of internal variable theory requires that such evolution equations be not in contradiction with the entropy production inequality. This provides an example where concepts issued originally from the microscopic world are constrained to satisfy a macroscopic principle. The *conformation* of macromolecules (Kuhn and Kuhn, 1945 ; Cerf, 1957) is a good candidate for an internal variable in the phenomenological description of polymers and polyelectrolyte solutions. This is what is exposed in the subsequent developments on account of various couplings with electric entities. The reader shall find elsewhere a comparison between the present approach, the kinetic theory of polymeric solutions, and other phenomenological approaches that make abstraction of any conformation or use other types of internal variables (Drouot, 1987 ; Drouot and Maugin, 1988b ; Maugin and Drouot, 1988b). The model, *per se*, of solutions of polyelectrolytes is described in Section 2 where special attention is paid to the notions of conformation, charge densities and electric currents present in the solution. The (simple) general field equations are also recalled with the peculiarity that, in general, the stress tensor of the fluid mixture is not symmetric. The main thermodynamical equation in that section is the statement of the Clausius-Duhem inequality. Section 3 is devoted to thermodynamical processes and the conformation enters only at that point since it is an internal variable which is to represent one of the dissipative mechanisms present in the mixture, the others being electric conduction, heat conduction, electric relaxation and species production through chemical reactions. In that respect the dissipation inequality (3.11) is the fundamental equation which is indeed exploited to obtain simple representations of all dissipative processes (with the exception of the diffusion of macroions which is not accounted for in the present approach). Additional effects and couplings are considered later on in Section 5. It is remarkable that up to so-called gyroscopic (i.e, non-entropy producing) terms, the whole thermodynamics of the solution follows from the datum of the free energy function and a dissipation potential. This is perfectly in line with the spirit of the continuum thermodynamics involving internal variables as it is nowadays applied for describing a large variety of fluid and solid behaviors and many relaxation processes (Maugin, 1987). Section 4 gives a few applications to the discussion of conformations taken by polyelectrolyte macromolecules in various instances (when they are "free", acted upon by flows, electric fields, or when the pH of the solution is changed by an additive). Section 5 considers additional phenomena such as diffusion and migration of macromolecules in the solvent and new couplings such as those arising between conformation and a local vorticity field related to turbulence. In Section 6 a short digression to specialized cases such as liquid crystals is given. The final Section 7 emphasizes the general mathematical structure (evolution problem) that emerges from the present considerations. In all the thermodynamical model, naîve at it is (it may also be complicated at will), provides an insight in many bevaviors, classical or less obvious. It is hoped that further elaborations on its basis will contribute to improve the understanding of the behavior of solutions of macromolecules on a macroscopic scale of observation.

2. - THERMODYNAMICAL MODEL FOR MACROMOLECULE SOLUTIONS

A. - Conformation

It is well known that polyelectrolytes consist of macromolecules whose monomers contain an ionic group (Oosawa, 1970). Very interesting phenomena are related to polyelectrolytes in solution. Indeed, when dissolved in a fluid carrier the macromolecules are dissociated into polyvalent *macroions* and a large number of small ions of opposite charge, so-called *counterions* (Figure 1). The macroions have a structure which heavily depends on the interactions between ionized groups. Consequently the "conformation" (a concept to be made precise herein after) and the "volume" of macromolecules in solution may be strongly influenced by the spatial distribution of counterions and the action of an external electric field.

Various phenomena concerning polyelectrolytes in solution are intimately connected with the structure that these macromolecules take in the dissolved state. Long molecular chains are randomly coiled in the solution. Such a structure is maintained by molecules with hydrogen-bonded helical structures in solvents that have little

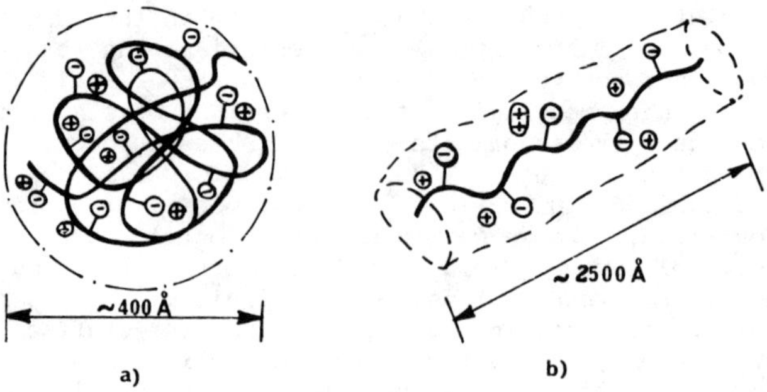

Figure 1: Conformations of polyelectrolytes in solution :
(a) contracted random coil (spherical conformation at low pH ;
(b) extended shape (rod-like or cylindrical conformation at high pH)
[after Drouot and Maugin, 1985].

tendency to form hydrogen bonds between solvent and solute molecules. In dealing with theoretical considerations of macromolecular solutions it is advantageous to model the macromolecules as "material" particles undergoing deformations. Two plausible representations are a sphere and a long thin rod ; the former might be a good representation of globular proteins (which have a compact and symmetrical structure at least in water solution), the latter of helical rod-shaped molecules. In fact, due to mathematical convenience and the generality of the model, dissolved molecules are often represented as ellipsoids of revolution. Thus the deformable configuration of macromolecules could be described by a symmetric second-order tensor **C**. As to randomly coiled molecules we may still describe this configuration by a symmetric tensor, still noted **C**, which then is defined via an ensemble average over all possible entanglements of the polymer chain (Figure 1). We call *conformation* this second-order tensor which is to be considered as a field (i.e. a rather smooth function of time and spatial position). It is not to be confused with the "configuration" tensor of Leonov's (1976) theory of polymeric liquids (which is none other than an Eulerian finite-strain tensor directly defined in terms of the macroscopic flow of the solution -see Giesekus (1986) and other authors for this confusion (Grmela and Carreau, 1987). In tensorial language we may set (tr = trace)

$$\mathbf{C} = \frac{1}{3}\kappa\,\mathbf{1} + \mathbf{K} \quad , \quad \mathrm{tr}\,\mathbf{K} = 0 \quad , \quad \kappa = \mathrm{tr}\,\mathbf{C} \tag{2.1}$$

where κ is the "dilatation" part of **C** and **K** is its deviatic (or traceless) contribution that describes pure shears of the conformation (i.e., deviations from a spherical overall shape) ; **1** is the unit dyadic. For "volume preserving" conformations κ = const. and we need consider **K** only. In general **C** is related to the statistical description by (Bird *et al*, 1977)

$$\mathbf{C} = \langle \mathbf{R} \otimes \mathbf{R} \rangle \quad , \quad (\langle \mathbf{R} \rangle = \mathbf{0} \text{ at rest}) \ , \tag{2.2}$$

where **R** is the directed distance between extreme monomers in the main, $\otimes$ indicates the tensorial product and $\langle .. \rangle$ denotes the ensemble average over all realizations. For instance, in polyox with repeated motif $-O-CH_2-CH_2-$ of length b one has at rest $\kappa = \mathrm{tr}\,\mathbf{C} = R_0^2$ and $R_0 \simeq \sqrt{nb}$. In polyelectrolytes R_0 is of the order of hundreds of angströms at neutral pH. At equilibrium in a neutral mixture **K** = **0** and **C** =

(R_0^2 **1**/3). We shall consider that **C** or, equivalently κ and **K**, are *internal* thermodynamical state variables whose evolution equations describe the conformational relaxation towards equilibrium.

B. - **Mixture and chemical reactions**

The polyelectrolyte solution is modelled as a mixture, the solvent being the carrier fluid. The solute merely consists of the polyelectrolyte molecules, or rather macroions. The different types of ions are dissolved into the fluid. For the sake of simplicity the solvent is supposed to undergo isochoric motions. The motion of the different types of ions gives rise to electric current densities which need not be conserved because of the possible ionization and recombination reactions. The volume charge density as a whole is assumed to vanish. The different types of ions are described through a set of charge densities q_α, $\alpha = 1,2,\ldots$, one of these, q_m, being the charge density of macroions. So the polyelectrolyte solution is modelled as a reacting mixture of several fluid components. However, in order to avoid the complexity inherent in the standard theory of fluid mixtures we introduce simplifying approximations which, while appearing physically sound, reduce the number of effective unknown fields. Indeed, let ρ_c, ρ_m be the mass densities of the carrier fluid and the macroion "fluid". Owing to the assumption of isochoric motions we have ρ_c, ρ_m = const. and hence the total mass density $\rho = \rho_c + \rho_m$ is constant (the mass of counterions is discarded). The corresponding velocity fields $\mathbf{v}$ and $\mathbf{v}_m$ satisfy $\nabla.\mathbf{v} = \nabla.\mathbf{v}_m = 0$, ∇ denoting the spatial gradient operator. We ascribe material properties to the carrier fluid and the macroion *only*. Thus ρ is the mass density of the mixture and we disregard the partial pressures of the different ions on the fluid. Finally, the local electric properties are described through the electric polarization per unit volume (effective density of electric dipoles) **P**, or unit mass, $\pi = \mathbf{P}/\rho$, and Maxwell's equations are considered within the framework of quasi-electrostatics (optics, however, requires the use of the full set of Maxwell's equations ; cf. Drouot and Maugin, 1987). the carrier fluid in itself is regarded as electrically neutral.

The balances of electric charge for the ions α in the solution are written as

$$\partial q_\alpha/\partial t + \nabla.(q_\alpha \mathbf{v}_\alpha) = \Upsilon_\alpha \quad , \quad \alpha = 1,2,\ldots \quad , \tag{2.3}$$

Υ_α being the charge supply and $\mathbf{v}_\alpha$ the velocity of the α-th type of ion. Conservation of charge implies by summation over α that

$$\sum_\alpha \Upsilon_\alpha = 0 \ . \tag{2.4}$$

Moreover, the charge supplies Υ_α are not independent of one another since there can occur n distinct chemical reactions so that

$$\Upsilon_\alpha = \sum_{r=1}^{n} \nu_{\alpha r} \dot{\Lambda}_r \tag{2.5}$$

where the $\nu_{\alpha r}$'s are stoichiometric coefficients and $\dot{\Lambda}_r$ are reaction progress rates. Let $\mathfrak{J}_\alpha = q_\alpha(\mathbf{v}_\alpha - \mathbf{v})$. In view of the condition $\nabla.\mathbf{v} = 0$, we have

$$\nabla.(q_\alpha \mathbf{v}_\alpha) = \nabla.\mathfrak{J}_\alpha + (\mathbf{v}.\nabla)q_\alpha \ ,$$

and eqn. (2.3) transforms to

$$\dot{q}_\alpha + \nabla.\mathfrak{J}_\alpha = \Upsilon_\alpha \, , \tag{2.6}$$

where a superimposed dot denotes the usual *material* time derivative (relative the whole mixture). We emphasize that macroions are carried with the fluid and thus

$$\mathfrak{J}_m = 0 \quad , \quad \dot{q}_m = \Upsilon_m \, . \tag{2.7}$$

C. - **Field equations**

On account of the above-enunciated working hypotheses the general thermomechanical equations of the fluid mixture are those of a globally charge free, but electrically polarized, electricity and heat conducting, incompressible viscous fluid. The local equations of motion and the local statement of the energy equation and of the second principle of thermodynamics are therefore taken directly from the general theory of continua with electromagnetic interactions (Maugin, 1980 ; the conformation does not intervene at all at this point since it is an internal variable). The equations of linear and angular momenta for the whole solution are considered in the form (the divergence is taken on the second index of tensors)

$$\rho \, \dot{\mathbf{v}} = \nabla.\mathbf{t} + \mathbf{f} + \mathbf{f}^{em} \, , \tag{2.8}$$

$$\mathbf{t} = \sigma + (\mathbf{P} \otimes \mathcal{E})_A \quad \Longrightarrow \quad \mathbf{t}_A = (\mathbf{P} \otimes \mathcal{E})_A \tag{2.9}$$

with

$$\mathbf{f}^{em} = (\mathbf{P}.\nabla)\mathcal{E} \tag{2.10}$$

Here $\mathbf{t}$ is Cauchy's (nonsymmetric) stress tensor, $\mathbf{f}$ is a mechanical body force per unit volume, $\mathbf{f}^{em}$ is the electrostatic form of the ponderomotive force and σ is the symmetric mechanical *intrinsic* stress tensor. The subscript A denotes the operation of anti- (or skew) symmetrization, so that $\mathbf{P} \times \mathcal{E}$ is the electric couple if $\mathcal{E}$ is the electric field in a frame co-moving with the element of fluid mixture.

Let $\mathfrak{J} = \sum_\alpha \mathfrak{J}_\alpha$ the total current density in the solution, e the internal energy density, $\mathbf{q}$ the heat flux, $\mathbf{S}$ the entropy flux, θ the thermodynamical temperature ($\theta > 0$, $\inf \theta = 0$), η the entropy density, and h the heat supply per unit mass. The local energy equation and the local statement of the second principle of thermodynamics read (Maugin, 1980)

$$\rho \dot{e} = \mathrm{tr}\,(\sigma\mathbf{D}) + \mathfrak{J}.\mathcal{E} - \rho\mathbf{E}^L.D_J\pi - \nabla.\mathbf{q} + \rho h \tag{2.11}$$

and

$$\rho\dot{\eta} \geqslant \theta^{-1}\,\rho h - \nabla.\mathbf{S} \, . \tag{2.12}$$

In eqn. (2.11) $(\mathfrak{J}.\mathcal{E})$ is the Joule term and $\mathbf{E}^L$ is the so-called *local* electric field which satisfies the "balance" equation

$$\mathbf{E}^L + \mathcal{E} = \mathbf{0}. \tag{2.13}$$

We have set (T = transpose)

$$D_J\pi = \dot{\pi} - \Omega \times \pi \quad , \quad \Omega = \frac{1}{2}\nabla \times \mathbf{v} \quad , \quad \mathbf{D} = \frac{1}{2}[\nabla\mathbf{v} + (\nabla\mathbf{v})^T] \ , \tag{2.14}$$

i.e., Ω is the vorticity vector (calculated from the **v**-field), **D** is the strain-rate tensor, and $D_J\pi$ is the so-called Jaumann (co-rotational) derivative of π. The form of eqn. (2.11) follows from the frame indifference of the energy equation. While $\mathcal{E}$ is the field that appears in Maxwell's equations, $\mathbf{E}^L$ is a material quantity for which one needs a constitutive equation in terms of the electric polarization. It happens that (2.13) is a simplified form of a more general equation which here gives $\mathbf{E}^L = -\mathcal{E}$ (see Maugin, 1980). In general

$$\mathbf{S} = \frac{1}{\theta}\mathbf{q} + \mathbf{k} \ , \tag{2.15}$$

where **k** is the extra flux of entropy which shall be shortly specified (usually **k** = **0** but this is *not* the case when diffusion of species is present).

Introducing the Helmholtz free energy ψ per unit mass by $\psi = e - \eta\theta$, we can also combine eqns. (2.11) and (2.12) on account of (2.15) to obtain the so-called *Clausius-Duhem inequality*

$$-\rho(\dot{\psi} + \eta\dot{\theta}) + \mathrm{tr}(\sigma\mathbf{D}) + \mathcal{J}.\mathcal{E} - \rho\mathbf{E}^L.D_J\pi + \theta\nabla.\mathbf{k} - \mathbf{q}.\nabla(\ln\theta) \geqslant 0 \ . \tag{2.16}$$

This is the fundamental thermodynamical inequality in the present approach. The next section is devoted to exploiting this inequality. Because of the incompressibility **D** is reduced to its deviatoric part and, by duality, only the deviatoric part of σ intervenes in eqn. (2.16). A Lagrange multiplier p, called the mechanical pressure, needs to be introduced to preserve the kinematical constraint tr **D** = ∇.**v** = 0.

3. - THERMODYNAMICAL PROCESSES

A. - Dissipation inequality

In harmony with the spirit of internal variable theory one assumes that the thermodynamical potential ψ has the following functional dependence

$$\psi = \psi(\theta, q_\alpha, \pi, \kappa, \mathbf{K}) \ , \tag{3.1}$$

where κ and K are the elements of the canonical decomposition of the (symmetric) conformation tensor **C**. We can *define* the following partial derivatives (*forces* in the language of the theory of irreversible processes)

$$\bar{\eta} = -\partial\psi/\partial\theta \ , \tag{3.2}$$

$$\mu_\alpha = \partial\psi/\partial q_\alpha \ , \tag{3.3}$$

$$\mathbf{E}^{Lr} = -\partial\psi/\partial\pi \ , \tag{3.4}$$

$$f = -\rho\,\partial\psi/\partial\kappa \ , \tag{3.5}$$

$$\mathcal{F} = -\rho\,\partial\psi/\partial\mathbf{K} \ . \tag{3.6}$$

In addition we set

$$\mathbf{E}^R = \mathbf{E}^L - \mathbf{E}^{Lr} \tag{3.7}$$

and make the (reasonable) choice

$$\mathbf{k} = -\frac{\rho}{\theta}\sum_{\alpha} \mu_\alpha \mathcal{J}_\alpha . \tag{3.8}$$

Finally, we assume that for a spatially uniform state (i.e., $\nabla\mathbf{v} = \mathbf{0}$, $\mathcal{E} = -\nabla\varnothing = \mathbf{0}$, $\nabla\theta = 0$, where $\varnothing$ is the electrostatic potential) entropy has its definition of *thermostatics* when all rates of internal variables (κ, K, Λ_r) are null ; by virtue of the axiom of local accompanying state (Germain 1973 ; Germain *et al*, 1983) we can state that this definition

$$\eta = \bar{\eta} = -\partial\psi/\partial\theta \tag{3.9}$$

still holds true slightly outside equilibrium and thus the inequality (2.16) -after computing $\dot{\psi}$, taking account of the invariance of ψ by rotations, expressed by (invariance under infinitesimal rotations of a scalar-valued function of a vector field π and a symmetric second-order tensor $\mathbf{K}$)

$$2\left(\frac{\partial\psi}{\partial K_{ki}} K_{kj} - \frac{\partial\psi}{\partial K_{kj}} K_{ki}\right) + \frac{\partial\psi}{\partial\pi_i}\pi_j - \frac{\partial\psi}{\partial\pi_j}\pi_i = 0 , \tag{3.10}$$

and accounting for eqns. (2.6), (2.15), (3.3) through (3.8) -transforms to the following residual *dissipation inequality*

$$\mathrm{tr}(\sigma\mathbf{D}) + \left[f\dot{\kappa} + \mathrm{tr}(\mathcal{F}D_J\mathbf{K})\right] - \mathbf{E}^R.D_J\mathbf{P} + \sum_\alpha \mathcal{J}_\alpha.\mathcal{E}_\alpha - \sum_{r=1}^{n} R_r\dot{\Lambda}_r - \mathbf{S}.\nabla\theta \geqslant 0 , \tag{3.11}$$

where we have set (in Cartesian tensor notation)

$$\begin{aligned}(D_J\mathbf{K})_{ij} &= \dot{K}_{ij} - (\Omega_{ik}K_{kj} + \Omega_{jk}K_{ki}) ,\\ \Omega_{ij} &= \frac{1}{2}(v_{i,j} - v_{j,i}) ,\end{aligned} \tag{3.12}$$

and

$$\mathcal{E}_\alpha = \mathcal{E} - \nabla\bar{\mu}_\alpha , \quad R_r = \sum_\alpha \bar{\mu}_\alpha \nu_{\alpha r} , \tag{3.13}$$

$$\bar{\mu}_\alpha = \rho\mu_\alpha = \partial\psi/\partial q_\alpha , \quad \mathbf{P} = \rho\pi , \tag{3.14}$$

while accounting for the incompressibility so that ($\Psi = \rho\psi$)

$$\begin{aligned}S &= \rho\eta = -\partial\Psi/\partial\theta , \quad \mathbf{E}^{Lr} = -\partial\Psi/\partial\mathbf{P} ,\\ f &= -\partial\Psi/\partial\kappa , \quad \mathcal{F} = -\partial\Psi/\partial K .\end{aligned} \tag{3.15}$$

All dissipative processes can now be studied on the basis of eqn. (3.11) and the theory of irreversible processes since (3.11) is written in the standard bilinear form

$$\sum_i X_i \; Y_i \geqslant 0 \; .$$

Thermodynamical equilibrium is defined by three groups of conditions :

$$S_{eq} = - \, (\partial\Psi/\partial\theta)_{eq} \quad , \quad (\bar{\mu}_\alpha)_{eq} = (\partial\Psi/\partial q_\alpha)_{eq} \; , \tag{3.16}$$

$$\nabla \mathbf{v} = \mathbf{0} \quad , \quad \nabla \varnothing = \mathbf{0} \quad , \quad \nabla \bar{\mu}_\alpha = 0 \quad , \quad \nabla \theta = 0 \; , \tag{3.17}$$

and

$$f(\kappa_{eq}) = 0 \quad , \quad \mathcal{F}(\mathbf{K}_{eq}) = \mathbf{0} \quad , \quad (\mathbf{E}^{L})_{eq} = (\mathbf{E}^{Lr})_{eq} \quad , \quad R_r = 0 \quad , \tag{3.18}$$

of which the second corresponds to a spatially uniform state while eqns. (3.18) define equilibrium values of κ, **K**, **P** and Λ_r once the conditions (3.16) are fulfilled. A linear theory usually implies that $\mathbf{K}_{eq} = \mathbf{0}$ and $(\Lambda_r)_{eq} = 0$. But $\kappa_{eq} \neq 0$ and $(3.17)_2$ and (2.13) combined with $(3.18)_3$ do not necessarily imply that $\mathbf{P}_{eq} = \mathbf{0}$ since $(\mathbf{E}^{Lr})_{eq}$ may be strongly nonlinear in **P** and the equation $(\mathbf{E}^{Lr})_{eq} = \mathbf{0}$ could have non-zero **P** solutions (case of a ferroelectric behavior).

B. - **Dissipative processes**

Equations (3.17) and (3.18) clearly single out two different types of dissipative processes through their limit condition corresponding to thermodynamical equilibrium. The former are the usual dissipative processes of the theory of irreversible processes (de Groot and Mazur, 1962), that correspond, at equilibrium, to spatially uniform states of the basic *observable* variables of thermo-electro-mechanics, while the latter represent *relaxation* processes governed by evolution laws and of which reaction kinetics have probably offered historically the first example. Electric and conformational relaxation belong to this last class while shear viscosity, electric conduction and thermal conduction belong to the former. A third class of dissipative processes is not represented here ; there are of the plasticity type in that they present *no* time scale while their formulation *do* involve some rates (Maugin, 1987).

The various parities in time and tensorial orders of the *fluxes* and *forces* involved in the inequality (3.11) make that one can immediately envisage certain couplings, as also uncouplings, between various dissipative effects. Equivalently to the linear equations of the theory of irreversible processes one can also assume the existence of a *Rayleigh dissipative function* that is homogeneous of degree *two* in the fluxes. That is, on account of the various parities and the overall isotropy of the fluid solution, one has the following dissipation potential :

$$\Phi = \frac{1}{2}\left\{2\eta_v \,\mathrm{tr}\, \mathbf{D}^2 + \zeta \,\mathrm{tr}\, \mathcal{F}^2 + \xi\, f^2 + \sum_\alpha \mathcal{J}_\alpha . \mathbf{R}_\alpha . \mathcal{J}_\alpha + \right.$$

$$\left. + B(\nabla\theta)^2 + 2\sum_\alpha A_\alpha\, \mathcal{J}_\alpha . \nabla\theta + \nu(\mathbf{E}^R)^2 + \sum_{\ell, r=1}^{n} R_\ell \xi_{\ell r} R_r \right\}, \qquad (3.19)$$

so that the general formulas

$$\mathbf{S} = -\,\partial\Phi/\partial(\nabla\theta) \quad , \quad \mathcal{E}_\alpha = \partial\Phi/\partial\mathcal{J}_\alpha \quad , \quad \sigma^{ir} = \partial\Phi/\partial\mathbf{D} \qquad (3.20)$$

and

$$\left.\begin{aligned} D_J\mathbf{K} &= \partial\Phi/\partial\mathcal{F} \quad , \quad \dot{\kappa} = \partial\Phi/\partial f \; , \\ D_J\mathbf{P} &= -\,\partial\Phi/\partial\mathbf{E}^R \quad , \quad \dot{\Lambda}_r = -\,\partial\Phi/\partial R_r \; , \end{aligned}\right\} \qquad (3.21)$$

yield for this special choice

$$\left.\begin{aligned} \mathbf{S} &= -\,B\,\nabla\theta - \sum_\alpha A_\alpha\, \mathcal{J}_\alpha \; , \\ \mathcal{E}_\alpha &= \mathbf{R}_\alpha . \mathcal{J}_\alpha + A_\alpha \nabla\theta \; , \end{aligned}\right\} \qquad (3.22)$$

$$\sigma^{ir} = 2\,\eta_v\, \mathbf{D} \qquad (3.23)$$

and

$$D_J\mathbf{K} = -\,\zeta\, \frac{\partial\Psi}{\partial\mathbf{K}} \quad , \quad \dot{\kappa} = -\,\xi\, \frac{\partial\Psi}{\partial\kappa} \; , \qquad (3.24)$$

$$D_J\mathbf{P} = \nu\left(\mathcal{E} - \frac{\partial\Psi}{\partial\mathbf{P}}\right) \quad , \quad \dot{\Lambda}_r = -\sum_{k=1}^{n}\sum_\alpha \left(\frac{\partial\Psi}{\partial q_\alpha}\, \nu_{\alpha k}\, \xi_{kr}\right) . \qquad (3.25)$$

The first two of these represent the coupled *Fourier-Ohm laws*. They can be rewritten in the following form on account of eqns. (2.15) and (3.8) :

$$\mathbf{q} = -\,(\theta B)\nabla\theta - \sum_\alpha \left(\theta A_\alpha - \frac{\partial\Psi}{\partial q_\alpha}\right)\mathcal{J}_\alpha \; , \qquad (3.26)$$

$$\mathcal{E} - \nabla\left(\frac{\partial\Psi}{\partial q_\alpha}\right) = \mathbf{R}_\alpha . \mathcal{J}_\alpha + A_\alpha \nabla\theta \quad , \quad \alpha = 1, 2, \ldots \; . \qquad (3.27)$$

The positive definiteness of Φ requires that the following inequalities be satisfied :

$$\left.\begin{array}{llll} \eta_v \geqslant 0 \ , & \zeta \geqslant 0 \ , & \xi \geqslant 0 & , \ \nu \geqslant 0 \ , \\ B \geqslant 0 \ , & A_\alpha^2 \leqslant B\,\bar{R}_\alpha \ , & \xi_{\ell r} \text{ positive definite} & , \end{array}\right\} \tag{3.28}$$

where $\bar{R}_\alpha$ is the smallest eigenvalue of $\mathbf{R}_\alpha$.

Since dissipative entities can also depend on the *state* variables (e.g., **K**) it is clear that eqn. (3.19) is only to be considered as one of the simplest examples used for illustrative purposes. In particular, one can envisage conduction processes which involve **K** and will, therefore, be *anisotropic*.

C. - Gyroscopic terms

It is clear that there still remains to specify the functional dependence of the free energy Ψ for a complete expression of eqns. (3.23) through (3.27). However, this is not all because we can notice that the expressions of σ and $D_J\mathbf{K}$, which can parametrically depend on the state variables present in Ψ, and thus on **K** directly, in addition to through $\mathcal{F}$ (in the case of $D_J\mathbf{K}$), can present *coupled* parts which are not ruled out by parity but have the peculiarity that when considered together they do not produce any entropy. Such contributions, which may be called *gyroscopic* because they correspond to forces orthogonal to their dual velocities in a generalized sense (Maugin, 1980), were first noticed by D. Lhuillier. Examples of these are

$$\begin{aligned} \sigma^G &= \lambda\mathcal{F} - \beta(\mathbf{K}\mathcal{F} + \mathcal{F}\mathbf{K}) \ , \\ \hat{\mathbf{K}}^G &= \lambda\mathbf{D} + \beta(\mathbf{K}\mathbf{D} + \mathbf{D}\mathbf{K}) \ . \end{aligned} \tag{3.29}$$

These two expressions are linear affine in $\mathcal{F}$ and **D**, respectively, depending parametrically on **K**, and they satisfy *together* the following generalized orthogonality property :

$$\mathrm{tr}(\sigma^G\mathbf{D} + \mathcal{F}\hat{\mathbf{K}}^G) = 0 \ , \tag{3.30}$$

so that thermodynamics imposes restrictions neither on the sign nor on the magnitude of the coefficients λ and β which appear *simultaneously* in both equations (3.29). It can be shown that λ is related to the osmotic pressure of the polymeric solution (Drouot and Maugin, 1983 ; Doi and Edwards, 1978) while a plausible value of β may be found by studying the stability of a special flow problem influencing **K** (e.g., Lhuillier, 1981) ; it is found that $0 < \beta < 1$). Expressions such as (3.29) must be added to the right-hand side of eqns. (3.23) and $(3.24)_1$ without contradicting continuum thermodynamics. For instance, $(3.24)_1$ is replaced by

$$D_J\mathbf{K} = -\zeta\,\frac{\partial\Psi}{\partial\mathbf{K}} - \lambda\mathbf{D} + \beta(\mathbf{K}\mathbf{D} + \mathbf{D}\mathbf{K}) \ . \tag{3.31}$$

D. - Free energy function

As an example for the free energy Ψ we may consider the following expression (k_B : Boltzmann's constant)

$$\Psi = -\frac{1}{2}(k_B\theta)\,\mathrm{Ln}(\det \mathbf{C}) - q^{eff}\left(\frac{1}{\sqrt{\kappa}} - \nu_e I_3^{-\frac{1}{2}}\right) +$$
$$+ \frac{1}{2}\, e_1^*(\varepsilon_{cr} - \varepsilon)(\kappa - \kappa_0)^2 + \frac{1}{4}\gamma(\kappa - \kappa_0)^4 +$$
$$+ \Big(\alpha_2 I_2 + \alpha_2' I_2^2 + \alpha_3 I_3 + \alpha_3' I_3^2 + \alpha_3'' I_3^3 +$$
$$+ \alpha_0 I_3 I_4 + \alpha_4 I_4 + \alpha_6 I_6 + \alpha_6' I_6^2\Big) + \bar{\Psi}(\theta, q_\alpha) \ , \qquad (3.32)$$

where the first line represents *singular* contributions, the second line accounts for dilatational effects of the conformation with a coupling with an electric field ε which may take a critical value ε_{cr} (see below) and the long contribution withing parentheses gathers *regular* contributions in the traceless tensor **K**, the electric polarization **P** and the electroelastic couplings between these two.The α's, (α')'s and (α'')'s are coefficients and the invariants I_α are defined by

$$I_2 = \mathbf{P}^2 \ , \ I_3 = \mathrm{tr}\,\mathbf{K}^2 \ , \ I_4 = \mathbf{P.KP} \ ,$$
$$I_5 = \mathbf{P.K}^2\mathbf{P} \ , \ I_6 = \mathrm{tr}\,\mathbf{K}^3 \ . \qquad (3.33)$$

The first term in (3.32) accounts for *Brownian agitation* (Grmela and Carreau, 1987) while q^{eff} being some effective electric charge, the contribution in $1/\sqrt{\kappa}$ accounts for *Coulombic electrostatic interactions*, that in $I_3^{-\frac{1}{2}}$ will account for the *mechanical-chemical* effect with coefficient ν_e, and the coefficient α_4 accounts for the *electrostriction effect*. Coulombic interactions were considered by Dunlap and Leal (1984) in a dumbbell model, as also by Carreau *et al* (1986 ; also Ait-Kadi *et al*, 1988) in a model using the conformation **C**. The other terms were introduced by the authors in co-operation with A. Morro from Genova (Morro, *et al*, 1987). Other types of conformational elasticity contributions can be introduced to account for the finite extensibility of molecular chains (so-called FENE-models ; see Bird *et al*, 1977). One needs to account for this last type in strongly elongational flows. A rather simple electro-elastic expression for Ψ is

$$\Psi^{ee} = \alpha_2 I_2 + \alpha_3 I_3 + \alpha_4 I_4 \ , \qquad (3.34)$$

which contains no nonlinearities and only the lowest order electroelastic coupling.

4. - APPLICATIONS

Here we consider a sequence of physical situations in which eqns. $(3.24)_2$, (3.31) and (3.32) can be used to illustrate the influence of the flow and external fields on the conformation of macromolecules and the stability of the obtained equilibria once the transients have died out. That is, the macromolecules of polyelectrolytes are envisioned as *test particles* in the sense that they are acted upon by external stimuli while their own action on the flow and fields is negligible so that, in effect, we are not interested in the alteration in the rheological properties of the solutions as a result of the presence of macromolecules, effect which is to be minute in very dilute solutions although we know that a slight addition of polymers may sometimes have a drastic effect on some flows (drag reduction for instance). In particular, we focus on *induced* changes of conformation and their coupling with electrical properties.

A. - Equilibrium conformations

A phase of the polyelectrolyte solution in which κ = const., **K** = 0, **P** = 0 is said to be *isotropic*. Obviously when the solution suffers a prescribed shear flow, whether or not in presence of an applied electric field, the conformation of macromolecules typically takes an ellipsoid shape with a principal axis more or less aligned with the shear direction (Drouot and Maugin, 1985 ; at 45 degrees for a rigid particle in a weak shear flow ; Jeffery, 1922). With an additional electric field switched on, we may have a competition effect between flow and field. The natural question then arises of the possible existence of *natural* or *forced anisotropic* phases and their stability once stationary conditions are reached. Obviously, the essential factor in the discussion then is the degree of the free energy function Ψ in the invariants I_2 and I_3 since, with constant pH, it is the last two lines of the expression (3.32) which are relevant. Stability then is investigated through the minimum condition of the free energy -or another potential (modified free energy)- (second variation criterion). A few instructive cases are as follows.

a) Absence of electric phenomena and external forcing

Considering energies such as (**P** = 0)

$$\Psi = \alpha_3 I_3 + \alpha_3' I_3^2 + \alpha_3'' I_3^3 + \alpha_6 I_6 + \alpha_6' I_6^2$$

it is shown by direct computation of $\partial\Psi/\partial\mathbf{K}$ and $\partial^2\Psi/\partial K_{ij}\partial K_{k\ell}$ that although one may have possible (nonspherical) axisymmetric solutions for **K**, only the isotropic phase is stable (Drouot and Maugin, 1988a).

b) Electrically polarizable microstructure

One may consider the simple expression (3.34) in the absence of flow and electric field so that the equilibrium equations read

$$\partial\Psi/\partial\mathbf{K} = 0 \quad , \quad \partial\Psi/\partial\mathbf{P} = 0 \ . \qquad (4.1)$$

Electromechanical couplings occur only through electrostriction (α_4). It is worth mentioning that an energy of the form (3.34) for describing conformations of biological molecules has been suggested by Frölich (1973) and investigated by Scott (1984) in connection with solitons of electrical polarization. The analysis of eqns. (4.1) and of the second variation of Ψ allows one to conclude that the only stable equilibria correspond to *spherical* conformations (**K** = 0) and *zero* polarization. The same result can be proved at the cost of some lengthy algebra for more general expressions of Ψ where the degree in $|\mathbf{P}|$ and $|\mathbf{K}|$ is increased (Drouot and Maugin, 1988a).

c) Action of an applied field

Then eqns. (4.1) are replaced by

$$\partial\Psi/\partial\mathbf{K} = 0 \quad , \quad (\partial\Psi/\partial\mathbf{P}) - \mathcal{E} = 0 \ . \qquad (4.2)$$

Upon considering the energy (3.34) the analysis shows that eqns. (4.2) admit axisymmetric solutions **K** with one eigenvector collinear with **P** and $\mathcal{E}$. The component E_3 of $\mathcal{E}$ along this direction must satisfy the inequality

$$|E_3| < \frac{4}{3}\left(\alpha_2\alpha_3^2\right)^{1/2}/\alpha_4^2 \quad , \quad \alpha_2, \alpha_3 > 0 \ , \qquad (4.3)$$

for the corresponding nonspherical conformations to turn out to be stable. The weaker is the electromechanical coupling (α_4), the larger will be the allowed magnitude of E_3.

B. - Flow-induced conformational phase transition

Equation (3.31) can formally be rewritten as

$$D_J K + \zeta \frac{\partial \tilde{\Psi}}{\partial K} = 0 \tag{4.4}$$

with

$$\tilde{\Psi} = \psi - \mathrm{tr}(\mathbf{M}\ \mathbf{D}) \tag{4.5}$$

$$\mathbf{M}(\mathbf{K}) = \frac{1}{\zeta}(\beta \mathbf{K}^2 - \lambda \mathbf{K}) \ .$$

In a controlled stationary irrotational flow eqn. (4.4) reduces to

$$\partial\tilde{\Psi}/\partial \mathbf{K} = 0 \tag{4.6}$$

and this equilibrium, which is *not* a thermodynamical one, is the basis of an argument very similar to that used in flow-induced isotropic-nematic transitions in nematogens (Sluckin, 1981). It also resorts to the notion of potential for systems under external forcing (Keizer, 1985). Here $\tilde{\Psi}$ is a kind of modified Landau free energy where **D** plays the role of the applied force. The stability is still defined by the second variation of $\tilde{\Psi}$ since the latter is only linear in **D**. Stable axisymmetric conformations can be obtained for the significant case α_3 and $\alpha_6 > 0$, $\alpha_3' = 0$ with the stability condition

$$D < 2 \left(\frac{\zeta}{\lambda}\right) \frac{\alpha_3^2}{\alpha_6} \ , \tag{4.7}$$

where D is the component of **D** along the main principal axis of **K**, the latter having the same eigenvectors as **D**. The smaller is the osmotic pressure (or λ) the higher is the upper bound for the flow guaranteeing stability (Drouot and Maugin, 1988a).

C. - Flow-induced electric polarization

Accounting for (4.6) but with $\beta = 0$, $\alpha_3' = \alpha_3'' = \alpha_0 = \alpha_6 = \alpha_6' = 0$ but with $\alpha_4 \neq 0$, we can obtain *stable electrically polarized axisymmetric conformations* provided that

$$D < -2 \left(\frac{\zeta}{\lambda}\right) \frac{\alpha_3^2}{\alpha_4} \ . \tag{4.8}$$

There will appear an electric polarization only for microstructures with a *negative* electromechanical coupling α_4 (since both D and λ are positive).

The result (4.7) and (4.8) give evidence of a flow-induced conformational phase transition as well as a flow-induced electric polarization through a change in conformation (Drouot and Maugin, 1988a ; Drouot, Maugin and Morro, 1987).

D. - **Mechano-chemical effect**

Now we consider the solvent (carrier fluid) at rest and a vanishing electric field. We focus the interest on deviations from a spherical coil (and thus $\kappa = \kappa_0$ = fixed constant for all practical purposes). The effect of the charge $q_m < 0$ that we select for the q^{eff} in eqn. (3.32) is thought to be dominant over that of electric polarization. At the simplest level of approximation the free energy then is viewed as resulting from the deformation energy of macromolecules and the interaction between that deformation and electric charge. From (3.31) one therefore extracts the relevant contribution

$$\Psi^{me} = \alpha_3 I_3 + q_m \nu_e I_3^{-\frac{1}{2}} . \tag{4.9}$$

Equation (4.6) with D = 0 yields equilibrium conformations $\mathbf{K} \neq \mathbf{0}$ given by

$$\ell^3 \equiv I_3^{3/2} = \frac{\nu_e}{2\alpha_3} q_m . \tag{4.10}$$

Since $\ell > 0$ and $q_m < 0$ one must have $\nu_e < 0$. Let ℓ_e the value of ℓ given by eqn. (4.10). Then we can show in absence of flow that the *relaxation equation* (4.4) gives after tensorial contraction with **K** (Morro *et al*, 1985) :

$$\dot{\ell} = - \frac{1}{\tau(\ell,\ell_e)} (\ell - \ell_e) , \tag{4.11}$$

where the conformation dependent relaxation time is given by

$$\tau = \ell^2/2\zeta \alpha_3 \left(\ell^2 + \ell\ell_e + \ell_e^2\right) .$$

Thus ℓ tends eventually to ℓ_e. Now the interaction between macroions and solvent is taken into account by observing that the number of dissociated groups in the polyanionic macroion increases (hence q_m decreases) with increasing the pH value, p, of the solution, namely $dq_m/dp < 0$. In view of (4.10) at equilibrium one has $\partial I_3/\partial q_m < 0$ and thus

$$\partial I_3/\partial p > 0 \quad \text{or} \quad \partial\ell/\partial p > 0 . \tag{4.12}$$

This means that the polyanionic macroion elongates with increasing pH (and vice versa) and this is the content of the mechano-chemical effect (Volkenshtein, 1983, pp. 91-94).

E. - **Electrically induced conformational phase transition**

Without entering the detailed biophysical mechanism which allow for the occurence of sudden volume changes, we borrow from the previous analysis and, by analogy with Landau's theory of phase transition (cf. Landau and Lifshitz, 1984 ; Maugin *et al*, 1990) we set up a very simple scheme for describing the volume changes. We are thus concerned with the internal variable κ = tr **C** rather than **K** and we concentrate on the second line in the expression (3.32) of the free energy function. There $e_1^* > 0$ is a pure constant (bulk elasticity of conformations) and ε is the externally controllable parameter (magnitude of electric field or induced electric polarization when the solution is placed between the plates of a condenser) which causes the transition. The corresponding elastic energy is minimum at $\kappa = \kappa_0$ for $\varepsilon < \varepsilon_{cr}$. After relaxation eqn. $(3.24)_2$ produces the equilibrium values via $\partial\Psi/\partial\kappa = 0$, i.e.,

$$\kappa = \kappa_0 \quad \text{or} \quad \kappa - \kappa_0 = \frac{\left[e_1^* (\varepsilon_{cr} - \varepsilon)\right]^{1/2}}{\gamma} . \tag{4.13}$$

This implies that $\gamma > 0$ if $\varepsilon < \varepsilon_{cr}$. It follows from the second of eqns. (4.13) that

$$\frac{d\kappa}{d\varepsilon} = - \left[\frac{e_1^*}{\gamma(\varepsilon_{cr} - \varepsilon)}\right]^{1/2} , \tag{4.14}$$

which diverges to $-\infty$ for ε tending to ε_{cr}. This implies that although the transition is continuous, κ changes significantly near the critical value ε_{cr} of ε [for $\varepsilon \geqslant \varepsilon_{cr}$ we apply the first of (4.13)]. Equation $(4.13)_2$ could be used to fit the experimental data in connection with the collapse of polyelectrolytes in certain circumstances (e.g. Tanaka *et al*, 1982) where ε is the magnitude of the electric field and $\varepsilon_{cr} \simeq 0.3$ volt/cm).

5. - MORE INVOLVED MODELING

The complexity of dissipative models of solutions of macromolecules is reflected in the length and complexity of the left-hand side of the dissipative inequality (3.11). Clearly, this inequality does not contain mechanism such as *molecular migration* in nonhomogeneous flows and interactions between conformation and vorticity at a small scale. These can be dealt with following works by the authors and co-workers.

A. - Diffusion and migration of macromolecules

In recent years much attention has been devoted to macromolecular behaviors, such as *molecular migration*, in flows where the velocity gradients vary spatially over the flow domain of interest (nonhomogeneous flows). It seems that macromolecules in a nonhomogeneous flow tend to minimize their configurational entropy by migrating toward less non-homogeneous flow regions thus escaping the more highly oriented configurations induced by higher velocity gradients. Aubert and Tirrell (1980) have shown that the bead-spring theory (see Bird *et al*, 1977) for dilute solutions of polymers predicts a diffusion of macromolecules for nonhomogeneous flows. At the bottom of their mathematical model is a force (and then a diffusion flux) proportional to the second gradient ($\nabla \nabla v_L$) of the velocity of the liquid solvent. This diffusion phenomena has sometimes been considered to explain wall effects in laminar flows of polymeric fluids (Tirrell and Malone, 1977 ; Drouot and Maugin, 1983). Subsequent refinements were performed by Brunn (1983) by modeling macromolecules as nonlinear elastic dumbbells. While microscopic theories are fruitful for pointing out new mechanisms or more detailed mimics, macroscopic theories have the advantage of a simpler and more compact scheme. Moreover, a macroscopic theory allows framing the scheme within a *thermodynamic context*, which results in the selection of physically admissible phenomena.

A macroscopic (and thermodynamic) investigation of *polymer diffusion* in nonhomogeneous flows was developed by Drouot and Maugin (1983) by modeling the polymer solution as a fluid with a *vectorial* internal variable. They also allowed the constitutive properties of the solution to be described by second order velocity gradients, and deduced a cross-streamline migration effect. They showed that, accordingly, a second gradient model deprived of the internal variable is inadequate for a satisfactory account of polymer diffusion. More recently, Morro *et al*

(1990) have set up a more refined scheme where the polymer solution is modeled as a binary mixture. Essentially, the macromolecules constitute the dispersed phase (with properties indexed d) while the solvent is a classical liquid (index L). Then the internal variable (a tensor **C**, see eqn. (2.1)) is ascribed to the dispersed phase and the second-order velocity gradient to the liquid. Without entering details, it is worth mentioning the main thermodynamical equations of this modeling where inertial properties of counter-ions are discarded and the charge of macro-ions and the number of counter-ions remain constant in time. This means that the mixture consists of two *non-reacting* constituents, each one assumed incompressible for simplicity (but the mixture as a whole is compressible). Then with neglect of heat conduction and electric polarization, the *reduced dissipation inequality* (3.11) is replaced by

$$\mathrm{tr}(\sigma_L \mathbf{D}_L + \mathcal{F}\, D_J \mathbf{K}) + \mathcal{J}.\mathcal{E} + \mathbf{m}.(\mathbf{v}_L - \mathbf{v}_d) + \mathbf{c}_L.(\Omega_L - \Omega_d) \geqslant 0 \,, \tag{5.1}$$

where Ω_L and Ω_d are vorticity vectors related to liquid and dispersed phases, and **m** and $\mathbf{c}_L$ are the exchanges of linear and angular momenta between phases. In this modeling the extra flux of entropy **k** is given by (compare to expression (3.8))

$$\mathbf{k} = \lambda(\mathbf{v}_d - \mathbf{v}_L) \,, \tag{5.2}$$

where λ is an arbitrary function of time which contributes terms $\lambda\theta$ and $-\lambda\theta$ to partial pressures p_d and p_L, respectively. In eqn. (5.1) dissipative mechanisms are described by $\mathbf{m}$, $\mathbf{c}_L$, $\mathbf{D}_L$, $\mathcal{F}$ and $\mathcal{J}$. Then thermodynamic schemes for modeling diffusion fluxes proportional to $(\mathbf{K} : \nabla\nabla)\mathbf{v}_L$ can be elaborated which satisfy the dissipation inequality (5.1). The exploitation of the latter shows that a $\nabla\nabla\mathbf{v}_L$ -dependent model can produce coupled dissipative constitutive equations of the following type (Morro *et al*, 1990 ; energy quadratic in **K**) :

$$\begin{aligned} D_J \mathbf{K} &= \kappa_1 \bar{\delta}\, \mathbf{K} + \kappa_2 \mathbf{D}_L - \alpha(\nabla\nabla\mathbf{v}_L).\mathbf{m} + \dots \,, \\ \mathbf{v}_L - \mathbf{v}_d &= \nu_1 \mathbf{m} + \alpha\, \bar{\delta}(\mathbf{K} : \nabla\nabla)\mathbf{v}_L + \tau(\mathbf{D}_L : \nabla\nabla)\mathbf{v}_L + \dots \,, \\ \sigma_L &= 2\eta_v\, \mathbf{D}_L - \kappa_2\, \bar{\delta}\, \mathbf{K} - \tau(\nabla\nabla\mathbf{v}_L).\mathbf{m} + \dots \,, \end{aligned} \tag{5.3}$$

where κ_1, κ_2, ν_1, η_v, τ, α, and $\bar{\delta}$ are macroscopic material parameters. The second of eqns. (5.3) contains the looked for phenomenon. This must be compared to results of other works. For instance Lhuillier (1983), by using a classical theory of mixtures, arrived at an expression for the relative acceleration containing terms which, in the present notation, read $\nabla.(\mathcal{F}\mathbf{K} + \mathbf{K}\mathcal{F})$ and $\nabla.\sigma_L$. So he provided an expression for the force which causes migration via the divergence of the stress, and then through a scheme conceptually akin to the one above but without introducing a $\nabla\nabla\mathbf{v}_L$ -dependence. An analogous procedure was developed by Sekhon *et al* (1982). The procedure of Tirrell and Malone (1977), although essentially microscopic in character and without thermodynamical analysis, resembles more closely the above-presented $\nabla\nabla\mathbf{v}_L$ -dependent model.

B. - Coupling between vorticity and conformation

Recently, Grmela and Maugin (1990) have examined the thermodynamical admissibility of couplings between the conformation of macromolecules **C**, macroscopic fluid motion of velocity **v** and fields of vortices. The latter are a priori described independently of the mean fluid motion by means of an intrinsic spin density $\bar{\omega}$ (axial vector changing sign under time reversal). Through rotational invariance one is led to introducing a *relative spin density* $\omega = \bar{\omega} - \Omega$, where $\Omega = \frac{1}{2} \nabla \times \mathbf{v}$. This relative spin measures the deviation of intrinsic spin from the local vorticity. With neglect of heat and electricity conductions and with an energy density $\Psi = \tilde{\Psi}(\theta, C, D, \omega)$, one can deduce a reduced *dissipation inequality* in the form (Grmela and Maugin, 1990)

$$\mathrm{tr}\left(\tilde{\sigma}\, A^{(1)} + B\, A^{(2)} + \mathcal{F}\, D_c C\right) + A . D_c\, \omega \geqslant 0 \,, \tag{5.4}$$

where we have set

$$\begin{aligned} \tilde{\sigma} &= \sigma + (A \otimes \omega)_s + (\mathcal{F}C + C\mathcal{F}) + (BD + DB) \,, \\ A^{(1)} &\equiv D \,, \\ A^{(2)} &\equiv \dot{D} - (\nabla v . D + D . \nabla v) \,, \\ D_c \omega &\equiv \dot{\omega} - (\nabla v) . w \,, \\ D_c C &\equiv \dot{C} - (\nabla v . C + C . \nabla v) \,, \end{aligned} \tag{5.5}$$

and

$$\mathcal{F} = -\frac{\partial \tilde{\Psi}}{\partial C} \,, \quad A = -\frac{\partial \tilde{\Psi}}{\partial \omega} \,, \quad B = -\frac{\partial \tilde{\Psi}}{\partial D} \,. \tag{5.6}$$

Here convected time derivatives (see, e.g., Maugin, 1988, p. 84) are used instead of Jaumann co-rotational derivatives. The introduction of the new stress tensor $\tilde{\sigma}$ shows that the stress will in general contain *non-Newtonian* contributions (those involving B). At *thermodynamical equilibrium*, among other constraints we shall have

$$D_c \omega = 0 \tag{5.7}$$

and if $\bar{\omega} = \Omega$ initially, then this property is conserved according to (5.7). This is the essence of Helmholtz' theorem in perfect fluids. *Outside of equilibrium*, the dissipation inequality (5.4), which still is of the form $(3.11)_b$, can be exploited as in previous cases. In particular, *evolution* equations can be obtained both for C and ω. These two will couple resulting, for instance, in *dissipative* constitutive equations of the type (Grmela and Maugin, 1990 ; here $B \equiv 0$)

$$\begin{aligned} \sigma &= \mathcal{N}(C,\omega) . D + \check{\sigma}(C,\omega) \,, \\ D_c \omega &= -\, T_\omega^{-1}(D,C,\omega) . \omega \,, \\ D_c C &= -\, T_C^{-1}(D,C,\omega) : C + \kappa_2 D + \mathcal{K}(D,\omega) \,, \end{aligned} \tag{5.8}$$

where $\mathcal{N}$ and T_C^{-1} are fourth-order tensors, and $\check{\sigma}$, T_ω^{-1} and $\mathcal{K}$ are second-order tensors of which the last one is homogeneous of degree two in ω. The latter dependence and that of T_ω^{-1} on C are the looked for couplings between conformational behavior and the vorticity evolution. Satisfaction of eqn. (5.4) imposes that the coefficients present in eqns. (5.8) be not all independent.

The above modeling can be refined by introducing a *weakly nonlocal* behavior by assuming that the energy density depends as well on the additional gradients $\nabla\nabla v$, ∇C and $\nabla\omega$ (Grmela and Maugin, 1990). This allows one to replace the pure evolution equation $(5.8)_2$ by a *nonlinear evolution-diffusion* equation such as

$$D_c\omega = - T_\omega^{-1}(C,\omega).\omega + \mathcal{D}(C,\omega) : \nabla\nabla\omega , \qquad (5.9)$$

where $\mathcal{D}$ is a fourth-order tensor of diffusion coefficients depending markedly on **C** (the dependence on **D** has been discarded for simplicity). Equation (5.9) is a *nonlinear anisotropic* generalization of the well-known vorticity-diffusion equation of simple Newtonian viscous fluids. We refer the reader to original works for further elaboration of this modeling and its relation to the *metriplectic dynamics* of complex fluids (see below Sect. 7).

6. - LIQUID CRYSTALS AND OTHER COMPLEX FLUIDS

Most of the above-given development applies to a large class of *complex fluids*. By this we understand those fluids endowed with a microstructure whose influence manifests at the macroscopic scale, for instance, through improved rheological properties, a marked anisotropy, and peculiar optical behaviors. Liquid crystals and bubbly emulsions provide two classes of such complex fluids. In the end the strategy reduces to an astute choice of the internal variables (e.g., **C**) and whether a *weakly nonlocal* behavior (as briefly mentioned in Section 5) manifests itself, imposing thus the introduction of spatial gradients of these internal variables. For instance, liquid crystals and their orientational and relaxation properties (cf. de Gennes, 1974) can be well described by a thermodynamical theory involving one *vectorial* internal variable (the so-called *director* field) as shown by Drouot and Maugin (1983, Section 5) on the condition that inertial properties of directors be neglected, an assumption justified in most practical cases. The same description also applies, but without short-range ordering, hence no weak nonlocality, to dilute polymer solutions in strong elongational flows (Piau and Doremus, 1984 ; Cartalos and Piau, 1985). This *nematogen* behavior can also be reproduced by considering directly the *distribution function* as the (scalar) internal variable and accounting then for the Lipshitz or Onsager entropy term (Grmela and Chlon Ly, 1987). Short-range order in the isotropic phase of nematics and Cholesterics (de Gennes, 1971) together with the accompanying *Kerr effect* (Drouot and Maugin, 1985, 1987) enters the framework of the thermodynamical theory with one tensor (conformational) internal variable **C**. The strong analogy with the theory of *dilute emulsions* and *bubbly liquids* was noted by Lhuillier and Ouibrahim (1980, cf. Maugin and Drouot, 1983, Sect. 4).

7. - CONCLUSION

The deductive thermodynamical scheme presented exhibits a flexibility, internal coherence, and power of generalization which are typical features of physico-mathematical theories. Not only does it help in conceptual constructs and clarifying logical relationships, but it also helps in formulating couplings in a systematic manner and finding out limitations on, or relations between, various material coefficients relating to involved dissipative processes. In addition, it brings to the foreground a mathematical structure of utmost interest in contemporary mathematical physics, namely, a *metriplectic structure*. Indeed, the internal-variable formulation of complex behaviors in irreversible thermodynamics emphasizes the role of *evolution* and *evolution-diffusion equations*. But all equations, including dynamical balance laws, can be cast in such a form. Thus such a formalism yields systems of evolution equations which are *constrained* to satisfy the second law of thermodynamics. These systems, in spite of their obvious complexity (for instance, in the case mentioned in Paragraph 5.B) can nonetheless be written in the following general form (Grmela, 1986, 1990) :

$$\frac{\partial \mathcal{A}}{\partial t} = \{\mathcal{A}, \mathcal{G}_c\}_{cq} + \{\mathcal{A}, \mathcal{G}_D\}_D \ , \tag{7.1}$$

where $\mathcal{A}$ can be either *observable* or *internal* state variables, $\{.,.\}_{eq}$ is a Poisson bracket associated with the structure of the state space, $\mathcal{G}_c$ is the generator functional (e.g., the total free energy), $\{.,.\}_D$ is a Lie-Poisson bracket, and $\mathcal{G}_D$ the associated generator functional (e.g., global dissipation potential).

In the absence of dissipative processes (no D-contribution in the right-hand side of eqn. (7.1)), (7.1) exhibits a *Hamiltonian structure* which imposes the very form of thermodynamically recoverable contributions to the constitutive equations. As the Poisson bracket is skewsymmetric, while $\{.,.\}_D$ is symmetric after linearization, and

$$\{\mathcal{F},\mathcal{F}\}_D \leqslant 0 \tag{7.2}$$

for any functional $\mathcal{F}$, we have necessarily

$$\frac{\partial \mathcal{G}}{\partial t} \leqslant 0 \tag{7.3}$$

in a dissipative process ; and this is a statement of the second law of thermodynamics. Thus the "analytical" structure of balance laws and evolution constitutive equations is conserved, even outside equilibrium while the second law of thermodynamics is enforced. This can be achieved for various levels of description, i.e., with internal variables having a more or less microscopic significance (Grmela, 1990). The only difficulty that remains is the explicit construction of the brackets present in eqn. (7.1), as their definition varies from one peculiar theory to another one, even though basic properties of brackets are shared by them all (see, e.g., Grmela, 1986, and Grmela and Maugin, 1990).

Summary

A continuum modeling of dilute solutions of polymers and polyelectrolytes is presented that relies on the thermodynamics of continua involving *internal* variables (here the "conformation" of electrically polarized macromolecules). A deductive thermodynamical scheme is presented that exhibits the flexibility, internal coherence, power of generalization and typical feature of physico-mathematical theory of the present approach. In particular, its very structure allows one to study in the most convenient manner the competition between flow and applied electric fields, the stability of conformations taken by macromolecules being acted upon by external stimuli in the presence of electromechanical couplings, and to give rather simple thermodynamical accounts of the mechano-chemical effect and conformational phase transitions. Possible fruitful generalizations accounting for diffusion and eventual interactions between vortex fields and conformation are also given on the same basis. Most of the elements presented apply to many complex fluids (liquid crystals, emulsions).

Acknowledgments

For many enlightening discussions and help in clarifying many points we are indebted to D. Lhuillier (LMM, Paris), M. Grmela (Montreal), and A. Morro (Genova).

REFERENCES

Ait-Kadi, M., M. Grmela, and P.J. Carreau. 1988. A Rheological Equation of State for Dilute Polymer Solutions with Applications to Polyelectrolytes. *Rheol. Acta, 24* : 241-254.

Aubert, J., and M. Tirrell. 1980. Macromolecules in Nonhomogeneous Velocity Gradients Fields. *J. Chem. Phys. 72* : 2694-2701.

Bampi, F. and A. Morro. 1984. Non-equilibrium Thermodynamics : A Hidden Variable Approach. In *Recent Developments in Non-equilibrium Thermodynamics*, ed. J. Casas-Vasquez, D. Jou and G. Lebon, 211-232. Berlin, Springer-Verlag.

Bird, R.B., R.C. Amstrong, and D. Hassager. 1977. *Dynamics of Polymer Liquids*, vol. I. John Wiley : New York.

Brunn, P.O. 1983. Hydrodynamically Induced Cross Stream Migration of Dissolved macromolecules (modelled as nonlinearly elastic dumbbells). *Int. J. Multiphase Flows 9* : 187-232.

Carreau, P.J., M. Grmela, and A. Ait-Kadi. 1986. A Conformational model for polymer solutions (polyelectrolytes). *Commun. 2nd Conference of European Rheologists*, Prague, June 17-20, 1986.

Cartalos U., and J.M. Piau. 1985. Modèle vectoriel pour les solutions de polymères. Comparaison avec un modèle en haltère. *J. Méca. Théor. Appl.* Special issue 1985, : 101-140.

Cerf, R. 1957. La macromolécule en chaîne dans un champ hydrodynamique. Théorie générale, propriétés dynamo-optiques. *J. Polym. Sci. 23* : 125-150.

Doi, M. and S.F. Edwards. 1978. Dynamics of rod-like macromolecules in concentrated solutions-Part II. *J. Chem. Soc. Faraday Trans.* II. *74* : 918-932.

Drouot, R. 1987. Modélisation thermodynamique des solutions diluées de polymères et polyélectrolytes. *D. Sci. Thesis*, Université Pierre et Marie Curie, Paris.

Drouot, R., and G.A. Maugin. 1983. Phenomenological theory for polymer diffusion in non-homogeneous velocity gradient flows. *Rheol. Acta. 22* : 336-347.

Drouot, R., and G.A. Maugin. 1985. Continuum modelling of polyelectrolytes in solution. *Rheol. Acta 24* : 474-487.

Drouot, R., and G.A. Maugin. 1987. Optical and hydrodynamical regimes in dilute solutions of polyelectrolytes. *Rheol. Acta. 26* : 350-357.

Drouot, R., and G.A. Maugin. 1988a. Polyelectrolytes in solution : Equilibrium conformations of the microstructure without and with external fields. *Int. J. Engng. Sci. 26* : 225-241.

Drouot, R., and G.A. Maugin. 1988b. Phénomélogie et thermodynamique du modèle conformationnel des solutions de polymères. In *Rheology of Heterogeneous fluids*, ed. D. Scrivener. 8.1-8.8, I.M.F., Strasbourg.

Drouot, R., G.A. Maugin, and A. Morro. 1987. Anisotropic equilibrium conformations in solutions of polymers and polyelectrolytes. *J. Non-Newtonian Fluid. Mech. 23* : 295-304.

Dunlap, P.N., and L.G. Leal. 1984. The charged dumbbell model for dilute polyelectrolyte solutions in strong flows. *Rheol. Acta 23* : 271-294.

Frölich, H. 1973. *J. Coll. Phenomena, 1* : 101.

de Gennes, P.G. 1971. Short range order in the isotropic phase of nematics and cholesterics. *Mol. Cryst. Liq. Cryst. 12* : 193-214.

de Gennes, P.G. 1974. *The physics of liquid crystals.* London : Oxford University Press.

Giesekus, H. 1986. Constitutive models of polymer fluids : Toward a unified approach. In *Trends in applications of Pure mathematics to Mechanics*, ed. E. Kröner and K. Kirchgässner, 331-348. Springer-Verlag, Berlin.

Grmela, M. 1986. Bracket formulation of diffusion-convection equations. *Physica 21D* : 179-212.

Grmela, M. 1990. Hierarchy of Descriptions in thermodynamics. In *Continuum Models and Discrete Systems*, ed. G.A. Maugin, vol. 1, Longman , London.

Grmela, M., and P.J. Carreau. 1987. Conformation tensor rheological models. *J. Non-Newtonian Fluid. Mech. 23* : 271-294.
Grmela, M., and Chon Ly. 1987. Shear flow induced structural changes in polymeric liquid crystals. *phys. Lett., A120* : 281-285.
Grmela, M., and G.A. Maugin. 1990. Vorticity dynamics in complex fluids. (in preparation).
de Groot, S.R., and P. Mazur. 1962. *Non-equilibrium Thermodynamics*. North-Holland, Amsterdam (reprint, Dover, New York, 1986).
Jeffery, G.B. 1922. The motion of ellipsoidal particles immersed in a viscous fluid. *Proc. Roy. Soc. Lond. A102* : 161-179.
Keizer, J. 1985. Heat, work and the thermodynamic temperature at non-equilibrium steady states. *J. Chem. Phys. 82* : 2751-2771.
Kestin, J., and J. Bataille. 1975. L'interprétation physique de la thermodynamique rationnelle. *J. Mécanique 14* : 365-386.
Kuhn, W., and H. Kuhn. 1945. Die Abhörigheit der Viskosität von Strömungsgefälle bei hochverdünnter Suspensionen und Lösungen. *Helv. Chim. Acta 28* : 97-127.
Landau, L.D., and E.M. Lifshitz. 1984. *Electrodynamics of continua*. Pergamon Press, Oxford. Revised edition.
Leonov, A.I. 1976. Non-equilibrium thermodynamics and rheology of viscoelastic polymer media. *Rheol. Acta 21* : 683-691.
Lhuillier, D. 1981. Molecular models and the Taylor stability of dilute polymer solutions. *J. Non-Newtonian Fluid Mech. 9* : 329-337.
Lhuillier, D. 1983. Phenomenology of polymer migration. *J. Physique 44* : 303-309.
Lhuillier, D. and A. Ouibrahim. 1980. A Thermodynamical model for solutions of deformable macromolecules. *J. Mécanique 19* : 1-17.
Maugin, G.A. 1980. The principle of virtual power in continuum mechanics : application to coupled fields. *Acta Mechanica 35* : 1-70.
Maugin, G.A. 1987. Thermodynamique à variables internes et applications. In *Seminar : Thermodynamics of Irreversible Processes*, Institut Français du Pétrole, Rueil-Malmaison, France.
Maugin, G.A. 1988. *Continuum Mechanics of Electromagnetic Solids*. Amsterdam : North-Holland .
Maugin, G.A., and R. Drouot. 1983. Internal variables and the thermodynamics of macromolecules solutions. *Int. J. Engng. Sci. 21* : 705-724.
Maugin, G.A., and R. Drouot. 1988a. Stability of conformations of polyelectrolytes in flows and applied electric fields. In *Progress and Trends in Rheology* II, ed. H. Giesekus, 94-96. Darmstadt : Steinkopf Verlag.
Maugin, G.A., and R. Drouot. 1988b. Thermodynamic modelling of polymers in solutions. In *Constitutive laws and microstructure*, ed. D.R. Axelrad and W. Muschik, 136-161. Berlin : Springer-Verlag.
Maugin G.A., B. Collet, R. Drouot, and J. Pouget. 1990. *Nonlinear Electromechanical Couplings*. Manchester, U.K. : Manchester University Press.
Morro, A., R. Drouot, and G.A. Maugin. 1985. Thermodynamics of polyelectrolyte solutions in an electric field. *J. Non-equilibrium Thermodyn. 10* : 131-144.
Morro, A., R. Drouot, and G.A. Maugin. 1987. A continuum model for polyelectrolytes in solution. In *Continuum models of discrete systems* (5), ed. A.J.M. Spencer, 139-144. Rotterdam : A.A. Balkema.
Morro, A., G.A. Maugin and R. Drouot. 1990. Diffusion in polyelectrolyte solutions. *Rheologica Acta* (submitted for publication).
Oosawa, F. 1971. *Polyelectrolytes*. New York : Marcel Dekker.
Piau, J.M., and P. Doremus. 1984. Constitutive equations for dilute polymer solutions in strong elongational flows. *Rheol. Acta 23* : 465-476.
Scott, A.C. 1984. Biological solitons. In *Dynamical Problems in Soliton Systems*, ed. S. Takeno, 224-235. Berlin : Springer-Verlag.
Sekhon, G. R.C. Amstrong, and M.S. Jhon. 1982. The origin of polymer migration in nonhomogeneous flow field. *J. Polym. Sci. Polym. Phys. 20* : 947-952.

Sluckin, T.J. 1981. Influence of flow on the isotropic-nematic transition in polymer solutions : a thermodynamical approach. *Macromolecules 14* : 1676-1680.
Tanaka, T., I. Nishio, S.T. Sun, and S. Ueno-Nishio. 1982. Collapse of goels in an electric field. *Science 21B* : 467-469.
Tirrell, M. and M.F. Malone. 1977. Stress-induced diffusion of macromolecules. *J. Polym. Sci. 15* : 1569-1583.
Volkenshtein, M.V. 1983. *Biophysics*. Moscow : Mir Publishers.

Thermodynamics and Extremum Principles in Viscoelasticity

A. Morro
DIBE, University
Viale Causa 13, 16145 Genova, Italy

Abstract. The paper deals with linearly viscoelastic bodies (solids and fluids) and has a twofold purpose. First, to provide a comprehensive analysis of the thermodynamics of viscoelasticity. Second, to show how the thermodynamic restrictions ensure the validity of extremum principles for the dynamics of the body. To this end, the thermodynamic scheme is based on general notions of state, process, cycle, approximate cycle. Then the restrictions are derived; in essence, for solids a negative half-range sine transform of the Boltzmann function turns out to be necessary and sufficient for the validity of the second law. The recourse to the internal variable model is also made to derive the expression of the free energy. Then extremum principles are set up just on the basis of the thermodynamic restriction. Though in various ways, such principles are shown to hold for the Laplace transform of the displacement field. Analogous properties are shown to hold for fluids, both compressible and incompressible.

1. INTRODUCTION

Within the recent, far-reaching developments of thermodynamics, materials with fading memory have played a prominent role. The thermodynamics of materials with fading memory was developed in the sixties as a remarkable chapter of Rational Thermodynamics. It was Coleman (1964) who first provided a mathematically rigorous theory of thermodynamics of materials with fading memory. In this approach the Clausius-Duhem inequality is taken to be the expression of the second law of thermodynamics and then a systematic method yields the mathematical consequences on the constitutive relations. The approach by Day (1972), developed at the end of the sixties, has the conceptual advantage that the entropy does not enter into the starting formulation but, rather, it enters the theory as a derived concept. A paper by Coleman and Owen (1974), somewhat reminiscent of a previous paper by Noll (1972), exhibits a general theory of thermodynamics where each system is a pair of sets; the set of states and that of processes. A fundamental concept of the theory is that of an action and the second law is taken as the assertion that a given action has the Clausius property.

Aside from the developments of more and more general theories of thermodynamics, recently particular attention has been devoted to specific thermodynamic contexts and material models with the apparent aim of arriving at more definite results. This trend is emphasized by various aspects. Foremost among them is the elaboration of appropriate thermodynamic frameworks such as the Extended irreversible thermodynamics and the Thermodynamics of materials with internal variables. Though framed within the general theory of materials with fading memory, this work develops a thermodynamic analysis of a specific model, namely linear viscoelasticity.

A number of reasons motivate the interest in linear viscoelasticity where the stress tensor $\mathbf{T}$ is given by the history of the infinitesimal strain tensor $\mathbf{E}$ in the form

$$\mathbf{T}(t) = \mathbf{G}_0\,\mathbf{E}(t) + \int_0^\infty \mathbf{G}'(s)\,\mathbf{E}(t-s)\,ds$$

which, to a large extent, it is a realistic model of material behaviour. It exhibits the simplicity peculiar to linear systems but still maintains the difficulties due to the (fading) memory property. So the interest into the physical aspects of the model is strengthened by that into the mathematical descriptions in terms of memory functionals. Further, particular models of linear viscoelasticity can be framed within simpler thermodynamic schemes thus allowing suggestive contrasts between corresponding procedures and results.

The subject presented in this article is based on recent researches carried out by the author[1]. Starting from a general scheme of thermodynamics of simple materials, precise definitions and statements are given for states, processes, cycles, and laws of thermodynamics. It is observed that, in linear viscoelasticity, no information is lost by considering the second law for cycles rather than for approximate cycles. Then the thermodynamic theory is developed for solids and fluids. It is shown that, for solids, the essence of the dissipative character is summarized by the negative definiteness of the half-range Fourier sine transform of the kernel $\mathbf{G}'$. Really, it is a new result that such negative definiteness yields all conditions derived so far via various procedures. What is more, such condition has been proved to be sufficient for the validity of the second law. As an aside, an interesting connection is made with the model of materials with internal variables by deriving the form of the free energy for a solid. Next, in a similar way, necessary and sufficient conditions for the validity of the second law are derived for fluids. The role of the thermodynamic restrictions is emphasized by investigating extremum principles for solids and fluids. Essentially the extremum properties are framed within the context of the Laplace transform of the pertinent dynamic problem. It is shown how the extremum property and the uniqueness of the solution is guaranteed by the thermodynamic restrictions.

[1]Mainly joint researches with M. Fabrizio.

NOTATION. As far as possible the customary notation of continuum thermodynamics and linear viscoelasticity has been used throughout. The meaning of some mathematical symbols is given as follows. tr denotes the trace, det the determinant, Lin the set of second- or fourth-order tensors, Lin^+ the set of tensors with positive determinant, Sym the set of symmetric tensors, sym the symmetric part, a superposed dot the time derivative, ∇ the gradient. Moreover, $\mathbb{R}$, $\mathbb{R}^+$, $\mathbb{R}^{++}$, $\mathbb{R}^-$ are the sets of reals, positive reals, strictly positive reals, negative reals. $\Re$ and $\Im$ denote the real and imaginary part of a complex number, $\Re^+$ and $\Re^{++}$ stand for the complex numbers with positive and strictly positive real part. Usually the complex variable is denoted by p and α, β are the real and imaginary parts. Unfortunately, p denotes also the pressure of the fluid but this should not cause any ambiguity.

2. STATES, PROCESSES, AND LAWS OF THERMODYNAMICS

At the bottom, any thermodynamic theory is based on the notions of state and process. Here we provide these notions by following a general scheme of thermodynamics of simple materials which traces back to Coleman and Owen (1974) and has been re-examined by Fabrizio and Morro (1990).

For any material point $\mathbf{X}$ in a body $\mathcal{B}$ and time t in a suitable interval in $\mathbb{R}$ we define the *configuration* as the pair

$$C(\mathbf{X},t) = (\mathbf{F}(\mathbf{X},t),\, \theta(\mathbf{X},t)) \tag{2.1}$$

of the deformation gradient $\mathbf{F}$ and the absolute temperature θ. Whenever no ambiguity can arise, the dependence on $\mathbf{X}$ is understood and not written. Though inessential, we let $\det \mathbf{F} > 0$ and then $\mathbf{F} \in \mathrm{Lin}^+$. Owing to mass conservation, the mass density ρ is related to the reference mass density ρ_0 by $\rho = \rho_0/\det \mathbf{F}$, which yields ρ once $\mathbf{F}$ is known.

Let E be an open, connected subset of $\mathrm{Lin} \times \mathbb{R} \times V$. A map $P : [0, d_P) \to E$, piecewise continuous on $[0, d_P)$, is called *process* of duration $d_P \in \mathbb{R}^+$. Each restriction of P to an interval $[t_1, t_2) \subset [0, d_P)$ is called *segment* of P. In particular, we denote by P_t the restriction of P to the interval $[0, t)$, $t < d_P$. Letting P_1 and P_2 be two processes we define the process $P_1 * P_2$ as

$$P_1 * P_2 = \begin{cases} P_1(t), & t \in [0, d_{P_1}) \\ P_2(t - d_{P_1}), & t \in [d_{P_1}, d_{P_1} + d_{P_2}); \end{cases}$$

the process $P_1 * P_2$ is called *composition* of P_2 with P_1. Here we consider *thermokinetic* processes defined, for every $t \in [0, d_P)$, as

$$P(t) = (\mathbf{L}(t),\, \dot{\theta}(t),\, \mathbf{g}(t)). \tag{2.2}$$

The symbol $\mathcal{C}$ denotes the set of all configurations which are accessible for the body $\mathcal{B}$ at the point $\mathbf{X}$ under consideration. The set Π of all accessible processes is such that if $P \in \Pi$ then every segment of P is in Π. Moreover, if $P_1, P_2 \in \Pi$ then $P_1 * P_2 \in \Pi$.

Any process $P \in \Pi$ can be viewed as the *input* of $\mathcal{B}$. The *output* U of $\mathcal{B}$ is an element of $\mathrm{Sym} \times \mathbb{R} \times V$ and is defined as

$$U = (\mathbf{T},\, h,\, \mathbf{q}) \tag{2.3}$$

where $\mathbf{T}$ is the Cauchy stress tensor, h the total heat supply, and $\mathbf{q}$ the heat flux vector. Any state σ is an element of a metric space Σ. We denote by $\varrho : \Sigma\Diamond\Pi \rightarrow \Sigma$ the ***evolution function*** (or state-transformation function). Its domain $\Sigma\Diamond\Pi$ is a subset of $\Sigma \times \Pi$ such that any element is the pair (σ, P) of a state σ and a process P which can occur by starting from σ. The function ϱ is such that if $(\sigma, P_1) \in \Sigma\Diamond\Pi$ and $(\varrho(\sigma, P_1), P_2) \in \Sigma\Diamond\Pi$ then $(\sigma, P_1 * P_2) \in \Sigma\Diamond\Pi$ and $\varrho(\sigma, P_1 * P_2) = \varrho(\varrho(\sigma, P_1), P_2)$. The function $\hat{U} : \Sigma \times E \rightarrow \mathrm{Sym} \times \mathbb{R} \times V$ maps the pair (σ, P) of the state σ and the process P at a time t into the output U at the time t. Specifically, $\hat{U}$ consists of the triplet $\hat{\mathbf{T}}, \hat{h}, \hat{\mathbf{q}}$ such that

$$\hat{\mathbf{T}} : \Sigma \times E \rightarrow \mathrm{Sym}, \quad \hat{h} : \Sigma \times E \rightarrow \mathbb{R}, \quad \hat{\mathbf{q}} : \Sigma \times E \rightarrow V. \tag{2.4}$$

In words, $U(t)$ depends on the state $\sigma(t)$ and the present value of the input, $P(t)$. To avoid any ambiguity we write

$$P(t) = (\mathbf{L}_P(t),\ \dot{\theta}_P(t),\ \mathbf{g}_P(t)) \tag{2.5}$$

where the subscript P is a reminder that the time dependence for $\mathbf{L}, \dot{\theta}$, and $\mathbf{g}$ is that induced by the process P.

As an example, consider a material with internal (or hidden) variables; such a model is employed extensively in this book.[2] Let $\boldsymbol{\xi}$ be the set of internal variables; they may represent any collection of scalars, vectors, and tensors. We write the constitutive equations as

$$\mathbf{T} = \mathbf{T}(\mathbf{F}, \theta, \boldsymbol{\xi}), \qquad h = a(\mathbf{F}, \theta, \boldsymbol{\xi})\dot{\theta} + \mathbf{B}(\mathbf{F}, \theta, \boldsymbol{\xi}) \cdot \mathbf{L}, \qquad \mathbf{q} = \mathbf{q}(\mathbf{F}, \theta, \mathbf{g}, \boldsymbol{\xi})$$

along with the growth equation for $\boldsymbol{\xi}$,

$$\dot{\boldsymbol{\xi}} = \mathbf{f}(\mathbf{F}, \theta, \mathbf{g}, \boldsymbol{\xi}).$$

Here the state σ is the triplet $(\mathbf{F}, \theta, \boldsymbol{\xi})$ and the process is $P(t) = (\mathbf{L}_P(t), \dot{\theta}_P(t), \mathbf{g}_P(t))$. The evolution function ϱ determines the current state $\sigma(t)\rho(\sigma^i, P_t)$ in terms of the initial state σ^i, the segment P_t, and the (given) growth function $\mathbf{f}$. Letting $\sigma^i = (\mathbf{F}_0, \theta_0, \boldsymbol{\xi}_0)$, formally we determine σ as the solution to the Cauchy problem

$$\frac{d}{dt}\mathbf{F}(t) = \mathbf{L}_P(t)\mathbf{F}(t), \quad \mathbf{F}(0) = \mathbf{F}_0, \tag{2.6}$$

$$\frac{d}{dt}\theta(t) = \dot{\theta}_P(t), \quad \theta(0) = \theta_0, \tag{2.7}$$

and

$$\frac{d}{dt}\boldsymbol{\xi}(t) = \mathbf{f}(\mathbf{F}_P(t),\ \theta_P(t),\ \mathbf{g}_P(t),\ \boldsymbol{\xi}(t)), \quad \boldsymbol{\xi}(0) = \boldsymbol{\xi}_0,$$

with $t \in [0, d_P)$, where $\mathbf{F}_P, \theta_P$ stands for the solutions to (2.6), (2.7).

A second example, related to the subject of this article, deals with materials with fading memory[3]. The output function is characterized by the constitutive equations

$$\mathbf{T} = \hat{\mathbf{T}}(\mathbf{F}^t, \theta^t), \quad h = \hat{a}(\mathbf{F}^t, \theta^t)\dot{\theta} + \hat{\mathbf{B}}(\mathbf{F}^t, \theta^t) \cdot \mathbf{L}, \quad \mathbf{q} = \hat{\mathbf{q}}(\mathbf{F}^t, \theta^t, \mathbf{g}). \tag{2.8}$$

[2]For the reader desiring a detailed account and applications of the model we mention Morro $(1985)_1$ and the references therein.
[3]See, e. g., Day (1972).

Here the state σ is the pair $(\mathbf{F}^t, \theta^t)$ of the histories $\mathbf{F}^t, \theta^t$ which are elements of the fading memory space

$$H = \{\mathbf{F}^t, \theta^t : \mathbb{R}^+ \to \mathrm{Lin}^+ \times \mathbb{R}^+; \int_0^\infty [|\mathbf{F}^t(s)|^2 + |\theta^t(s)|^2]k(s)ds < \infty\}$$

where k is the usual influence function. It is a routine matter to determine the evolution function ϱ.

Given a process $P \in \Pi$ let

$$\mathcal{D}(P) := \{\sigma \in \Sigma; (\sigma, P) \in \Sigma \Diamond \Pi\}.$$

A pair (σ, P), $\sigma \in \mathcal{D}(P)$, is called a *cycle* if

$$\varrho(\sigma, P) = \sigma.$$

In words, the final state produced by the process P coincides with the initial state. In such a case we say also that the process P is cyclic.

The mass density ρ is determined by the state σ, indeed the dependence on σ is through $\det\mathbf{F}$ only. To save writing, henceforth we write $\rho(t)$ as an abbreviation for $\rho(\sigma(t))$.

First Law. *For every cycle $(\sigma, P) \in \Sigma \Diamond \Pi$ the equality*

$$\int_0^{d_P} [\hat{h}(\sigma(t), P(t)) + \frac{1}{\rho(t)} \hat{\mathbf{T}}(\sigma(t), P(t)) \cdot \mathbf{L}(t)]dt = 0$$

holds with $\sigma(t) = \varrho(\sigma, P_t)$.

Let $e : \Sigma \Diamond \Pi \to \mathbb{R}$ be defined by

$$e(\sigma, P) = \int_0^{d_P} [\hat{h}(\sigma(t), P(t)) + \frac{1}{\rho(t)} \hat{\mathbf{T}}(\sigma(t), P(t)) \cdot \mathbf{L}(t)]dt. \tag{2.9}$$

Given $\sigma_0, \sigma \in \Sigma$ it is convenient to consider the set

$$\mathcal{E}(\sigma_0, \sigma) := \{e(\sigma_0, P), P \in \Pi : \varrho(\sigma_0, P) = \sigma\}.$$

As a consequence of the first law it follows that *for any $\sigma_0, \sigma \in \Sigma$, the set $\mathcal{E}(\sigma_0, \sigma)$ is bounded and consists of the only value $e(\sigma_0, \bar{P}), \bar{P} \in \Sigma$ being any process such that $\hat{\varrho}(\sigma_0, \bar{P}) = \sigma$.*

A function $A : \Sigma \to \mathbb{R}$ is said to be a *potential* for $e(\sigma, P)$ if

$$A(\sigma_2) - A(\sigma_1) = e(\sigma_1, P)$$

for every pair of states $\sigma_1, \sigma_2 \in \Sigma$ and process $P \in \Pi$ such that $\hat{\varrho}(\sigma_1, P) = \sigma_2$.

It is a further consequence of the first law that *there exists a potential $\epsilon : \Sigma \to \mathbb{R}$ for $e(\sigma, P)$;* ϵ being a potential for $e(\sigma, P)$ means that for any pair of states $\sigma_1, \sigma_2 \in \Sigma$ and any process $P \in \Pi$, with $\varrho(\sigma_1, P) = \sigma_2$,

$$\epsilon(\sigma_2) - \epsilon(\sigma_1) = \int_0^{d_P} [\hat{h}(\sigma(t), P(t)) + \frac{1}{\rho(t)} \hat{\mathbf{T}}(\sigma(t), P(t)) \cdot \mathbf{L}(t)]dt$$

where $\sigma(t) = \varrho(\sigma_1, P_t)$. Given $\sigma_0 \in \Sigma$ and $P \in \Pi$, the function

$$\epsilon(\sigma) = e(\sigma_0, P), \tag{2.10}$$

where $\sigma = \varrho(\sigma_0, P)$, turns out to be the pertinent potential.

On the basis of (2.9) and (2.10) we regard the function $\epsilon(\sigma)$ as the *internal energy* (density). Of course the internal energy ϵ is determined to within an additive constant. This arbitrariness can be removed by regarding the internal energy at a state σ_0 as the reference value.

By the existence of the potential ϵ, at any time t where P is continuous,

$$\dot{\epsilon}(\sigma(t)) = \hat{h}(\sigma(t), P(t)) + \frac{1}{\rho(t)}\hat{\mathbf{T}}(\sigma(t), P(t)) \cdot \mathbf{L}(t).$$

By the definition of h, viz.

$$\rho h = -\nabla \cdot \mathbf{q} + \rho r, \tag{2.11}$$

we have the relation

$$\rho\dot{\epsilon} = -\nabla \cdot \mathbf{q} + \mathbf{T} \cdot \mathbf{L} + \rho r \tag{2.12}$$

which identifies ϵ with the internal energy of standard continuum thermomechanics.

The content of the second law is given the following form.
Second law. *For every cycle* $(\sigma, P) \in \Sigma\diamond\Pi$ *the inequality*

$$\int_0^{d_P} [\frac{\hat{h}(\sigma(t), P(t))}{\theta(t)} - \frac{1}{\rho(t)\theta^2(t)}\hat{\mathbf{q}}(\sigma(t), P(t)) \cdot \mathbf{g}(t)]dt \le 0$$

holds with $\sigma(t) = \varrho(\sigma, P_t)$.

To derive consequences of the second law it is convenient to consider the function $s : \Sigma\diamond\mathbb{R}$, defined by

$$s(\sigma, P) = \int_0^{d_P} [\frac{\hat{h}(\sigma(t), P(t))}{\theta(t)} - \frac{1}{\rho(t)\theta^2(t)}\hat{\mathbf{q}}(\sigma(t), P(t)) \cdot \mathbf{g}(t)]dt,$$

and the set

$$\mathcal{S}(\sigma_0, \sigma) = \{s(\sigma_0, P), P \in \Pi : \varrho(\sigma_0, P) = \sigma\}.$$

Moreover it is convenient to introduce the notion of upper potential.

A function $B : \Sigma \to \mathbb{R}$ is said to be an *upper potential* for $s(\sigma, P)$ if

$$B(\sigma_2) - B(\sigma_1) \ge s(\sigma_1, P)$$

for every pair of states $\sigma_1, \sigma_2 \in \Sigma$ and process $P \in \Pi$ such that $\varrho(\sigma_1, P) = \sigma_2$.

It is a consequence of the second law that *there exists at least one upper potential for* $s(\sigma, P)$; that is

$$\eta^s(\sigma) = \sup\{s(\sigma_0, P), P \in \Pi : \varrho(\sigma_0, P) = \sigma\}.$$

The upper potential $\eta^s(\sigma)$ is called *entropy*. It is worth remarking that in general the upper potential $\eta^s(\sigma)$ is not uniquely determined. We can fix the value of $\eta^s(\sigma)$ at a reference state σ_0 and denote the upper potential by $\eta(\sigma)$. Yet even $\eta(\sigma)$ is not uniquely determined and we have $\eta^s(\sigma) \geq \eta(\sigma)$.

For completeness and useful reference we give the explicit statement of the second law when the processes under consideration can be regarded as isothermal. Observe that, by (2.11) and (2.12),

$$\rho h = \rho \dot{\epsilon} - \mathbf{T} \cdot \mathbf{L}.$$

Substitution yields

$$s(\sigma, P) = \int_0^{d_P} [\frac{\dot{\hat{\epsilon}}(\sigma(t), P(t))}{\theta(t)} - \frac{1}{\rho(t)} \hat{\mathbf{T}}(\sigma(t), P(t)) \cdot \mathbf{L}(t) - \frac{1}{\rho(t)\theta^2(t)} \hat{\mathbf{q}}(\sigma(t), P(t)) \cdot \mathbf{g}(t)] dt.$$

If the process P is cyclic and θ is constant and uniform, whence $\mathbf{g} = 0$, then the only significant contribution is given by $\mathbf{T} \cdot \mathbf{L}$. Hence we have the following statement.

Second law for isothermal processes. *For every cycle* $(\sigma, P) \in \Sigma \diamond \Pi$ *the inequality*

$$\int_0^{d_P} \frac{1}{\rho(t)} \hat{\mathbf{T}}(\sigma(t), P(t)) \cdot \mathbf{L}(t) \geq 0$$

holds with $\sigma(t) = \varrho(\sigma, P_t)$ *and* $P(t) = (\mathbf{L}(t), 0, \mathbf{0})$.

For materials with fading memory cycles are quite rare in that it usually happens that upon a process the material gets to a state which, though close to the initial one, is different from it. Then, to make the thermodynamic theory widely applicable, a broader class of processes are considered as the counterpart of cyclic processes. Such is the case of approximately-cyclic processes, and then of approximate cycles, which are characterized as follows.

Let $\mathcal{O}_\nu(\sigma)$ stand for the ν-neighbourhood of σ consisting of the set of elements $\bar{\sigma}$ such that $\|\bar{\sigma} - \sigma\| < \nu$. A pair $(\sigma, P) \in \Sigma \times \Pi$ is called an ν-approximate cycle if $\varrho(\sigma, P) \in \mathcal{O}_\nu(\sigma)$. The extension of the second law to approximate cycles can then be given the following form.

Second law (approximate cycles). *For every* $\varepsilon > 0$ *there exists* $\nu_\varepsilon > 0$ *such that*

$$\int_0^{d_P} [\frac{\hat{h}(\sigma(t), P(t))}{\theta(t)} + \frac{1}{\rho(t)\,\theta^2(t)} \hat{\mathbf{q}}(\sigma(t), P(t)) \cdot \mathbf{g}(t)]\, dt < \varepsilon$$

for any state σ *and process* P *with* $\varrho(\sigma, P) \in \mathcal{O}_{\nu_\varepsilon}(\sigma)$.

For later convenience it is worth introducing the *dissipation* functional $\hat{\gamma} : \Sigma \times E \to \mathbb{R}$ as defined by

$$\hat{\gamma}(\sigma(t), P(t)) = \frac{1}{\theta(t)} \hat{h}(\sigma(t), P(t)) - \frac{1}{\rho(t)\theta^2(t)} \hat{\mathbf{q}}(\sigma(t), P(t)) \cdot \mathbf{g}(t).$$

By (2.11) and (2.12) the value of the dissipation, γ, is given by

$$\gamma = \frac{1}{\theta} \dot{\epsilon} - \frac{1}{\rho\theta} \mathbf{T} \cdot \mathbf{L} + \frac{1}{\rho\theta^2} \mathbf{q} \cdot \mathbf{g}.$$

The statement of the second law simplifies when, as is the case in linear viscoelasticity, isothermal processes are involved. First, the contribution to the dissipation due to heat conduction vanishes. Second, the integral of $\dot{\epsilon}/\theta$ in $[0, d)$ gives $[\hat{\epsilon}(\sigma(d)) - \hat{\epsilon}(\sigma(0))]/\theta$. Provided only that the functional $\hat{\epsilon}$ is continuous, we can have $|\hat{\epsilon}(\sigma(d)) - \hat{\epsilon}(\sigma(0))|$ as small as we please by considering a small enough neighbourhood $\mathcal{O}_\nu(\sigma)$ of the initial state σ. Then in the case of isothermal processes the second law for approximate cycles becomes the following

Work inequality. *For every $\varepsilon > 0$ there exists $\nu_\varepsilon > 0$ such that*

$$\int_0^d \frac{1}{\rho(t)} \hat{\mathbf{T}}(\mathbf{E}^t) \cdot \dot{\mathbf{E}}(t)\, dt > -\varepsilon$$

for every ν_ε-approximate cycle.

There are cases where the set of cycles is quite large, relative to the set of approximate cycles, in the sense that no additional information arises from the second law for approximate cycles, beyond that from the second law for cycles. Of course this is not known a priori. As we shall see in a moment, the linear theory of viscoelasticity is a case where this feature occurs.

3. THERMODYNAMICS OF LINEAR VISCOELASTICITY

For the sake of definiteness, henceforth we apply the thermodynamic approach in connection with viscoelastic solids and fluids. Solids are modelled as follows. Let $\mathbf{E}$ be the infinitesimal strain tensor, namely

$$\mathbf{E} = \text{sym} \frac{\partial \mathbf{u}}{\partial \mathbf{x}}$$

where $\mathbf{x}$ is the present position (identified with $\mathbf{X}$). The state of the body, σ, is then taken as the history of $\mathbf{E}$, i. e.

$$\sigma(t) = (\mathbf{E}(t),\ {}_r\mathbf{E}^t)$$

where $\mathbf{E}(t)$ denotes the present value and ${}_r\mathbf{E}^t$ the past history, ${}_r\mathbf{E}^t(s) = \mathbf{E}(t-s)$, $s \in \mathbb{R}^{++}$. Owing to the linear approximation we let $\det \mathbf{F} = 1$, and then ρ be constant, and identify the (first) Piola-Kirchhoff stress tensor $\mathbf{S}$ with the Cauchy stress tensor $\mathbf{T}$. The power of the stress $\mathbf{T} \cdot \mathbf{D}$ or $\mathbf{S} \cdot \dot{\mathbf{F}}$ can then be identified with $\mathbf{T} \cdot \dot{\mathbf{E}}$. We let $\mathbf{T}$ be given by the constitutive functional of linear viscoelasticity, namely

$$\mathcal{T}(\mathbf{E}^t) = \mathbf{G}_0 \mathbf{E}(t) + \int_0^\infty \mathbf{G}'(s) \mathbf{E}(t-s)\, ds,$$

where $\mathbf{G}_0 \in \text{Lin(Sym)}$ and $\mathbf{G}' : \mathbb{R}^+ \to \text{Lin(Sym)}$. As usual,

$$\mathbf{G}(s) = \mathbf{G}(0) + \int_0^s \mathbf{G}'(u)\, du$$

is called *relaxation function*. Of course $\mathbf{G}$ is allowed to depend also on the position $\mathbf{x}$ but, as usual, such dependence will be understood and not written.

The fourth-order tensor $\mathbf{G}_0 := \mathbf{G}(0)$ is called ***instantaneous elastic modulus*** and governs the response to instantaneous changes in strain. The limit of $\mathbf{G}(s)$, as $s \to \infty$, is supposed to exist and

$$\mathbf{G}_\infty = \lim_{s\to\infty} \mathbf{G}(s)$$

is called ***equilibrium elastic modulus*** in that the constancy of the history $\mathbf{E}^t$, $\mathbf{E}^t(s) = \mathbf{E}, s \in \mathbb{R}^+$, gives

$$\mathbf{T}(t) = \mathbf{G}_\infty \mathbf{E}.$$

We assume that

$$\mathbf{G}_\infty > 0.$$

Consistent with the model of viscoelastic solid, whereby the power $\mathbf{T} \cdot \mathbf{D}$ is approximated with $\mathbf{T} \cdot \dot{\mathbf{E}}$, we restate the second law through the work inequality by saying that, for each $\varepsilon > 0$, there exists $\nu_\varepsilon > 0$ such that[4]

$$\int_0^d \mathcal{T}(\mathbf{E}^t) \cdot \dot{\mathbf{E}}(t)\, dt > -\varepsilon \tag{3.1}$$

for any ν_ε-approximate cycle. Upon substitution of the constitutive functional $\mathcal{T}$ it follows from (3.1) that

$$\int_0^d \dot{\mathbf{E}}(t) \cdot \mathbf{G}_0 \mathbf{E}(t)\, dt + \int_0^d \dot{\mathbf{E}}(t) \cdot \int_0^\infty \mathbf{G}'(s)\mathbf{E}^t(s)\, ds > -\varepsilon. \tag{3.2}$$

Of course, in the case of cycles the inequality (3.2) is replaced by

$$\int_0^d \dot{\mathbf{E}}(t) \cdot \mathbf{G}_0 \mathbf{E}(t)\, dt + \int_0^d \dot{\mathbf{E}}(t) \cdot \int_0^\infty \mathbf{G}'(s)\mathbf{E}^t(s)\, ds \geq 0. \tag{3.3}$$

The second law of thermodynamics requires that (3.3) hold for any cycle and that equality hold if and only if $\mathbf{E}^t$ is a constant history.

The functional $\mathcal{T}$ is compatible with thermodynamics if and only if (3.2) is satisfied for approximate cycles. As a particular case, compatibility with thermodynamics means that (3.3) holds for cycles. Then we begin by deriving thermodynamic restrictions on $\mathcal{T}$.

Besides being important for easy correlation with experiments, time-harmonic variation proves to be especially suited to the derivation of thermodynamic restrictions. Accordingly we consider oscillatory strain-tensor evolutions of the form

$$\tilde{\mathbf{E}}(t) = \mathbf{E}_1 \cos\omega t + \mathbf{E}_2 \sin\omega t, \tag{3.4}$$

where $\omega \in \mathbb{R}^{++}$, $\mathbf{E}_1, \mathbf{E}_2 \in \mathrm{Sym}$. The corresponding process $\tilde{P} \in \Pi$ may be expressed as

$$\tilde{P}(t) = -\omega\mathbf{E}_1 \sin\omega t + \omega\mathbf{E}_2 \cos\omega t, \qquad t \in [0, d), \tag{3.5}$$

[4]Henceforth d stands for the duration of the process under consideration.

where $d = 2\pi m/\omega$, m being any positive integer. Accordingly, it follows at once that any current state

$$\tilde{\sigma}(t) = (\tilde{\mathbf{E}}(t), {}_r\tilde{\mathbf{E}}^t), \tag{3.6}$$

and the process P constitute a cycle in that

$$\varrho(\tilde{\sigma}(t), \tilde{P}) = \tilde{\sigma}(t).$$

Then $\tilde{\sigma}(t), \tilde{P}$ must satisfy the second law in the form (3.3).

The application of the inequality (3.3) to any cycle $(\tilde{\sigma}(t), \tilde{P})$ yields

$$\begin{aligned} \mathbf{E}_1 \cdot [\mathbf{G}_0^T - \mathbf{G}_0]\,\mathbf{E}_2 - \int_0^\infty [\mathbf{E}_1 \cdot \mathbf{G}'(s)\mathbf{E}_1 + \mathbf{E}_2 \cdot \mathbf{G}'(s)\mathbf{E}_2] \sin\omega s \, ds \\ - \int_0^\infty \mathbf{E}_1 \cdot [\mathbf{G}'(s) - \mathbf{G}'^T(s)]\mathbf{E}_2 \cos\omega s \, ds > 0 \end{aligned} \tag{3.7}$$

for every $\omega \in \mathbb{R}^{++}$ and every $\mathbf{E}_1, \mathbf{E}_2 \in \mathrm{Sym}$. This inequality in turn yields some simple, important results by considering the limiting cases $\omega \to \infty$, $\omega \to 0$. They consist in the symmetry of $\mathbf{G}_0$, i.e.

$$\mathbf{G}_0 = \mathbf{G}_0^T, \tag{3.8}$$

and the symmetry of $\mathbf{G}_\infty$,

$$\mathbf{G}_\infty = \mathbf{G}_\infty^T. \tag{3.9}$$

Hence (3.7) reduces to

$$\int_0^\infty [\mathbf{E}_1 \cdot \mathbf{G}'(s)\mathbf{E}_1 + \mathbf{E}_2 \cdot \mathbf{G}'(s)\mathbf{E}_2] \sin\omega s \, ds + \int_0^\infty \mathbf{E}_1 \cdot [\mathbf{G}'(s) - \mathbf{G}'^T(s)]\mathbf{E}_2 \cos\omega s \, ds < 0 \tag{3.10}$$

for every $\omega \in \mathbb{R}^{++}$ and every $\mathbf{E}_1, \mathbf{E}_2 \in \mathrm{Sym}$.

Let $\mathbf{G}'_s$ on $\mathbb{R}^+$ be the half-range Fourier sine transform of $\mathbf{G}'$, namely

$$\mathbf{G}'_s(\omega) = \int_0^\infty \mathbf{G}'(u) \sin\omega u \, du.$$

The inequality (3.10) implies that, in Sym,

$$\mathbf{G}'_s(\omega) < 0, \qquad \omega \in \mathbb{R}^{++}, \tag{3.11}$$

while $\mathbf{G}'_s(0) = 0$ is identically true. As a consequence of (3.11) we obtain

$$\mathbf{G}_0 - \mathbf{G}(s) > 0, \quad s \in \mathbb{R}^{++}. \tag{3.12}$$

This is easily shown as follows. Upon extending the definition of $\mathbf{G}'(u)$ to $\mathbb{R}^-$ by letting $\mathbf{G}'(-u) = -\mathbf{G}'(u)$ we can write the Fourier sine transform as

$$\mathbf{G}'_s(\omega) = \int_0^\infty \sin\omega u \, \mathbf{G}'(u) \, du$$

and the inversion formula

$$\mathbf{G}'(u) = \frac{2}{\pi}\int_0^\infty \sin\omega u\, \mathbf{G}'_s(\omega)\, d\omega.$$

Hence the obvious integration with respect to u gives

$$\mathbf{G}(u) - \mathbf{G}_0 = \frac{2}{\pi}\int_0^\infty \frac{1-\cos\omega u}{\omega}\, \mathbf{G}'_s(\omega)\, d\omega, \qquad u \in \mathbb{R}^+,$$

whence the desired result.

By (3.12) it follows that, in Sym,

$$\mathbf{G}'(0) < 0, \tag{3.13}$$

$$\mathbf{G}_0 - \mathbf{G}_\infty \geq 0. \tag{3.14}$$

In essence these results have already appeared in the literature though derived through various approaches and at different times. Here we mention the main pertinent references. The symmetry of the instantaneous elastic modulus (3.8) was proved by Coleman $(1964)_2$ to follow from the second law in the form of the Clausius-Duhem inequality. The symmetry of the equilibrium elastic modulus (3.9) was obtained by Day (1971) via the Clausius inequality. Apart from the inequality being strict, (3.11) was first derived by Graffi (1928) in the case of isotropic materials by requiring that energy be dissipated in a period of a sinusoidal strain function $\mathbf{E}(t) = \mathbf{E}\sin\omega t$. This connection is emphasized in Leitman and Fisher (1973), §13, where the energy dissipated in one period $[0, d]$, $d = 2\pi/\omega$, is shown to be

$$\int_0^d \mathcal{T}(\mathbf{E}^t)\cdot\dot{\mathbf{E}}(t)dt = -\pi\mathbf{E}\cdot\mathbf{G}'_s(\omega)\mathbf{E}.$$

Incidentally, that is why $-\mathbf{G}'_s(\omega)$ is often called the *loss modulus*. The inequality (3.13) for the initial derivative of the relaxation function was proved first by Bowen and Chen (1973), by having recourse to discontinuous histories, in the one-dimensional case via the Clausius-Duhem inequality. The same result was proved in Morro $(1985)_2$ with C^∞ histories in the three-dimensional case. The inequality (3.14) traces back to Gurtin and Sternberg (1962) and Coleman $(1964)_2$.
Seemingly, the inequality (3.12) appeared in Fabrizio and Morro (1988) for the first time but is in a sense related to a previous result by Day (1971) who showed that, as a consequence of a dissipativity condition, the relaxation function satisfies the condition

$$\mathbf{G}_0 - \mathbf{G}_\infty \geq \pm[\mathbf{G}(s) - \mathbf{G}_\infty]. \tag{3.15}$$

To show the connection, observe that the limit $u \to \infty$ in the expression for $\mathbf{G}_0 - \mathbf{G}(u)$ gives

$$\mathbf{G}_\infty - \mathbf{G}_0 = \frac{2}{\pi}\int_0^\infty \frac{1}{\omega}\, \mathbf{G}'_s(\omega)\, d\omega.$$

Hence

$$\mathbf{G}(u) - \mathbf{G}_\infty = -\frac{2}{\pi}\int_0^\infty \frac{\cos\omega u}{\omega}\, \mathbf{G}'_s(\omega)\, d\omega.$$

Then the obvious inequalities

$$\int_0^\infty \frac{1}{\omega}|\mathbf{E}\cdot\mathbf{G}'_s(\omega)\mathbf{E}|d\omega \geq \int_0^\infty \frac{|\cos\omega u|}{\omega}|\mathbf{E}\cdot\mathbf{G}'_s(\omega)\mathbf{E}|d\omega \geq \left|\int_0^\infty \frac{\cos\omega u}{\omega}\mathbf{E}\cdot\mathbf{G}'_s(\omega)\mathbf{E}\, d\omega\right|$$

for any $\mathbf{E} \in \text{Sym}$ provide (3.15).

The set of inequalities derived in the previous section comprises the thermodynamic inequalities already known. What is more, they are also sufficient for the second law to hold in the general form of the Clausius property. The proof, quite involved, is given in Fabrizio and Morro (1988).

The determination of the free energy for linearly-viscoelastic materials traces back to Volterra (1940). The subject was reinvestigated by Graffi (1974) who showed that the expression given by Volterra satisfies the thermodynamic restrictions derived by Coleman (1964) for the free energy of materials with memory. As shown by Morro (1990) and Morro and Vianello (1990), there is an intrinsic non-uniqueness in the determination of the free energy even if the thermodynamic restrictions are taken into account. Materials with internal variables may describe linear viscoelasticity when the relaxation function is of exponential type. In such a case the free energy can be determined uniquely and this is outlined as follows.

The state of the body is the pair $(\mathbf{E}, \boldsymbol{\xi})$, $\boldsymbol{\xi} \in \text{Sym}$ representing the internal variables. Letting a superposed ring denote the trace-free part, we consider an explicit model in the form

$$\mathbf{T} = \kappa(\text{tr}\mathbf{E})\mathbf{1} + 2\mu \mathring{\mathbf{E}} + \boldsymbol{\xi}, \tag{3.16}$$

$$\dot{\boldsymbol{\xi}} = -\upsilon\boldsymbol{\xi} - \kappa_\xi(\text{tr}\mathbf{E})\mathbf{1} - 2\mu_\xi \mathring{\mathbf{E}}\,. \tag{3.17}$$

The quantities $\upsilon, \kappa, \kappa_\xi, \mu, \mu_\xi$ are supposed to be real constants. Indeed, κ and μ can be viewed as the instantaneous bulk and shear elasticities and then we let $\kappa, \mu > 0$. Moreover we let $\upsilon > 0$. Incidentally, the integration of (3.17) yields

$$\begin{aligned}\boldsymbol{\xi}(t) =& \boldsymbol{\xi}(t_0)\exp[-\upsilon(t-t_0)] \\ &- \kappa_\xi\{\int_{t_0}^t \exp[-\upsilon(t-s)](\text{tr}\mathbf{E}(s))ds\}\mathbf{1} - 2\mu_\xi \int_{t_0}^t \exp[-\upsilon(t-s)]\, \mathring{\mathbf{E}}\,(s)ds.\end{aligned} \tag{3.18}$$

The limit $t_0 \to -\infty$ provides the model of linear viscoelasticity with an exponential-type relaxation function.

By following the procedure developed by Coleman et al (1982) we can show that, as a consequence of the second law of thermodynamics, the free energy function $\tilde{\psi}$, the stress function $\tilde{\mathbf{T}}$ and the evolution function $\tilde{\boldsymbol{\xi}}$ of $\mathbf{E}$ and $\boldsymbol{\xi}$ satisfy

$$\tilde{\mathbf{T}} = \rho \frac{\partial \tilde{\psi}}{\partial \mathbf{E}} \tag{3.19}$$

and

$$\frac{\partial \tilde{\psi}}{\partial \boldsymbol{\xi}} \cdot \tilde{\boldsymbol{\xi}} \le 0. \tag{3.20}$$

It is a routine matter to show that (3.19) and (3.20) follow also by assuming a statement of the second law in terms of the Clausius-Duhem inequality. We now apply (3.19) and (3.20) to the functions (3.17) and (3.18). Since ρ is regarded as constant it is convenient to consider the free energy per unit volume $\Psi = \rho\psi$. By (3.19) and (3.17) we have

$$\Psi = \frac{1}{2}\kappa(\mathrm{tr}\mathbf{E})^2 + \mu\,\mathring{\mathbf{E}}\cdot\mathring{\mathbf{E}} + \frac{1}{3}(\mathrm{tr}\mathbf{E})(\mathrm{tr}\boldsymbol{\xi}) + \mathring{\mathbf{E}}\cdot\mathring{\boldsymbol{\xi}} + \Psi^*(\boldsymbol{\xi}) \tag{3.21}$$

where Ψ^* is an arbitrary function on Sym. Restrictions on Ψ^* are placed by (3.20). In view of (3.18) substitution gives

$$[\frac{1}{3}(\mathrm{tr}\mathbf{E})\mathbf{1} + \mathring{\mathbf{E}} + \frac{\partial\Psi^*}{\partial\boldsymbol{\xi}}]\cdot[\upsilon\frac{1}{3}(\mathrm{tr}\boldsymbol{\xi})\mathbf{1} + \upsilon\,\mathring{\boldsymbol{\xi}} + \kappa_\xi(\mathrm{tr}\mathbf{E})\mathbf{1} + 2\mu_\xi\,\mathring{\mathbf{E}}] \geq 0. \tag{3.22}$$

The exploitation of (3.22) is based on the following result[5].

Let $\mathbf{A},\mathbf{B} \in$ Sym, with $\mathrm{tr}\mathbf{A} = \mathrm{tr}\mathbf{B} = 0$, $\mathbf{F} \in C^1(\mathrm{Sym}, \mathrm{Sym})$, $a, b \in \mathbb{R}$ and

$$\Phi(a,b,\mathbf{A},\mathbf{B}) := [a\mathbf{1} + \mathbf{A} + \mathbf{F}(b\mathbf{1}+\mathbf{B})]\cdot[b\mathbf{1} + \mathbf{B} + \gamma a\mathbf{1} + \nu\mathbf{A}], \qquad \gamma,\nu \in \mathbb{R}.$$

Then $\Phi \geq 0$ for arbitrary values of a, b and $\mathbf{A},\mathbf{B}$ only if

$$\mathbf{F}(b\mathbf{1}+\mathbf{B}) = \frac{b}{\gamma}\mathbf{1} + \frac{1}{\nu}\mathbf{B} \tag{3.23}$$

This is shown by observing that Φ is minimal at $b = -\gamma a$, $\mathbf{B} = -\nu\mathbf{A}$. Then $\partial\Phi/\partial(b\mathbf{1}+\mathbf{B})$ must vanish whenever $b = -\gamma\, a$ and $\mathbf{B} = -\nu\,\mathbf{A}$. This condition gives

$$a\mathbf{1} + \mathbf{A} + \mathbf{F}(b\mathbf{1}+\mathbf{B}) = 0 \qquad \text{whenever} \qquad b = -\gamma\, a, \quad \mathbf{B} = -\nu\,\mathbf{A}$$

whence (3.23). Substitution from (3.23) yields

$$\Phi(a,b,\mathbf{A},\mathbf{B}) = 3\gamma(a + \frac{1}{\gamma}b)^2 + \nu(\mathbf{A} + \frac{1}{\nu}\mathbf{B})^2$$

and hence $\Phi \geq 0$ if and only if $\gamma, \nu > 0$. The exploitation of (3.22) is immediate once we make the identifications

$$a = \frac{1}{3}\mathrm{tr}\mathbf{E}, \quad \mathbf{A} = \mathring{\mathbf{E}}, \quad b = \frac{1}{3}\mathrm{tr}\boldsymbol{\xi}, \quad \mathbf{B} = \mathring{\boldsymbol{\xi}}, \quad \gamma = 3\frac{\kappa_\xi}{\upsilon}, \quad \nu = 2\frac{\mu_\xi}{\upsilon}.$$

We get

$$\frac{\partial\Psi^*}{\partial\boldsymbol{\xi}} = \frac{\upsilon}{9\kappa_\xi}(\mathrm{tr}\boldsymbol{\xi})\mathbf{1} + \frac{\upsilon}{2\mu_\xi}\mathring{\boldsymbol{\xi}}, \tag{3.24}$$

$$\kappa_\xi > 0 \qquad \mu_\xi > 0. \tag{3.25}$$

Integration of (3.24) gives

$$\Psi^* = \frac{1}{2}[\frac{\upsilon}{9\kappa_\xi}(\mathrm{tr}\boldsymbol{\xi})^2 + \frac{\upsilon}{2\mu_\xi}\mathring{\boldsymbol{\xi}}\cdot\mathring{\boldsymbol{\xi}}].$$

[5] This result generalizes a simpler case investigated by E. Dunn and the author during a conversation in Galway, March 1988.

Substitution into (3.21) and some rearrangement yield

$$\Psi = \frac{1}{2}(\kappa - \frac{\kappa_\xi}{v})(\mathrm{tr}\mathbf{E})^2 + (\mu - \frac{\mu_\xi}{v})\,\overset{\circ}{\mathbf{E}} \cdot \overset{\circ}{\mathbf{E}}$$
$$+ \frac{\kappa_\xi}{2v}(\mathrm{tr}\mathbf{E} + \frac{v}{3\kappa_\xi}\mathrm{tr}\boldsymbol{\xi})^2 + \frac{\mu_\xi}{v}(\overset{\circ}{\mathbf{E}} + \frac{v}{2\mu_\xi}\overset{\circ}{\boldsymbol{\xi}})^2. \tag{3.26}$$

We have thus determined explicitly the expression of the free energy. As it must be, (3.26) complies with a general property of materials with internal variables. By (3.17) each strain $\mathbf{E} = \frac{1}{3}(\mathrm{tr}\mathbf{E})\mathbf{1} + \overset{\circ}{\mathbf{E}}$ corresponds to an equilibrium state with

$$\mathrm{tr}\boldsymbol{\xi} = -\frac{3\kappa_\xi}{v}\mathrm{tr}\mathbf{E}, \qquad \overset{\circ}{\boldsymbol{\xi}} = -\frac{2\mu_\xi}{v}\,\overset{\circ}{\mathbf{E}}\,.$$

Then, by (3.25), for each strain $\mathbf{E}$ the function $\Psi(\mathbf{E}, \boldsymbol{\xi})$ has a strict minimum at the equilibrium state.

It follows at once that the result (3.26) coincides with Day's free energy Ψ_d. What is more, we cannot arrive at an expression like Ψ_g on the basis of the internal variable model in that no internal variable term can provide $\mathbf{E}(t) - \mathbf{E}(s)$. On assuming the monotone-decreasing property of the relaxation function, Graffi (1982) proved that $\Psi_d \le \Psi_g$. This result can also be derived as a consequence of the minimal property of Ψ_d, as shown by Morro and Vianello (1990).

Fluids are defined as materials for which the symmetry group is the full unimodular group. In the linear approximation, and following a standard notation in fluid-dynamics, we write the constitutive equation as

$$\mathbf{T}(t) = -p(\rho)\,\mathbf{1} + 2\int_0^\infty \mu(s)\mathbf{D}^t(s)\,ds + \int_0^\infty \lambda(s)[\mathrm{tr}\,\mathbf{D}^t(s)]\,ds\,\mathbf{1}. \tag{3.27}$$

Quite naturally in this case the fading memory space consists of all histories $\mathbf{D}^t$ with finite k-norm. It is convenient to write (3.27) in terms of tr$\mathbf{D}$ and the trace-free part $\overset{\circ}{\mathbf{D}}$. We have

$$\mathbf{T}(t) = -p(\rho)\,\mathbf{1} + 2\int_0^\infty \mu(s)\,\overset{\circ}{\mathbf{D}}{}^t(s)\,ds + \int_0^\infty \kappa(s)[\mathrm{tr}\mathbf{D}^t(s)]ds\,\mathbf{1} \tag{3.28}$$

where $\kappa(s) = \lambda(s) + \frac{2}{3}\mu(s)$ is the bulk relaxation function. Incidentally, the well-known Navier-Stokes model of viscous fluids follows as a limiting case when $\mu(s)$ and $\lambda(s)$ - or $\kappa(s)$ - are Dirac's delta-functions of the form $\mu\delta(s)$, $\lambda\delta(s)$.

To complete this section about thermodynamics of viscoelasticity, we examine the model of viscoelastic fluid characterized by the constitutive functional $\hat{\mathbf{T}}(\rho, \mathbf{D}^t)$ defined by (3.27), or (3.28), the state σ is the pair of the mass density ρ and the history of $\mathbf{D}$. Again we confine the attention to isothermal processes and then we can state the second law of thermodynamics by saying that, for each $\epsilon > 0$, there exists $\nu_\epsilon > 0$ such that

$$\int_0^d \hat{\mathbf{T}}(\rho, \mathbf{D}^t) \cdot \mathbf{D}(t)\,dt > -\epsilon$$

for any ν_ϵ-approximate cycle. Really, we investigate the

Second law for cycles. *For every cycle the condition*

$$\int_0^d \hat{\mathbf{T}}(\rho, \mathbf{D}^t) \cdot \mathbf{D}(t)\, dt \geq 0 \tag{3.29}$$

holds; equality holds if and only if the process is reversible, which occurs when $\mathbf{D}^d$ *is the zero-history.*

By paralleling the procedure developed for viscoelastic solids we might show that the validity of the second law for cycles implies the validity of the second law for approximate cycles. That is why we examine (3.29) only. Let $\sigma(0) = (\rho(0), \mathbf{D}^0)$ be the initial state (at $t = 0$). The process $P = \mathbf{D}$ on $[0, d)$ determines the state σ at any time $t \in [0, d)$. While $\mathbf{D}^0$ and $\mathbf{D}(\tau)$, as $\tau \in [0, d)$, determine $\mathbf{D}^\tau$ for any $\tau \in [0, d)$, we have

$$\rho(t) = \rho(0) \exp[- \int_0^t \mathrm{tr}\mathbf{D}(\tau)\, d\tau]. \tag{3.30}$$

Then $\rho(d) = \rho(0)$ if and only if the integral of $\mathrm{tr}\mathbf{D}$ on the interval $[0, d)$ vanishes.

Let the subscript c denote the half-range Fourier cosine transform. The constitutive equation (3.28) complies with the second law for cycles (3.29) if and only if

$$\mu_c(\omega) > 0, \qquad \kappa_c(\omega) > 0, \qquad \omega \in \mathbb{R}^{++}. \tag{3.31}$$

For, by generalizing the proof given by Fabrizio (1989), observe preliminarily that, letting H be the integral of p/ρ, we have

$$\int_0^d -p(\rho(0) \exp[- \int_0^t \mathrm{tr}\mathbf{D}(\tau)\, d\tau])\, \mathrm{tr}\mathbf{D}(t)\, dt = \int_0^d \frac{dH(\rho(t))}{dt}\, dt.$$

Hence the integral along any cycle of $p(\rho)\mathrm{tr}\mathbf{D}$ vanishes. Now consider a time-harmonic stretching function $\tilde{\mathbf{D}}(t), t \in \mathbb{R}$, defined as

$$\tilde{\mathbf{D}}(t) = \mathbf{D}_1 \cos \omega t + \mathbf{D}_2 \sin \omega t, \qquad \omega \in \mathbb{R}^{++}, \quad \mathbf{D}_1, \mathbf{D}_2 \in \mathrm{Sym}.$$

The duration d is taken to be $2\pi/\omega$ times any positive integer. The integral of $\tilde{\mathbf{D}}$ on $[0, d)$ vanishes and hence, by (3.30), $\rho(d) = \rho(0)$. Since $\tilde{\mathbf{D}}^d = \tilde{\mathbf{D}}^0$, then $\sigma(t) = (\rho(t), \tilde{\mathbf{D}}^t)$, as t runs over $[0, d)$, constitutes a cycle. Substitution into (3.29) yields

$$2 \int_0^d \{ \int_0^\infty \mu(s) [\overset{\circ}{\mathbf{D}}_1 \cdot \overset{\circ}{\mathbf{D}}_1 \cos \omega(t-s) \cos \omega t + \overset{\circ}{\mathbf{D}}_2 \cdot \overset{\circ}{\mathbf{D}}_2 \sin \omega(t-s) \sin \omega t$$
$$+ \overset{\circ}{\mathbf{D}}_1 \cdot \overset{\circ}{\mathbf{D}}_2 (\cos \omega(t-s) \sin \omega t + \sin \omega(t-s) \cos \omega t)] ds \} dt$$

$$+ \int_0^d \{ \int_0^\infty \kappa(s) [(\mathrm{tr}\mathbf{D}_1)^2 \cos \omega(t-s) \cos \omega t + (\mathrm{tr}\mathbf{D}_2)^2 \sin \omega(t-s) \sin \omega t$$
$$+ \mathrm{tr}\mathbf{D}_1 \mathrm{tr}\mathbf{D}_2 (\cos \omega(t-s) \sin \omega t + \sin \omega(t-s) \cos \omega t)] ds \} dt \geq 0$$

whence, by changing the order of integration, we have

$$\mu_c(\omega)\, (\overset{\circ}{\mathbf{D}}_1 \cdot \overset{\circ}{\mathbf{D}}_1 + \overset{\circ}{\mathbf{D}}_2 \cdot \overset{\circ}{\mathbf{D}}_2) + \kappa_c(\omega)\, [(\mathrm{tr}\mathbf{D}_1)^2 + (\mathrm{tr}\mathbf{D}_2)^2] > 0, \qquad \omega \in \mathbb{R}^{++},$$

for any non-zero $\mathbf{D}_1, \mathbf{D}_2 \in \mathrm{Sym}$. Hence it follows the results (3.31). To show that (3.31) is sufficient for the validity of (3.29) along cycles, observe that $(\sigma(0), P)$ on $[0, d)$ constitutes a cycle if and only if the function $\mathbf{D}$ is periodic with vanishing mean value, in the period $[0, d)$. The corresponding history $\mathbf{D}^t$ can be represented through its Fourier series provided only that $\mathbf{D}$ is piecewise continuous. More conveniently we write the Fourier series representation as

$$\mathbf{D}^t(s) = \sum_{h=1}^{\infty}\{\overset{\circ}{\mathbf{A}}_h \cos h\omega(t-s) + \overset{\circ}{\mathbf{B}}_h \sin h \sin(t-s) + \frac{1}{3}[\mathrm{tr}\mathbf{A}_h \cos h\omega(t-s) + \mathrm{tr}\mathbf{B}_h \sin h\omega(t-s)]\mathbf{1}\}$$

where $\mathbf{A}_h$, $\mathbf{B}_h \in \mathrm{Sym}$, $\omega = 2\pi/d$. We know already that the integral of $p(\rho)\mathrm{tr}\mathbf{D}$ along a cycle vanishes. Then the work w performed along a cycle, namely the left-hand side of (3.29), takes the form

$$w = 2\int_0^d \int_0^\infty \mu(s) \sum_{h,k=1}^{\infty} [\overset{\circ}{\mathbf{A}}_h \cos h\omega(t-s) + \overset{\circ}{\mathbf{B}}_h \sin h\omega(t-s)] \cdot [\overset{\circ}{\mathbf{A}}_k \cos k\omega t + \overset{\circ}{\mathbf{B}}_k \sin k\omega t]\, ds\, dt$$

$$+\frac{1}{3}\int_0^d \int_0^\infty \kappa(s) \sum_{h,k=1}^{\infty} [\mathrm{tr}\mathbf{A}_h \cos h\omega(t-s) + \mathrm{tr}\mathbf{B}_h \sin h\omega(t-s)][\mathrm{tr}\mathbf{A}_k \cos k\omega t + \mathrm{tr}\mathbf{B}_k \sin k\omega t]\, ds\, dt.$$

Term by term integration with respect to t gives

$$w = \frac{\pi}{\omega}\sum_{k=1}^{\infty}\{[2\,\mu_c(k\omega)[\overset{\circ}{\mathbf{A}}{}_k^2 + \overset{\circ}{\mathbf{B}}{}_k^2] + \kappa_c(k\omega)[(\mathrm{tr}\mathbf{A}_k)^2 + (\mathrm{tr}\mathbf{B}_k)^2]\}.$$

The validity of (3.31) implies that $w > 0$ for any non-zero history $\mathbf{D}^t$ whence $\hat{\mathbf{T}}(\rho, \mathbf{D}^t)$ meets the second law for cycles.

We may say that the thermodynamic analysis of the viscoelastic fluid parallels that of the viscoelastic solid. Formally, though, there is a difference in that the power is $\mathcal{T}(\mathbf{E}^t) \cdot \dot{\mathbf{E}}$ for the viscoelastic solid and $\hat{\mathbf{T}}(\rho, \mathbf{D}^t) \cdot \mathbf{D}$ for the viscoelastic fluid. As shown above, this difference is reflected in the thermodynamic restrictions for the relaxation functions. For, while in viscoelastic solids thermodynamics requires the *negative* definiteness of the half-range Fourier *sine* transform of $\mathbf{G}(s)$, in viscoelastic fluids thermodynamics requires the *positive* definiteness of the half-range Fourier *cosine* transform of $\mu(s)$ and $\kappa(s)$.

For the sake of completeness we consider the *incompressible* viscoelastic fluid namely (3.27), or (3.28), subject to the constraint

$$\mathrm{tr}\mathbf{D} = 0. \tag{3.32}$$

One might think that the incompressible fluid can be regarded as a particular case. It is really so but we have to remark that, while (3.32) demands that ρ be constant, a parameter say, the pressure p is no longer a function of ρ but is an unknown function of time. We can prove that the viscoelastic incompressible fluid complies with the second law for cycles (3.29) if and only if

$$\mu_c(\omega) > 0, \qquad \omega \in \mathbb{R}^{++}. \tag{3.33}$$

This is so because the contribution of the pressure p to the work is

$$-\int_0^d p(t)\,\mathrm{tr}\mathbf{D}(t)\,dt$$

and then it vanishes whatever the dependence of p on t is. Accordingly the previous proof can be applied step by step with the restrictions $\mathrm{tr}\mathbf{A}_h$, $\mathrm{tr}\mathbf{B}_h = 0$ and the result (3.33) follows.

4. EXTREMUM PRINCIPLES FOR VISCOELASTIC SOLIDS

Usually variational formulations consist in a proper functional being stationary at a solution in an appropriate class of functions. Yet a critical point of a functional need not be an extremum point for that functional. In fact extremum principles are decisively more useful than stationary principles especially in connection with the development of approximation schemes. Whenever nonequilibrium phenomena are concerned, variational formulations are difficult to obtain. Things are obviously worse if we aim at extremum principles. In the case of linear elastodynamics, a minimum principle was established by Benthien and Gurtin (1970) for a mixed boundary-value problem; they showed that the Laplace transform of the energy functional is minimal at the solution. An improvement is due to Reiss (1978) who elaborated a method whereby Benthien-Gurtin's functional can be transformed back to the original space-time domain while preserving the extremum property of the functional.

By paralleling in part Fabrizio et al (1989), in this section we investigate some extremum (minimum) properties and show how the extremum property is related to the class of functions (displacements) under consideration. Meanwhile we emphasize that the validity of the extremum properties is crucially related to the thermodynamic restrictions on the relaxation function.

Preliminarly, though, it is worth introducing some pertinent notation and definitions. Let $\mathbf{u}(\mathbf{x},t)$ be known in $\Omega \times \mathbb{R}^-$. Then

$$\mathbf{T}_0(t) = \int_t^\infty \mathbf{G}'(s)\nabla\mathbf{u}(t-s)\,ds$$

is known for every $t \in \mathbb{R}^+$. Let

$$A\mathbf{u} := \rho\ddot{\mathbf{u}} - \nabla\cdot(\mathbf{G}_0\nabla\mathbf{u} + \mathbf{G}' * \nabla\mathbf{u}) - \rho\mathbf{f}$$

where $\rho\mathbf{f} = \rho\mathbf{b} + \nabla\cdot\mathbf{T}_0$, $\mathbf{b}$ being the body force, and $*$ denotes the convolution in time. Moreover let the boundary $\partial\Omega$ of Ω consist of two disjoint sets $\partial\Omega_u, \partial\Omega_T$ such that

$$\partial\Omega_u \cup \partial\Omega_T = \partial\Omega, \qquad \partial\Omega_u \cap \partial\Omega_T = \emptyset.$$

Consider the initial-boundary value problem (P)

$$A\mathbf{u} = 0 \quad \text{on} \quad \Omega\times\mathbb{R}^{++}, \tag{4.1}$$

$$\begin{cases} \mathbf{u} = \hat{\mathbf{u}} & \text{on} \quad \partial\Omega_u \times \mathbb{R}^+, \\ \mathbf{T}\mathbf{n} = \hat{\mathbf{t}} & \text{on} \quad \partial\Omega_T \times \mathbb{R}^+, \end{cases} \tag{4.2}$$

$$\mathbf{u}(0) = \mathbf{u}_0, \qquad \dot{\mathbf{u}}(0) = \dot{\mathbf{u}}_0 \quad \text{on} \quad \bar{\Omega}. \tag{4.3}$$

By a kinematically admissible displacement field we mean a vector field on $\Omega \times \mathbb{R}^+$ that satisfies the displacement boundary condition $(4.2)_1$. Denote by $\mathcal{U}$ the set of all kinematically admissible displacements. Let $\Re^{++}$ be the set of complex numbers p with strictly-positive real part. We denote by $\mathcal{U}_L$ the set of all $\mathbf{u} \in \mathcal{U}$ that possess Laplace transform $\mathbf{u}_L$ with respect to time in the complex half-plane $\Re^{++}$. We assume that $\mathbf{T}_0, \mathbf{b}, \hat{\mathbf{u}}, \hat{\mathbf{t}}$ possess Laplace transform in $\Re^{++}$. Of course $\mathbf{u}_L$ must comply with

$$\mathbf{u}_L(\mathbf{x}, p) = \hat{\mathbf{u}}_L(\mathbf{x}, p) \qquad \text{on} \quad \partial\Omega_u \times \Re^{++}. \tag{4.4}$$

In connection with the problem P, application of the Laplace transform operator to (4.1) and account of (4.3) give

$$A_p \mathbf{u}_L := \rho p^2 \mathbf{u}_L - \nabla \cdot [(\mathbf{G}_0 + \mathbf{G}'_L)\nabla \mathbf{u}_L] = \mathbf{g}, \tag{4.5}$$

where

$$\mathbf{g} = \rho \mathbf{b}_L + \nabla \cdot (\mathbf{T}_0)_L + \rho(p\mathbf{u}_0 - \dot{\mathbf{u}}_0).$$

The counterpart of the problem P in the Laplace-transform context, say P_L, is then given by

$$A_p \mathbf{u} = \mathbf{g} \qquad \text{on} \quad \partial\Omega \times \Re^{++} \tag{4.6}$$

along with the boundary conditions (4.4) and

$$[(\mathbf{G}_0 + \mathbf{G}'_L)\nabla \mathbf{u}_L]\mathbf{n} = (\hat{\mathbf{t}}_0)_L \qquad \text{on} \quad \partial\Omega_T \times \Re^{++}, \tag{4.7}$$

where $\hat{\mathbf{t}}_0 = \hat{\mathbf{t}} - \mathbf{T}_0\mathbf{n}$.

By generalizing a procedure which traces back to Reiss we establish a minimum principle in terms of an appropriate weight function. To this purpose it is convenient to consider some space functions as follows. For the time being let H be any Hilbert space and define L^2_L as

$$L^2_L(\mathbb{R}^+; H) = \{\mathbf{u} \in L^2_L : \mathbf{u}_L(p) \in H, \ p \in \Re^{++}\}.$$

Moreover let

$$H^1_L(\mathbb{R}^+; L^2(\Omega)) = \{\mathbf{u} \in H^1_{loc}(\mathbb{R}^+; L^2(\Omega)) : \mathbf{u}_L(p) \in L^2(\Omega), \ p \in \Re^{++}\}.$$

Consider a function $Y \in C(\mathbb{R}^+, \mathbb{R}^+)$, with a finite limit at infinity, such that

$$y(t) := \int_0^\infty Y(s) \exp(-st) ds$$

exists for any $t \in \mathbb{R}^+$. In terms of y we define the bilinear form $(\cdot,\cdot)_y$ on $L^2_L(\mathbb{R}^+; L^2(\Omega))$ as

$$(v, w)_y = \int_0^\infty \int_0^\infty y(t+\tau) \int_\Omega v(\mathbf{x}, t) w(\mathbf{x}, \tau) \, dx \, d\tau \, dt;$$

if (both) v and w are vectors or tensors then the product vw is meant as the scalar product. By the definition of y we have

$$(v,w)_y = \int_0^\infty \int_\Omega Y(s) v_L(\mathbf{x},s) w_L(\mathbf{x},s)\, dx\, ds. \tag{4.8}$$

This shows that the bilinear form $(\cdot,\cdot)_y$ is positive semidefinite. Of course $(\cdot,\cdot)_y$ is well defined on L_L^2 when $Y \in C_0^\infty$. As before, to save writing, the dependence on $\mathbf{x} \in \Omega$ is understood and not written.

We now show that the mixed problem P can be given a variational formulation in terms of $(\cdot,\cdot)_y$. Consider the functional

$$\begin{aligned}\Lambda_y(\mathbf{u}) = {} & \frac{1}{2}(\nabla\mathbf{u}, \mathbf{G}_0\nabla\mathbf{u} + \mathbf{G}' * \nabla\mathbf{u})_y + \frac{1}{2}\rho(\dot{\mathbf{u}},\dot{\mathbf{u}})_y - \rho(\mathbf{u},\mathbf{f})_y \\ & - \int_0^\infty \int_0^\infty \int_{\partial\Omega_T} y(t+\tau)\mathbf{u}(t)\cdot\hat{\mathbf{t}}_0(\tau)\, d\tau\, dt\, da \\ & - \int_0^\infty y(t) \int_\Omega \rho\{\dot{\mathbf{u}}(t)\cdot[\mathbf{u}(0)-\mathbf{u}_0] - \mathbf{u}(t)\cdot\dot{\mathbf{u}}_0\}\, dx\, dt + y(0)\int_\Omega \rho\mathbf{u}(0)\cdot[\frac{1}{2}\mathbf{u}(0)-\mathbf{u}_0]\, dx.\end{aligned}$$

where $\rho\mathbf{f} = \rho\mathbf{b} + \nabla\cdot\mathbf{T}_0$ and $\hat{\mathbf{t}}_0 = \hat{\mathbf{t}} - \mathbf{T}_0\mathbf{n}$ are supposed to belong to $L_L^2(\mathbb{R}^+; L^2(\Omega))$ and $L_L^2(\mathbb{R}^+; L^2(\partial\Omega))$, respectively. Owing to the occurrence of the convolution $\mathbf{G}' * \nabla\mathbf{u}$ it is worth observing that, by definition,

$$\int_0^\infty y(t+\tau)(v*w)(\tau)\, d\tau = \int_0^\infty y(t+\tau)\int_0^\tau v(\tau-\sigma)\, w(\sigma)\, d\sigma\, d\tau.$$

The change of variables $\lambda = \tau - \sigma, \quad \eta = \sigma$ yields

$$\int_0^\infty y(t+\tau)\,(v*w)(\tau)\, d\tau = \int_0^\infty \int_0^\infty y(t+\lambda+\eta)\, v(\lambda)\, w(\eta)\, d\lambda\, d\eta. \tag{4.9}$$

The spaces

$$\mathcal{P} = \{\mathbf{u} \in L_L^2(\mathbb{R}^+; H^1(\Omega)) \cap H_L^1(\mathbb{R}^+; L^2(\Omega)) : \mathbf{u}\,|_{\partial\Omega_u} = \hat{\mathbf{u}}\},$$

$$\mathcal{P}_0 = \{\mathbf{h} \in L_L^2(\mathbb{R}^+; H^1(\Omega)) \cap H_L^1(\mathbb{R}^+; L^2(\Omega)) : \mathbf{h}\,|_{\partial\Omega_u} = 0\},$$

are the natural setting for a minimum principle. To establish this result we need a condition of thermodynamic character, namely

$$\mathbf{G}_0 + \mathbf{G}_L'(s) > 0, \quad s \in \mathbb{R}^+, \qquad \text{in Sym}, \tag{4.10}$$

which follows from (3.11); for the proof see Giorgi (1989) or Fabrizio et al (1989).

If $\mathbf{u}^0 \in \mathcal{P}$ is a solution to the mixed problem P with $\mathbf{f} \in L_L^2(\mathbb{R}^+; L^2(\Omega))$ and $\hat{\mathbf{t}}_0 \in L_L^2(\mathbb{R}^+; L^2(\partial\Omega))$ then, for every function $Y \in C_0^\infty(\mathbb{R}^+)$ the functional $\Lambda_y(\mathbf{u})$ has a minimum at $\mathbf{u}^0$ on $\mathcal{P}$.

To prove this property let $\mathbf{u}^0 \in \mathcal{P}$ and represent any element $\mathbf{u} \in \mathcal{P}$ as $\mathbf{u}^0 + \mathbf{h}$, where $\mathbf{h} \in \mathcal{P}_0$. Then by (4.9) we have

$$\begin{aligned}\Lambda_y(\mathbf{u}^0+\mathbf{h}) - \Lambda_y(\mathbf{u}^0) = &(\nabla\mathbf{h}, \mathbf{G}_0\nabla\mathbf{u}^0 + \mathbf{G}' * \nabla\mathbf{u}^0)_y + \frac{1}{2}(\nabla\mathbf{h}, \mathbf{G}_0\nabla\mathbf{h} + \mathbf{G}' * \nabla\mathbf{h})_y \\ &+ \rho(\dot{\mathbf{h}}, \dot{\mathbf{u}}^0)_y + \frac{1}{2}\rho(\dot{\mathbf{h}}, \dot{\mathbf{h}})_y - \rho(\mathbf{h}, \mathbf{f})_y \\ &- \int_0^\infty \int_0^\infty y(t+\tau) \int_{\partial\Omega_T} \mathbf{h}(t)\cdot\hat{\mathbf{t}}_0(\tau)\, da\, d\tau\, dt \\ &- \rho\int_0^\infty y(t)\int_\Omega \{[\dot{\mathbf{u}}^0(t) + \dot{\mathbf{h}}(t)]\cdot\mathbf{h}(0) + [\mathbf{u}^0(0) - \mathbf{u}_0]\cdot\dot{\mathbf{h}} - \dot{\mathbf{h}}(t)\cdot\dot{\mathbf{u}}_0\}\, dx\, dt \\ &+ \frac{1}{2}y(0)\rho\int_\Omega \mathbf{h}(0)\cdot\{\mathbf{h}(0) + 2[\mathbf{u}^0(0) - \mathbf{u}_0]\}\, dx\end{aligned}$$

Suitable integrations by parts and account that $\mathbf{u}^0$ is a solution to the mixed problem P lead to

$$\begin{aligned}\Lambda_y(\mathbf{u}^0+\mathbf{h}) - \Lambda_y(\mathbf{u}^0) = &\frac{1}{2}(\nabla\mathbf{h}, \mathbf{G}_0\nabla\mathbf{h} + \mathbf{G}' * \nabla\mathbf{h})_y + \frac{1}{2}\rho(\dot{\mathbf{h}}, \dot{\mathbf{h}})_y \\ &- \rho\int_0^\infty y(t)\int_\Omega \dot{\mathbf{h}}(t)\cdot\mathbf{h}(0)\, dx\, dt + \frac{1}{2}\rho y(0)\int_\Omega \mathbf{h}w(0)\cdot\mathbf{h}w(0)\, dx.\end{aligned}$$

Hence, by using again (4.9) we have

$$\Lambda_y(\mathbf{u}^0+\mathbf{h}) - \Lambda_y(\mathbf{u}^0) = \frac{1}{2}\int_\Omega\int_0^\infty Y(s)[\nabla\mathbf{h}_L(s)\cdot(\mathbf{G}_0 + \mathbf{G}'_L)\nabla\mathbf{h}_L(s) + \rho s^2\mathbf{h}_L(s)\cdot\mathbf{h}_L(s)]\, ds\, dx.$$

By the thermodynamic inequality (4.10) we have $\Lambda_y(\mathbf{u}^0+\mathbf{h}) \geq \Lambda_y(\mathbf{u}^0)$ for any $\mathbf{h} \in \mathcal{P}_0$ and the result is proved. The converse also holds.

If the function $\mathbf{w} \in \mathcal{P}$ minimizes the functional Λ_y on $\mathcal{P}$, for every $Y \in C_0^\infty(\mathbb{R}^+)$, then $\mathbf{w}$ is the unique solution to the problem P with $\mathbf{f} \in L_L^2(\mathbb{R}^+; L^2(\Omega))$ and $\hat{\mathbf{t}}_0 \in L_L^2(\mathbb{R}^+; L^2(\partial\Omega))$.

This is proved as follows. Let $\mathbf{w} \in \mathcal{P}$ be a function which minimizes Λ_y in $\mathcal{P}$ for every $Y \in C_0^\infty(\mathbb{R}^+)$. Then, for every $\mathbf{h} \in \mathcal{P}_0$ it follows that

$$\begin{aligned}0 = d\Lambda_y(\mathbf{w} \mid \mathbf{h}) = &(\nabla\mathbf{h}, \mathbf{G}_0\nabla\mathbf{w}\mathbf{G}' * \nabla\mathbf{w})_y + \rho(\dot{\mathbf{h}}, \dot{\mathbf{w}})_y - (\mathbf{h}, \mathbf{f})_y \\ &- \int_0^\infty\int_0^\infty y(t+\tau)\int_{\partial\Omega_T} \hat{\mathbf{t}}_0(\tau)\cdot\mathbf{h}(t)da\, d\tau\, dt \\ &- \rho\int_0^\infty y(t)\int_\Omega \{\dot{\mathbf{h}}(t)\cdot[\mathbf{w}(0) - \mathbf{u}_0] + \mathbf{h}(0)\cdot\dot{\mathbf{w}}(t)\mathbf{h}(t)\cdot\dot{\mathbf{u}}_0\}dx\, dt \\ &+ y(0)\rho\int_\Omega [\mathbf{w}(0) - \mathbf{u}_0]\cdot\mathbf{h}(0)dx.\end{aligned}$$

Let $\mathcal{T}_0$ denote the stress functional of $\mathbf{E}^t$ with $\mathbf{T}_0(t)$ given. Upon suitable integrations by parts we obtain

$$\int_0^\infty\int_0^\infty y(t+\tau)\{\int_\Omega [\rho\ddot{\mathbf{w}}(t) - \nabla\cdot\mathcal{T}_0(\nabla\mathbf{w}^t) - \rho\mathbf{b}(t)]\cdot\mathbf{h}(\tau)dx + \int_{\partial\Omega_T}[\mathcal{T}_0(\nabla\mathbf{w}^t)\mathbf{n} - \hat{\mathbf{t}}(t)]\cdot\mathbf{w}(\tau)da\}d\tau\, dt$$

$$-\rho\int_0^\infty y(t)\int_\Omega \{\dot{\mathbf{w}}(0) - \dot{\mathbf{u}}_0]\cdot\mathbf{h}(t) + [\mathbf{w}(0) - \mathbf{u}_0]\cdot\dot{\mathbf{h}}(t)\}dx\, dt + \rho y(0)\int_\Omega [\mathbf{w}(0) - \mathbf{u}_0]\cdot\mathbf{h}(0)dx = 0$$

whence, by (4.9), we have

$$\int_0^\infty Y(s)\{\int_\Omega [L_s(\mathbf{w})_L(s) - \mathbf{g}(s)]\cdot \mathbf{h}_L(s)\,dx + \int_{\partial\Omega_T} [(\mathbf{G}_0 + \mathbf{G}'_L(s))\nabla w_L(s))\mathbf{n} - (\hat{\mathbf{t}}_0)_L(s)]\cdot \mathbf{h}_L(s)\,da\}ds = 0 \tag{4.11}$$

Let $\mathbf{h}(\mathbf{x},t)$ take the form

$$\mathbf{h}(\mathbf{x},t) = \mathbf{h}_1(\mathbf{x})h_2(t)$$

where $\mathbf{h}_1 \in H_0^1(\Omega), h_2 \in H^1(\mathbb{R}^+)$. By the arbitrariness of $\mathbf{h}_1$ we obtain

$$\int_0^\infty Y(s)[L_s\mathbf{w}_L(s) - \mathbf{g}(s)](h_2)_L(s)ds = 0$$

a. e. in Ω. Hence the arbitrariness of $Y(h_2)_L$ on $\mathbb{R}^+$ gives

$$A_s\mathbf{w}_L(s) - \mathbf{g}(s) = 0 \tag{4.12}$$

a. e. in $\Omega\times\mathbb{R}^+$. The arbitrariness of $Y\mathbf{h}_L$ on $\partial\Omega_T\times\mathbb{R}^+$ and (4.12) make (4.11) to reduce to

$$[(\mathbf{G}_0 + \mathbf{G}'_L(s))\nabla\mathbf{w}(s)]\mathbf{n} = (\hat{\mathbf{t}}_0)_L(s) \tag{4.13}$$

a. e. in $\partial\Omega_T\times\mathbb{R}^+$. Accordingly, $\mathbf{w}_L$ is a solution to the problem P_L for almost every $s\in\mathbb{R}^+$. By $\mathbf{G}_\infty > 0$ and $\mathbf{G}'_s(\omega) < 0$ it follows the invertibility of $\mathbf{G}_0 + \mathbf{G}'_L(p)$ in Sym for any complex p with $\alpha \geq 0$. Then the operator

$$A_p = \rho p^2 - \nabla\cdot[(\mathbf{G}_0 + \mathbf{G}'_L(p))\nabla], \qquad \alpha \geq 0,$$

is strongly elliptic in $H^1(\Omega)$. As a consequence, $\mathbf{w}_L$ is the unique solution to P_L. Now, let $\mathbf{v}$ be the solution to the problem P which exists and is unique in $\mathcal{P}$. Its Laplace transform $\mathbf{v}_L$ exists for any $p\in\Re^{++}$ and is a solution to the problem P_L. This implies that $\mathbf{v}_L(\alpha)$ and $\mathbf{w}_L(p)$ coincide when $p=\alpha$. Since the Laplace transform $\mathbf{w}_L(p)$ of the minimizing function $\mathbf{w}$ and $\mathbf{v}_L(p)$ are analytic functions which coincide on $\mathbb{R}^{++}$ then they must coincide in the complex half-plane $\Re^{++}$. Hence, by the uniqueness of the Laplace transform, the minimizing function $\mathbf{w}$ is the unique solution to the problem P in $\mathcal{P}$.

A further minimum principle can be established through the direct use of the Laplace transform. The solution $\mathbf{w}^0(p)$ to the problem P_L is obviously a stationary point of the functional

$$\mathcal{L}_p(\mathbf{w}) = \int_\Omega\{\frac{1}{2}\rho p^2\mathbf{w}(p)\cdot\mathbf{w}(p) + \frac{1}{2}\nabla\mathbf{w}(p)\cdot[\mathbf{G}_0+\mathbf{G}'_L(p)]\nabla\mathbf{w}(p) - \mathbf{g}(p)\cdot\mathbf{w}(p)\}dx - \int_{\partial\Omega_T}(\hat{\mathbf{t}}_0)_L\cdot\mathbf{w}(p)\,da$$

which is parameterized by the complex quantity p both explicitly and through $\mathbf{w}(p)$. Letting

$$\Gamma_p(\mathbf{u}) = \mathcal{L}_p(\mathbf{u}_L(p))$$

we can regard Γ_p as a functional on $\mathcal{U}_L$ parameterized by the complex quantity p. Indeed, a solution $\mathbf{u}^0\in\mathcal{U}_L$ to the problem P_L is a stationary point of Γ_p for any $p\in\Re^{++}$. Now the stationary point is shown to be a minimum of the functional Γ_p.

If a function $\mathbf{u}^0 \in \mathcal{U}_L$ is a solution to the problem P then, for any $\alpha \in \mathbb{R}^+$, the functional Γ_α on $\mathcal{U}_L$ has a strict minimum at $\mathbf{u}^0$.

To prove this result let $\mathbf{u}^0 \in \mathcal{U}_L$ be a solution to P. Then $\mathbf{u}_L^0(p)$, $p \in \Re^{++}$, satisfies (4.4), (4.6), (4.7) and is a stationary point of $\mathcal{L}_p$. Moreover, letting $\mathbf{u}^0 + \lambda \mathbf{h} \in \mathcal{U}_L$ as $\lambda \in [0, \bar{\lambda}]$, $\bar{\lambda} > 0$ we have

$$d^2\mathcal{L}_p(\mathbf{u}_L^0(p)|\mathbf{h}_L(p)) = \int_\Omega [\rho p^2 \mathbf{h}_L(p) \cdot \mathbf{h}_L(p) + \nabla \mathbf{h}_L(p) \cdot (\mathbf{G}_0 + \mathbf{G}'_L(p)) \nabla \mathbf{h}_L(p)] dx.$$

Let p be real, that is $p = \alpha \in \mathbb{R}^{++}$. Then $\mathbf{h}_L(\alpha), \nabla \mathbf{h}_L(\alpha)$, and $\mathbf{G}'_L(\alpha)$ are real-valued as well as $\mathcal{L}_\alpha$. Meanwhile, by (4.10), we have $\mathbf{G}_0 + \tilde{\mathbf{G}}'(\alpha) > 0$ and hence, for any non-zero admissible $\mathbf{h}_L$, it follows that

$$d^2\mathcal{L}_\alpha(\mathbf{u}_L^0(\alpha)|\mathbf{h}_L(\alpha) > 0.$$

The vanishing of $\mathbf{h}_L(\alpha)$ on $\mathbb{R}^{++}$ implies the vanishing of $\mathbf{h}$. Then we have

$$d^2\Gamma_\xi(\mathbf{u}^0|\mathbf{h}) > 0$$

for any non-zero admissible $\mathbf{h}$ and for any $\alpha \in \mathbb{R}^{++}$, which completes the proof.

The converse holds too. Let $\mathcal{Q} = \{\mathbf{u} \in \mathcal{U}_L : \mathbf{u}_L(p) \in H^1(\Omega), p \in \Re^{++}\}$. If Γ_α, $\alpha \in \mathbb{R}^{++}$, has a strict minimum in $\mathcal{Q}$ at $\mathbf{u}^0$ then $\mathbf{u}^0$ is a solution to the problem P. For the proof see Fabrizio et al (1989).

Since a convolution occurs in the stress-strain relation, it seems natural to ask about the possibility of setting up extremum principles by using the convolution as a non-degenerate form. In general, this possibility is precluded by the fact that the pertinent operator is not self-adjoint and the convolution is not a definite form. Yet an extremum principle occurs by having recourse to a class separable displacement fields in the form

$$\mathbf{u}(\mathbf{x}, t) = [\mathbf{u}^0(\mathbf{x}) + \mathbf{w}(\mathbf{x})]h(t) \tag{4.14}$$

where $\mathbf{u}^0, \mathbf{w} \in C^2(\Omega, V)$ and $\mathbf{u}^0$ is a solution to the mixed problem P while $h \in C^2(\mathbb{R}, \mathbb{R})$. We denote by $\mathcal{U}'$ the set of (kinematically admissible) displacement fields characterized by (4.14). Quite easily we can show that the functional

$$\Phi(\mathbf{u}) = \int_0^t \int_\Omega (\frac{1}{2} \nabla \dot{\mathbf{u}} \cdot \mathbf{G} * \nabla \dot{\mathbf{u}} - \rho \mathbf{f} \cdot \dot{\mathbf{u}})\, dx\, d\tau - \int_0^t \int_{\partial \Omega_T} \hat{\mathbf{t}} \cdot \dot{\mathbf{u}}\, da\, d\tau.$$

has a minimum in $\mathcal{U}'$ at any solution $\mathbf{u}^0(\mathbf{x})h(t)$ to the quasi-static version of the equation of motion

$$\nabla \cdot (\mathbf{G} * \nabla \dot{\mathbf{u}}) + \rho \mathbf{f} = 0$$

with the boundary conditions of P. Owing to the limited extent of this principle it is not worth going through the details. We observe only that the minimum property is guaranteed by the dissipation principle

$$\int_{-\infty}^{\infty} \mathcal{T}(\mathbf{E}^\tau) \cdot \dot{\mathbf{E}}(\tau)\, d\tau \geq 0, \tag{4.15}$$

for any history $\mathbf{E}^{r}$, which in linear viscoelasticity is shown to be a consequence of the second law of thermodynamics.

It is of interest to observe that the relaxation functions that satisfy the dissipation principle make a correspondent functional convex. Since the dissipation principle is in fact a consequence of the second law, along with the positive definiteness of $\mathbf{G}_\infty$, we can express the result by saying that thermodynamic restrictions imply the convexity of a proper functional.

Let H be a convex set of a normed space, e. g. the fading memory space $\mathcal{H}$ of histories of the velocity $\dot{\mathbf{u}}$ and the velocity gradient $\nabla\dot{\mathbf{u}}$. Let $\dot{\xi} = (\dot{\mathbf{u}}, \nabla\dot{\mathbf{u}})$ be any element of H. Indeed, it is the history of $\dot{\xi}$ which is the element of H but, for formal simplicity, we write functionals on H in the form $\Psi(\dot{\xi})$.

A functional Ψ is said convex if, for each pair of elements $\dot{\xi}_1, \dot{\xi}_2 \in H$, and each $\lambda \in (0,1)$ the inequality

$$\Psi(\lambda\dot{\xi}_1 + (1-\lambda)\dot{\xi}_2) \leq \lambda\Psi(\dot{\xi}_1) + (1-\lambda)\Psi(\dot{\xi}_2)$$

holds. As is well known this inequality is equivalent to each of the conditions

$$\Psi(\dot{\xi}_2) - \Psi(\dot{\xi}_1) - d\Psi(\dot{\xi}_1|\dot{\xi}_2 - \dot{\xi}_1) \geq 0,$$

$$d^2\Psi(\dot{\xi}_1|\dot{\xi}_2 - \dot{\xi}_1) \quad \text{positive definite.}$$

By examining the second-order Gateaux derivative $d^2\Psi$ now we show that the functional

$$\Psi(\dot{\xi}) = \frac{1}{2}\int_{-\infty}^{t}\int_{-\infty}^{\tau} \nabla\dot{\mathbf{u}}(\tau)\cdot\mathbf{G}(\tau-\sigma)\nabla\dot{\mathbf{u}}(\sigma)\,d\sigma\,d\tau - \rho\int_{-\infty}^{t}\mathbf{b}(\tau)\cdot\dot{\mathbf{u}}(\tau)\,d\tau$$

is convex.

As it stands, the functional Ψ is scarcely useful because of the non-symmetric role of the tensors $\nabla\dot{\mathbf{u}}$ in the first integral $(\tau \geq \sigma)$. Here we recover the symmetry by extending the definition of $\mathbf{G}$ to $\mathbb{R}$ in the form

$$\mathbf{G}(\tau) = \mathbf{G}(|\,\tau\,|), \qquad \tau \in \mathbb{R}.$$

Then we can write

$$\Psi(\mathbf{u}) = \frac{1}{4}\int_{-\infty}^{t}\int_{-\infty}^{t} \nabla\dot{\mathbf{u}}(\tau)\cdot\mathbf{G}(|\,\tau-\sigma\,|)\nabla\dot{\mathbf{u}}(\sigma)\,d\sigma\,d\tau - \rho\int_{-\infty}^{t}\mathbf{b}(\tau)\cdot\dot{\mathbf{u}}(\tau)\,d\tau.$$

Hence, letting $\mathbf{u} + \nu\mathbf{w} \in H$, $\nu \in [0,\bar{\nu}], \bar{\nu} > 0$, we have

$$d^2\Psi(\nabla\dot{\mathbf{u}} \mid \nabla\dot{\mathbf{w}}) = \frac{1}{2}\int_{-\infty}^{t}\int_{-\infty}^{t} \nabla\dot{\mathbf{w}}(\tau)\cdot\mathbf{G}(|\,\tau-\sigma\,|)\nabla\dot{\mathbf{w}}(\sigma)\,d\sigma\,d\tau$$

where we have written $\nabla\dot{\mathbf{u}}$ and $\nabla\dot{\mathbf{w}}$ in the argument of $d^2\Psi$ to emphasize the independence of $\dot{\mathbf{u}}$ and $\dot{\mathbf{w}}$. Then we can write

$$d^2\Psi(\nabla\dot{\mathbf{u}} \mid \nabla\dot{\mathbf{w}}) = \int_{-\infty}^{t} \nabla\dot{\mathbf{w}}(\tau)\cdot\mathcal{T}(\nabla\mathbf{w}^{\tau})\,d\tau.$$

This proves that the dissipation principle, and then the second law of thermodynamics, implies the convexity of the functional Ψ.

5. EXTREMUM PRINCIPLES FOR VISCOELASTIC FLUIDS

The minimum principles elaborated so far for viscoelastic solids have their non-trivial counterpart for viscoelastic fluids. Not to be much too repetitive we are content with examining the analogue of the formulation via a weight function. Both incompressible and compressible fluids are considered.

For formal convenience we write the constitutive equation (3.27) as

$$\mathbf{T}(t) = -p(t)\,\mathbf{1} + \int_0^\infty \mathbf{M}(s)\,\mathbf{L}(t-s)\,ds$$

where $\mathbf{M} : \mathrm{I\!R}^+ \to \mathrm{Lin(Sym)}$ and, in indicial notation,

$$M_{ijkl} = \mu(\delta_{ik}\,\delta_{jl} + \delta_{il}\,\delta_{jk}) + \lambda\,\delta_{ij}\,\delta_{kl}.$$

We take the history of the velocity $\mathbf{v}$ to be known up to time $t = 0$ for any $\mathbf{x} \in \Omega$ and let $\mathbf{v}$ be given at the boundary $\partial\Omega$ for any time $t \geq 0$. If $\mathbf{v}$ is known in $\Omega \times \mathrm{I\!R}^-$ then we can write

$$\mathbf{T}(t) = -p\,\mathbf{1} + \int_0^t \mathbf{M}(s)\,\mathbf{L}(t-s)\,ds + \mathbf{T}_0(t)$$

where

$$\mathbf{T}_0(t) = \int_t^\infty \mathbf{M}(s)\,\mathbf{L}(t-s)\,ds$$

is known. The boundary $\partial\Omega$ of Ω is taken to consist of two disjoint subsets $\partial\Omega_v$, $\partial\Omega_T$, namely

$$\partial\Omega_v \cup \partial\Omega_T = \partial\Omega, \qquad \partial\Omega_v \cap \partial\Omega_T = \emptyset.$$

For definiteness we begin by considering the incompressible fluid. Consistent with the linear approximation inherent in the model of viscoelasticity, we identify $\dot{\mathbf{v}}$ with $\partial\mathbf{v}(\mathbf{x},t)/\partial t$. As usual we omit writing the dependence of the fields on the position $\mathbf{x}$. Then we consider the dynamics of the viscoelastic fluid subject to initial-boundary data through the following problem ($\mathrm{P_I}$)

$$\rho\dot{\mathbf{v}}(t) = -\nabla p(t) + 2\int_0^t \mu(s)\Delta\mathbf{v}(t-s)\,ds + \rho\mathbf{f}, \tag{5.1}$$

$$\nabla\cdot\mathbf{v} = 0 \quad \text{on} \quad \Omega \times \mathrm{I\!R}^+, \tag{5.2}$$

$$\begin{cases} \mathbf{v} = 0 & \text{on} \quad \partial\Omega_v \times \mathrm{I\!R}^+, \\ \mathbf{T}\mathbf{n} = \hat{\mathbf{t}} & \text{on} \quad \partial\Omega_T \times \mathrm{I\!R}^+, \end{cases} \tag{5.3}$$

$$\mathbf{v}(0) = \mathbf{v}_0 \quad \text{on} \quad \bar{\Omega}, \tag{5.4}$$

where $\rho\mathbf{f} = \rho\mathbf{b} + \nabla\cdot\mathbf{T}_0$ and $\mathbf{b}$, as usual, is the body force. Consistency of (5.1)-(5.4) requires that $\mathbf{v}_0$ is divergence-free. Any solution to the problem $\mathrm{P_I}$ is a pair of vector and scalar fields $\{\mathbf{v}, p\}$ on

$\Omega \times \mathbb{R}^+$ satisfying (5.1) to (5.4). Of course the initial velocity field $\mathbf{v}_0$ is required to satisfy the incompressibility constraint (5.2). Moreover, letting

$$J_v^1(\Omega) = \{\mathbf{v} \in H^1(\Omega) : \nabla \cdot \mathbf{v} = 0 \text{ on } \Omega, \quad \mathbf{v} = 0 \text{ on } \partial\Omega_v\},$$

we require that

$$\mathbf{v} \in H^1(\mathbb{R}^+; J_v^1(\Omega)), \qquad p \in L^2(\mathbb{R}^+; L^2(\Omega)).$$

For later convenience we recall that the shear relaxation function μ is required to satisfy the thermodynamic inequality (3.33), namely the half-range Fourier cosine transform $\mu_c(\omega)$ is strictly positive for any $\omega \in \mathbb{R}^+$. This in turn implies the following result.

If $\mu_c(\omega) > 0$, $\omega \in \mathbb{R}^+$, then

$$\mu_L(\alpha) > 0, \quad \alpha \in \mathbb{R}^{++}. \tag{5.5}$$

The proof is immediate. Letting $\alpha \in \mathbb{R}^{++}$, consider the Laplace transform

$$\mu_L(\alpha) = \int_0^\infty \exp(-\alpha s)\mu(s)ds.$$

By the inversion formula for Fourier transforms we have

$$\mu(s) = \frac{2}{\pi}\int_0^\infty \mu_c(\omega)\cos\omega s\, d\omega,$$

a. e., and then

$$\mu_L(\alpha) = \frac{2}{\pi}\int_0^\infty\int_0^\infty \exp(-\alpha s)\mu_c(\omega)\cos\omega s\, ds\, d\omega.$$

On interchanging the order of integration we have

$$\mu_L(\alpha) = \frac{2}{\pi}\int_0^\infty \frac{\alpha\mu_c(\omega)}{\alpha^2+\omega^2}d\omega, \qquad \alpha \in \mathbb{R}^{++},$$

whence (5.5).

As in §4 we consider the function y with $Y \in C_0^\infty(\mathbb{R}^+)$ and make use of the bilinear form $(v,w)_y$ for any pair of functions $v, w \in L_L^2(\mathbb{R}^+; L^2(\Omega))$. Then we consider the functional

$$\Upsilon_y(\mathbf{v},p) = \tfrac{1}{2}[\rho(\dot{\mathbf{v}},\mathbf{v})_y + (\mathbf{M} * \nabla\mathbf{v}, \nabla\mathbf{v})_y] - \rho(\mathbf{f},\mathbf{v})_y$$
$$+\rho\int_0^\infty y(t)\int_\Omega [\tfrac{1}{2}\mathbf{v}(0) - \mathbf{v}_0]\cdot\mathbf{v}(t)\,dx\,dt - \int_0^\infty\int_0^\infty y(t+\tau)\int_{\partial\Omega_T}\mathbf{v}(\tau)\cdot\hat{\mathbf{t}}_0(t)\,d\tau\,dt\,da$$

where $\hat{\mathbf{t}}_0 = \hat{\mathbf{t}} - \mathbf{T}_0\mathbf{n}$. It is assumed that $\mathbf{f} \in L_L^2(\mathbb{R}^+; L^2(\Omega))$ and $\hat{\mathbf{t}}_0 \in L_L^2(\mathbb{R}^+; L^2(\partial\Omega_T))$.

In order to prove the desired minimum property for the functional Υ_y we make use of the following two properties. First, if the functions $u, \dot{u}, w, \dot{w}$ on $\mathbb{R}^+$ are Laplace transformable on $\mathbb{R}^+$ then

$$\int_0^\infty\int_0^\infty y(t+\tau)\dot{u}(t)w(\tau)\,dt\,d\tau = \int_0^\infty\int_0^\infty y(t+\tau)u(t)\dot{w}(\tau)\,dt\,d\tau$$
$$-\int_0^\infty y(t)u(0)w(t)\,dt + \int_0^\infty y(t)u(t)w(0)\,dt. \tag{5.6}$$

The simple proof is based on elementary properties of the Laplace transform.

Second, if $\mathbf{h} \in C^1(\Omega \times \mathbb{R}^+)$ is such that $\nabla \cdot \mathbf{h} = 0$ on $\Omega \times \mathbb{R}^+$ and $\mathbf{h} \cdot \mathbf{n} = 0$ on $\partial\Omega_v \times \mathbb{R}^+$ then, for any scalar field $q \in C^1(\Omega \times \mathbb{R}^+)$,

$$\int_\Omega \nabla q(t) \cdot \mathbf{h}(\tau)\, dx = \int_{\partial\Omega_T} q(t)\, \mathbf{h}(\tau) \cdot \mathbf{n}\, da, \qquad t, \tau \in \mathbb{R}^+. \tag{5.7}$$

This is easily seen by observing that because $\nabla \cdot \mathbf{h} = 0$ then $\nabla q(t) \cdot \mathbf{h}(\tau) = \nabla \cdot [q(t)\, \mathbf{h}(\tau)]$. The divergence theorem and the vanishing of $\mathbf{h} \cdot \mathbf{n}$ on $\partial\Omega_v$ provide the result.

We are now able to prove two theorems about the minimum of the functional Υ_y; of course it is understood that μ meets the thermodynamic inequality (3.31) and then μ_L meets (5.5). To this end it is convenient to consider the space

$$\mathcal{Q}_v = \{\mathbf{v} \in L^2_L(\mathbb{R}^+; J^1_v(\Omega)) \cap H^1_L(\mathbb{R}^+; L^2(\Omega))\}.$$

If $\mathbf{v}^0 \in \mathcal{Q}_v$ is a solution to the problem $\mathrm{P_l}$ then, for every function $Y \in C_0^\infty(\mathbb{R}^+)$, the functional $\Upsilon_y(\mathbf{v})$ has a minimum at $\mathbf{v}^0$ on $\mathcal{Q}_v$.

For, let $\mathbf{v}^0 \in \mathcal{Q}_v$ and represent any element $\mathbf{v} \in \mathcal{Q}$ as $\mathbf{v}^0 + \mathbf{h}$, where $\mathbf{h} \in \mathcal{Q}_v$. Then we can write

$$\begin{aligned}
\Upsilon_y(\mathbf{v}^0 + \mathbf{h}) - \Upsilon_y(\mathbf{v}^0) = {} & \tfrac{1}{2}[\rho(\dot{\mathbf{v}}^0, \mathbf{h})_y + \rho(\dot{\mathbf{h}}, \mathbf{v}^0)_y + \rho(\dot{\mathbf{h}}, \mathbf{h})_y - 2\rho(\mathbf{f}, \mathbf{h})_y \\
& + (\mathbf{M} * \nabla\mathbf{h}, \nabla\mathbf{v}^0)_y + (\mathbf{M} * \nabla\mathbf{v}^0, \nabla\mathbf{h})_y + (\mathbf{M} * \nabla\mathbf{h}, \nabla\mathbf{h})_y] \\
& + \rho \int_0^\infty y(t) \int_\Omega [\tfrac{1}{2}\mathbf{v}^0(0) \cdot \mathbf{h}(t) + \tfrac{1}{2}\mathbf{h}(0) \cdot (\mathbf{v}^0(t) + \mathbf{h}(t)) - \mathbf{v}_0 \cdot \mathbf{h}(t)] dx\, dt \\
& + \int_0^\infty \int_0^\infty y(t+\tau) \int_{\partial\Omega_T} \hat{\mathbf{t}}_0(t) \cdot \mathbf{h}(\tau)\, da\, d\tau\, dt.
\end{aligned}$$

By virtue of (5.6) and (5.2) we have

$$\begin{aligned}
\Upsilon_y(\mathbf{v}^0 + \mathbf{h}) - \Upsilon_y(\mathbf{v}^0) = {} & \rho(\dot{\mathbf{v}}^0, \mathbf{h})_y + (\mathbf{M} * \nabla\mathbf{v}^0, \nabla\mathbf{h})_y - \rho(\mathbf{f}, \mathbf{h})_y \\
& + \rho \int_0^\infty y(t) \int_\Omega \{[\mathbf{v}^0(0) - \mathbf{v}_0] \cdot \mathbf{h}(t) + \tfrac{1}{2}\mathbf{h}(0) \cdot \mathbf{h}(t)\} dx\, dt \\
& - \int_0^\infty \int_0^\infty y(t+\tau) \int_{\partial\Omega_T} \hat{\mathbf{t}}_0(t) \cdot \mathbf{h}(\tau)\, da\, d\tau\, dt \\
& + \tfrac{1}{2}[\rho(\dot{\mathbf{h}}, \mathbf{h})_y + (\mathbf{M} * \nabla\mathbf{h}, \nabla\mathbf{h})_y].
\end{aligned}$$

Observe that

$$\begin{aligned}
\int_\Omega (\mathbf{M} * \nabla\mathbf{v}^0)(t) \cdot \nabla\mathbf{h}(\tau) dx = {} & \int_{\partial\Omega} \mathbf{n} \cdot (\mathbf{M} * \nabla\mathbf{v}^0)(t)\, \mathbf{h}(\tau) da \\
& - \int_\Omega (\mu * \Delta\mathbf{v}^0)(t) \cdot \mathbf{h}(\tau) dx.
\end{aligned}$$

Because $\mathbf{v}^0$ is the solution to the problem (5.1)-(5.4), the vanishing of $\mathbf{h}(0)$ and $\mathbf{v}^0(0) - \mathbf{v}_0$, the use of (5.7), and account of the boundary conditions (5.3) yield

$$\Upsilon_y(\mathbf{v}^0 + \mathbf{h}) - \Upsilon(\mathbf{v}^0) = \tfrac{1}{2}[\rho(\dot{\mathbf{h}}, \mathbf{h})_y + (\mathbf{M} * \nabla\mathbf{h}, \nabla\mathbf{h})_y].$$

Now, $(\mathbf{M} * \nabla \mathbf{h}, \nabla \mathbf{h})_y = (2\mu * \mathrm{sym}\nabla \mathbf{h}, \mathrm{sym}\nabla \mathbf{h})_y$. Hence, by the choice of the weight function y, we obtain

$$\Upsilon_y(\mathbf{v}^0 + \mathbf{h}) - \Upsilon_y(\mathbf{v}^0) = \frac{1}{2}\int_0^\infty Y(\alpha) \int_\Omega [\rho\, \alpha\, |\mathbf{h}_L(\alpha)|^2 + 2\,\mu_L(\alpha)\, |(\mathrm{sym}\nabla \mathbf{h})_L(\alpha)|^2] dx\, d\alpha.$$

In view of (5.5) we conclude that $\Upsilon_y(\mathbf{v}^0 + \mathbf{h}) - \Upsilon_y(\mathbf{v}^0)$ vanishes only at $\mathbf{h} = 0$, that is at $\mathbf{v} = \mathbf{v}^0$. Really, since Y may have a compact support we can conclude that $\mathbf{h}_L = 0$ on the pertinent subinterval. However, real Laplace transforms can be considered as restrictions to the real axis of analytic functions on the complex plane. The fact that analytic functions vanishing on an interval of the real axis vanish on $\mathbb{R}^+$ too provides the vanishing of $\mathbf{h}_L$ on $\Omega \times \mathbb{R}^+$ and then that of $\mathbf{h}$ on $\Omega \times \mathbb{R}^+$. The second theorem, providing the sufficiency property, can be written as follows.

If $\mathbf{v}^0 \in \mathcal{Q}_v$ and

$$\Upsilon_y(\mathbf{v}^0) \le \Upsilon_y(\mathbf{v}), \qquad \mathbf{v} \in \mathcal{Q}_v,$$

then $\mathbf{v}^0$ is the unique solution to the problem (5.1)-(5.4).

To show that this is so, observe that any $\mathbf{v} \in \mathcal{Q}_v$ can be expressed as $\mathbf{v} = \mathbf{v}^0 + \eta\mathbf{h}$ with $\mathbf{h} \in \mathcal{Q}_v$, $\eta \in \mathbb{R}$. By direct calculation we have

$$\begin{aligned}
&\Upsilon_y(\mathbf{v}^0 + \eta\mathbf{h}) - \Upsilon_y(\mathbf{v}^0) = \\
&\qquad \eta\{\tfrac{1}{2}[\rho(\dot{\mathbf{v}}^0, \mathbf{h})_y + \rho(\dot{\mathbf{h}}, \mathbf{v}^0)_y + (\mathbf{M} * \nabla \mathbf{v}^0, \nabla \mathbf{h})_y + (\mathbf{M} * \nabla \mathbf{h}, \nabla \mathbf{v}^0)_y - \rho(\mathbf{f}, \mathbf{h})_y] \\
&\qquad\quad + \rho \int_0^\infty y(t) \int_\Omega [\tfrac{1}{2}\mathbf{v}^0(0) - \mathbf{v}_0] \cdot \mathbf{h}(t)\, dx\, dt \\
&\qquad\quad - \int_0^\infty \int_0^\infty y(t+\tau) \int_{\partial\Omega_T} \hat{\mathbf{t}}_0(t) \cdot \mathbf{h}(\tau) d\tau\, dt\, da\} \\
&\qquad + \tfrac{1}{2}\eta^2\{\rho(\dot{\mathbf{h}}, \mathbf{h})_y + (\mathbf{M} * \nabla \mathbf{h}, \nabla \mathbf{h})_y + \rho \int_0^\infty y(t) \int_\Omega \mathbf{h}(0) \cdot \mathbf{h}(t)\, dx\, dt\}.
\end{aligned}$$

Accordingly, Υ_y is minimal at $\mathbf{v}^0$ only if the coefficient of η vanishes while that of η^2 is non-negative. Then by means of (5.6) and integrations by parts we obtain

$$\begin{aligned}
&(\rho\dot{\mathbf{v}}^0 - \mu * \Delta \mathbf{v}^0 - \rho\mathbf{f}, \mathbf{h})_y + \rho \int_0^\infty y(t) \int_\Omega [\mathbf{v}^0(0) - \mathbf{v}_0] \cdot \mathbf{h}(t)\, dx\, dt \\
&\qquad + \int_0^\infty \int_0^\infty y(t+\tau) \int_{\partial\Omega_T} [-\hat{\mathbf{t}}_0(t) + \mathbf{n}(\mathbf{M} * \nabla \mathbf{v}^0)(t)] \cdot \mathbf{h}(\tau)\, da\, d\tau\, dt = 0,
\end{aligned} \tag{5.8}$$

$$(\rho\dot{\mathbf{h}}, \mathbf{h})_y + (\mathbf{M} * \nabla \mathbf{h}, \nabla \mathbf{h})_y + \int_0^\infty y(t) \int_\Omega \mathbf{h}(0) \cdot \mathbf{h}(t) dx\, dt \ge 0. \tag{5.9}$$

The left-hand side of (5.9) can be written as

$$\int_0^\infty Y(\alpha) \int_\Omega [\rho\, \alpha\, |\mathbf{h}_L(\alpha)|^2 + 2\,\mu_L(\alpha)\, |\mathrm{sym}\nabla \mathbf{h}_L(\alpha)|^2] d\alpha\, dx$$

and shows that, by (5.5), Υ_y has a strict minimum at $\mathbf{v}^0$, namely the minimum is unique. In view of (5.7), the divergence-free condition on $\mathbf{h}$ and the vanishing of $\mathbf{h}$ on $\partial\Omega_v$ allows us to add the (zero) quantity

$$\int_\Omega \nabla p(t) \cdot \mathbf{h}(\tau) dx - \int_{\partial\Omega_T} p(t)\mathbf{h} \cdot \mathbf{n} da$$

on the left-hand side of (5.8). Hence (5.8) yields

$$\rho\,\alpha\mathbf{v}_L^0 + \nabla p_L - 2\,\mu_L\,\Delta\mathbf{v}_L^0 - \rho\mathbf{f}_L - \rho\mathbf{v}_0 = 0$$

along with

$$(\mathbf{t}_0)_L = -p_L\,\mathbf{n} + \mathbf{n}(\mathbf{M}_L\nabla\mathbf{v}_L^0) \quad \text{on} \quad \partial\Omega_T.$$

So $\mathbf{v}_L^0$ is the unique solution to the Laplace transformed problem and so $\mathbf{v}^0$ is the unique solution to the problem (5.1)-(5.4).

Now let the fluid be compressible. We take the balance equations to be given by their linear approximations and the history of $\mathbf{v}$ and ρ to be known up to time $t = 0$ for any $\mathbf{x} \in \Omega$. Accordingly we let ρ_0 be the average mass density and set $\varphi = \rho - \rho_0$. Then, letting again $\dot{} = \partial/\partial t$, we write

$$\rho_0\dot{\mathbf{v}} = -p_\rho\nabla\varphi + \nabla\cdot\int_0^t \mathbf{M}(s)\nabla\mathbf{v}(t-s)\,ds + \rho_0\mathbf{f}, \tag{5.10}$$

$$\dot{\varphi} + \rho_0\nabla\cdot\mathbf{v} = 0, \tag{5.11}$$

where

$$\rho_0\mathbf{f} = \rho_0\mathbf{b} + \nabla\cdot\int_t^\infty \mathbf{M}(s)\nabla\mathbf{v}(t-s)\,ds$$

and p_ρ stands for the reference value $p_\rho(\rho_0)$. The boundary data (5.3) hold along with the initial data

$$\mathbf{v}(0) = \mathbf{v}_0, \qquad \varphi(0) = \varphi_0 \qquad \text{on} \quad \bar{\Omega}. \tag{5.12}$$

Owing to mass conservation φ is subject to the constraint

$$\int_\Omega \varphi\,dx = 0 \quad \text{on} \quad \mathbb{R}^+. \tag{5.13}$$

On the whole (5.10), (5.11), (5.3), (5.12), and (5.13) constitute the dynamic problem $\mathrm{P_f}$.

It is a straighforward matter to verify that the functional

$$\begin{aligned}\Xi_y(\mathbf{v},\varphi) = {} & \tfrac{1}{2}[\rho_0(\dot{\mathbf{v}},\mathbf{v})_y - \frac{p_\rho}{\rho_0}(\dot{\varphi},\varphi)_y + (\mathbf{M} * \nabla\mathbf{v},\nabla\mathbf{v})_y] + p_\rho(\nabla\varphi,\mathbf{v})_y - \rho_0(\mathbf{f},\mathbf{v})_y \\ & - \int_0^\infty\int_0^\infty y(t+\tau)\int_{\partial\Omega_T}\mathbf{v}(t)\cdot[p_\rho\varphi(\tau)\mathbf{n} + \hat{\mathbf{t}}_0(\tau)]\,da\,d\tau\,dt \\ & + \int_0^\infty y(t)\int_\Omega[\rho_0(\tfrac{1}{2}\mathbf{v}(0) - \mathbf{v}_0)\cdot\mathbf{v}(t) - \frac{p_\rho}{\rho_0}(\tfrac{1}{2}\varphi(0) - \varphi_0)\varphi(t)]\,dx\,dt\end{aligned}$$

is stationary at the solution $\mathbf{v},\varphi$ to the problem $\mathrm{P_f}$. Yet, owing to the term $(\nabla\varphi,\mathbf{v})_y$ and the $-$ sign in front of $(\dot{\varphi},\varphi)_y$ we cannot find a definite form for the possible extremum. To remove the difficulty we consider, in a sense, a functional for one equation only but account for the other one as a constraint.

In the space

$$\mathcal{J}_v^1(\Omega) = \{\mathbf{v} \in H^1(\Omega) : \int_\Omega \nabla\cdot\mathbf{v}\,dx = 0, \quad \mathbf{v} = 0 \text{ on } \partial\Omega_v\},$$

look at the functional

$$\Theta_y(\mathbf{v},\varphi) = \tfrac{1}{2}[\rho_0(\dot{\mathbf{v}},\mathbf{v})_y + (\mathbf{M} * \nabla\mathbf{v}, \nabla\mathbf{v})_y + \frac{p_\rho}{\rho_0}(\dot{\varphi},\varphi)_y] - \rho_0(\mathbf{f},\mathbf{v})_y$$
$$+ \rho_0 \int_0^\infty y(t) \int_\Omega [\tfrac{1}{2}\mathbf{v}(0) - \mathbf{v}_0]\cdot\mathbf{v}(t)\, dx\, dt - \int_0^\infty \int_0^\infty y(t+\tau) \int_{\partial\Omega_T} \mathbf{v}(\tau)\cdot\hat{\mathbf{t}}_0(t)\, da\, d\tau\, dt$$
$$- \frac{p_\rho}{\rho_0} \int_0^\infty y(t) \int_\Omega [\tfrac{1}{2}\varphi(0) - \varphi_0]\varphi(t)\, dx\, dt$$

where $\hat{\mathbf{t}}_0 = \hat{\mathbf{t}} - \mathbf{T}_0\mathbf{n}$ and

$$\mathbf{T}_0(t) = -p(\rho_0)\mathbf{1} + \int_t^\infty \mathbf{M}(s)\mathbf{L}(t-s)\, ds.$$

We can show that if $\mathbf{v}^0 \in \bar{\mathcal{Q}}_v = \{\mathbf{v} \in L^2_L(\mathbb{R}^+; \bar{J}^1_v(\Omega)) \cap H^1_L(\mathbb{R}^+; L^2(\Omega))\}$, and $\mathbf{v}^0, \varphi^0$ is a solution to the problem $\mathrm{P_f}$ then, for every function $Y \in C_0^\infty(\mathbb{R}^+)$, the functional $\Theta_y(\mathbf{v},\varphi)$ has a minimum at $\mathbf{v}^0$ on $\bar{\mathcal{Q}}_v$.

Represent any element $\mathbf{v} \in \bar{\mathcal{Q}}_v$ as $\mathbf{v}^0 + \mathbf{h}$ and, correspondingly, φ as $\varphi^0 + \xi$. Then we can write

$$\Theta_y(\mathbf{v}^0 + \mathbf{h}, \varphi^0 + \xi) - \Theta_y(\mathbf{v}^0,\varphi^0) = \tfrac{1}{2}[\rho_0(\dot{\mathbf{v}^0},\mathbf{h})_y + \rho_0(\dot{\mathbf{h}},\mathbf{v}^0)_y + \rho_0(\dot{\mathbf{h}},\mathbf{h})_y - 2\rho_0(\mathbf{f},\mathbf{h})_y$$
$$+ (\mathbf{M} * \nabla\mathbf{h}, \nabla\mathbf{v}^0)_y + (\mathbf{M} * \nabla\mathbf{v}^0, \nabla\mathbf{h})_y + (\mathbf{M} * \nabla\mathbf{h}, \nabla\mathbf{h})_y$$
$$+ \frac{p_\rho}{\rho_0}(\dot{\varphi^0},\xi)_y + \frac{p_\rho}{\rho_0}(\dot{\xi},\varphi^0)_y + \frac{p_\rho}{\rho_0}(\dot{\xi},\xi)_y]$$
$$+ \rho \int_0^\infty y(t) \int_\Omega [(\tfrac{1}{2}\mathbf{v}^0(0) - \mathbf{v}_0)\cdot\mathbf{h}(t) + \tfrac{1}{2}\mathbf{h}(0)\cdot(\mathbf{v}^0(t) + \mathbf{h}(t))]dx\, dt$$
$$- \int_0^\infty \int_0^\infty y(t+\tau) \int_{\partial\Omega_T} \hat{\mathbf{t}}_0(t)\cdot\mathbf{h}(\tau)\, da\, d\tau\, dt$$
$$- \frac{p_\rho}{\rho_0} \int_0^\infty y(t) \int_\Omega [\tfrac{1}{2}\xi(0)(\varphi^0(t) + \xi(t)) + (\tfrac{1}{2}\varphi^0(0) - \varphi_0)\xi(t)]dx\, dt.$$

By virtue of (5.6) and (4.9) and use of (5.11) in connection with ξ we have

$$\Theta_y(\mathbf{v}^0 + \mathbf{h}, \varphi^0 + \xi) - \Theta_y(\mathbf{v}^0,\varphi^0) = \rho_0(\dot{\mathbf{v}^0},\mathbf{h})_y - p_\rho(\nabla\cdot\mathbf{h},\varphi^0)_y + (\mathbf{M} * \nabla\mathbf{v}^0, \nabla\mathbf{h})_y - \rho_0(\mathbf{f},\mathbf{h})_y$$
$$+ \rho_0 \int_0^\infty y(t) \int_\Omega \{[\mathbf{v}^0(0) - \mathbf{v}_0]\cdot\mathbf{h}(t) + \tfrac{1}{2}\mathbf{h}(0)\cdot\mathbf{h}(t)\}dx\, dt$$
$$- p_\rho\rho_0 \int_0^\infty y(t) \int_\Omega \{[\varphi^0(0) - \varphi_0]\xi(t) + \tfrac{1}{2}\xi(0)\xi(t)\}dx\, dt$$
$$- \int_0^\infty \int_0^\infty y(t+\tau) \int_{\partial\Omega_T} \hat{\mathbf{t}}_0(t)\cdot\mathbf{h}(\tau)\, da\, d\tau\, dt$$
$$+ \tfrac{1}{2}[\rho_0(\dot{\mathbf{h}},\mathbf{h})_y + \frac{p_\rho}{\rho_0}(\dot{\xi},\xi)_y + (\mathbf{M} * \nabla\mathbf{h}, \nabla\mathbf{h})_y].$$

Integration by parts, the divergence theorem, and some rearrangement yield

$$(\mathbf{M} * \nabla\mathbf{v}^0, \nabla\mathbf{h})_y - p_\rho(\nabla\cdot\mathbf{h},\varphi^0)_y = (-\nabla\cdot\mathbf{M} * \nabla\mathbf{v}^0 + p_\rho\nabla\varphi^0, \mathbf{h})_y$$
$$+ \int_0^\infty \int_0^\infty y(t+\tau) \int_{\partial\Omega} [\mathbf{n}\,\mathbf{M} * \nabla\mathbf{v}^0 - p_\rho\varphi^0\mathbf{n}](t)\cdot\mathbf{h}(\tau)\, da\, d\tau\, dt.$$

Because $\mathbf{v}^0, \varphi^0$ is the solution to the problem P_f, the vanishing of $\mathbf{h}(0)$, $\xi(0)$ and $\mathbf{v}^0(0)-\mathbf{v}_0, \varphi^0(0)-\varphi_0$, and account of the boundary conditions (5.3) yield

$$\Theta_y(\mathbf{v}^0+\mathbf{h}, \varphi^0+\xi) - \Theta(\mathbf{v}^0, \varphi^0) = \tfrac{1}{2}[\rho(\dot{\mathbf{h}}, \mathbf{h})_y + (\mathbf{M} * \nabla\mathbf{h}, \nabla\mathbf{h})_y + \frac{p_\rho}{\rho_0}(\dot{\xi}, \xi)_y].$$

Hence, by the choice of the weight function y and the definition of $\mathbf{M}$, we obtain

$$\Theta_y(\mathbf{v}^0+\mathbf{h}, \varphi^0+\xi) - \Theta_y(\mathbf{v}^0, \varphi^0) =$$
$$\frac{1}{2}\int_0^\infty Y(\alpha)\int_\Omega [\rho_0\alpha|\mathbf{h}_L(\alpha)|^2 + 2\mu_L(\alpha)|(\mathrm{sym}\,\nabla\overset{\circ}{\mathbf{h}})_L\,(\alpha)|^2 + \kappa_L(\alpha)|\nabla\cdot\mathbf{h}_L(\alpha)|^2 + \frac{p_\rho}{\rho_0}\alpha|\xi_L(\alpha)|^2]\,dx\,d\alpha.$$

Since the thermodynamic inequalities $\mu_c, \kappa_c > 0$ imply $\mu_L, \kappa_L > 0$ we conclude that $\Theta_y(\mathbf{v}^0+\mathbf{h}, \varphi^0+\xi) - \Theta_y(\mathbf{v}^0, \varphi^0)$ vanishes only at $\mathbf{h}, \xi = 0$, that is at $\mathbf{v} = \mathbf{v}^0$, $\varphi = \varphi^0$, provided only that $p_\rho > 0$. This is a standard, usually understood, assumption on the function p. So for the compressible fluid too an extremum principle is available which holds just because of the thermodynamic restrictions (3.31).

The converse holds too as can be seen by paralleling the previous proofs.

ACKNOWLEDGMENTS

The research leading to this work was supported in part by the Italian Ministry of Education through the 40% Research Project "Problemi di evoluzione nei fluidi e nei solidi".

REFERENCES

Benthien, G. and M.E. Gurtin. 1970. A principle of minimum transformed energy in linear elastodynamics. *J. Appl. Mech.* 37: 1147-1149.

Bowen, R. and P.J. Chen. 1973. Thermodynamic restriction on the initial slope of the stress relaxation function. *Arch. Rational Mech. Anal.* 51: 278-284.

Coleman, B.D. 1964. Thermodynamics of materials with memory. *Arch. Rational Mech. Anal.* 17: 1-46.

Coleman, B.D. 1964. On thermodynamics, strain impulses, and viscoelasticity. *Arch. Rational Mech. Anal.* 17: 230-254.

Coleman, B.D., M. Fabrizio, and D.R. Owen. 1982. On the thermodynamics of second sound in dielectric crystals. *Arch. Rational Mech. Anal.* 80: 135-158.

Coleman, B.D. and D.R. Owen. 1974. A mathematical foundation for thermodynamics. *Arch. Rational Mech. Anal.* 54: 1-104.

Day, W.A. 1971. Restrictions on relaxation functions in linear viscoelasticity. *Q. Jl Mech. Appl. Math.* 24: 487-497.

Day, W.A. 1972. *The thermodynamics of simple materials with fading memory,* Springer, Berlin.

Fabrizio, M. 1989. Proprietà e restrizioni costitutive per fluidi viscosi con memoria. *Atti Sem. Mat. Fis. Univ. Modena* 37: 429-446.

Fabrizio, M., C. Giorgi, and A. Morro. 1989. Minimum principles, convexity, and thermodynamics in viscoelasticity. *Continuum Mech. Thermodyn.* 1: 197-211.

Fabrizio, M. and A. Morro. 1988. Viscoelastic relaxation functions compatible thermodynamics, *J. Elasticity* 19: 63-75.

Fabrizio, M. and A. Morro. 1990. *Mathematical problems in linear viscoelasticity,* SIAM publications, to appear.

Giorgi, C. 1989. Alcune conseguenze delle restrizioni termodinamiche per mezzi viscoelastici lineari. *Quaderno n. 6/89.* Università Cattolica, Dipartimento di Matematica, Brescia.

Graffi, D. 1928. Sui problemi dell'ereditarietà lineare. *Nuovo Cimento* 5: 53-71.

Graffi, D. 1974. Sull'espressione dell'energia libera nei materiali visco-elastici lineari. *Ann. Mat. Pura Appl.* 98: 273-279.

Graffi, D. 1982. Sull'espressione analitica di alcune grandezze termodinamiche nei materiali con memoria. *Rend. Sem. Mat. Univ. Padova* 68: 17-29.

Gurtin, M.E. and E. Sternberg. 1962. On the linear theory of viscoelasticity. *Arch. Rational Mech. Anal.* 11: 291-356.

Leitman, M.J. and G.M.C. Fisher. 1973. The linear theory of viscoelasticity. In *Encyclopedia of Physics,* ed. C. Truesdell, VIa/3, Springer, Berlin.

Morro, A. 1985. Relaxation phenomena via hidden variable thermodynamics. In *Thermodynamics and constitutive equations,* ed. G. Grioli, Springer, Berlin.

Morro, A. 1985. Negative semi-definiteness of the viscoelastic attenuation tensor via the Clausius-Duhem inequality. *Arch. Mech.* 37: 255-259.

Morro, A. 1990. *Topics in the theory of materials with memory.* Mediterranean Press, Rende.

Morro, A. and M. Vianello. 1990. Minimal and maximal free energy for materials with memory. *Boll. Un. Mat. Ital.,* to appear.

Noll, W. 1972. A new mathematical theory of simple materials. *Arch. Rational Mech. Anal.* 48: 1-50.

Reiss, R. 1978. Minimum principles for linear elastodynamics. *J. Elasticity* 8: 35-45.

Volterra, V. 1940. Energia nei fenomeni ereditarii. *Acta Pontificiae Acad. Scient.* 4: 115-128.

Extended Thermodynamics—A Theory with Hyperbolic Field Equations

Ingo Müller
Physikalische Ingenieurwissenschaft
TU Berlin, Germany

ABSTRACT

Obeying the strict injunctions of the editors, I have hereby prepared a review of my personal work on Extended Thermodynamics which began 25 years ago and still continues today. I discuss how the desire for having finite speeds for heat waves and shear waves has eventually led to a systematic theory that is in full agreement with the kinetic theory of gases. In order to achieve this goal, many formerly firmly believed ingredients of thermodynamics had to be eliminated, viz.

- material frame indifference as a general principle,
- continuity of temperature at thermometric walls, and
- universal relation between the fluxes of entropy and heat.

What has developed is a theory of viscosity and heat conduction in which the thermal and caloric equations of state, and the measured transport coefficients determine all unknown coefficients. This theory has so simple a mathematical structure that in the future it may allow mathematicians to find a relation between the existence of entropy and the natural requirements - such as uniqueness and well-posedness - placed on the solutions of the field equations of thermodynamics.

1. Motivation - Cattaneo's Work.

There has always been a consensus that the parabolic equations of heat conduction, shear diffusion, and mass diffusion were necessarily flawed in that they predicted infinite speeds. Also, upon reflection it became clear that the classical constitutive laws of Fourier, Navier-Stokes, and Fick were to blame for this situation. Few people cared, however, since these laws were perfectly well-suited for the every-day problems of engineers and physicists. Only a few natural philosophers and theoretical physicists were concerned.

One of the concerned was C. Cattaneo. In 1948 he wrote a paper entitled "Sulla conduzione del calore" (Cattaneo 1948) in which he proposed a modification of Fourier's law. This modification is based on kinetic arguments. The reasoning proceeds as follows:

Cattaneo first reviews an old argument of Maxwell's which gives a molecular interpretation of heat conduction. To understand the gist of that argument, we may look at a volume element of linear dimensions of the mean free path and set up a temperature gradient across the element, see Figure 1. A particle moving upwards will, in the mean, carry more energy that a particle moving downwards. Therefore, there is a net flux of energy upwards, i.e., opposite to the temperature gradient associated with the passing of the pair of particles across the middle layer. Thus, we get the

$$\text{"stationary" Fourier law} \qquad q_i = \kappa_1 \frac{\partial T}{\partial x_i} \,. \tag{1.1}$$

Cattaneo changed the argument slightly. He argued that there is a time lag between the start of the particles at their point of departure and the time of passage through the middle layer. If the temperature changes in time, it is thus clear that the heat flux at a certain time depends on the temperature gradient at an earlier time. Therefore, is seems reasonable to write the

$$\text{"instationary" Fourier law} \qquad q_i = -\kappa_1 \frac{\partial T}{\partial x_i} + \kappa_2 \left(\frac{\partial T}{\partial x_i}\right)^{\cdot} . \tag{1.2}$$

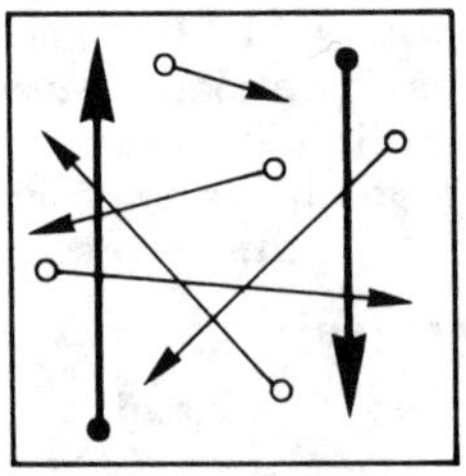

Figure 1: On Fourier's law

If the equations (1) or (2) are inserted into the energy balance for a body at rest, i.e.,

$$\varrho \dot{u} + \frac{\partial q_i}{\partial x_i} = 0 \qquad \text{with} \qquad u = cT + u_0 \,, \tag{1.3}$$

we obtain the differential equations of heat conduction

$$\dot{T} = \frac{\kappa_1}{\varrho c} \Delta T \qquad \text{with (1.1) and} \tag{1.4}$$

$$\ddot{T} = \frac{\kappa_1}{\varrho c} \Delta T - \frac{\kappa_2}{\varrho c} (\Delta T)^{\cdot} \qquad \text{with (1.2)} \,. \tag{1.5}$$

Both of these equations are parabolic and both predict the spreading of heat pulses with infinite speed. Cattaneo must have noticed this, because he proceeds to modify equation (1.2) by passing through a sequence of steps as follows:

$$q_i = -\kappa_1 \left(1 - \sigma \frac{d}{dt}\right) \frac{\partial T}{\partial x_i} \qquad \text{with} \qquad \sigma = \frac{\kappa_2}{\kappa_1} \tag{1.6}$$

$$\frac{1}{1-\sigma\frac{d}{dt}} q_i = -\kappa_1 \frac{\partial T}{\partial x_i} \tag{1.6}$$

$$q_i + \sigma \dot{q}_i = -\kappa_1 \frac{\partial T}{\partial x_i} .$$

The last equation is often called the Cattaneo equation; it is again a modification of Fourier's law. The above transition from (1.2) to (1.6) is difficult to justify rigorously, and therefore, I simply quote what Cattaneo had to say about it:

> Nel risultato ottenuto approfitteremo della piccolezza del parametro σ per trascurare il termine che contiene a fattore il suo quadrato, consevando peraltro il termine in cui σ compare a primo grado. Naturalmente, per delimitare la portata delle conseguenze che stiamo per trarre, converrà precisare un pò meglio le condizioni in cui tale approssimazione è lecita. Allo scopo ammetteremo espli citamente che il fenomeno di conduzione calorifica avvenga nell'-intorno di uno stato stazionario o, in altri termini, che durante il suo svolgersi si mantengano abbastanza piccole le derviate temporali delle varie grandezze in giuoco.

If we combine (1.6) and the energy equation (1.3), we obtain

$$\frac{\kappa_2}{\kappa_1} \ddot{T} + \dot{T} = -\frac{\kappa_1}{\varrho c} \Delta T . \tag{1.7}$$

This is the telegraph equation, it is hyperbolic - if $\kappa_2 > 0$ - and predicts the propagation of heat pulses at a finite speed.

We conclude that - innocent as the transition from (1.2) to (1.6) may seem - it is all important for finite speeds, and it changes a parabolic equation into a hyperbolic one.

The mathematical daring of Cattaneo's argument is obvious, but that is not the only criticism against equation (1.6). In fact, physicists do not agree with Cattaneo's equation, because it does not remain invariant under a change of frame. We will come back to this point later in Section 3.

2. "Erweiterte Thermodynamik" - Müller's Work of 1966. (Müller 1966, 1967)

Nevertheless, Cattaneo had pointed the way, and his work motivated Müller to try and derive generalizations of the laws of Fourier and Navier-Stokes from the principles of Linear Irreversible Thermodynamics, a theory universally known as TIP, or, Thermodynamics of Irreversible Processes.

Let us first recall how TIP derives the laws of Fourier and Navier-Stokes. Our starting point is the Gibbs equation

$$\dot{s} = \frac{1}{T}\left(\dot{u} - \frac{p}{\varrho^2}\dot{\varrho}\right) . \tag{2.1}$$

The balance equations of mass and internal energy, viz.

$$\dot{\varrho} = -\varrho \frac{\partial v_i}{\partial x_i} \tag{2.2}$$

$$\varrho\dot{u} = -\frac{\partial q_i}{\partial x_i} + t_{ij}\frac{\partial v_i}{\partial x_j} \tag{2.2}$$

are used to eliminate $\dot{s}$ and $\dot{u}$ between (2.1) and (2.2). This leads to the equation

$$\varrho\dot{s} + \frac{\partial}{\partial x_i}\left(\frac{q_i}{T}\right) = q_i\left(-\frac{1}{T^2}\frac{\partial T}{\partial x_i}\right) + \frac{1}{T}t_{\langle ij\rangle}\frac{\partial v_{\langle i}}{\partial x_{j\rangle}} + \frac{1}{T}\left(\frac{1}{3}t_{ii}+p\right)\frac{\partial v_i}{\partial x_i} \tag{2.3}$$

which we may interpret as the balance of entropy. This interpretation implies that

$\Phi_i = {}^{q_i}/_T$ is the entropy flux, and (2.4)

$\sigma = -\frac{q_i}{T^2}\frac{\partial T}{\partial x_i} + \frac{1}{T}t_{\langle ij\rangle}\frac{\partial v_{\langle i}}{\partial x_{j\rangle}} + \frac{1}{T}\left(\frac{1}{3}t_{ii}+p\right)\frac{\partial v_i}{\partial x_i}$ is the entropy production. (2.5)

Inspection shows that the entropy production is a sum of products of

<u>thermodynamic fluxes</u>	and	<u>thermodynamic forces</u>
heat flux q_i		temperature gradient $\frac{\partial T}{\partial x_i}$
stress deviator $t_{\langle ij\rangle}$		deviatoric velocity gradient $\frac{\partial v_{\langle i}}{\partial x_{j\rangle}}$
non-equilibrium pressure $\pi = \frac{1}{3}t_{ii}+p$		divergence of velocity $\frac{\partial v_i}{\partial x_i}$. (2.6)

The entropy production is assumed to be non-negative, and this inequality is ensured by the following linear relations between forces and fluxes.

$$q_i = -\kappa\frac{\partial T}{\partial x_i}\,, \qquad \kappa > 0\,, \tag{2.7}$$

$$t_{\langle ij\rangle} = 2\mu\frac{\partial v_{\langle i}}{\partial x_{j\rangle}}\,, \qquad \mu > 0\,,$$

$$\pi = \lambda\frac{\partial v_i}{\partial x_i}\,, \qquad \lambda > 0\,. \tag{2.8}$$

These are the laws of Fourier and Navier-Stokes with κ being the heat conductivity and μ and λ the shear- and bulk-viscosity, respectively.

Müller noticed that the same argument can be used to derive the Cattaneo equation (1.6), if we only assume that s - which heretofore was considered a function of u and ϱ - depends on q_i as well as on u and ϱ. He further argued that if s depends on q_i, there is no reason it should not also depend on $t_{\langle ij\rangle}$ and π. Of course these dependencies must be such that s is a scalar and that s has a maximum in equilibrium, where q_i, $t_{\langle ij\rangle}$, and π all vanish. Restricting attention to quadratic terms, Müller therefore allowed s to depend on the scalars q_iq_i, $t_{\langle ij\rangle}\,t_{\langle ij\rangle}$, and π^2 with negative coefficients a, b, and c. Thus, he generalized the Gibbs equation (2.1) to read

$$\dot{s} = \frac{1}{T}\left(\dot{u} - \frac{p}{\varrho^2}\dot{\varrho} + 2aT\,\dot{q}_iq_i + 2bT\,t_{\langle ij\rangle}\,\dot{t}_{\langle ij\rangle} + 2cT\,\pi\dot{\pi}\right)\,. \tag{2.9}$$

Again, (2.2) is used, and we obtain an entropy balance of the form

$$\begin{aligned}
\varrho\dot{s} + \frac{\partial}{\partial x_i}\left(\frac{q_i}{T} + \underline{K\, t_{\langle ij\rangle}\, q_j + L\pi q_i}\right) &= \\
= q_i\left(-\frac{1}{T^2}\frac{\partial T}{\partial x_i} + 2\varrho a\, \dot{q}_i + \underline{K\frac{\partial t_{\langle ij\rangle}}{\partial x_j} + L\,\frac{\partial \pi}{\partial x_i}}\right) &+ \\
+ t_{\langle ij\rangle}\left(\frac{1}{T}\frac{\partial v_{\langle i}}{\partial x_{j\rangle}} + 2\varrho b\, \dot{t}_{\langle ij\rangle} + \underline{K\,\frac{\partial q_{\langle i}}{\partial x_{j\rangle}}}\right) &+ \\
+ \pi\left(\frac{1}{T}\frac{\partial v_i}{\partial x_i} + 2\varrho c\, \dot{\pi} + \underline{L\,\frac{\partial q_i}{\partial x_i}}\right) &.
\end{aligned} \tag{2.10}$$

(For now we ignore the underlined terms. We will presently discuss these terms.) Once more, we assume that the entropy production on the right hand side is non-negative and ensure this by postulating linear relations between the fluxes and the (modified) forces.

$$\begin{aligned}
q_i &= -\kappa\left\{\frac{\partial T}{\partial x_i} - 2\varrho\, T^2 a\, \dot{q}_i - \underline{KT^2\,\frac{\partial t_{\langle ij\rangle}}{\partial x_j} - LT^2\,\frac{\partial \pi}{\partial x_i}}\right\}, \\
t_{\langle ij\rangle} &= \mu\left\{\frac{\partial v_{\langle i}}{\partial x_{j\rangle}} + 2\varrho\ Tb\ \dot{t}_{\langle ij\rangle} + \underline{KT\,\frac{\partial q_{\langle i}}{\partial x_{j\rangle}}}\right\}, \\
\pi &= \lambda\left\{\frac{\partial v_i}{\partial x_i} + 2\varrho\ Tc\ \dot{\pi} + \underline{LT\,\frac{\partial q_i}{\partial x_i}}\right\}.
\end{aligned} \tag{2.11}$$

Thus indeed, the Fourier law is now replaced by $(2.11)_1$ which - still disregarding the underlined terms - is equivalent to the Cattaneo equation (1.6). Also in Müller theory corresponding changes were made in the Navier-Stokes equations, see $(2.8)_{2,3}$ as compared to $(2.11)_{2,3}$. Therefore, the theory provides hyperbolic equations not only for heat conduction, but also for shear diffusion. The close analogy between the equations for heat conduction and shear diffusion has been illustrated by Müller (Müller 1987, p. 245 ff).

Having made the assumption that s depend on q_i, $t_{\langle ij\rangle}$ and π, Müller searched for corroboration in the kinetic theory of gases and - more specificly - in Grad's 13 moment theory (Grad 1949). Grad had proposed a non-equilibrium phase density of the form

$$f = f_{\text{MAXWELL}}\left(1 - \frac{1}{2p}(t_{ij}+p\delta_{ij})\left[\frac{m}{kT}\left(C_iC_j - \frac{1}{3}C^2\delta_{ij}\right)\right] - \frac{1}{p}\,\frac{m}{kT}q_iC_i\left(1 - \frac{1}{5}\,\frac{mC^2}{kT}\right)\right). \tag{2.12}$$

If this is used to calculate s from

$$s = -\frac{k}{\varrho}\int f\ \ln f\ d\underset{\sim}{C} \tag{2.13}$$

it turns out that s indeed depends on q_i and t_{ij}. The kinetic theory even provides numbers for the coefficients a, b, and c in (2.9), viz.

$$a = -\frac{1}{5\ p^2\ T}, \qquad b = -\frac{1}{4\ p\ \varrho}, \qquad c = 0\ . \tag{2.14}$$

But it also turns out that the entropy flux

$$\Phi_i = - k \int C_i \; f \ln f \; d\underset{\sim}{C} \qquad (2.15)$$

is not equal to q_i/T. Rather, we obtain

$$\Phi_i = \frac{q_i}{T} + \frac{2}{5\,p\,T} t_{\langle ij \rangle} q_j \; . \qquad (2.16)$$

In his phenomenological theory, Müller therefore assumed

$$\Phi_i = \frac{q_i}{T} + K\, t_{\langle ij \rangle} q_j + L\, \pi\, q_i \qquad (2.17)$$

and added the divergence of $+ K t_{\langle ij \rangle} q_j + L \pi q_i$ to both side of the entropy balance (2.10). Restricting the attention to the quadratic terms, he obtained the underlined terms in (2.10) and, a forteriori, in (2.11).

Thus, within the framework of TIP, Müller had achieved two things:

- Hyperbolic equations for heat conduction and shear diffusion,
- Compatibility with Grad's 13 moment method.

Müller called this theory "Erweiterte Thermodynamik", which later was translated to Extended Thermodynamics. It was presented in his dissertation (Müller 1966), and a shortened version was published in the *Zeitschrift für Physik* (Müller 1967).

The theory proved useful in calculating the dispersion relation for plane harmonic waves of small amplitude and, hence, phase speeds $\vartheta_{Ph}(\omega) = \omega/k$ as quotients of frequency ω and wave number k. The phase speed in the limit $\omega \to \infty$ gives the fastest speed of propagation, and Müller calculated this speed to be equal to

$$\vartheta_{Ph}(\infty) = 1.65 \; \vartheta(0) \qquad (2.18)$$

using the values from (2.14) for a, b, c, and $K = + 2/(5pT)$, $L = 0$ as dictated by a comparison of (2.16), (2.17). $\vartheta(0)$ is the normal sound speed, i.e., it stands for $\sqrt{\kappa(k/m)T}$, where κ is the ratio of specific heats.

Note that the constitutive relations (2.11) contain the five unknown coefficients a, b, c, K, and L <u>in addition to</u> the transport coefficients κ, μ, and ν which occur also in ordinary thermodynamics of irreversible processes, see (2.7), (2.8).

3. The Question of Invariance.

It is obvious that neither the Cattaneo equation (1.6) nor Müller's equations can be solid physical laws, because they are not properly invariant. Indeed, $\dot{q}_i$ and $\dot{t}_{\langle ij \rangle}$ in (1.6) and (2.11) are not objective tensors, while all other terms in those equations are such. Therefore, even if the equations hold true in an inertial frame, they do not hold true in a non-inertial frame.

Here again, it is useful to turn to the kinetic theory in order to direct further research. Among the equations of transfer for the moments

$$p_{i_1 i_2 \ldots i_n} = m \int C_{i_1} C_{i_2} \ldots C_{i_n} \, f \, d\underset{\sim}{C} \tag{3.1}$$

of the distribution function, the first 13 read

$$\dot{\varrho} + \varrho \frac{\partial v_i}{\partial x_i} = 0 \; ,$$

$$\varrho\left[\dot{v}_i - 2\, W_{ik}(v_k - \dot{b}_k) + W_{ik}^2 (x_k - b_k) + \dot{W}_{ik}(x_k - b_k) + \ddot{b}_i\right] - \frac{\partial t_{ij}}{\partial x_j} = \varrho \, f_i \; ,$$

$$\varrho \dot{u} + \frac{\partial q_i}{\partial x_i} = t_{ij} \frac{\partial v_i}{\partial x_j} \; , \tag{3.2}$$

$$t_{\langle ij \rangle} = -\frac{2p}{\sigma}\left\{\frac{\partial v_{\langle i}}{\partial x_{j\rangle}} - \frac{1}{2p}\left[\dot{t}_{\langle ij\rangle} + 2t_{\langle n\langle i\rangle}\frac{\partial v_{j\rangle}}{\partial x_n} - 4t_{\langle n\langle i\rangle}W_{j\rangle n}\right] + \frac{2}{5p}\frac{\partial q_{\langle i}}{\partial x_{j\rangle}}\right\} + \boxed{\frac{1}{\sigma} t_{\langle ij\rangle} \frac{\partial v_n}{\partial x_n}}$$

$$q_i = \frac{1}{\tau}\frac{5}{2}\frac{k}{m}p\left\{\frac{\partial T}{\partial x_i} + \frac{2}{5}\frac{m}{k}\frac{1}{p}\left[\dot{q}_i + q_n\frac{\partial v_i}{\partial x_n} - 2q_n W_{in}\right] - \frac{2}{5}\frac{T}{p}\frac{\partial t_{\langle in\rangle}}{\partial x_n}\right\} + \boxed{\frac{1}{\tau}\left(\frac{7}{5}q_i\frac{\partial v_n}{\partial x_n} + \frac{4}{5}q_n\frac{\partial v_i}{\partial x_n} - \frac{7}{2}\frac{k}{m}t_{\langle iq\rangle}\frac{\partial T}{\partial x_q} - \frac{1}{\varrho}t_{\langle ik\rangle}\frac{\partial t_{kn}}{\partial x_n}\right)} .$$

Closure of this system of equations has been achieved by calculating the highest moments from the Grad phase density (2.12). The collision integrals have been calculated by use of the interaction potential for Maxwellian molecules. σ and τ are inverse relaxation times.

The first three equations (3.2) are the well-known balance equations for mass, momentum, and energy in a non-inertial frame. W_{ij} is the matrix of angular velocity of that frame with respect to an inertial frame, and b_i is the distance vector between the origins of the frames.

We ignore the framed terms in $(3.2)_{4,5}$, because they are non-linear. That leaves only the curly bracket terms on the right-hand sides where comparison with the phenomenological equations $(2.11)_{1,2}$ shows a significant similarity. Indeed, the only difference between the two sets of equations is - apart from the fact that now most coefficients are explicit - that $\dot{q}_i$ and $\dot{t}_{\langle ij\rangle}$ is now replaced by the objective tensors

$$\underbrace{\dot{q}_i + q_n\frac{\partial v_i}{\partial x_n} - 2q_n W_{in}} \qquad \dot{t}_{\langle ij\rangle} + 2t_{\langle n\langle i\rangle}\frac{\partial v_{j\rangle}}{\partial x_n} - 4t_{\langle n\langle i\rangle}W_{j\rangle n} \; . \tag{3.3}$$

We conclude that the non-invariance of the equations (1.6) and (2.11) of Cattaneo and Müller that was mentioned earlier must be corrected by joining the braced terms in (3.3) to $\dot{q}_i$ and $\dot{t}_{\langle ij\rangle}$ in (2.11). But, of course, at this stage this suggestion comes from the kinetic theory; there is nothing contained within TIP that would suggest such an addition. Indeed, if it were only for removing the non-invarience of equations (1.6) and (2.11), we could have replaced $\dot{t}_{\langle ij\rangle}$ and $\dot{q}_i$ by

$$\dot{t}_{\langle ij\rangle} - 2t_{\langle n\langle i\rangle}\frac{\partial v_{j\rangle}}{\partial x_n} \quad \text{and} \quad \dot{q}_i - q_n\frac{\partial v_i}{\partial x_n} \; , \tag{3.4}$$

the so-called Jaumann derivatives of $t_{\langle ij \rangle}$ and q_i which are objective tensors. However, the kinetic theory does not permit this alternative; it insists on inertial terms in equations $(3.2)_{4,5}$, namely, the terms with $\underset{\sim}{W}$. Later in Section 4 we shall discuss these terms and realize that they are eminently sensible.

Comparison of equations $(2.11)_{1,2}$ with $(3.2)_{4,5}$ permits the identification of the coefficients a, b, c, and K,L appropriate for a mon-atomic gas. In fact, we obtain the values of a,b, and c as already given in (2.14), and now we also obtain a value for K, viz.,

$$K = \frac{2}{5}\frac{1}{pT} . \tag{3.5}$$

4. Rational and Extended Thermodynamics as Systematic Phenomenological Theories of Mon-atomic Ideal Gases.

Until the mid-1960's, TIP was the only systematic thermodynamic theory, but at that time another theory emerged which came to be known as "Rational Thermodynamics", (Truesdell 1969). This theory was more effective in grasping the essential ingredients of thermodynamics, distinquishing them as either basic objectives, universal laws, or material properties. However, this theory does not provide new results - at least not for simple viscous - heat conducting fluids. Instead, it provided new and more systematic ways to derive the old results.

Liu and Müller (Liu and Müller 1983) have applied the methods of Rational Thermodynamics to derive a version of Extended Thermodynamics of a mon-atomic gas far superior to the version presented heretofore (see also Müller 1985, Chapter 13). In particular, this new theory determines the coefficients a, b, c, K, and L explicitly, as well as providing the properly invariant laws without recourse to the kinetic theory of gases.

The best way to explain this new theory is to proceed in a systematic manner, by which the tenets of Rational Thermodynamics of viscous, heat-conducting gases are set in juxtaposition to those of Extended Thermodynamics of such gases.

Rational Thermodynamics	Extended Thermodynamics
The basic objective is the determination of	
5 fields, viz., the fields	13 fields, viz., the fields
F - mass density	F - mass density
F_i - momentum density	F_i - momentum density
F_{ii} - 2x energy density	F_{ij} - momentum flux density
	F_{ijj} - 2x energy flux density.

(4.1)

In order to determine these fields, we need field equations, and what we have are equations of balance

$$
\begin{array}{lll}
\text{mass} & \dfrac{\partial F}{\partial t} + \dfrac{\partial F_r}{\partial x_r} = 0 \\[2ex]
\text{momentum} & \dfrac{\partial F_i}{\partial t} + \dfrac{\partial F_{ir}}{\partial x_r} = 0 \\[2ex]
\text{energy} & \dfrac{\partial F_{ii}}{\partial t} + \dfrac{\partial F_{iir}}{\partial x_r} = 0
\end{array}
\qquad \Bigg\| \qquad
\begin{array}{lll}
\text{mass} & \dfrac{\partial F}{\partial t} + \dfrac{\partial F_r}{\partial x_r} = 0 \\[2ex]
\text{momentum} & \dfrac{\partial F_i}{\partial t} + \dfrac{\partial F_{ir}}{\partial x_r} = 0 \\[2ex]
\text{momentum flux} & \dfrac{\partial F_{ij}}{\partial t} + \dfrac{\partial F_{ijr}}{\partial x_r} = S_{\langle ij\rangle} \\[2ex]
\text{energy flux} & \dfrac{\partial F_{ijj}}{\partial t} + \dfrac{\partial F_{ijjr}}{\partial x_r} = S_{irr}
\end{array}
\qquad (4.2)
$$

The balance equations of Rational Thermodynamics are all conservation laws, while those of Extended Thermodynamics have productions, viz. $S_{\langle ij\rangle}$ and S_{ijj}. The trace of the balance of momentum flux is the conservation law of energy. This assumption and the supposed symmetry of all tensors in (4.2) reduces the theory described here to a thermodynamic theory of mon-atomic ideal gases.

Obviously, neither set of equations (4.2) is a set of field equations for the fields (4.1). They must be supplemented by constitutive relations for

$$
F_{\langle ij\rangle},\ F_{irr} \qquad \Big\| \qquad F_{\langle ijr\rangle},\ F_{ijjr},\ S_{\langle ij\rangle},\ S_{ijj}\ . \qquad (4.3)
$$

For a viscous, heat-conducting gas, the constitutive equations assume the form

$$
\begin{array}{l}
F_{\langle ij\rangle} = \hat{F}_{\langle ij\rangle}(F, F_k, F_{ll}, F_{,r}, F_{k,r}, F_{ll,r}) \\[1ex]
F_{irr} = \hat{F}_{irr}\ (\text{———————————})
\end{array}
\qquad \Bigg\| \qquad
\begin{array}{l}
F_{\langle ijr\rangle} = \hat{F}_{\langle ijr\rangle}(F, F_k, F_{ij}, F_{ijj}) \\[1ex]
F_{ijjr} = \hat{F}_{ijjr}\ (\text{—————}) \\[1ex]
S_{\langle ij\rangle} = \hat{S}_{\langle ij\rangle}\ (\text{—————}) \\[1ex]
S_{ijj} = \hat{S}_{ijj}\ (\text{—————})\ .
\end{array}
\qquad (4.4)
$$

Note that in Rational Thermodynamics the constitutive relations are not truly local ones, due to the fact that gradients of the basic fields appear. On the other hand, in Extended Thermodynamics the constitutive relations are local. For that reason the field equations of Extended Thermodynamics are quasi-linear first order differential equations.

In both cases, it is true that if the constitutive functions were known, we could formulate field equations by eliminating the constitutive quantities (4.3) between (4.2) and (4.4). Every solution of the field equations is called a <u>thermodynamic process</u>.

In reality, of course, the constitutive functions are not known. Therefore, it is the task of the constitutive theory to determine these functions, or at least, to reduce them to a few coefficients which might then be measured. The tools of the constitutive theory are universal physical principles, above all:

- the entropy principle
- the principle of relativity.

The entropy principle assumes that the entropy inequality

$$
\frac{\partial h}{\partial t} + \frac{\partial \varphi_k}{\partial x_k} \geqslant 0 \qquad (4.5)
$$

holds for all thermodynamic processes, i.e., for all solutions of the field equations. The entropy density h and the entropy flux φ_k are considered as constitutive quantities such that we set

$$
\begin{array}{l|l}
h = \hat{h}\,(F, F_k, F_{ll}, F_{,r}, F_{k,r}, F_{ll,r}) & h = \hat{h}\,(F, F_k, F_{ij}, F_{ijj})\;, \\
\varphi_k = \hat{\varphi}_k\,(\text{———————}) & \varphi_k = \hat{\varphi}_k\,(\text{————})\;.
\end{array}
\qquad (4.6)
$$

In a manner of speaking, we may say that the fields (4.1), which must satisfy the inequality, are constrained by the requirement that they be solutions of the field equations (4.2) and (4.4). According to Liu's results (Liu 1972), we may take care of this constraint by using Lagrange multipliers Λ. Indeed, the new inequality

$$
\frac{\partial h}{\partial t} + \frac{\partial \varphi_k}{\partial x_k} - \Lambda\left(\frac{\partial F}{\partial t} + \frac{\partial F_r}{\partial x_r}\right) - \Lambda^{i}\left(\frac{\partial F_i}{\partial t} + \frac{\partial F_{ir}}{\partial x_r}\right) - \Lambda^{jj}\left(\frac{\partial F_{ii}}{\partial t} + \frac{\partial F_{iir}}{\partial x_r}\right) \geq 0 \;.
$$

$$
\frac{\partial h}{\partial t} + \frac{\partial \varphi_k}{\partial x_r} - \Lambda\left(\frac{\partial F}{\partial t} + \frac{\partial F_r}{\partial x_r}\right) - \Lambda^{i}\left(\frac{\partial F_i}{\partial t} + \frac{\partial F_{ir}}{\partial x_r}\right) - \Lambda^{ij}\left(\frac{\partial F_{ij}}{\partial t} + \frac{\partial F_{ijr}}{\partial x_r} - S_{\langle ij\rangle}\right) - \Lambda^{ijj}\left(\frac{\partial F_{ikk}}{\partial t} + \frac{\partial F_{ikkr}}{\partial x_r} - S_{ikk}\right) \geq 0 \;. \qquad (4.7)
$$

must hold for <u>all</u> fields. The Λ's are constitutive functions of the same types as (4.4) and (4.6). Insertion of the constitutive relations (4.4) and (4.6) into the inequalities (4.7) results in inequalities on whose left-hand side we have linear functions of the derivatives

$$
\begin{array}{l|l}
\dfrac{\partial F}{\partial t}\;\dfrac{\partial F_i}{\partial t}\;\dfrac{\partial F_{ii}}{\partial t}\;\dfrac{\partial F_{,r}}{\partial t}\;\dfrac{\partial F_{i,r}}{\partial t}\;\;\dfrac{\partial F_{ii,r}}{\partial t}\;, & \dfrac{\partial F}{\partial t}\;\dfrac{\partial F_i}{\partial t}\;\dfrac{\partial F_{ij}}{\partial t}\;\dfrac{\partial F_{ijj}}{\partial t}\;, \\
F_{,rs}, F_{k,rs}, F_{k,rs} F_{iirs} & F_{,r}, F_{i,r}, F_{ij,r}, F_{ijj,r}\;.
\end{array}
\qquad (4.8)
$$

Since the inequality must hold for abitrary values of these derivatives we obtain the following conditions on the constitutive functions

$$
\frac{\partial h}{\partial F} = \Lambda, \quad \frac{\partial h}{\partial F_i} = \Lambda^{i}, \quad \frac{\partial h}{\partial F_{ii}} = \Lambda^{ii}
$$

h − independent of $F_{,r}, F_{i,r}, F_{ii,r}$

$$
\frac{\partial \varphi_{(k}}{\partial F_{,r)}} = \Lambda^{j}\frac{\partial F_{j(k}}{\partial F_{,r)}} + \Lambda^{ii}\frac{\partial F_{jj(k}}{\partial F_{,r)}}
$$

$$
\frac{\partial \varphi_{(k}}{\partial F_{i,r)}} = \Lambda^{j}\frac{\partial F_{j(k}}{\partial F_{i,r)}} + \Lambda^{ii}\frac{\partial F_{jj(k}}{\partial F_{i,r)}}
$$

$$
\frac{\partial \varphi_{(k}}{\partial F_{ll,r)}} = \Lambda^{j}\frac{\partial F_{j(k}}{\partial F_{ll,r)}} + \Lambda^{ii}\frac{\partial F_{jj(k}}{\partial F_{ll,r)}}
$$

$$
\frac{\partial h}{\partial F} = \Lambda, \quad \frac{\partial h}{\partial F_i} = \Lambda^{i}, \quad \frac{\partial h}{\partial F_{ij}} = \Lambda^{ij}, \quad \frac{\partial h}{\partial F_{ijj}} = \Lambda^{ijj}
$$

$$
\frac{\partial \varphi_k}{\partial F} = \qquad + \Lambda^{ij}\frac{\partial F_{ijk}}{\partial F} + \Lambda^{ijj}\frac{\partial F_{ijjk}}{\partial F}
$$

$$
\frac{\partial \varphi_k}{\partial F_r} = \Lambda\delta_{kr} \qquad + \Lambda^{ij}\frac{\partial F_{ijk}}{\partial F_r} + \Lambda^{ijj}\frac{\partial F_{ijjk}}{\partial F_r}
$$

$$
\frac{\partial \varphi_k}{\partial F_{rs}} = \Lambda^{i}\delta_{i(r}\delta_{s)k} \qquad + \Lambda^{ij}\frac{\partial F_{ijk}}{\partial F_{rs}} + \Lambda^{ijj}\frac{\partial F_{ijjk}}{\partial F_{rs}}
$$

$$
\frac{\partial \varphi_k}{\partial F_{rss}} = \qquad + \Lambda^{ij}\frac{\partial F_{ijk}}{\partial F_{rss}} + \Lambda^{ijj}\frac{\partial F_{ijjk}}{\partial F_{rss}} \qquad (4.9)
$$

There remains an inequality in each of these cases whose left-hand side represents the entropy production

$$\left.\begin{array}{l}
\left(\dfrac{\partial\varphi_k}{\partial F} \qquad -\Lambda^j\dfrac{\partial F_{jk}}{\partial F} -\Lambda^{jj}\dfrac{\partial F_{iik}}{\partial F}\right)\dfrac{\partial F}{\partial x_k} + \\[2ex]
+\left(\dfrac{\partial\varphi_k}{\partial F_i} -\Lambda\delta_{ki}-\Lambda^j\dfrac{\partial F_{jk}}{\partial F_i} -\Lambda^{jj}\dfrac{\partial F_{iik}}{\partial F_i}\right)\dfrac{\partial F_i}{\partial x_k} + \\[2ex]
+\left(\dfrac{\partial\varphi_k}{\partial F_{ii}} \qquad -\Lambda^i\dfrac{\partial F_{jk}}{\partial F_{ii}} -\Lambda^{jj}\dfrac{\partial F_{iik}}{\partial F_{ii}}\right)\dfrac{\partial F_{ii}}{\partial x_k} \geq 0
\end{array}\right\|\quad \Lambda^{ij}\, S_{\langle ij\rangle} + \Lambda^{ijj}\, S_{inn} \geq 0\ . \tag{4.10}$$

This exhausts the consequences of the entropy principle. But clearly we shall have to be more specific. This will be possible after the formulation of the principle of relativity.

It the present non-relativistic context, the principle of relativity may be stated in two parts:

i.) The balance equations (4.2) and (4.5) are invariant under Galilei transformations

$$x_i^* = x_i + V_i\ t\ . \tag{4.11}$$

ii.) The constitutive functions are invariant under Euclidean transformations

$$x_i^* = O_{ij}(t)\ x_j + b_i(t)\ , \tag{4.12}$$

where O_{ij} is an orthogonal matrix.

Ruggeri (Ruggeri 1989) has shown that the postulated Galilei invariance leads to the following decompositions of the densities, fluxes, and productions in (4.2) and (4.5), which make the velocity dependencies explicit as follows:

$$\begin{aligned}
F &= m \\
F_i &= \qquad + m\, v_i \\
F_{ij} &= m_{ij} \qquad\qquad + m\, v_i v_j \\
F_{ijk} &= m_{ijk} + 3\, m_{(ij}v_{k)} \qquad\qquad + m\, v_i v_j v_k \\
F_{ijjk} &= m_{ijjk} + 4\, m_{(ijk}v_{j)} + 6\, m_{(ij}v_j v_{k)} \qquad + m\, v_i v_j v_j v_k \\
S_{\langle ij\rangle} &= s_{\langle ij\rangle}, \qquad S_{irr} = s_{irr} + 2\, s_{\langle ij\rangle}\, v_j\ , \\
\varphi_i &= \Phi_i + h\, v_i\ .
\end{aligned} \tag{4.13}$$

So as not to disrupt the simple formal structure of the theory, a synthetic notation is maintained. Later, we shall drop this notation, and introduce a more conventional one, whereby

m		ϱ, the mass density
$m_{\langle ij\rangle}$		$-\ t_{\langle ij\rangle}$, with $t_{\langle ij\rangle}$ as stress deviator
m_{ii}	is denoted by	$2\varrho u = 3p$ with u as specific internal energy, and p as pressure.
m_{ijj}		$2q_i$, with q_i as heat flux. (4.14)

The m's, s's, and h, Φ_k are non-convective tensors, i.e. they are independent of velocity. Therefore, the constitutive relations (4.4) and (4.6) may be written in a somewhat simpler form, viz.

$$\begin{array}{lll|lll}
m_{\langle ij\rangle} & = & \hat{m}_{\langle ij\rangle}(m, m_{ll}, m_{,r}, v_{k,r}, m_{ll,r}) & m_{\langle ijk\rangle} & = & \hat{m}_{\langle ijk\rangle}(m, m_{ij}, m_{ijj}) \\
m_{ijj} & = & \hat{m}_{ijj}\,(\text{———}) & m_{ijjk} & = & \hat{m}_{ijjk}\,(\text{———}) \\
 & & & s_{\langle ij\rangle} & = & \hat{s}_{\langle ij\rangle}\,(\text{———}) \\
 & & & s_{ijj} & = & \hat{s}_{ijj}\,(\text{———}) \\
h & = & \hat{h}\,(\text{———}) & h & = & \hat{h}\,(\text{———}) \\
\Phi_k & = & \hat{\Phi}_k\,(\text{———}) & \Phi_k & = & \hat{\Phi}_k\,(\text{———})\,.
\end{array} \qquad (4.15)$$

Once the decomposition (4.13) is known, it is an easy matter to write the balance equations (4.2) in a form appropriate to a non-inertial frame. In such a frame the equations read

$$\begin{array}{l|l}
\dfrac{\partial F}{\partial t} + \dfrac{\partial F_r}{\partial x_r} = 0 & \dfrac{\partial F}{\partial t} + \dfrac{\partial F_r}{\partial x_r} \\
\dfrac{\partial F_i}{\partial t} + \dfrac{\partial F_{ir}}{\partial x_r} = F\, i_i^o + 2\, F_k W_i & \dfrac{\partial F_i}{\partial t} + \dfrac{\partial F_{ir}}{\partial x_r} = F\, i_i^o + 2F_k W_{ik} \\
\dfrac{\partial F_{ii}}{\partial t} + \dfrac{\partial F_{iir}}{\partial x_r} = F_i\, i_i^o & \dfrac{\partial F_{ij}}{\partial t} + \dfrac{\partial F_{ijr}}{\partial x_r} - S_{\langle ij\rangle} = 2F_{(i} i_{j)}^o + 4F_{k(j} W_{i)k} \\
 & \dfrac{\partial F_{ijj}}{\partial t} + \dfrac{\partial F_{ijjr}}{\partial x_r} - S_{ijj} = 3F_{(ij} i_{j)}^o + 2F_{kjj} W_{ik}.
\end{array} \qquad (4.16)$$

Here W_{ij} is the matrix of angular velocity of the non-inertial frame with respect to an inertial one and i_j^o is the part of the inertial acceleration that is independent of velocity, viz.

$$i_j^o = -\,2W_{jk}\,\dot{b}_k - W_{jk}^2(x_k - b_k) + \dot{W}_{jk}(x_k - b_k) + \ddot{b}_j\,. \qquad (4.17)$$

All constitutive quantities (4.15) are now objective tensors. Therefore, the requirement of Euclidean invariance of the constitutive functions implies that non-objective tensors may not occur as variables, and that the functions must be isotropic. In particular, this means that in $(4.15)_1$, $v_{k,r}$ may not occur with all of its components, but only with its symmetric part which we shall denote by $d_{kr} = v_{(k,r)}$. There are representation theorems for isotropic functions which we shall not use in their whole generality, since we intend to restrict our attention to processes close to equilibrium.

Equilibria are processes in which the entropy production vanishes. For the variables, this means that we must have*)

$$\left.\frac{\partial F}{\partial x_j}\right|_E = 0, \quad \left.\frac{\partial F_i}{\partial x_j}\right|_E = 0, \quad \left.\frac{\partial F_{ii}}{\partial x_j}\right|_E = 0 \quad \Bigg| \quad m_{\langle ij\rangle}\big|_E = 0, \quad m_{iij}\big|_E = 0\,. \qquad (4.18)$$

---

*) As far as Rational Thermodynamics is concerned this characterisation of equilibrium is slightly overstated, because we may have density gradients and rigid rotations in equilibrium. I stick with it for brevity of the argument and because in the end it will not make a difference.

In processes close to equilibrium, we shall therefore neglect higher order terms in the quantities (4.18). Specifically, we write linear representations for the isotropic functions $\hat{\mathfrak{m}}$, $\hat{\mathfrak{s}}$, and quadratic representations for $\hat{h}$, $\hat{\Phi}$. [Note that the entropy production contains only quadratic terms. For consistency, we must therefore consider quadratic terms in $\hat{h}$ and $\hat{\Phi}_k$]. The linear and quadratic representations read

$$
\begin{array}{ll|ll}
m_{\langle ij\rangle} & = -2\mu\, d_{ij} & m_{\langle ijk\rangle} & = 0 \\
m_{ijj} & = a\, m_{ii} + b\, m_{ll,i} & m_{ijjk} & = \beta\, \delta_{ij} + \gamma\, m_{\langle ik\rangle} \\
 & & s_{\langle ij\rangle} & = \sigma\, m_{\langle ij\rangle} \\
 & & s_{ijj} & = \tau\, m_{ijj} \\
h & = h_0(m, m_{ll}) \quad {}^{*)} & h & = h_0 + h_1 m_{\langle ij\rangle} m_{\langle ij\rangle} + h_2 m_{ijj} m_{ikk} \\
\Phi_k & = \Phi_k(m, m_{ll}, m_{,r}, d_{kr}, m_{ll,r}) \quad {}^{**)} & \Phi_k & = \varphi_1 m_{kll} + \varphi_2 m_{\langle ki\rangle} m_{ill}.
\end{array}
\qquad (4.19)
$$

where all coefficients may only depend on m and m_{ll}. The relations (4.19) will now be subjected to the conditions (4.9) and (4.10) from the entropy principle. First of all we shall identify the Lagrange multipliers. Rather obviously we have from $(4.9)_1$

$$
d\,h_0 = \Lambda^{ii}\, dm_{ii} + \Lambda_I\, dm \qquad \Big| \qquad d\,h_0 = \Lambda_I^{ii}\, dm_{ii} + \Lambda_I\, dm \qquad (4.20)
$$

where the index I denotes the velocity independent parts of the Lagrange multipliers. Comparison with the Gibbs equation of equilibrium thermodynamics, viz.

$$
d\,h_0 = \frac{1}{T}\, d(\varrho u) - \frac{g}{T}\, d\varrho \qquad (4.21)
$$

leads us to identify Λ^{ii}, or Λ_I^{ii}, and Λ_I as follows

$$
\begin{array}{ll}
\Lambda_I^{ii} = {}^{1}/_{2T} & T \text{ - absolute temperature,} \\
\Lambda_I = -{}^{g}/_{T} & g \text{ - specific free enthalpy.}
\end{array}
\qquad (4.22)
$$

Once the absolute temperature has been identified, it is convenient and customary to introduce it as a variable, instead of m_{ll}. This will subsequently be done.

In the case of Rational Thermodynamics the consequences of (4.9) are derived in a few lines, but in Extended Thermodynamics the derivation requires several pages. For the calculations the reader is referred to Liu & Müller 1983 or to Müller 1985. Here we only write the results. For the entropy density and the entropy flux we obtain

$$
\begin{array}{ll|ll}
h & = h_0(\varrho, T) & h & = h_0(\varrho, T) - \dfrac{1}{4pT} m_{\langle ij\rangle} m_{\langle ij\rangle} - \dfrac{2}{5pT} \dfrac{1}{\gamma - 5\frac{p}{\varrho}} q_i q_i \\
\Phi_k & = \dfrac{q_k}{T} & \Phi_k & = \dfrac{q_k}{T} + \dfrac{2}{5pT} t_{\langle kl\rangle} q_l \,.
\end{array}
$$

---

*) By $(4.9)_1$, h is independent of $m_{,r}$, $v_{i,r}$, $m_{ll,r}$. Note that this is exactly true in Rational Thermodynamics and is not the result of the approximation.

**) There is no need to decompose Φ_k, because it will presently be seen to be universally related to m_{llk} in Rational Thermodynamics.

The entropy production (4.10) assumes the form

$$t_{\langle ik\rangle} \frac{\partial v_{\langle i}}{\partial x_{k\rangle}} + q_k \frac{\partial \frac{1}{T}}{\partial x_k} \geq 0 \quad \Bigg[\,\Bigg] \quad - \frac{\sigma}{4pT} t_{\langle ij\rangle} t_{\langle ij\rangle} - \frac{2}{5pT} \frac{\tau}{\gamma - 5\frac{p}{\varrho}} q_i q_i \geq 0 \,. \qquad (4.24)$$

The coefficients a, b, and ß, γ in (4.19) turn out to be related by the equations

$$a = - b\left(\frac{\partial \varrho u}{\partial \varrho}\right)_T \quad \Bigg[\,\Bigg] \quad \beta = \frac{5}{7} p\gamma + C \qquad (4.25)$$

and we have the inequalities

$$b\left(\frac{\partial u}{\partial T}\right)_\varrho \leq 0\,, \quad \mu \geq 0 \quad \Bigg[\,\Bigg] \quad \sigma \leq 0\,, \quad \frac{\tau}{\gamma - 5\frac{p}{\varrho}} \leq 0\,. \qquad (4.26)$$

C in (4.25) is a constant of integration.

The entropy ought to have a maximum in equilibrium. Therefore inspection of (4.23) implies

$$\gamma - 5\frac{p}{\varrho} > 0 \qquad \text{hence by (4.26)} \qquad \sigma \leq 0\,, \quad \tau \leq 0\,. \qquad (4.27)$$

The Gibbs equation (4.21) may be written in the alternative form

$$d\left(\frac{h_0}{\varrho}\right) = \frac{1}{T}\left(du - \frac{p}{\varrho^2}\, d\varrho\right)\,.$$

It implies the intergrability condition

$$\left(\frac{\partial u}{\partial \varrho}\right)_T = \frac{1}{\varrho^2}\left(p - T\left(\frac{\partial p}{\partial T}\right)_\varrho\right)$$

or, since $u = \frac{3p}{2\varrho}$ (see (4.14))

$$\frac{3}{2}\frac{\partial \ln p}{\partial \ln \varrho} + \frac{\partial \ln p}{\partial \ln T} = \frac{5}{2}\,,$$

whence follows that the thermal equation of state must have the form

$$p(\varrho, T) = T^{5/2} F\left(\frac{\varrho}{T^{3/2}}\right) \qquad (4.28)$$

where F is an arbitrary function. This result holds in Rational and Extended Thermodynamics. In Extended Thermodynamics the exploitation of the conditions also identifies the coefficient γ in terms of the function F. We obtain

$$\gamma = 7\frac{T}{F}\left(\int \frac{FF'}{z}\, dz + c\right)\,. \qquad (4.29)$$

z stands for $\frac{\varrho}{T^{3/2}}$ and c is a constant of integration.

In summary we write the equations for the stress deviator and the heat flux, the latter with $\kappa = -\varrho\, b \left(\frac{\partial u}{\partial T}\right)_\varrho$

$$t_{\langle ij\rangle} = 2\mu\, d_{\langle ij\rangle} \qquad \Bigg\| \qquad t_{\langle ij\rangle} = \frac{2p}{\sigma}\left\{\frac{\partial v_{\langle i}}{\partial x_{j\rangle}} - \frac{1}{2p}\left[\dot{t}_{\langle ij\rangle} + 2t_{\langle n\langle i\rangle}\frac{\partial v_{j\rangle}}{\partial x_n} - 4t_{\langle n\langle i\rangle}W_{j\rangle n}\right] + \frac{2}{5p}\frac{\partial q_{\langle i}}{\partial x_{j\rangle}}\right\} + \frac{1}{\sigma}t_{\langle ij\rangle}\frac{\partial v_n}{\partial x_n}$$

$$q_i = -\kappa\frac{\partial T}{\partial x_i} \qquad \Bigg\| \qquad q_i = \frac{1}{\tau}\left\{\frac{1}{2}\frac{\partial \frac{5}{7}p\gamma}{\partial x_i}\right\} - \frac{5}{2}\frac{p}{\varrho}\frac{\partial p}{\partial x_i} + \left[\dot{q}_i + q_n\frac{\partial v_i}{\partial x_n} - 2q_nW_{in}\right] - \left(\frac{\gamma}{2} - \frac{5}{2}\frac{p}{\varrho}\right)\frac{\partial t_{\langle ij\rangle}}{\partial x_j}\Bigg\}$$

$$+ \frac{1}{\tau}\left(\frac{7}{5}q_i\frac{\partial v_k}{\partial x_k} + \frac{4}{5}q_k\frac{\partial v_{(k}}{\partial x_{i)}} - t_{\langle ij\rangle}\frac{\partial\frac{\gamma}{2}}{\partial x_j} - \frac{1}{\varrho}t_{\langle ik\rangle}\frac{\partial t_{\langle kn\rangle}}{\partial x_n} + \frac{1}{\varrho}t_{\langle ij\rangle}\frac{\partial p}{\partial x_j}\right\} . \tag{4.30}$$

In Rational Thermodynamics these results come about by combining (4.19) and (4.25). In Extended Thermodynamics they represent the balance equation (4.16) for the fluxes of momentum and energy; the decompositions (4.13) have been used and (4.19), (4.25) were introduced.

Remarkable about $(4.30)_2$ is the fact that these equations are so similar to the last two among the equations (3.2) which came from the kinetic theory. In fact the equations for $t_{\langle ij\rangle}$ are entirely identical. In particular, the equations $(4.30)_2$ contain the inertial terms in exactly the same way as in the kinetic theory. It will be remembered that the old version of Extended Thermodynamics could not predict the presence of those terms.

The only difference between $(4.30)_2$ and the equations (3.2) appears in the presence of the coefficient γ in the equation for q_i. By (4.29) that coefficient may be calculated from the function $F(z)$ which determines the thermal equation of state, see (4.28).

As was observed before, the present theory refers to mon-atomic ideal gases. For such gases the thermal equation of state has been determined centuries ago to read

$$p = \varrho\frac{k}{m}T \qquad \text{hence} \qquad F(z) = \frac{k}{m}z . \tag{4.31}$$

By (4.29) this gives

$$\gamma = 7\frac{k}{m}T + 7c\frac{m}{k}\frac{T^{5/2}}{\varrho} . \tag{4.32}$$

There is some uncertainty about the meaning of the constant c. <u>If we set it equal to zero</u> and introduce γ into $(4.30)_2$ we obtain

$$t_{\langle ij\rangle} = -\frac{2p}{\sigma}\left\{\frac{\partial v_{\langle i}}{\partial x_{j\rangle}} - \frac{1}{2p}\left[\dot{t}_{\langle ij\rangle} + 2t_{\langle n\langle i\rangle}\frac{\partial v_{j\rangle}}{\partial x_n} - 4t_{\langle n\langle i\rangle}W_{j\rangle n}\right] + \frac{2}{5p}\frac{\partial q_{\langle i}}{\partial x_{j\rangle}}\right\} + \frac{1}{\sigma}t_{\langle ij\rangle}\frac{\partial v_n}{\partial x_n}$$

$$q_i = \frac{1}{\tau}\frac{5}{2}\frac{k}{m}p\left\{\frac{\partial T}{\partial x_i} + \frac{2}{5}\frac{m}{k}\frac{1}{p}\left[\dot{q}_i + q_n\frac{\partial v_i}{\partial x_n} - 2q_nW_{in}\right] - \frac{2}{5}\frac{T}{p}\frac{\partial t_{\langle ij\rangle}}{\partial x_j}\right\} + \tag{4.33}$$

$$+\frac{1}{\tau}\left(\frac{7}{5}q_i\frac{\partial v_k}{\partial x_k}+\frac{4}{5}q_k\frac{\partial v_{(k}}{\partial x_{i)}}-\frac{7}{2}\frac{k}{m}t_{\langle ij\rangle}\frac{\partial T}{\partial x_j}-\frac{1}{\varrho}t_{\langle ik\rangle}\frac{\partial t_{\langle kn\rangle}}{\partial x_n}+\frac{1}{\varrho}t_{\langle ij\rangle}\frac{\partial p}{\partial x_j}\right). \tag{4.33}$$

These are exactly the equations (3.2) of the kinetic theory derived entirely from macroscopic arguments. Thus the new version of Extended Thermodynamics has indeed determined the coefficients that were left open in the old theory and which were denoted by a, b, c, and K, L there.

We conclude that Extended Thermodynamics catches the essentials of the much more sophisticated Boltzmann theory on the level appropriate to the Grad approximation. It is also clear what must be done, if we wish to describe a fluid other than a mon-atomic gas:

i.) it must no longer be assumed that the pressure is related to the in-ternal energy density by $p=(^2/_3)\varrho u$,

ii.) the tensors in (4.2) must no longer be assumed symmetric.

iii.) more balance equations may have to be added to (4.2)

Kremer (Kremer 1985, 1989) and Dreyer (Dreyer 1990) have investigated such cases.

But let us stay with mon-atomic gases for the moment. The thermal equation of state (4.31) is not the only one describing such gases. Indeed, at low temperature and high density degeneration occurs and equilibrium statistical mechanics dictates a different thermal equation of state

$$p(\varrho,T)=\frac{2}{3}\frac{k}{m}\frac{1}{A}T^{5/2}I_4^{\pm}(\alpha)\quad\text{with}\quad A\frac{\varrho}{T^{3/2}}=I_2^{\pm}(\alpha) \tag{4.34}$$

where

$$I_n^{\pm}(\alpha)=\int_0^\infty\frac{x^n dx}{e^{x^2+\alpha}\pm 1} \tag{4.35}$$

and where A is some constant involving m, k, and h, the atomic mass and the constants of Boltzmann and Planck.

The + sign refers to Fermi gases and the – sign refers to Bose gases. Liu and Müller (1983) have investigated the cases of full degeneracy and calculated the fastest propagation speeds to be

$$\vartheta_{Ph}^{Fermi}(\infty)=1.34\vartheta(0)\quad\text{and}\quad\vartheta_{Ph}^{Bose}(\infty)=2.52\vartheta(0) \tag{4.36}$$

All speeds are finite, of course in this theory.

Greco and Müller (Greco and Müller 1984) have applied the equations of Extended Thermodynamics to a Bose gas with Bose-Einstein condensation. In this way they recovered Landau's theory of superfluidity by assigning the constant c the lowest value that it may assume by virtue of the inequality $(4.27)_1$. This argument is somewhat speculative and it may well deserve more attention.

The most natural question to be asked of Extended Thermodynamics is this: How does it relate to Rational Thermodynamics? To answer that question Liu and Müller (1983) have proposed an iterative scheme which proceeds as follows: In the first step all quantities on the right hand side of (4.33) are replaced by their equilibrium values. In particular, $t_{\langle ij\rangle}$ and q_i

are set equal to zero on the right hand sides of (4.33) and these equations thus determine first iterates $t_{<ij>}^{(1)}$ and $q_i^{(1)}$ for the stress deviator and the heat flux. Then first iterates are subsequently introduced into the right hand sides and second iterates are calculated. The first iterates read

$$t_{<ij>}^{(1)} = -\frac{2p}{\sigma} d_{<ij>} ,$$
$$q_i^{(1)} = \frac{5}{2\tau}\frac{k}{m}\frac{\partial T}{\partial x_i} . \tag{4.37}$$

These are the laws of Navier-Stokes and Fourier and comparison with $(4.30)_1$ allows us to determine the relation between σ,τ and μ,κ, the viscosity and heat conductivity. We have

$$\mu = -\frac{2p}{\sigma} \qquad \text{and} \qquad \kappa = -\frac{5}{2\tau}\frac{k}{m} . \tag{4.38}$$

The second iterates are more complex. They read

$$t_{<ij>}^{(2)} = 2\mu\frac{\partial v_{<i}}{\partial x_{j>}} - 2\frac{\mu}{p}\left[\frac{d}{dt}\left(\mu\frac{\partial v_{<i}}{\partial x_{j>}}\right) + \mu\frac{\partial v_{<k}}{\partial x_{i>}}\left(\frac{\partial v_{[j}}{\partial x_{k]}} - 2W_{jk}\right) + \mu\frac{\partial v_{<k}}{\partial x_{j>}}\left(\frac{\partial v_{[i}}{\partial x_{k]}} - 2W_{ik}\right)\right] - $$
$$- \frac{10}{3}\frac{\mu^2}{p}\frac{\partial v_l}{\partial x_l}\frac{\partial v_{<i}}{\partial x_{j>}} + \frac{4}{5}\frac{\mu}{p}\frac{\partial}{\partial x_{<j}}\left(\kappa\frac{\partial T}{\partial x_{i>}}\right)$$
$$\tag{4.39}$$
$$q_i^{(2)} = -\kappa\frac{\partial T}{\partial x_i} - \frac{1}{\tau}\left[\frac{d}{dt}\left(\kappa\frac{\partial T}{\partial x_i}\right) + \kappa\frac{\partial T}{\partial x_k}\left(\frac{\partial v_{[i}}{\partial x_{k]}} - 2W_{ik}\right)\right] - 2\frac{\kappa}{\tau}\frac{\partial v_l}{\partial x_l}\frac{\partial T}{\partial x_i} - \frac{1}{\tau}\frac{k}{m}T\frac{\partial}{\partial x_j}\left(\mu\frac{\partial v_{<i}}{\partial x_{j>}}\right) +$$
$$+ \frac{1}{\tau}\frac{\mu}{\varrho}\frac{\partial p}{\partial x_j}\frac{\partial v_{<i}}{\partial x_{j>}} .$$

This iterative scheme is purely formal, of course, but it does produce iterates with increasing powers of μ and κ which makes higher iterates smaller. The scheme is akin to the Maxwellian iteration of the kinetic theory in which higher iterates have increasingly high powers of the mean free path.

Inspection of (4.39) shows that the constitutive functions which relate $t_{<ij>}$ and q_i to the fields of ϱ, v_i, and T depend on frame in this second iteration – and all further ones, – because the matrix W_{ik} appears. This fact contradicts the principle of material frame indifference as stated in Truesdell and Noll (1965). It is easy though to understand that violation in the following manner: For simplicity we consider the special case of a gas at rest and with T = 0. In that case $(4.39)_2$ reduces to

$$q_i = -\kappa\left(\frac{\partial T}{\partial x_i} + \frac{2}{\tau}W_{ij}\frac{\partial T}{\partial x_j}\right) \tag{4.40}$$

and now the frame dependence is obvious: In an inertial frame, where W_{ij} is zero, the heat flux is parallel to $\partial T/\partial x_i$, whereas in a non-inertial frame the heat flux contains a term perpendicular to $\partial T/\partial x_i$. We consider a cylinder with a gas at rest and a radial temperature gradient, as in Figure 2 and focus attention on a small volume element of the linear dimensions of

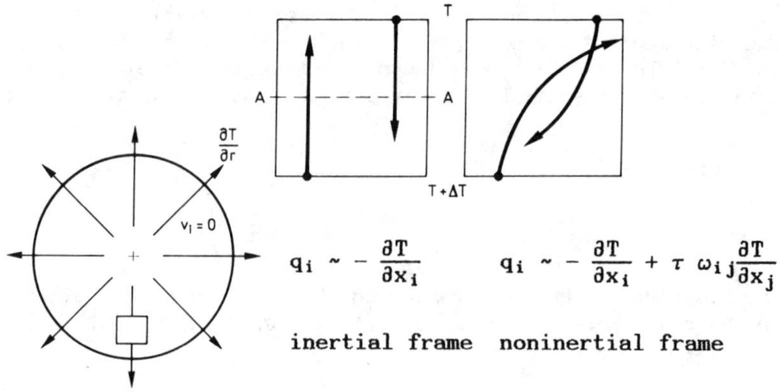

Figure 2: On the frame dependence of the heat flux.

the mean free path. In the middle of Figure 2 we see this element enlarged and two particles which - upon passage through the middle layer A-A - carry an energy flux upward, opposite to $\partial T/\partial x_i$ as it must be according to (4.40) in an inertial frame. Now, however, we take the cylinder to a non-inertial frame and consider the same situation. The paths of the particles are now curved by the Coriolis force as shown on the right hand side of Figure 2. The two particles carry energy not only upwards but also to the right, i.e. in the direction perpendicular to $\partial T/\partial x_i$ and to the angular velocity, and this is exactly what equation (4.40) implies.

This violation of the principle of material frame indifference was first discussed by Müller (1972) in the context of the kinetic theory of gases. Extended Thermodynamics helps us to see this violation in the proper perspective. Indeed, material frame dependence in ordinary thermodynamics reflects the explicit frame dependence of the basic equations of balance (4.16) or of the field equations (4.30), while the constitutive relations of extended thermodynamics are frame indifferent. Thus it is only in ordinary thermodynamics that material frame indifference becomes a problem.

There is one more interesting feature in Extended Thermodynamics and that concerns the form (4.23) of the entropy flux. In contrast to the usual assumption the entropy flux is not equal to q_i/T, rather it contains an additional explicit non-linear contribution of considerable consequence for the interpretation of temperature. Let us consider:

In thermodynamics we concentrate on temperature as a variable, because temperature is measurable and it is measurable, because it is continuous at a thermometric wall. Let us consider what becomes of this proposition in the light of (4.23): A thermometric wall is a wall that does not contribute to the processes on either side and, in particular, it does not produce entropy. Therefore the normal component of the entropy flux is continuous on such a wall, i.e.

$$\Phi_i^+ e_i = \Phi_i^- e_i$$

where e_i is a unit normal to the wall. If Φ_i were equal to q_i/T, this condition would immediately entail

$$T^+ = T^- ,$$

since $q_i e_i$ is continuous. Thus temperature is continuous. But this argument relies on the relation $\Phi_i = q_i/_T$. If instead we accept (4.23) we obtain for the ratio of T^+ and T^-

$$\frac{T^+}{T^-} = \frac{1 + \dfrac{2}{5}\dfrac{\overset{+}{t}_{\langle ij\rangle}\,\overset{+}{q}_j\, e_i}{p^+\,\dfrac{k}{m}\,q_i\, e_i}}{1 + \dfrac{2}{5}\dfrac{\overset{-}{t}_{\langle ij\rangle}\,\overset{-}{q}_j\, e_i}{p^-\,\dfrac{k}{m}\,q_i\, e_i}} \ .$$

Thus, if we are not in equilibrium - or close to it - temperature is not continuous at a wall and therefore not easily measurable. This fact detracts much from the usefulness of temperature as a variable and we shall have to be aware of that fact in non-equilibrium thermodynamics.

5. Relativistic Extended Thermodynamics.

Understandably relativists were much more concerned about infinite speeds of propagation than were classical physicists. Therefore Müller (1966) already formulated the equations of Extended Thermodynamics relativistically. This went largely unnoticed until Israel (1976) published the equations again. Relativists took note then and the subject is known to them simply as "Relativistic Thermodynamics" without the epithet "extended". I shall not describe those early versions of Müller and Israel, because physically they contain no more than Section 2 of this review, except that the notation is four-dimensional. Rather I shall describe the systematic theory of relativistic gases that extends the new version of Extended Thermodynamics described in the previous section into relativity. That version was first published by Liu, Müller, Ruggeri (1986) and here I present an abbreviated version concentrating on the results.

The principal objective of relativistic extended thermodynamics is the determination of the 14 fields

$$\begin{aligned} &A^A \ \text{- particle flux vector,} \\ &A^{AB} \ \text{- energy-momentum tensor,} \end{aligned} \tag{5.1}$$

in all events x^D of a body. The energy-momentum tensor is assumed symmetric so that it has 10 independent components. For the determination of the fields (5.1) we need field equations and these are formed by the conservation laws of particle number and energy-momentum, viz.

$$A^A{}_{,A} = 0 \tag*{(5.2)$_1$}$$

$$A^{AB}{}_{,B} = 0 \tag*{(5.2)$_2$}$$

and by the equations of balance of fluxes

$$A^{ABC}{}_{,C} = I^{AB} \ . \tag*{(5.2)$_3$}$$

A^{ABC} is the completely symmetric tensor of fluxes and I^{AB} is its production density. The set (5.2) consists of 15 equations for only 14 fields. Therefore we assume

$$I^A{}_A = 0 \qquad \text{and} \qquad A^A{}_A{}^C \approx A^C \tag{5.3}$$

so that the trace of $(5.2)_3$ reduces to $(5.2)_1$. We are then left with only 14 independent equations (5.2). The factor of proportionality in $(5.3)_2$ is chosen to be equal to m^2c^2, because that is the value it has in the kinetic theory of relativistic gases.

The set of equations (5.2) is not a set of field equations for the basic fields (5.1), because two new quantities appear in them: The flux tensor A^{ABC} and the flux production I^{AB}. In this situation we let ourselves be guided by the arguments of non-relativistic continuum mechanics and thermodynamics and assume that A^{ABC} and I^{AB} are constitutive quantities. Such quantities are related to the basic fields in a materially dependent manner by constitutive functions. In particular, when the constitutive relations are of the general form

$$\begin{aligned} A^{ABC} &= \hat{A}^{ABC}\ (A^M,\ A^{MN})\ , \\ I^{AB} &= \hat{I}^{AB}\ (A^M,\ A^{MN})\ , \end{aligned} \tag{5.4}$$

we say that they characterize a viscous heat-conducting gas.

If the constitutive functions $\hat{A}^{ABC}$ and $\hat{I}^{AB}$ were known, we could eliminate A^{ABC} and I^{AB} between (5.2) and (5.4) and obtain an explicit set of field equations for the 14 fields A^A and A^{AB}. Each solution of that set of field equations is called a <u>thermodynamic process.</u>

In reality of course the constitutive functions $\hat{A}^{ABC}$ and $\hat{I}^{AB}$ are not known and therefore the thermodynamicist tries to determine them or at least to restrict their generality. In these efforts the thermodynamicist is guided by universal principles to which the constitutive functions must conform. The most important such principles are

i.) the entropy principle,

ii.) the requirement of hyperbolicity,

iii.) the principle of relativity.

The entropy principle postulates the existence of the entropy flux vector h^A which is a constitutive quantity so that in a viscous, heat-conducting gas we have

$$h^A = \hat{h}^A\ (A^M,\ A^{MN})\ . \tag{5.5}$$

Moreover, the principle requires that h^A satisfy the entropy inequality

$$h^A{}_{,A} \geq 0\ . \tag{5.6}$$

for all thermodynamic processes.

Hyperbolicity is a condition on the character of the field equations. It requires that these equations are symmetric hyperbolic. This property guarantees that Cauchy's initial value problem is well-posed and that all wave speeds are finite.

The principle of relativity asumes that the field equations and the entropy inequality are invariant under space-time transformations

$$\overset{*}{x}{}^A = \overset{*}{x}{}^A\ (x^B)\ . \tag{5.7}$$

Since the balance equations (5.2) are tensor equations which are naturally invariant, the principle requires that the constitutive functions be invariant. If G and G^* stand for A^{ABC}, I^{AB}, and h^A in the two frames connected by the transformation (5.7), the principle of relativity can be stated as follows

$$G = \hat{G}\,(A^M, A^{MN}) \qquad \text{and} \qquad G^* = \hat{G}\,(\overset{*}{A}{}^M, \overset{*}{A}{}^{MN}) \;. \tag{5.8}$$

Note that $\hat{G}$ is the same function in both equations.

The exploitation of the above restrictive principles is the subject of the thermodynamic constitutive theory. The details of this theory are explained in Liu, Müller and Ruggeri (1986) and in a forthcoming book by Müller and Ruggeri. Here I shall only list the results and indicate certain special cases of interest.

It has proved convenient to introduce the 4-velocity U^A, with $U^A U_A = c^2$ and the projector $h_{AB} = (^1/_{c^2})U_A U_B - g_{AB}$. We use these to decompose the particle flux vector, the energy-momentum tensor and the entropy flux vector by the equations

$$\begin{aligned} A^A &= n\,U^A \,, \\ A^{AB} &= t^{\langle AB\rangle} + (p(n,e)+\pi)\,h^{AB} + \frac{1}{c^2}(U^A q^B + U^B q^A) + \frac{e}{c^2}U^A U^B \,, \\ h^A &= hU^A + \Phi^A \,. \end{aligned} \tag{5.9}$$

The advantage of this decompostion is that the different parts have suggestive meaning as follows

n	–	number density	
$t^{\langle AB\rangle}$	–	stress deviator	
$p + \pi$	–	pressure	
q^A	–	heat flux	(5.10)
e	–	energy density	
h	–	entropy density	
Φ^A	–	entropy flux	

The stress deviator $t^{\langle AB\rangle}$, the heat flux q^A and the dynamic pressure π vanish in equilibrium.

We present here the results of the constitutive theory whose derivation can be found in Liu, Müller and Ruggeri 1986.

To within second order terms in $t^{\langle AB\rangle}$, q^A, and π the flux tensor A^{ABC} and the flux production I^{AB} have the forms

$$\begin{aligned} A^{ABC} = {} & \left(\frac{1}{2c^2}\frac{\Gamma_1'}{T} + C_1^\pi\,\pi\right) U^A U^B U^C - \\ & - \left(\frac{1}{2}\frac{\Gamma_1}{T} + \frac{c^2}{6}\,C_1^\pi\,\pi\right)(g^{AB}U^C + g^{BC}U^A + g^{CA}U^B) + \\ & + C_3\left(g^{AB}q^C + g^{BC}q^A + g^{CA}q^B\right) - \frac{6}{c^2}C_3\left(U^A U^B q^C + U^B U^C q^A + U^C U^A q^B\right) + \\ & + C_5\left(t^{\langle AB\rangle}U^C + t^{\langle BC\rangle}U^A + t^{\langle CA\rangle}U^B\right) \,, \end{aligned} \tag{5.11}$$

$$I^{AB} = B_1^{\pi}\, \pi g^{AB} - \frac{4}{c^2} B_1^{\pi}\, \pi U^A U^B + B_3 t^{\langle AB\rangle} + B_4 (q^A U^B + q^B U^A)\ . \tag{5.12}$$

Also the entropy flux vector h^A to within third order terms in $t^{\langle AB\rangle}$, q^A and π assumes the form

$$\begin{aligned} h^A = & \left(h|_E + A_1^{\pi^2} \pi^2 + A_1^q q_E q^E + A_1^t t_{\langle EF\rangle} t^{\langle EF\rangle}\right) U^A + \\ & + \left(A_2^0 + A_2^{\pi}\, \pi\right) q^A + A_3^0 t^{\langle AE\rangle} q_E\ . \end{aligned} \tag{5.13}$$

$h|_E$ is the equilibrium entropy density which satisfies the Gibbs equation

$$d\left(\frac{h|_E}{n}\right) = \frac{1}{T}\left[d\left(\frac{e}{n}\right) + p\, d\left(\frac{1}{n}\right)\right]\ , \tag{5.14}$$

where T is the absolute temperature.

The coefficients C in the flux tensor (5.11) and A in the entropy flux vector (5.13) are related to the three functions p, Γ_1, and Γ_2 of the two variables

fugacity $\alpha = -\frac{1}{T}\frac{1}{n}(e - Th|_E + p)$ and absolute temperature T

in the following manner:

$$C_1^{\pi} = -\frac{3}{c^2}\frac{1}{T}\, \frac{\begin{bmatrix} -\ddot{p} & \dot{p} - \dot{p}' & \dot{\Gamma}_1 \\ \dot{p} - \dot{p}' & p' - p'' & \Gamma_1' - \Gamma_1 \\ \dot{\Gamma}_1 & \Gamma_1' - \Gamma_1 & \frac{5}{3}\Gamma_2 \end{bmatrix}}{\begin{bmatrix} -\ddot{p} & \dot{p} - \dot{p}' & \dot{\Gamma}_1 \\ \dot{p} - \dot{p}' & p' - p'' & \Gamma_1' - \Gamma_1 \\ -\dot{p} & -p' & \frac{5}{3}\Gamma_1 \end{bmatrix}}\ , \tag{5.15$_1$}$$

$$C_3 = -\frac{1}{2}\frac{1}{T}\, \frac{\begin{bmatrix} \dot{p} & -\dot{\Gamma}_1 \\ \Gamma_1 & \Gamma_2 \end{bmatrix}}{\begin{bmatrix} \dot{p} & -\dot{\Gamma}_1 \\ p' & \Gamma_1 - \Gamma_1' \end{bmatrix}}\ , \tag{5.15$_2$}$$

$$C_5 = -\frac{1}{2}\frac{1}{T}\frac{\Gamma_2}{\Gamma_1}\ , \tag{5.15$_3$}$$

$$A_1^{\pi^2} = -\frac{c^2}{6}\, C_1^{\pi}\, \frac{\begin{bmatrix} \ddot{p} & p - p' \\ \dot{p} - \dot{p}' & p' - p'' \end{bmatrix}}{\begin{bmatrix} -\ddot{p} & \dot{p} - \dot{p}' & \dot{\Gamma}_1 \\ \dot{p} - \dot{p}' & p' - p'' & \Gamma_1' - \Gamma_1 \\ -\dot{p} & -p' & \frac{5}{3}\Gamma_1 \end{bmatrix}}\ , \tag{5.16$_1$}$$

$$A_1^q = -\frac{1}{2c^2}\,\frac{-\frac{1}{T}\Gamma_1' - 10\,C_3\,\dot{p}}{\begin{bmatrix}\dot{p} & -\dot{\Gamma}_1\\ p' & \Gamma_1-\Gamma_1'\end{bmatrix}}\,, \tag{5.16$_2$}$$

$$A_1^t = -\frac{1}{T}\,C_5\,\frac{1}{\Gamma_1}\,, \tag{5.16$_3$}$$

$$A_2^o = \frac{1}{T} \tag{5.16$_4$}$$

$$A_2^\pi = -\frac{\frac{1}{T}\begin{bmatrix}-\ddot{p} & \dot{\Gamma}_1\\ \dot{p}-\dot{p}' & \Gamma_1'-\Gamma_1\end{bmatrix} - \frac{10}{3}\,C_3\begin{bmatrix}-\ddot{p} & \dot{p}-\dot{p}'\\ \dot{p}-\dot{p}' & p'-p''\end{bmatrix}}{\begin{bmatrix}-\ddot{p} & \dot{p}-\dot{p}' & \dot{\Gamma}_1\\ \dot{p}-\dot{p}' & p'-p'' & \Gamma_1'-\Gamma_1\\ \dot{p} & -p' & \frac{5}{3}\Gamma_1\end{bmatrix}} = -\frac{\frac{1}{T}\dot{\Gamma}_1 - \frac{c^2}{3}C_1^\pi\,\dot{p}}{\begin{bmatrix}\dot{p} & -\dot{\Gamma}_1\\ p' & \Gamma_1-\Gamma_1'\end{bmatrix}} \tag{5.16$_5$}$$

$$A_3^o = \frac{-\frac{1}{T}\dot{\Gamma}_1 - 2\,C_5\,\dot{p}}{\begin{bmatrix}\dot{p} & -\dot{\Gamma}_1\\ p' & \Gamma_1-\Gamma_1'\end{bmatrix}} = -\,C_3\,\frac{2}{\Gamma_1}\,. \tag{5.16$_6$}$$

Dots and primes denote derivatives with respect to α and $\ln T$ respectively.

Fugacity and absolute temperature are the natural variables of statistical mechanics of a gas as we shall see later. Also they are easier to measure than the original variable n and e. However, if we wish to switch between the sets of variables (α,T) and (n,e), this can be done by use of the formulae

$$n = -\frac{1}{T}\,\dot{p} \qquad\qquad \text{and} \qquad\qquad e = p' - p \tag{5.17}$$

provided that the thermal equation of state $p = p(\alpha,T)$ is known.

The results (5.15) and (5.16) seem to indicate that the three functions $p(\alpha,T)$, $\Gamma_1(\alpha,T)$ and $\Gamma_2(\alpha,T)$ determine the flux tensor A^{ABC} and the entropy flux vector h^A. However, the functions $\Gamma_1(\alpha,T)$, $\Gamma_2(\alpha,T)$ are not independent of $p(\alpha,T)$, because we have

$$-\frac{1}{2}\Gamma_1' + 3\Gamma_1 = m^2c^2\,\dot{p}\,, \qquad -\frac{1}{2}\Gamma_2' + 4\Gamma_2 = -\,m^2c^2\,\dot{\Gamma}_1 \tag{5.18}$$

so that Γ_1 and Γ_2 can be related to p by integration. We obtain from (5.18)

$$\begin{aligned}\Gamma_1 &= T^6\left[-\,2m^2c^2\int\frac{\dot{p}}{T^7}\,dT + A_1(\alpha)\right]\\ \Gamma_2 &= T^8\left\{2m^2c^2\int\frac{1}{T^3}\left[-\,2m^2c^2\int\frac{\ddot{p}}{T^7}\,dT + A_1(\alpha)\right]dT + A_2(\alpha)\right\}\end{aligned} \tag{5.19}$$

and we conclude that Γ_1 and Γ_2 may be determined from $p(\alpha,T)$; each to within a function of the single variable α, the fugacity.

It follows from (5.15), (5.16) that all coefficients of the flux tensor A^{ABC} and of the entropy flux vector h^A can be determined to within two functions of α, if only the thermal equation of state $p = p(\alpha,T)$ is known.

The requirement of hyperbolicity furnishes conditions on the coefficients of A^{ABC} and h^A which in effect restrict the thermal equation of state $p = p(\alpha,T)$ and the functions $A_1(\alpha)$, $A_2(\alpha)$. Most of these restrictions are not easy to interpret but we mention some which are:

$$\begin{bmatrix} \frac{1}{T}\ddot{p} & -\frac{1}{2T}(\dot{p}-\dot{p}') \\ -\frac{1}{2T}(\dot{p}-\dot{p}') & \frac{1}{4T}(p''-p') \end{bmatrix} - \text{positive definite } A_1^{\pi^2} < 0,\ A_1^q > 0,\ A_1^t < 0 . \tag{5.20}$$

Inspection of (5.13) shows that the inequalities $(5.20)_{2,3,4}$ are necessary for the entropy density $h=(^1/_{c^2})U_A h^A$ to have a maximum in equilibrium. The positive definiteness of the matrix $(5.20)_1$ ensures that the compressibility $^1/_n(^{\partial n}/\partial p)_T$ and the specific heat $(^{\partial e}/\partial T)_n$ are positive. These are two well-known "thermodynamic stability conditions", i.e. requirements on the constitutive functions which guarantee that the entropy of a body in homogeneous equilibrium is bigger than in a non-homogenous equilibrium (e.g. see Müller 1985, p. 22 ff.)

There is a residual inequality whose left hand side is a sum of squares of the dynamic pressure π, the heat flux and the stress deviator. Therefore there are three dissipative "mechanisms" in the gas, due to dynamic pressure, heat flux and stress deviator, because the entropy production is a quadratic form in these variables. Necessary and sufficient conditions for the entropy production to be non-negative are given by the inequalities

$$B_1^{\pi}\frac{\begin{bmatrix} -\ddot{p} & \dot{p}-\dot{p}' \\ \dot{p}-\dot{p}' & p'-p'' \end{bmatrix}}{\begin{bmatrix} -\ddot{p} & \dot{p}-\dot{p}' & \dot{\Gamma}_1 \\ \dot{p}-\dot{p}' & p'-p'' & \Gamma_1'-\Gamma_1 \\ -\dot{p} & -p' & \frac{5}{3}\Gamma_1 \end{bmatrix}} \geqslant 0, \quad B_4\frac{\dot{p}}{\begin{bmatrix} \dot{p} & -\dot{\Gamma}_1 \\ p' & \Gamma_1-\Gamma_1' \end{bmatrix}} \geqslant 0, \quad B_3\frac{1}{\Gamma_1} \leqslant 0 \tag{5.21}$$

which provide restrictions for the coefficients B in the flux production.

It is possible to identify the coefficients B_1^{π}, B_3, B_4 in the flux production by relating them to the bulk viscosity, the shear viscosity and the heat conductivity. These are the transport coefficients in ordinary thermodynamics which can in principle be measured. In order to find that relation we proceed as in non-relativistic Extended Thermodynamics with an iteration of which, in the present case, we consider only the first step. On the left hand side of the field equations

$$\begin{aligned} &(nU^A)_{,A} = 0 , \\ &\left(t^{\langle AB\rangle} + (p+\pi)\, h^{AB} + \frac{1}{c^2}(U^A q^B + U^B q^A) + \frac{e}{c^2}\, U^A U^B\right)_{,B} = 0 , \\ &A^{ABC}{}_{,C} = B_1^{\pi}\, \pi g^{AB} - \frac{4}{c^2}B_1^{\pi}\pi U^A U^B + B_3 t^{\langle AB\rangle} + B_4(q^A U^B + q^B U^A) , \end{aligned} \tag{5.22}$$

we set π, q^A and $t^{\langle AB\rangle}$ equal to zero. In particular this means setting A^{ABC} equal to

$$A^{ABC}\big|_E = \frac{1}{2c^2}\frac{\Gamma_1'}{T}\,U^AU^BU^C - \frac{1}{2}\frac{\Gamma_1}{T}(g^{AB}U^C + g^{BC}U^A + g^{CA}U^B)\ . \tag{5.23}$$

We then decompose the energy-momentum balance and the flux balance into temporal and spatial parts and obtain

$$\begin{aligned}
0 &= -\ddot{p}\,\frac{d\alpha}{d\tau} - (\dot{p}'-\dot{p})\,\frac{d\ln T}{d\tau} - \dot{p}\,U^A{}_{,A}\ ,\\
0 &= -\frac{\dot{\Gamma}_1}{2T}\frac{d\alpha}{d\tau} + (p''-p')\,\frac{d\ln T}{d\tau} + p'\,U^A{}_{,A}\ ,\\
0 &= -\frac{\Gamma_1}{2T}\,h^A_M\,\alpha_{,A} + p'\,h^A_M\,(\ln T)_{,A} - \frac{p'}{c^2}\frac{dU_M}{d\tau}\ ,\\
-3B_1^\pi\,\pi &= \frac{\dot{\Gamma}_1'-3\dot{\Gamma}_1}{2T}\frac{d\alpha}{d\tau} + \frac{\Gamma_1''-4\Gamma_1'+3\Gamma_1}{2T}\frac{d\ln T}{d\tau} + \frac{\Gamma_1'-\Gamma_1}{2T}\,U^A{}_{,A}\ ,\\
-B_4q_M &= -\frac{\dot{\Gamma}_1}{2T}\,h^A_M\,\alpha_{,A} - \frac{\Gamma_1'-\Gamma_1}{2T}\,h^A_M(\ln T)_{,A} + \frac{1}{c^2}\frac{\Gamma_1'-\Gamma_1}{2T}\frac{dU_M}{d\tau}\ ,\\
B_3t_{\langle MN\rangle} &= -\frac{\Gamma_1}{2T}\,h^B_M\,h^C_N\,U_{\langle B,C\rangle}\ .
\end{aligned} \tag{5.24}$$

The equations (5.17) have been used here to replace $\dot{n}$, e, n' and e' by derivatives of p. The symbol $^{dR}/_{d\tau}$ stands for the material time derivative $U^AR_{,A}$. Elimination of the derivatives $d\alpha/d\tau$, $d\ln T/d\tau$ between $(5.24)_{1,2,4}$ and of the derivatives $h^A{}_M\alpha_{,A}$ between $(5.24)_{3,5}$ allows us to determine π, q_M and $t_{\langle MN\rangle}$ from $(5.24)_{4-6}$ as follows

$$\pi = -\frac{1}{2T}\frac{1}{B_1^\pi}\,\frac{\begin{bmatrix}\ddot{p} & \dot{p}-\dot{p}' & \dot{\Gamma}_1\\ \dot{p}-\dot{p}' & p'-p'' & \Gamma_1'-\Gamma_1\\ -\dot{p} & -p' & \frac{5}{3}\Gamma_1\end{bmatrix}}{\begin{bmatrix}-\ddot{p} & \dot{p}-\dot{p}'\\ \dot{p}-\dot{p}' & p'-p''\end{bmatrix}}\left[U^A{}_{,A}\right],$$

$$q_M = -\frac{1}{2T}\frac{1}{B_4}\,\frac{\begin{bmatrix}\dot{p} & -\dot{\Gamma}_1\\ \dot{p}' & \Gamma_1-\Gamma_1'\end{bmatrix}}{\dot{p}}\left[h^A_M\left[(\ln T)_{,A} - \frac{1}{c^2}\frac{dU_A}{d\tau}\right]\right], \tag{5.25}$$

$$t_{\langle MN\rangle} = -\frac{1}{2T}\frac{1}{B_3}\,\Gamma_1\left[h^B_M\,h^C_N\,U_{\langle B,C\rangle}\right].$$

The equations $(5.25)_{2,3}$ are the relativistic analogues to the phenomenological equations of Fourier and Navier-Stokes. Equation $(5.25)_1$ gives a linear relation between the dynamic pressure and the divergence of velocity. In analogy to the nomenclature of ordinary irreversible thermodynamics we may call

$$\pi,\ q_M,\ t_{<MN>} \quad \text{and} \quad \left[U^A{}_{,A}\right],\ \left[h^A_M\, T_{,A} - \frac{T}{c^2}\frac{dU_M}{d\tau}\right],\ \left[h^B_M\, h^C_N\, U_{<B,C>}\right] \tag{5.26}$$

thermodynamic forces and fluxes respectively. The factors of proportionality between these forces and fluxes are called transport coefficients. There are three of them, known as the bulk viscosity λ, the heat conductivity κ and the shear viscosisity μ. We have

$$\lambda = \frac{1}{2T}\frac{1}{B_1^{\pi}} \frac{\begin{bmatrix} \ddot{p} & \dot{p}-\dot{p}' & \dot{\Gamma}_1 \\ \dot{p}-\dot{p}' & p'-p'' & \Gamma_1'-\Gamma_1 \\ -\dot{p} & -p' & \frac{5}{3}\Gamma_1 \end{bmatrix}}{\begin{bmatrix} -\ddot{p} & \dot{p}-\dot{p}' \\ \dot{p}-\dot{p}' & p'-p'' \end{bmatrix}} \geqslant 0\,,$$

$$\kappa = \frac{1}{2T^2}\frac{1}{B_4} \frac{\begin{bmatrix} \dot{p} & -\dot{\Gamma}_1 \\ \dot{p}' & \Gamma_1-\Gamma_1' \end{bmatrix}}{\dot{p}} \geqslant 0\,, \tag{5.27}$$

$$\mu = -\frac{1}{4T}\frac{1}{B_3}\Gamma_1 \geqslant 0\,.$$

By comparison with the inequalities (5.21) we conclude that the requirement of a non-negative entropy production determines the sign of the transport coefficients ν, μ and κ as indicated in (5.27).

Since in principle viscosity, heat conductivity and bulk viscosity can be measured, we may use the equations (5.27) to identify the coefficients B_1^{π}, B_3 and B_4.

Note that, by $(5.25)_2$ the "thermodynamic force" driving the heat flux is not just equal to the temperature gradient. Rather there is an additional acceleration term which was first derived by Eckart (1940). This term has the interesting consequence that a gas in equilibrium within a gravitational field has a non-uniform temperature. A discussion of this effect and related ones may be found in (Müller 1985, p. 427 ff). or elsewhere in the literature.

Liu, Müller, and Ruggeri proceed to derive more specific results by introducing special choices for the thermal equation of state $p = p(\alpha,T)$. They cover the whole spectrum of ideal gas properties from the non-degenerate, non-relativistic case to the fully degenerate, ultrarelativistic case.

6. The Mathematical Structure of Extended Thermodynamics.

The reader who has followed this review up to this point will have noticed that Extended Thermodynamics - as it has developed over 25 years - has become far more than a physically more or less well motivated theory with a finite speed of propagation of heat waves. It has become a general thermodynamic theory of quasi-linear hyperbolic field equations based on simple balance laws. It has merged into the mathematical theory of symmetric hyperbolic systems with a convex extension formulated by Friedrichs and Lax (1971). At the same time the theory has revealed more and more of its basically simple structure. Ruggeri and his co-workers (e.g. see Ruggeri 1983) have done much to "streamline" that structure. The present section offers a brief account of their view.

According to that view the objective of Extended Thermodynamics is the determination of N fields, generically denoted of $\underset{\sim}{u}(x^{\mathrm{D}})$, where $\underset{\sim}{u}$ is an N-vector. The necessary field equations are based on N equations of balance of the form

$$\underset{\sim}{F}^{A}{}_{,A} = \underset{\sim}{f} \; . \tag{6.1}$$

The comma denotes partial differentiation. The components of F^0 may be called densities and the components of F^a fluxes and we refer to F^A as four-dimensional fluxes. Their divergence in space-time is equal to the production vector $\underset{\sim}{f}$.

In order to give field equations for the fields $\underset{\sim}{u}$, the balance equations must be supplemented by constitutive equations. Such constitutive equations relate the fluxes $\underset{\sim}{F}^A$ and the productions $\underset{\sim}{f}$ to the fields $\underset{\sim}{u}$ in a materially dependent manner. In Extended Thermodynamics the constitutive equations are of the form

$$\underset{\sim}{F}^{A} = \hat{\underset{\sim}{F}}^{A}(\underset{\sim}{u}) \qquad \text{and} \qquad \underset{\sim}{f} = \hat{\underset{\sim}{f}}(\underset{\sim}{u}) \tag{6.2}$$

so that the fluxes $\underset{\sim}{F}^A$ and productions $\underset{\sim}{f}$ at one event depend only on the values of the fields $\underset{\sim}{u}$ at that event. We say that the constitutive relations are <u>local</u> in space time.

If the constitutive functions $\hat{\underset{\sim}{F}}^A$ and $\hat{\underset{\sim}{f}}$ were known explicitly, we could eliminate $\underset{\sim}{F}^A$ and $\underset{\sim}{f}$ between (6.1) and (6.2) and obtain an explicit set of field equations. Every solution of these field equations is called a <u>thermodynamic process.</u>

In reality of course the constitutive functions are not known and it is the task of the constitutive theory to determine these functions, or at least to reduce their generality. The tools of the constitutive theory are certain universal principles which we rely on from experience. The most important three among these principles are

i.) the entropy principle,

ii.) the requirement of convexity and causality,

iii.) the principle of relativity. *)

---

*) In non-relativistic theories this principle is sometimes known as the principle of material objectivity).

The most important part of the entropy principle is the entropy inequality

$$h^A{}_{,A} \leq 0 \tag{6.3}$$

which must hold for all thermodynamic processes. Here $-h^A$ is the entropy flux vector which is a constitutive quantity such that in Extended Thermodynamics we have

$$h^A = \hat{h}^A(\underset{\sim}{u}) \ . \tag{6.4}$$

The requirement of convexity and causality demands that the field equations be symmetric hyperbolic. This requirement ensures the well-posedness of Cauchy problems for the field equations and it ensures that all wave speeds are finite. If the entropy principle is accepted as stated above, the requirement of convexity and causality is equivalent to the statement that

$$-\ \hat{h}^0(\underset{\sim}{u}) \quad \text{is a concave function, i.e.} \quad \frac{\partial^2 \hat{h}^0}{\partial \underset{\sim}{u} \partial \underset{\sim}{u}} \ - \ \text{pos. def.} \tag{6.5}$$

The principle of relativity states that the field equations have the same form in all frames. In particular this implies that the constitutive functions must be invariant under transformations of frame. (In the non-relativistic case the statement of this principle is a little more complicated, because the balance equations are usually different in inertial and non-inertial frames).

For the exploitation of the entropy principle, we recall that the entropy inequality must hold for all thermodynamic processes, i.e. solutions of the field equations. In a manner of speaking the field equations provide constraints for the field which must satisfy the entropy inequality. It is possible to use Lagrange multipliers $\underset{\sim}{\Lambda}$ to eliminate such constraints. Indeed, the new inequality

$$h^A_{,A} - \underset{\sim}{\Lambda}(\underset{\sim}{F}^A_{,A} - \underset{\sim}{f}) \leq 0 \tag{6.6}$$

must hold for <u>all</u> fields $\underset{\sim}{u}$. The Lagrange multipliers are functions of $\underset{\sim}{u}$. We rewrite (6.6) in the form

$$(h^A - \underset{\sim}{\Lambda}\underset{\sim}{F}^A)_{,A} + \underset{\sim}{F}^A \underset{\sim}{\Lambda}_{,A} + \underset{\sim}{\Lambda}\ \underset{\sim}{f} \leq 0 \tag{6.7}$$

and introduce a new vector

$$h'^A = -\ h^A + \underset{\sim}{\Lambda} \cdot \underset{\sim}{F}^A \ , \tag{6.8}$$

which we call the vector potential for reasons that will soon become apparent. Thus (6.7) reads

$$-\ h'^A{}_{,A} + \underset{\sim}{F}^A\ \underset{\sim}{\Lambda}_{,A} + \underset{\sim}{\Lambda}\ \underset{\sim}{f} \leq 0 \ . \tag{6.9}$$

In the next step we introduce the Lagrange multipliers $\underset{\sim}{\Lambda}$ as new variables instead of $\underset{\sim}{u}$. The requirement of convexity and causality ensures that there is a one-to-one correspondence between the two sets, but we postpone proof of that statement. With the Λ's as variables $\underset{\sim}{F}^A$, $\underset{\sim}{f}$ and h^A, hence h'^A are all function of $\underset{\sim}{\Lambda}$:

$$\underset{\sim}{F}^A = \bar{\underset{\sim}{F}}^A(\underset{\sim}{\Lambda}) \ , \quad \underset{\sim}{f} = \bar{\underset{\sim}{f}}(\underset{\sim}{\Lambda}) \ , \quad h^A = \bar{h}^A(\underset{\sim}{\Lambda}) \ , \quad h'^A = \bar{h}'^A(\underset{\sim}{\Lambda}) \ . \tag{6.10}$$

Therefore - by use of the chain rule - we may write (6.9) in the form

$$\left(- \frac{\partial h'^A}{\partial \Lambda} + F^A\right) \Lambda_{,A} + \Lambda\, f \leq 0 . \tag{6.11}$$

This inequality must now hold for all fields Λ. Therefore it must hold in particular for all gradients $\Lambda_{,A}$ at an event. And since the left hand side of the inequality is explicitly linear in $\Lambda_{,A}$, the factor of that gradient must vanish lest the inequality be violated for some choice of $\Lambda_{,A}$. Thus we conclude

$$F^A = \frac{\partial h'^A}{\partial \Lambda} \quad \text{and with (6.8)} \quad h^A = - h'^A + \Lambda \frac{\partial h'^A}{\partial \Lambda} . \tag{6.12}$$

If follows that <u>the constitutive quantities F^A and h^A can be derived from a single vector,</u> viz. h'^A. This justifies the name "vector potential" for h'^A.

Of course there is a residual inequality, viz.

$$\Lambda\, f \leq 0 . \tag{6.13}$$

The relations (6.12) and (6.13) exhaust the consequences of the entropy principle and now we proceed to investigate the significance of these results for our primary constitutive quantities F^A and f. First F^A: The integrability condition for h'^A implies by (3.7) states that the four (N×N)-matrices

$$\frac{\partial F^A}{\partial \Lambda} \quad \text{be symmetric} . \tag{6.14}$$

These are obviously many severe restrictions on the functions $F^A(\Lambda)$. Next f: f is less severely restricted; all we know is that it must satisfy the inequality (6.13).

There is, however, a serious blemish on these results, because they hold for the Lagrange-multipliers as variables and the Lagrange multipliers - a priori - have no physical significance. Before we obtain "useful" results, i.e. results for F^A and f as functions of the chosen physically significant variables u, we must reverse the switch of variables from u to Λ. This is a most cumbersome task that cannot be approached within the present generic formulation. The reason is that the necessary arguments must rely on the tensorial character of the variables u and Λ in space-time which is different for different materials and - even for one material - it is different in a relativistic and non-relativistic description. There is presently much work on a general classification of the variables and field equations according to their tensorial character.

Let us return to the above statement that the fields u may always be replaced by the Lagrange multipliers Λ. Note that the inequality (6.6) may be written in the form

$$\left(\frac{\partial h^A}{\partial u} - \Lambda \frac{\partial F^A}{\partial u}\right) u_{,A} + \Lambda\, f \leq 0 \tag{6.15}$$

which must hold for all fields u. By an argument akin to the one that led from (6.11) to (6.12) we obtain

$$\frac{\partial h^A}{\partial \underset{\sim}{u}} = \underset{\sim}{\Lambda}\,\frac{\partial \underset{\sim}{F}^A}{\partial \underset{\sim}{u}}\ . \tag{6.16}$$

Now let us choose - without essential loss of generality - the fields $\underset{\sim}{u}$ to be equal to the densities $\underset{\sim}{F}^{\circ}$. Hence follows from (6.16) for $\alpha = 0$

$$\frac{\partial h^o}{\partial \underset{\sim}{u}} = \underset{\sim}{\Lambda} \tag{6.17}$$

and by differentiation with respect to $\underset{\sim}{u}$

$$\frac{\partial^2 h^o}{\partial \underset{\sim}{u}\,\partial \underset{\sim}{u}} = \frac{\partial \underset{\sim}{\Lambda}}{\partial \underset{\sim}{u}}\ . \tag{6.18}$$

Now by the requirement (6.5) of convexity and causality the left hand side of (6.18) is negative definite. Therefore the Jacobian of the transformation $\underset{\sim}{u} \rightleftarrows \underset{\sim}{\Lambda}$ is non-zero and the change of variables is possible.

Let us now show that the field equations (6.1) and (6.2) or $(6.10)_{1,2}$, form a symmetric hyperbolic system, if their solution satisfies the entropy inequality (6.3) with (6.4), or $(6.10)_3$, in which $-\hat{h}^{\circ}(\underset{\sim}{u})$ is a concave function.

Indeed, we combine (6.1) and (6.10) to give

$$\frac{\partial \underset{\sim}{F}^A}{\partial \underset{\sim}{\Lambda}}\,\underset{\sim}{\Lambda}_{,A} = \underset{\sim}{f} \qquad \text{or by } (6.12)_1 \qquad \frac{\partial^2 h'^A}{\partial \underset{\sim}{\Lambda}\,\partial \underset{\sim}{\Lambda}}\,\underset{\sim}{\Lambda}_{,A} = \underset{\sim}{f}\ . \tag{6.19}$$

Therefore the system of equations is symmetric, since the Hessian matrices $\frac{\partial^2 h'^A}{\partial \underset{\sim}{\Lambda}\,\partial \underset{\sim}{\Lambda}}$ are, of course, symmetric. And the system is symmetric hyperbolic, if $\frac{\partial^2 h'^A}{\partial \underset{\sim}{\Lambda}\,\partial \underset{\sim}{\Lambda}}$ is positive definite so that h'^o is a convex function of the components of $\underset{\sim}{\Lambda}$.

This latter condition, however, is satisfied, because of (6.5). For proof we assume - again without essential loss of generality - that $\underset{\sim}{u} = \underset{\sim}{F}^{\circ}$ holds. By $(6.12)_1$, (6.17) and (6.5) we then have

$$\frac{\partial^2 h'^o}{\partial \underset{\sim}{\Lambda}\,\partial \underset{\sim}{\Lambda}}\,\delta\underset{\sim}{\Lambda}\,\delta\underset{\sim}{\Lambda} = \delta\underset{\sim}{u}\,\delta\underset{\sim}{\Lambda} = \frac{\partial^2 h^o}{\partial \underset{\sim}{u}\,\partial \underset{\sim}{u}}\,\delta\underset{\sim}{u}\,\delta\underset{\sim}{u}\ . \tag{6.20}$$

The reason why we focus attention on symmetric hyperbolic systems lies in the convenient and desirable properties of existence, uniqueness and well-posedness of their solutions. We note that the symmetric hyperbolic character has several ingredients which are all satisfied in Extended Thermodynamics, viz.

- field equations of balance type,
- local constitutive relations,
- existence of an entropy inequality,
- convexity of the entropy density.

This is quite satisfactory.

And yet, one might have more far-reaching ambitions. Indeed, there is a persistant suspicion among thermodynamisists that the entropy principle should not be postulated a priori, but rather that it should follow from a reasonable and well-motivated requirement for the solution of the field equations.

In thermostatics Carathéodory (1909) has realized such a program: For the solution of the adiabatic energy equation he postulated inaccessibility of some states adjacent to the intial state and, starting from there, he derived the concept of entropy and the entropy inequality.

In thermodynamics an analogous argument is not yet known. In my paper "Entropy in Non-Equilibrium - A Challenge to Mathematicians" (Müller 1977) I have tried to advertise the view that such an argument is needed. So far no mathematician has seriously taken up that challenge.

7. Outlook.

As I have explained in the previous section the simple mathematical structure of Extended Thermodynamics may finally enable the thermodynamisist to ask the right question of the mathematician.

Also in Section 4 I have explained that Extended Thermodynamics puts a strong and convincing qualification upon the principle of material frame indifference which had been accepted in many quarters as a basic law of nature.

Moreover the theory shows - according to another argument presented in Section 4 - that the temperature may loose much of its usefulness as a variable in non-equilibrium processes.

Complete agreement with the kinetic theory of gases was achieved, at least with the Grad 13-moment version of the kinetic theory.

And all this was set in motion by the desire to have finite speeds of propagation. That goal was achieved; speeds were calculated, e.g. see (2.18) or (4.36). But what do they mean? Are they indeed the speeds of heat waves and shear waves?

They are not!

Indeed, in a most recent work Weiss (1990) has used the kinetic theory of gases to derive a whole hierarchy of extended theories differing by the number of moments that are taken as variables. Each theory should be better than the previous one with less moments. Weiss goes up to 20,000 moments and he finds that the maximum speed of propagation becomes bigger and bigger the more moments are considered. There is little doubt now that eventually - with infinitely many moments - the speed of propagation will be infinite.

On that anticlimatic note I should like to finish this review.

References.

Carathéodory, C. 1909, Untersuchungen über die Grundlagen der Thermodynamik. Math. Ann. 67.

Cattaneo, C. 1948, Sulla Conduzione del Calore., Atti del Seminari Mat. Fis. Univ. di Modena 3.

Dreyer, W. 1990, Molekulare Erweiterte Thermodynamik realer Gase (Hyperbolizität und Entropie). Habilitation thesis TU Berlin.

Eckart, C. 1940, The Thermodynamics of Irreversible Processes III: Relativistic Theory of the Simple Fluid. Phys. Rev. 58.

Friedrichs, K.O. 1971, Lax, P.D., System of Conservation Equations with a Convex Extension. Proc. Nat. Acad. Sci. USA 68.

Grad, H. 1949, On the Kinetic Theory of Rarefied Gases., Comm. Pure Appl. Math 2.

Greco, A., Müller, I. 1984, Extended Thermodynamics and Superfluidity. Arch. Rational Mech. Anal. 85.

Israel, W. 1976, Non-Stationary Irreversible Thermodynamics: A Causal Relativistic Theory. Annals of Physics 100.

Kremer, G.M. 1985, Zur Erweiterten Thermodynamik idealer und dichter Gase. Dissertation TU Berlin.

Kremer, G.M. 1989, Extended Thermodynamics of Molecular Ideal Gases., Cont. Mech. and Thermod. 1.

Liu, I-Shih 1972, Method of Lagrange Multipliers for Exploitation of the Entropy Principle. Arch. Rational Mech. Anal. 46.

Liu, I-Shih, Müller, I. 1983, Extended Thermodynamics of Classical and Degenerate Gases., Arch. Rational Mech. Anal. 83.

Liu, I-Shih, Müller, I., Ruggeri, T. 1986, Relativistic Thermodynamics of Gases. Annals of Physics 169.

Müller, I. 1966, Zur Ausbreitungsgeschwindigkeit von Störungen in kontinuierlichen Medien. Dissertation TH Aachen.

Müller, I. 1967, Zum Paradox der Wärmeleitungstheorie. Z. f. Physik, 198.

Müller, I. 1972, On the Frame Dependence of Stress and Heat Flux. Arch. Rational Mech. Anal. 45.

Müller, I. 1977, Entropy in Non-Equilibrium - A Challenge to Mathematicians. Proc. Symp. Trends in Appl. of Math. to Mechanics, Kozubnik, Poland.

Müller, I. 1985, Thermodynamics. Pitman London.

Müller, I. 1987, Kinetic Theory and Extended Thermodynamics., Proc. of Symp. on Kin. Th. and Ext. Thermod. Pitagora, Bologna.

Ruggeri, T. 1983, Symmetric Hyperbolic System of Conservative Equations for a Viscous Heat-Conducting Fluid. Acta Mechanica 47.

Ruggeri, T. 1989, Relativistic Extended Thermodynamics: General Assumptions and Mathematical Procedure. Springer Lecture Notes in Mathematics 1385.

Ruggeri, T. 1989, Galilean Invariance and Entropy Principle for Systems of Balance Laws. The Structure of Extended Thermodynamics. Cont. Mech. Thermod. 1.

Truesdell, C. 1969, Rational Thermodynamics. McGraw Hill.

Truesdell, C., Noll, W. 1965, The Non-Linear Field Theories of Mechanics. Handbuch der Physik III/3 Springer Berlin.

Weiss, W. 1990, Zur Hierarchie der Erweiterten Thermodynamik. Dissertation TU Berlin.

On Extended Thermodynamics of Ideal and Real Gases

G. M. Kremer
Departamento de Física
Universidade Federal do Paraná
Caixa Postal 19091, 81504 Curitiba, Brazil

Abstract

Ideal and real monatomic gases, mixtures of ideal monatomic gases and molecular gases are investigated within the framework of an extended thermodynamic theory.

For ideal and real monatomic gases a 14 field theory is presented where the 13 fields of mass density, velocity, pressure tensor and heat flux are supplemented by an additional scalar field. The scalar field is introduced in order to investigate gases with non-vanishing volume viscosities.

The extended thermodynamic of a mixture of ν ideal monatomic gases is based on a 13ν field theory. From this theory it is shown how the Onsager reciprocity relations in the presence of axial fields came out without the use of statistical arguments.

For molecular gases the intrinsic energy (rotational or vibrational energy of the molecules) and the corresponding intrinsic heat flux are taking as basic fields, resulting an extended thermodynamic theory with 17 scalar fields. In this theory, the transition from extended to ordinary thermodynamics leads to the determination of the so-called Eucken formulae. The dynamic pressure and the volume viscosity are also discussed.

1. INTRODUCTION

In order to avoid the so-called paradox of heat conduction in the Navier-Stokes and Fourier fluid, which predicts an infinite speed of propagation for disturbances of temperature, an extended thermodynamic theory for fluids was formulated by (Müller 1966) within the framework of linear irreversible thermodynamics. For the case of monatomic ideal gases, the theory was reformulated by (Liu and Müller 1983) within the framework of rational thermodynamics with Lagrange multipliers as a 13 field theory whose objective is the determination of the 13 scalar fields of density, velocity, pressure tensor and heat flux.

The aim of this review is to investigate, from an extended thermodynamic theory with Lagrange multipliers, ideal and real monatomic gases,molecular gases and mixtures of ideal monatomic gases.

In Section 2 an extended thermodynamic theory for real gases with 14 scalar fields is developed. A further scalar field, which is a contracted fourth order tensor, is added to the 13 variables in order to bring to the theory of viscous heat-conducting fluids the effect of the volume viscosity. The final linearized field equations are expressed in terms of eight scalar functions of density and temperature which are identified with the coefficients of shear and volume viscosities, thermal conductivity, three relaxation times and two thermal-viscous coefficients. All of them are, at least in principle, measurable quantities.

Section 3 deals with an extended thermodynamic theory of monatomic ideal gases with 14 scalar fields. The results of this theory follow directly from those of Section 2 as a limit case. The linearized field equations for ideal gases depend only on three relaxation times, provided that the thermal equation of state is given. Moreover, explicit results are formulated for classical ideal gas and for degenerate Bose and Fermi gases.

An extended thermodynamic theory for molecular gases is developed in Section 4. The objective of this theory is the determination of the 17 scalar fields of density, velocity, pressure tensor, heat flux, intrinsic energy and intrinsic heat flux. The intrinsic energy represents the rotational or the vibrational energy of the molecules. In this theory it turns out that all but 6 out of 34 constitutive coefficients can be related to the thermal and caloric equations of state. The remaining coefficients are identified either to the shear viscosity and thermal conductivity or with the coefficient of self-diffusion or with the absorption and dispersion of sound waves. The dynamic pressure and the volume viscosity are discussed and it is shown that in non-equilibrium the temperature is discontinuous at thermometric walls.

In Section 5 an extended thermodynamic theory for mixtures of ν monatomic ideal gases is discussed. The transition from extended to ordinary thermodynamics leads to constitutive equations for diffusion flux, heat flux and pressure tensor corresponding to the laws of Fick, Fourier and Navier-Stokes,respectively. From these laws it is shown,without the use of statistical arguments, how the Onsager reciprocity relations in the presence of axial fields came out of the phenomenological extended thermodynamics of a rotating mixture of ideal gases.

Cartesian notation for tensors with the usual summation convention is used. Parentheses denote symmetrization of all indices while angular parentheses and square brackets indicate traceless symmetrization and antisymmetrization, respectively.

2. EXTENDED THERMODYNAMICS OF REAL MONATOMIC GASES

The extended thermodynamics of real (or dense) monatomic gases as a theory of 13 fields of density, velocity, pressure tensor and heat flux was first formulated by (Kremer and Beevers 1984) and later developed by (Kremer 1985, Liu 1985, Kremer 1987b and Kremer and Müller 1987).

Recently the extended thermodynamic theory of real monatomic gases was reformulated by (Liu and Kremer 1989) as a 14 field theory. The objective of this theory is the determination of the 14 scalar fields of

$$\begin{aligned}&\text{density} \quad \varrho,\\ &\text{velocity} \quad v_i,\\ &\text{temperature} \quad T,\\ &\text{pressure tensor} \quad p_{ij} \quad \text{and}\\ &\text{heat flux} \quad q_i.\end{aligned} \tag{2.1}$$

A further scalar field was introduced in order to analyze gases with non- vanishing volume viscosity. This theory was based on an extended thermodynamic theory of monatomic ideal gases with 14 fields proposed by (Kremer 1986). On the other hand, the later was motivated by the relativistic version of extended thermodynamics of ideal gases (Liu, Müller and Ruggeri 1986) which is a natural 14 field theory with non- vanishing volume viscosity.

In this Section I will develope the theory for real gases along the same line as in (Kremer 1987b and Kremer and Müller 1987). The results of this theory, which I found to be more appropriate, do not have three shortcomings that have appeared in (Liu and Kremer 1989), namely

i) the specific internal energy ε has a term proportional to the divergence of velocity. Since this term does not appear in ordinary thermodynamics it cannot be present in an extended theory either;

ii) it was not possible to prove that the volume viscosity is non- negative in general;

iii) the final field equations are functions of nine coefficients that depend on the density and on the temperature. This does not agree with an extended theory based on linear irreversible thermodynamics proposed by (Müller 1966 and Müller 1967) in which there are only eight of such coefficients.

2.1 Thermodynamic Processes

For real monatomic gases the necessary field equations for the determination of the 14 fields (2.1) are based on the following balance equations

$$\frac{\partial \varrho}{\partial t} + \frac{\partial \varrho v_i}{\partial x_i} = 0, \tag{2.2}$$

$$\frac{\partial \varrho v_i}{\partial t} + \frac{\partial F_{ij}}{\partial x_j} = 0, \tag{2.3}$$

$$\frac{\partial G_{ij}}{\partial t} + \frac{\partial F_{ijk}}{\partial x_k} = P_{\langle ij \rangle}, \tag{2.4}$$

$$\frac{\partial G_{jji}}{\partial t} + \frac{\partial F_{jjik}}{\partial x_k} = P_{jji}, \tag{2.5}$$

$$\frac{\partial G_{iijj}}{\partial t} + \frac{\partial F_{iijjk}}{\partial x_k} = P_{iijj}. \tag{2.6}$$

The above equations are written in an inertial frame free of external supplies.

Eq. (2.2) is the balance equation of the mass density while Eq. (2.3) the balance equation of the momentum density with F_{ij} denoting the flux of the momentum density. Eq. (2.4) represents the balance of the energy tensor G_{ij}. G_{ij} is called the energy tensor since its trace is twice the energy density. G_{ij} differs from F_{ij} because in a real gas the pressure p is not related to the energy density $\varrho\varepsilon$. F_{ijk} is the flux of the energy tensor and its trace F_{jji} is twice the energy flux density. Eqs. (2.5) and (2.6) are the balance equations for the quantities G_{jji} and G_{iijj}, respectively. In thermodynamics they do not have proper names and I will call G_{jji} as moment vector and G_{iijj} as moment scalar. Hence F_{jjik} is the flux of the moment vector while F_{iijjk} the flux of the moment scalar.

In Eqs. (2.4), (2.5) and (2.6) $P_{\langle ij \rangle}$, P_{jji} and P_{iijj} are production terms. Eqs. (2.2), (2.3) and the trace of Eq. (2.4) do not have production terms since mass density, momentum density and energy density are conservative quantities.

For ideal monatomic gases the **G**'s are equal to the **F**'s and the balance equations (2.2)-(2.6) reduce to the 14 moment equations of the kinetic theory of monatomic ideal gases. In this case the flux in one equation is always equal to the density in the next equation and, in particular, it holds the relationship $3p = 2\varrho\varepsilon$.

While for ideal gases all tensors **F** are symmetric, this is no longer a safe assumption for real gases.

Basing on the requirement that the balance equations (2.2)-(2.6) are invariant under Galilean transformation, one can decompose the tensors **F** and **G** into a convective and a non-convective part (**p** and **g**) as follows

$$G_{ij} = \varrho v_i v_j + g_{ij}, \tag{2.7}$$
$$G_{jji} = \varrho v_j v_j v_i + 3g_{(jj}v_{i)} + g_{jji}, \tag{2.8}$$
$$G_{iijj} = \varrho v_i v_i v_j v_j + 6g_{(ii}v_j v_{j)} + 4g_{(iij}v_{j)} + g_{iijj}, \tag{2.9}$$
$$F_{ij} = \varrho v_i v_j + p_{ij}, \tag{2.10}$$
$$F_{ijk} = G_{ij}v_k + 2v_{(i}p_{j)k} + p_{ijk}, \tag{2.11}$$
$$F_{jjik} = G_{jji}v_k + 3v_{(j}v_j p_{i)k} + 3v_{(j}p_{ji)k} + p_{jjik}, \tag{2.12}$$
$$F_{iijjk} = G_{iijj}v_k + 4v_{(i}v_i v_j p_{j)k} + 6v_{(i}v_i p_{jj)k} + 4v_{(i}p_{ijj)k} + p_{iijjk}. \tag{2.13}$$

Insertion of Eqs. (2.7)-(2.13) into the balance equations (2.2)-(2.6) leads to a new system of balance equations that reads

$$\dot{\varrho} + \varrho\frac{\partial v_i}{\partial x_i} = 0, \tag{2.14}$$
$$\varrho\dot{v}_i + \frac{\partial p_{ij}}{\partial x_j} = 0, \tag{2.15}$$
$$\dot{g}_{ij} + g_{ij}\frac{\partial v_k}{\partial x_k} + \frac{\partial p_{\langle ij\rangle k}}{\partial x_k} + \frac{2}{3}\frac{\partial q_k}{\partial x_k}\delta_{ij} + 2p_{k(i}\frac{\partial v_{j)}}{\partial x_k} = P_{\langle ij\rangle}, \tag{2.16}$$
$$\dot{g}_i + g_i\frac{\partial v_k}{\partial x_k} + \frac{1}{2}\frac{\partial p_{ijjk}}{\partial x_k} + p_{\langle ij\rangle k}\frac{\partial v_j}{\partial x_k} + \frac{5}{3}q_k\frac{\partial v_i}{\partial x_k} - \frac{3}{2}\frac{g_{(ij}}{\varrho}\frac{\partial p_{j)k}}{\partial x_k} = Q_i, \tag{2.17}$$
$$\dot{g}_{iijj} + g_{iijj}\frac{\partial v_k}{\partial x_k} + \frac{\partial p_{iijjk}}{\partial x_k} + 4p_{ijjk}\frac{\partial v_i}{\partial x_k} - 8\frac{g_i}{\varrho}\frac{\partial p_{ik}}{\partial x_k} = G. \tag{2.18}$$

In the above equations the dot denotes the material time derivative and the following notations were introduced.

$$q_i = \frac{1}{2}p_{jji}, \qquad g_i = \frac{1}{2}g_{jji}, \tag{2.19}$$
$$p_{\langle ij\rangle} = \frac{1}{2}(p_{ij} + p_{ji}) - p\delta_{ij}, \qquad g_{\langle ij\rangle} = \frac{1}{2}(g_{ij} + g_{ji}) - \frac{2}{3}\varrho\varepsilon\delta_{ij}, \tag{2.20}$$
$$Q_i = \frac{1}{2}P_{jji} - P_{\langle ij\rangle}v_j, \qquad G = P_{iijj} - 4(P_{jji} - P_{\langle ij\rangle}v_j)v_i. \tag{2.21}$$

q_i is the heat flux, $p_{\langle ij\rangle}$ the pressure deviator and $g_{\langle ij\rangle}$ the energy deviator.

The system of balance equations (2.14)-(2.18) is closed by considering

$$g_{ij},\ p_{\langle ij\rangle k},\ g_i,\ p_{ijjk},\ g_{iijj},\ p_{iijjk},\ P_{\langle ij\rangle},\ Q_i \ \text{ and } \ G \tag{2.22}$$

as constitutives quantities which, at a given point and time, are related to the values of the fields (2.1) at that point and time in a manner dependent on material. The pressure tensor is still symmetric, because we are dealing with a non-polar material. g_{ij} and p_{ijjk} may be considered as symmetric, since in the approximation to be used, possible non-symmetric terms could not appear. But p_{ijk} can be taken as symmetric only in the first two indices.

Equivalently this system is closed by considering

$$p_{ij},\ q_i,\ p_{\langle ij\rangle k},\ p_{ijjk},\ p_{iijjk},\ P_{\langle ij\rangle},\ Q_i \quad \text{and} \quad G \tag{2.23}$$

as constitutives quantities that depend upon the 14 fields

$$\varrho,\ v_i,\ g_{ij},\ g_i \quad \text{and} \quad g_{iijj}. \tag{2.24}$$

The last version is more appropriate for the development of the constitutive theory and in the end it will be seen how to express the quantities in (2.22) as functions of the fields (2.1).

If the constitutive functions for the quantities in (2.23) were known explicitly, we should now have an explicit system of 14 field equations which are first order differential equations. Every solution of this system is called a *thermodynamic process.*

2.2 Constitutive Theory

The purpose of the constitutive theory is to restrict the generality of the constitutive functions.

The tools of the constitutive theory are universal principles and, above all *the entropy principle* and *the principle of material frame indifference.* I shall require also that the constitutive functions are restricted by *the criterion of thermodynamic stability.*

These principles will be exploited later since first I wish to discuss the meaning of equilibrium in this theory.

2.2.1 *Definition of equilibrium.* Equilibrium is defined as a process in which all productions vanish, so that there are 9 conditions

$$P_{\langle ij\rangle}\,|_E = 0, \qquad Q_i\,|_E = 0 \qquad \text{and} \qquad G\,|_E = 0. \tag{2.25}$$

These conditions imply the vanishing of the energy deviator $g_{\langle ij\rangle}$ and of the moment vector g_i but there must be one more relation as these are only 8 in number. Therefore in equilibrium the remaining variables g_{iijj}, ϱ and ε must be related. I write the equilibrium condition as

$$g_{iijj}\,|_E = g(\varrho, \varepsilon), \tag{2.26}$$

and the non-equilibrium condition as

$$g_{iijj} = g(\varrho, \varepsilon) + \Delta. \tag{2.27}$$

which gives, in Δ, an additional variable along with $g_{\langle ij\rangle}$ and g_i, that vanishes in equilibrium.

Hence the variables in the set (2.24) are replaced by

$$\varrho,\ v_i,\ g_{ij},\ g_i \quad \text{and} \quad \Delta. \tag{2.28}$$

2.2.2 *Entropy principle.*The entropy principle states that the inequality

$$\dot{h} + h\frac{\partial v_k}{\partial x_k} + \frac{\partial \phi_k}{\partial x_k} \geq 0 \tag{2.29}$$

holds for all thermodynamic processes. The entropy density h and the entropy flux ϕ_k are constitutive quantities that depend on the variables given in (2.28).

2.2.3 *Criterion of thermodynamic stability.* The criterion of thermodynamic stability requires that the inequality

$$\delta^2 h\,|_E = \frac{\partial^2 h}{\partial x_A \partial x_B}\bigg|_E \delta x_A \delta x_B < 0, \tag{2.30}$$

holds for all stable thermodynamic processes. x_A stands for $\{\varrho,\ v_i,\ g_{ij},\ g_i,\ \Delta\}$.

2.2.4 *Principle of material frame indiference.* The principle of material frame indifference requires that the constitutive functions of h and ϕ_i and of (2.23) are invariant under Euclidean transformations, or, in other words, the constitutive functions have the same form in an inertial and in a non-inertial frame. This implies that the constitutive quantities cannot depend on the velocity v_i, and that the constitutive functions must be isotropic with respect to arbitrary orthogonal transformations.

There are representations theorem for isotropic functions of which I write down the leading terms.

$$p_{ij} = (p + p_1\Delta + \underline{p_2\Delta^2 + p_3\, g_{\langle rs\rangle}\, g_{\langle rs\rangle} + p_4\, g_r\, g_r})\delta_{ij} + \chi\, g_{\langle ij\rangle}, \tag{2.31}$$

$$q_i = (\alpha_1 + \underline{\alpha_2\Delta})\, g_i + \underline{\alpha_3\, g_{\langle ij\rangle}\, g_j}, \tag{2.32}$$

$$p_{ijjk} = 2(\beta_o + \beta_1\Delta)\delta_{ik} + 2\gamma\, g_{\langle ik\rangle}, \tag{2.33}$$

$$p_{\langle ij\rangle k} = \nu\, g_{\langle i}\delta_{j\rangle k}, \tag{2.34}$$

$$p_{iijjk} = \xi\, g_k, \tag{2.35}$$

$$P_{\langle ij\rangle} = -\frac{1}{\tau_p}\, g_{\langle ij\rangle}, \tag{2.36}$$

$$Q_i = -\frac{1}{\tau_q} g_i, \tag{2.37}$$

$$G = -\frac{1}{\tau_\Delta}\Delta, \tag{2.38}$$

$$h = h_o + h_1\Delta + h_2\Delta^2 + \underline{h_3\Delta^3} + (h_4 + \underline{h_5\Delta})\, g_{\langle rs\rangle}\, g_{\langle rs\rangle} \\ +(h_6 + \underline{h_7\Delta})\, g_r\, g_r + \underline{h_8\, g_{\langle rs\rangle}\, g_{\langle sk\rangle}\, g_{\langle kr\rangle}} + \underline{h_9\, g_{\langle rs\rangle}\, g_r\, g_s}, \tag{2.39}$$

$$\phi_i = (\varphi_1 + \varphi_2\Delta)\, g_i + \varphi_3\, g_{\langle ir\rangle}\, g_r. \tag{2.40}$$

All coefficients may depend on the density ϱ and on the specific internal energy ε.

Essentially the constitutive theory to be discussed here is appropriated for processes close to equilibrium, where $g_{\langle ij\rangle}$, g_i and Δ vanish. This is why I shall confine interest to those parts of Eqs. (2.31)-(2.38) that are linear in $g_{\langle ij\rangle}$, g_i and Δ and to h and ϕ_i up to second order terms in $g_{\langle ij\rangle}$, g_i and Δ.

Note, however, that the underlined terms in Eqs. (2.31), (2.32) and (2.39) are quadratic and of third order. These terms affect the coefficients of lower order terms when the entropy inequality (2.29) is exploited. It may not be clear to the reader that I need only 10 such higher order terms but that can be proved. The proper and systematic way of doing this is by writing the full quadratic representations for p_{ij} through G and full third order representation for ϕ_i and then proceed the evaluation of the entropy inequality (2.29) with Lagrange multipliers. For more details one is refer to (Liu and Müller 1983), (Kremer 1985, 1986, 1987b, 1989a).

For ideal gases the pressure tensor p_{ij} is identified with the energy tensor g_{ij}, the heat flux q_i with the moment vector g_i and p_{ijjk} with g_{ijjk}. Hence, in this case, it follows from Eqs. (2.27) and (2.31)-(2.33) that

$$\chi = \alpha_1 = 1, \quad p_1 = p_2 = p_3 = p_4 = \alpha_2 = \alpha_3 = 0, \quad \beta_0 = \frac{g}{6}, \quad \beta_1 = \frac{1}{6}. \tag{2.41}$$

2.3 Results of the Constitutive Theory

From the exploitation of the entropy inequality with Lagrange multipliers it follows the residual inequality

$$\Sigma = -2h_4\frac{1}{\tau_p}\, g_{\langle rs\rangle}\, g_{\langle rs\rangle} - 2h_6\frac{1}{\tau_q}\, g_r\, g_r - (h_1 + 2h_2\Delta)\frac{1}{\tau_\Delta}\Delta \geq 0. \tag{2.42}$$

But Σ has a minimum in equilibrium, namely zero. Therefore it cannot have a linear term and I conclude that

$$h_1 = 0. \tag{2.43}$$

Further, this exploitation furnishes the equation

$$d\left(\frac{h_o}{\varrho}\right) = \frac{\partial}{\partial \varepsilon}\left(\frac{h_o}{\varrho}\right)\left[d\varepsilon - \frac{p}{\varrho^2}d\varrho\right]. \tag{2.44}$$

Comparision of Eq. (2.44) with the Gibbs equation of thermostatics identifies $\frac{\partial}{\partial \varepsilon}\left(\frac{h_o}{\varrho}\right)$ as the reciprocal of the absolute temperature T, i.e.,

$$\frac{\partial}{\partial \varepsilon}\left(\frac{h_o}{\varrho}\right) = \frac{1}{T} \qquad \text{hence} \qquad d\left(\frac{h_o}{\varrho}\right) = \frac{1}{T}\left[d\varepsilon - \frac{p}{\varrho^2}d\varrho\right]. \tag{2.45}$$

Since T is easily measurable, it is a more convenient variable than ε and I shall use it. In particular, I shall write the thermal and the caloric equations of state in the form

$$p = p(\varrho, T) \qquad \text{and} \qquad \varepsilon = \varepsilon(\varrho, T). \tag{2.46}$$

The Gibbs equation (2.45) implies the integrability condition

$$\frac{\partial \varepsilon}{\partial \varrho} = \frac{1}{\varrho^2}\left[p - T\frac{\partial p}{\partial T}\right]. \tag{2.47}$$

Moreover, from the exploitation of the entropy inequality one can obtain relations between the constitutive coefficients which read

$$2h_2\left[\varrho\frac{\partial g}{\partial \varrho} + \frac{T}{\varrho c_v}\frac{\partial p}{\partial T}\frac{\partial g}{\partial T} - (8\beta_o + g)\right] = \frac{p_1}{T}, \tag{2.48}$$

$$h_4 = -\frac{\chi}{4pT}, \qquad h_6 = -\frac{\chi\nu}{4pT(\gamma - \frac{5}{3}\varepsilon\chi)}, \tag{2.49}$$

$$h_2\left(\xi - \frac{\alpha_1}{\varrho c_v}\frac{\partial g}{\partial T}\right) = h_6\left(\beta_1 - \frac{5}{3}\varepsilon p_1\right), \tag{2.50}$$

$$\varphi_1 = \frac{\alpha_1}{T}, \qquad \varphi_2 = \frac{\alpha_2}{T} + 2h_6\left(\beta_1 - \frac{5}{3}\varepsilon p_1\right), \tag{2.51}$$

$$\varphi_3 = \frac{\alpha_3}{T} + 2h_6\left(\gamma - \frac{5}{3}\varepsilon\chi\right), \tag{2.52}$$

$$\frac{\alpha_1}{T} = -2h_6\left(\frac{\partial \beta_o}{\partial T} - \frac{5}{3}\varepsilon\frac{\partial p}{\partial T}\right), \qquad \frac{\partial \beta_o}{\partial \varrho} = \frac{5}{3}\varepsilon\frac{\partial p}{\partial \varrho}, \tag{2.53}$$

$$\frac{\partial h_6}{\partial \varrho}\left(\beta_1 - \frac{5}{3}\varepsilon p_1\right) - \frac{5}{3}h_6\frac{\partial(\varepsilon p_1)}{\partial \varrho} = \frac{h_2}{\varrho}\left(\varrho\frac{\partial \xi}{\partial \varrho} - 8\frac{\partial p}{\partial \varrho} - \frac{1}{c_v}\frac{\partial g}{\partial T}\frac{\partial \alpha_1}{\partial \varrho}\right), \tag{2.54}$$

$$\frac{\partial h_4}{\partial \varrho}\nu = h_6\left(\frac{\partial \gamma}{\partial \varrho} - \frac{5}{3}\varepsilon\frac{\partial \chi}{\partial \varrho} - \frac{1}{\varrho}\frac{\partial p}{\partial \varrho}\right), \tag{2.55}$$

$$\varrho\frac{\partial h_2}{\partial \varrho} + \frac{T}{\varrho c_v}\frac{\partial p}{\partial T}\frac{\partial h_2}{\partial T} + h_2\left(16\beta_1 - \frac{2p_1}{\varrho c_v}\frac{\partial g}{\partial T} + 1\right) + \frac{p_2}{T}$$
$$= 3h_3\varrho\left(\frac{\partial g}{\partial \varrho} + \frac{T}{c_v\varrho^2}\frac{\partial p}{\partial T}\frac{\partial g}{\partial T} - 8\beta_o - g\right), \tag{2.56}$$

$$-\frac{\alpha_2}{2T^2}+\frac{\partial h_6}{\partial T}\Big(\beta_1-\frac{5}{3}\varepsilon p_1\Big)-\frac{5}{3}h_6\frac{\partial(\varepsilon p_1)}{\partial T}$$
$$=\frac{h_2}{\varrho}\Big(\varrho\frac{\partial\xi}{\partial T}-8\frac{\partial p}{\partial T}-\frac{1}{c_v}\frac{\partial g}{\partial T}\frac{\partial\alpha_1}{\partial T}\Big)+h_7\Big(\frac{\partial\beta_o}{\partial T}-\frac{5}{3}\varepsilon\frac{\partial p}{\partial T}\Big), \tag{2.57}$$
$$-\frac{\alpha_3}{2T^2}+\nu\frac{\partial h_4}{\partial T}=h_9\Big(\frac{\partial\beta_o}{\partial T}-\frac{5}{3}\varepsilon\frac{\partial p}{\partial T}\Big)+h_6\Big(\frac{\partial\gamma}{\partial T}-\frac{5}{3}\varepsilon\frac{\partial\chi}{\partial T}-\frac{1}{\varrho}\frac{\partial p}{\partial T}\Big), \tag{2.58}$$
$$4h_4\chi+6ph_8=0,\qquad 2ph_9+\frac{1}{3}h_6(\nu+10\alpha_1)=0, \tag{2.59}$$
$$h_2\Big(8\gamma-\frac{\chi}{\varrho c_v}\frac{\partial g}{\partial T}\Big)+2ph_5+2p_1h_4=0. \tag{2.60}$$

In the above equations $c_v=\frac{\partial\varepsilon}{\partial T}$ is the specific heat at constant volume.

Ideally the above relations should enable us to determine all the constitutive coefficients as functions of ϱ and T, if only the thermal and caloric equations of state (2.46) are known. In the case of ideal gases this program can be carried out, and this will be shown in Section 3. But in the present case this seems to be impossible, since we have very many coefficients and only few equations.

2.4 Shear and Volume Viscosities and Thermal Conductivity

2.4.1 *Linear field equations.* The combination of the balance equations (2.14)-(2.18) with the constitutive equations (2.31)-(2.38) leads to a set of field equations in which I will replace the variables $g_{\langle ij\rangle}$, g_i and Δ by the more suggestive quantities: pressure deviator $p_{\langle ij\rangle}=\chi\, g_{\langle ij\rangle}$, heat flux $q_i=\alpha_1\, g_i$ and dynamic pressure $\varpi=p_1\Delta$. After linearization, it follows

$$\frac{\partial\varrho}{\partial t}+\varrho\frac{\partial v_i}{\partial x_i}=0, \tag{2.61}$$
$$\varrho\frac{\partial v_i}{\partial t}+\frac{\partial p}{\partial\varrho}\frac{\partial\varrho}{\partial x_i}+\frac{\partial p}{\partial T}\frac{\partial T}{\partial x_i}+\frac{\partial\varpi}{\partial x_i}+\frac{\partial p_{\langle ij\rangle}}{\partial x_j}=0, \tag{2.62}$$
$$\varrho c_v\frac{\partial T}{\partial t}+\frac{\partial q_k}{\partial x_k}+T\frac{\partial p}{\partial T}\frac{\partial v_i}{\partial x_i}=0, \tag{2.63}$$
$$\tau_p\frac{\partial p_{\langle ij\rangle}}{\partial t}+\frac{\tau_p\chi\nu}{\alpha_1}\frac{\partial q_{\langle i}}{\partial x_{j\rangle}}+2\tau_p p\chi\frac{\partial v_{\langle i}}{\partial x_{j\rangle}}=-p_{\langle ij\rangle}, \tag{2.64}$$
$$\tau_q\frac{\partial q_i}{\partial t}+\frac{\alpha_1\tau_q}{\chi}\Big(\gamma-\frac{5}{3}\varepsilon\chi\Big)\frac{\partial p_{\langle ij\rangle}}{\partial x_j}+\frac{\alpha_1\tau_q}{p_1}\Big(\beta_1-\frac{5}{3}\varepsilon\chi\Big)\frac{\partial\varpi}{\partial x_i}$$
$$+\frac{2\alpha_1^2p\tau_q}{\chi\nu T}\Big(\gamma-\frac{5}{3}\varepsilon\chi\Big)\frac{\partial T}{\partial x_i}=-q_i, \tag{2.65}$$
$$\tau_\Delta\frac{\partial\varpi}{\partial t}+p_1\tau_\Delta\Big(\frac{\xi}{\alpha_1}-\frac{1}{\varrho c_v}\frac{\partial g}{\partial T}\Big)\frac{\partial q_k}{\partial x_k}-\frac{p_1^2\tau_\Delta}{2h_2T}\frac{\partial v_k}{\partial x_k}=-\varpi. \tag{2.66}$$

In the derivation of Eqs. (2.65)-(2.66) I have made use of the restrictions (2.48)-(2.50) and (2.53). Also I have left out products of $p_{\langle ij\rangle}$, q_i, ϖ, $\frac{\partial v_i}{\partial x_j}$ and $\frac{\partial T}{\partial x_i}$ in the linearization. One may well consider to keep $\frac{\partial v_{[i}}{\partial x_{j]}}$ and $\frac{\partial v_k}{\partial x_k}$ in the equations, but for my purpose this makes no difference.

2.4.2 *Laws of Navier-Stokes and Fourier.* While in the present theory of extended thermodynamics $p_{\langle ij\rangle}$, ϖ and q_i are variables, they are constitutive functions in ordinary thermodynamics of viscous heat conducting fluids and they are given there by the linear laws of Navier-Stokes and Fourier. Here those laws are recovered from Eqs. (2.64), (2.65) and (2.66) by replacing $p_{\langle ij\rangle}$, ϖ and q_i on the left-hand side by their equilibrium values, viz. zero. Thus, it follows

$$2\tau_p p\chi \frac{\partial v_{\langle i}}{\partial x_{j\rangle}} = -p_{\langle ij\rangle}, \tag{2.67}$$

$$\frac{2\alpha_1^2 p\tau_q}{\chi\nu T}\left(\gamma - \frac{5}{3}\varepsilon\chi\right)\frac{\partial T}{\partial x_i} = -q_i, \tag{2.68}$$

$$-\frac{p_1^2\tau_\Delta}{2h_2 T}\frac{\partial v_k}{\partial x_k} = -\varpi, \tag{2.69}$$

i.e., linear relations between the pressure deviator and the gradient of velocity, between the heat flux and the gradient of temperature and between the dynamic pressure and the divergence of the velocity. Comparision of Eqs. (2.67)-(2.69) with the laws of Navier-Stokes and Fourier leads us to identify the shear viscosity μ, the thermal conductivity κ and the volume viscosity ς by the equations

$$\mu = \tau_p p\chi, \qquad \kappa = \frac{2\alpha_1^2 p\tau_q}{\chi\nu T}\left(\gamma - \frac{5}{3}\varepsilon\chi\right), \qquad \varsigma = -\frac{p_1^2\tau_\Delta}{2h_2 T}. \tag{2.70}$$

The argument leading to Eqs. (2.67)-(2.69) is the first step in an iterative procedure for the determination of $p_{\langle ij\rangle}$, ϖ and q_i that is akin to the well-known Maxwellian iteration in the kinetic theory of gases (Ikenberry and Truesdell 1956).

2.5 Results from the Inequalities

2.5.1 *Residual inequality.* The residual inequality (2.42) with $h_1 = 0$ can be written in terms of the variables $p_{\langle ij\rangle}$, q_i and ϖ as

$$\frac{1}{\mu}p_{\langle ij\rangle}p_{\langle ij\rangle} + \frac{2}{\kappa T}q_i q_i + \frac{2}{\varsigma}\varpi^2 \geq 0. \tag{2.71}$$

This implies the inequalities

$$\mu \geq 0, \qquad \kappa \geq 0, \qquad \varsigma \geq 0, \tag{2.72}$$

which show that the shear and volume viscosities and the thermal conductivity are all positive quantities.

2.5.2 *Thermodynamic stability criteria.* Insertion of the representation (2.39) of h into the inequality (2.30) it follows by the use of the Gibbs equation (2.45)

$$-\frac{1}{\varrho T}\frac{\partial p}{\partial \varrho}(\delta\varrho)^2 - \frac{\varrho}{T^2}c_v(\delta T)^2 + 2h_4\chi^2\delta p_{\langle ij\rangle}\delta p_{\langle ij\rangle} + 2h_6\alpha_1^2 q_i q_i + 2\frac{h_2}{p_1^2}(\delta\varpi)^2 < 0. \tag{2.73}$$

From this inequality one can conclude that

$$\frac{\partial p}{\partial \varrho} > 0, \qquad c_v > 0, \tag{2.74}$$

$$h_2 < 0, \qquad h_4 < 0, \qquad h_6 < 0. \tag{2.75}$$

The inequalities (2.74) are the well-known mechanical and thermal stability conditions that ensure the positiveness of the isothermal compressibility $\kappa_T = \frac{1}{\varrho}\frac{\partial \varrho}{\partial p}$ and of the specific heat at constant volume c_v.

In order to interpret the inequalities (2.75) let us combine Eqs. (2.49) and (2.70) with the inequalities (2.72) as follows

$$-\frac{\chi^2\tau_p}{4Th_4} = \mu \geq 0, \qquad -\frac{\alpha_1^2\tau_q}{2T^2h_6} = \kappa \geq 0, \qquad -\frac{p_1^2\tau_\Delta}{2T^2h_2} = \varsigma \geq 0. \tag{2.76}$$

Now from the inequalities (2.75) and (2.76) it is easy to conclude that all relaxation times of the pressure deviator, of the heat flux and of the dynamic pressure are all positive, i.e.

$$\tau_p > 0, \qquad \tau_q > 0, \qquad \text{and} \qquad \tau_\Delta > 0. \tag{2.77}$$

2.6 Final Field Equations

As we have seen, the field equations for ϱ, v_i, T, $p_{\langle ij\rangle}$, q_i and ϖ are given by Eqs. (2.61)-(2.66). In this section, I rewrite Eqs. (2.64)-(2.66) as

$$\tau_p\frac{\partial p_{\langle ij\rangle}}{\partial t} + 2\mu\sigma_1\frac{\partial q_{\langle i}}{\partial x_{j\rangle}} + 2\mu\frac{\partial v_{\langle i}}{\partial x_{j\rangle}} = -p_{\langle ij\rangle}, \tag{2.78}$$

$$\tau_q\frac{\partial q_i}{\partial t} + T\kappa\sigma_1\frac{\partial p_{\langle ij\rangle}}{\partial x_j} + T\kappa\sigma_2\frac{\partial \varpi}{\partial x_i} + \kappa\frac{\partial T}{\partial x_i} = -q_i, \tag{2.79}$$

$$\tau_\Delta\frac{\partial \varpi}{\partial t} + \varsigma\sigma_2\frac{\partial q_i}{\partial x_i} + \varsigma\frac{\partial v_i}{\partial x_i} = -\varpi, \tag{2.80}$$

where σ_1 and σ_2 are thermal-viscous coupling coefficients defined by

$$\sigma_1 = \frac{\nu}{2p\alpha_1}, \qquad \sigma_2 = \frac{\sigma_1 \chi}{p_1} \frac{(\beta_1 - \frac{5}{3}\varepsilon p_1)}{(\gamma - \frac{5}{3}\varepsilon\chi)}. \tag{2.81}$$

Hence the fields ϱ, v_i, T, $p_{\langle ij\rangle}$, q_i and ϖ can be determined from Eqs. (2.61)-(2.63) and (2.78)-(2.80) providing the thermal and caloric equations of state (2.46) and eight coefficients are known. Among these eight coefficients there are three transport coefficients, e.g. μ, ς and κ, three relaxation times, e.g. τ_p, τ_q and τ_Δ, and two thermal-viscous coupling coefficients, e.g. σ_1 and σ_2. All of them are, at least in principle, measurable.

The field equations obtained here agree with those derived by (Müller 1966 and Müller 1967) from an extended theory based on linear irreversible thermodynamics.

2.7 Entropy Density and Entropy Flux

Note that I have so far listed the form of the entropy density and of the entropy flux among the results. The entropy density and the entropy flux are given by Eqs. (2.39) and (2.40) and their quadratic expressions in terms of the pressure deviator $p_{\langle ij\rangle}$, heat flux q_i and dynamic pressure ϖ can be written as

$$h = h_o - \frac{\tau_\Delta}{2T\varsigma}\varpi^2 - \frac{\tau_p}{4T\mu}p_{\langle ij\rangle}p_{\langle ij\rangle} - \frac{\tau_q}{2T^2\kappa}q_i q_i, \tag{2.82}$$

$$\phi_i = \frac{1}{T}q_i - \frac{\sigma_2}{T}\varpi q_i - \frac{\sigma_1}{T}p_{\langle ij\rangle}q_j. \tag{2.83}$$

From Eq. (2.82) one can conclude that the coefficients of the non-equilibrium part of h depend on the three transport coefficients, μ, ς and κ and on the three relaxation times τ_p, τ_q and τ_Δ. Moreover, from Eq. (2.83) it follows that the entropy flux is equal to the heat flux divided by the temperature plus two terms that take into account thermal-viscous interactions.

3. EXTENDED THERMODYNAMICS OF IDEAL GASES WITH 14 FIELDS

In Section 2 it was said that the objective of extended thermodynamics of real gases is the determination of the 14 fields of density, velocity, temperature, pressure tensor and heat flux.

For ideal gases one has to reformulate this sentence by saying that the objective of extended thermodynamics of ideal gases (Kremer 1986) is the determination of the 14 fields of

$$
\begin{array}{lll}
\text{density} & \varrho, & \\
\text{velocity} & v_i, & \\
\text{pressure tensor} & p_{ij}, & \\
\text{heat flux} & q_i & \text{and} \\
\text{moment scalar} & p_{iijj}. &
\end{array}
\tag{3.1}
$$

The reason for this change is that, for ideal gases,, one has to identify the pressure tensor p_{ij} with the energy tensor g_{ij}, the heat flux q_i with the moment vector g_i and p_{ijjk} with g_{ijjk}. As a consequence, the trace of the pressure tensor is three times the pressure p which in turn is given by the thermal equation of state $p = p(\varrho, T)$.

From the results of the constitutive theory of Section 2 with the conditions (2.41), it follows that the linear field equations for ideal gases contain only three unknows functions of two variables and two constants, provided that the thermal equation of state is given. Explicit results are formulated for classical ideal gas and for degenerate gases of Bose and Fermi type.

3.1 Results of the Constitutive Theory

From the integrability condition of the Gibbs equation (2.47) with $3p = 2\varrho\varepsilon$ it follows a differential equation for p

$$\varrho\frac{\partial p}{\partial \varrho} + \frac{2}{3}T\frac{\partial p}{\partial T} - \frac{5}{3}p = 0, \tag{3.2}$$

whose general solution is

$$p = T^{5/2}F(z) \qquad \text{with} \qquad z = \frac{\varrho}{T^{3/2}}. \tag{3.3}$$

F is an arbitrary function. Eq. (3.3) represents the general form of the thermal equation of state whose explicit form must be taken from statistical mechanics of ideal gases.

Further, the relations between the constitutive coefficients (2.48)-(2.60) with the conditions (2.41) give explicit forms of the constitutive relations (2.33)-(2.35), (2.39) and (2.40), viz.

$$p_{ijjk} = \frac{1}{3}(g + \Delta)\delta_{ik} + 2\gamma p_{\langle ik\rangle}, \tag{3.4}$$

$$p_{\langle ij\rangle k} = \frac{4}{5}q_{\langle i}\delta_{j\rangle k}, \tag{3.5}$$

$$p_{iijjk} = \xi q_k, \tag{3.6}$$

$$h = \left[\frac{3}{2}\int\left(\frac{F'}{z} - \frac{5}{3}\frac{F}{z^2}\right)dz + a_1\right]\varrho + h_2\Delta^2 - \frac{1}{4pT}p_{\langle rs\rangle}p_{\langle rs\rangle} - \frac{4}{5pT}\left(\gamma - \frac{5p}{2\varrho}\right)^{-1}q_r q_r, \tag{3.7}$$

$$\Phi_k = \left[\frac{1}{T} - \frac{4\Delta}{15pT}\left(\gamma - \frac{5p}{2\varrho}\right)^{-1}\right]q_k - \frac{2}{5pT}p_{\langle kr\rangle}q_r, \tag{3.8}$$

where F' is the derivative of F with respect to its argument and

$$\gamma = \frac{7T}{2F}\Big(\int \frac{FF'}{z}dz + a_2\Big), \tag{3.9}$$

$$g = 15T^{7/2}\Big(\int \frac{FF'}{z}dz + a_2\Big), \tag{3.10}$$

$$h_2^{-1} = T^{11/2}\int\Big\{30F\Big[\frac{7}{F}\Big(\int \frac{FF'}{z}dz + a_2\Big) - \frac{5F}{z}\Big]\Big[\Big(\frac{7\big(\int \frac{FF'}{z}dz + a_2\big) - 3FF'}{F - 3F'z/5}\Big)' - \frac{4F'}{z}\Big]\Big\}dz + a_3T^{11/2}, \tag{3.11}$$

$$\xi = \frac{2}{3}\Big(\frac{\partial g}{\partial T}\Big/\frac{\partial p}{\partial T}\Big) - \frac{2}{15pT}\Big(\gamma - \frac{5p}{2\varrho}\Big)^{-1}\frac{1}{h_2}. \tag{3.12}$$

a_1, a_2 and a_3 are constants of integration.

One can conclude from Eqs. (3.4)-(3.12) that all coefficients in the representations (2.33)-(2.35), (2.39) and (2.40) have been reduced to the thermal equation, or rather to the function $F(\frac{\varrho}{T^{3/2}})$ which caracterizes the thermal equation of state. It is true though that three constants have also appeared of which a_1 is the additive entropy constant. This leaves only the coefficients τ_p, τ_q and τ_Δ, of the production terms $P_{\langle ij\rangle}$, Q_i and G in Eqs. (2.36)-(2.38), undetermined.

3.2 Specific Results

3.2.1 *Thermal equations of state for ideal gases.* In statistical mechanics the thermal equation of state of a monatomic ideal gas is given by (Huang 1963)

$$p = \frac{2k}{3m}\frac{T^{5/2}}{A_\pm}I_4^\pm(\alpha) \quad \text{with} \quad I_2^\pm(\alpha) = A_\pm\frac{\varrho}{T^{3/2}}. \tag{3.13}$$

$I_n^\pm(\alpha)$ is an integral defined by

$$I_n^\pm(\alpha) = \int_0^\infty \frac{x^n dx}{e^{(x^2+\alpha)} \pm 1}, \tag{3.14}$$

where the $\pm$ signs refer to Fermi and Bose gases respectively. $A_\pm$ is a constant which is different for fermions and bosons. The integrals $I_n^\pm$ satisfy the recurrence formula

$$\frac{dI_n^\pm(\alpha)}{d\alpha} = -\frac{n-1}{2}I_{n-2}^\pm(\alpha). \tag{3.15}$$

Comparision of Eq. (3.13) with the general expression (3.3) of the thermal equation of state shows that the function $F(z)$ is given by

$$F(z) = \frac{2k}{3m}\frac{1}{A_\pm} I_4^\pm(\alpha) \qquad \text{with} \qquad I_2^\pm(\alpha) = A_\pm z. \tag{3.16}$$

In general, α cannot be eliminated between (3.13) or (3.16) to give the thermal equation of state $p = \hat{p}(\varrho, T)$ or the form of $F(z)$. However, there are three limiting cases in which this can be done. These are the cases of the classical ideal gas and of the strongly degenerate Bose and Fermi gases which are characterized by

i) classical ideal gas: $\alpha \gg 1$

$$p = \varrho \frac{k}{m} T \qquad \text{i.e.} \qquad F(z) = \frac{k}{m} z, \tag{3.17}$$

ii) strongly degenerate Fermi gas: $\alpha \ll -1$

$$p = \nu_F \varrho^{5/3} \qquad \text{i.e.} \qquad F(z) = \nu_F z^{5/3}, \tag{3.18}$$

iii) strongly degenerate Bose gas: $\alpha \approx 0$

$$p = \nu_B T^{5/2} \qquad \text{i.e.} \qquad F(z) = \nu_B. \tag{3.19}$$

New constants ν_F and ν_B have been introduced in Eqs. (3.16) and (3.17) by the definitions

$$\nu_F = \frac{2k}{5m}(3A_+)^{2/3} \qquad \text{and} \qquad \nu_B = \frac{2k}{3m}\frac{1}{A_-} I_4^-(0). \tag{3.20}$$

Corresponding to the three limiting cases (3.17) thorugh (3.19) it follows

i) classical ideal gases:

$$I_n^\pm(\alpha) = \frac{n-1}{2} I_{n-2}^\pm(\alpha), \qquad I_2^\pm(\alpha) = \frac{\pi^{1/2}}{4} e^{-\alpha} = A_\pm \frac{\varrho}{T^{3/2}}, \tag{3.21}$$

ii) strongly degenerate Fermi gas:

$$I_n^+(\alpha) = \frac{1}{(n+1)}(-\alpha)^{\frac{n+1}{2}}, \qquad I_2^+(\alpha) = \frac{1}{3}(-\alpha)^{3/2}, = A_+ \frac{\varrho}{T^{3/2}}, \tag{3.22}$$

iii) strongly degenerate Bose gas:

$$I_0^-(0) = \infty, \quad I_2^-(0) \approx 1.16, \quad I_4^-(0) \approx 0.89, \quad I_6^-(0) \approx 1.87, \quad I_8^-(0) \approx 6.13. \tag{3.23}$$

Knowing $F(z)$ in general, by Eq. (3.16) one can proceed to write the conditions (3.4)-(3.8) in a specific form.

3.2.2 *The results for* $F(z) = (2k/3m)(I_4^{\pm}(\alpha)/A_{\pm})$. For $F(z)$ given by Eq. (3.16), Eqs. (3.9)-(3.12) assume the forms

$$\gamma = \frac{7}{5}\frac{k}{m}\frac{I_6^{\pm}(\alpha)}{I_4^{\pm}(\alpha)}T + \frac{21}{4}\frac{m}{k}\frac{A_{\pm}}{I_4^{\pm}(\alpha)}a_2 T, \tag{3.24}$$

$$g = 4\left(\frac{k}{m}\right)^2 \frac{I_6^{\pm}(\alpha)}{A_{\pm}}T^{7/2} + 15a_2 T^{7/2}, \tag{3.25}$$

$$\begin{aligned} h_2^{-1} = a_3 T^{11/2} + T^{11/2}\Bigg\{ &-144\left(\frac{k}{m}\right)^3 \frac{1}{A_{\pm}} I_8^{\pm}(\alpha) + \frac{560}{3}\left(\frac{k}{m}\right)^3 \frac{1}{A_{\pm}} \frac{I_6^{\pm}(\alpha) I_4^{\pm}(\alpha)}{I_2^{\pm}(\alpha)} + \Bigg[210 a_2 \\ &+20\left(\frac{k}{m}\right)^2 \frac{1}{A_{\pm}}\left(\frac{14}{5} I_6^{\pm}(\alpha) - \frac{10}{3}\frac{(I_4^{\pm}(\alpha))^2}{I_2^{\pm}(\alpha)}\right)\Bigg]\Bigg[\frac{14}{5}\frac{k}{m}\frac{I_6^{\pm}(\alpha)}{I_4^{\pm}(\alpha)} + \frac{21}{2}\frac{m}{k}\frac{A_{\pm} a_2}{I_4^{\pm}(\alpha)} \\ &-6\frac{k}{m}\frac{I_2^{\pm}(\alpha)}{I_0^{\pm}(\alpha)}\Bigg]\Bigg[1 - \frac{9}{5}\frac{(I_2^{\pm}(\alpha))^2}{I_0^{\pm}(\alpha) I_4^{\pm}(\alpha)}\Bigg]^{-1} + \frac{k}{m}a_2\left(1890\alpha + 700\frac{I_4^{\pm}(\alpha)}{I_2^{\pm}(\alpha)}\right)\Bigg\}. \end{aligned} \tag{3.26}$$

By Eq. (3.7) the specific entropy in equilibrium is given by

$$\eta_0 = \frac{k}{m}\alpha + \frac{5}{3}\frac{k}{m}\frac{I_4^{\pm}(\alpha)}{I_2^{\pm}(\alpha)} + a_1. \tag{3.27}$$

On the other hand, the value of ξ folllows from Eqs. (3.12), (3.25) and (3.26)

$$\begin{aligned} \xi = -\frac{T}{10}\Bigg[&a_3 - 144\left(\frac{k}{m}\right)^3 \frac{1}{A_{\pm}} I_8^{\pm}(\alpha) + \frac{560}{3}\left(\frac{k}{m}\right)^3 \frac{1}{A_{\pm}} \frac{I_6^{\pm}(\alpha) I_4^{\pm}(\alpha)}{I_2^{\pm}(\alpha)} + \frac{k}{m}a_2\Bigg(1890\alpha \\ &+700\frac{I_4^{\pm}(\alpha)}{I_2^{\pm}(\alpha)}\Bigg)\Bigg]\Bigg[\frac{21}{2}a_2 + \left(\frac{k}{m}\right)^2 \frac{1}{A_{\pm}}\left(\frac{14}{5} I_6^{\pm}(\alpha) - \frac{10}{3}\frac{(I_4^{\pm}(\alpha))^2}{I_2^{\pm}(\alpha)}\right)\Bigg]^{-1}. \end{aligned} \tag{3.28}$$

3.2.3 *The classical ideal gas.* Insertion of Eq. (3.21) into Eqs. (3.24)-(3.28) gives the values γ, g, h_2, η_0 and ξ for a classical ideal gas

$$\gamma = \frac{7}{2}\frac{k}{m}T + \frac{7}{2}\frac{m}{k}\frac{T^{5/2}}{\varrho}a_2, \tag{3.29}$$

$$g = 15\varrho\left(\frac{k}{m}\right)^2 T^2 + 15a_2 T^{7/2}, \tag{3.30}$$

$$\begin{aligned} h_2^{-1} = &-240\left(\frac{k}{m}\right)^3 \varrho T^4 + 105\frac{k}{m}a_2 T^{11/2}\left[18\,ln\frac{T^{3/2}}{\varrho} + 35\left(\frac{m}{k}\right)^2 a_2 \frac{T^{3/2}}{\varrho}\right] \\ &+\left[a_3 + 4200\frac{k}{m}a_2\left(\frac{9}{20}\,ln\frac{\pi^{1/2}}{4A_{\pm}} + 1\right)\right]T^{11/2} \end{aligned} \tag{3.31}$$

$$\eta_0 = \frac{k}{m}\Big[ln\frac{T^{3/2}}{\varrho} + \frac{5}{2} + ln\frac{\pi^{1/2}}{4A_\pm} + \frac{m}{k}a_1\Big], \tag{3.32}$$

$$\xi = \Big\{28\frac{k}{m}T - 63\Big(\frac{m}{k}\Big)\frac{T^{5/2}}{\varrho}a_2\,ln\frac{T^{3/2}}{\varrho} - \Big(\frac{m}{k}\Big)^2\frac{T^{5/2}}{\varrho}\Big[\frac{a_3}{30} + 35\frac{k}{m}a_2\Big(\frac{9}{5}\,ln\frac{\pi^{1/2}}{4A_\pm} + 1\Big)\Big]\Big\}\Big[1 + \frac{7}{2}a_2\Big(\frac{m}{k}\Big)^2\frac{T^{3/2}}{\varrho}\Big]. \tag{3.33}$$

There is a strong suspicion that the constants a_2 and a_3 are both zero, because in the kinetic theory of gases these constants do not appear. It is therefore appropriate to write down the final results for classical ideal gases without those constants. The results are obtained by insertion of Eqs. (3.29)-(3.33) into Eqs. (3.4) and (3.6)-(3.8).

$$p_{ijjk} = \frac{1}{3}\Big[15\varrho\Big(\frac{k}{m}\Big)^2T^2 + \Delta\Big]\delta_{ik} + 7\frac{k}{m}Tp_{\langle ik\rangle}, \tag{3.34}$$

$$p_{iijjk} = 28\frac{k}{m}Tq_k, \tag{3.35}$$

$$h = \frac{k}{m}\Big[ln\frac{T^{3/2}}{\varrho} + \frac{5}{2} + ln\frac{\pi^{1/2}}{4A_\pm} + \frac{m}{k}a_1\Big]\varrho - \frac{m^3}{240k^3\varrho T^4}\Delta^2 - \frac{1}{4pT}p_{\langle rs\rangle}p_{\langle rs\rangle} - \frac{m}{5kpT^2}q_rq_r, \tag{3.36}$$

$$\Phi_k = \Big(\frac{1}{T} - \frac{m}{15kpT^2}\Delta\Big)q_k - \frac{2}{5pT}p_{\langle kr\rangle}q_r. \tag{3.37}$$

3.2.4 *Strongly degenerate Fermi gas.* For a strongly degenerate Fermi gas insertion of Eq. (3.22) into Eqs. (3.24)-(3.28) leads to

$$\gamma = \frac{5}{2}\nu_F\varrho^{2/3} + \frac{7}{2}\frac{a_2T^{7/2}}{\nu_F\varrho^{5/3}}, \tag{3.38}$$

$$g = \frac{75}{7}\nu_F^2\varrho^{7/3} + 15a_2T^{7/2}, \tag{3.39}$$

$$h_2 = 0, \tag{3.40}$$

$$\eta_0 = a_1, \tag{3.41}$$

$$\xi = 35\nu_F\varrho^{2/3} - \frac{a_3}{105a_2}T. \tag{3.42}$$

Hence, the constitutive equations for a strongly degenerate Fermi gas are given by insertion of Eqs. (3.38)-(3.42) into Eqs. (3.4) and (3.6)-(3.8)

$$p_{ijjk} = \frac{1}{3}\Big(\frac{75}{7}\nu_F^2\varrho^{7/3} + 15a_2T^{7/2} + \Delta\Big)\delta_{ik} + \Big(5\nu_F\varrho^{2/3} + 7\frac{a_2T^{7/2}}{\nu_F\varrho^{5/3}}\Big)p_{\langle ik\rangle}, \tag{3.43}$$

$$p_{rrssk} = \Big(35\nu_F\varrho^{2/3} - \frac{a_3}{105a_2}T\Big)q_k, \tag{3.44}$$

$$h = a_1\varrho - \frac{1}{4\nu_F\varrho^{5/3}T}p_{\langle rs\rangle}p_{\langle rs\rangle} - \frac{2}{35a_2T^{9/2}}q_rq_r, \tag{3.45}$$

$$\Phi_k = \left(\frac{1}{T} - \frac{2\Delta}{105a_2T^{9/2}}\right)q_k - \frac{2}{5\nu_F\varrho^{5/3}T}p_{\langle kr\rangle}q_r. \tag{3.46}$$

3.2.5 *Strongly degenerate Bose gas.* The results for a strongly degenerate Bose gas follow by the substitution of the values (3.23) into Eqs. (3.24)-(3.28)

$$\gamma = \left(2.94\frac{k}{m} + \frac{7}{2}\frac{a_2}{\nu_B}\right)T, \tag{3.47}$$

$$g = \left(12.59\frac{k}{m}\nu_B + 15a_2\right)T^{7/2}, \tag{3.48}$$

$$h_2^{-1} = \left[a_3 + 1470\frac{a_2^2}{\nu_B}\left(1 + 1.68\frac{k}{m}\frac{\nu_B}{a_2}\right) - 449.48\left(\frac{k}{m}\right)^2\nu_B\right]T^{11/2}, \tag{3.49}$$

$$\eta_0 = 1.29\frac{k}{m} + a_1, \tag{3.50}$$

$$\xi = \frac{T}{105}\left[1032.33\left(\frac{k}{m}\right)^2\nu_B - 539.67\frac{k}{m}a_2 - a_3\right]\left[a_2 + 0.47\frac{k}{m}\nu_B\right]^{-1}. \tag{3.51}$$

Hence, it follows from Eqs. (3.4), (3.6)-(3.8) and (3.47)-(3.51) the constitutive equations for a strongly degenerate Bose gas

$$p_{ijjk} = \frac{1}{3}\left(12.59\frac{k}{m}\nu_BT^{7/2} + 15a_2T^{7/2} + \Delta\right)\delta_{ik} + \left(5.88\frac{k}{m}T + 7\frac{a_2}{\nu_B}T\right)p_{\langle ik\rangle} \tag{3.52}$$

$$p_{rrssk} = \frac{T}{105}\left[1032.33\left(\frac{k}{m}\right)^2\nu_B - 539.67\frac{k}{m}a_2 - a_3\right]\left[a_2 + 0.47\frac{k}{m}\nu_B\right]^{-1}q_k, \tag{3.53}$$

$$\begin{aligned} h = {} & \frac{k}{m}\left(1.29 + \frac{m}{k}a_1\right)\varrho + T^{-11/2}\left[a_3 + 1470\frac{a_2^2}{\nu_B}\left(1 + 1.68\frac{k}{m}\frac{\nu_B}{a_2}\right) - 449.48\left(\frac{k}{m}\right)^2\nu_B\right]^{-1}\Delta^2 \\ & - \frac{1}{4\nu_BT^{7/2}}p_{\langle rs\rangle}p_{\langle rs\rangle} - \frac{2}{5\nu_BT^{9/2}}\left(3.32\frac{k}{m} + 7\frac{a_2}{\nu_B}\right)^{-1}q_rq_r, \end{aligned} \tag{3.54}$$

$$\Phi_k = \left[\frac{1}{T} - \frac{2\Delta}{15\nu_BT^{9/2}}\left(3.32\frac{k}{m} + 7\frac{a_2}{\nu_B}\right)^{-1}\right]q_k - \frac{2}{5\nu_BT^{7/2}}p_{\langle kr\rangle}q_r. \tag{3.55}$$

3.3 On the Status of the Moment Scalar p_{iijj}

One aspect of the variable p_{iijj}, introduced here, is implicit in equation (2.69) which states that the first iterate of Δ vanishes. It is because of that that Δ does not play a role in ordinary thermodynamics, where as a rule the constitutive quantities are functions of ϱ, v_i and T and their first order spatial derivatives.

Another aspect of the variable Δ reveals itself in a relativistic theory. It was proved by (Dreyer and Weiss 1986) that the moment scalar p_{iijj} corresponds to the combination

$$8c^2\left[(e - nmc^2) - \frac{3}{2}p\right] \tag{3.56}$$

in relativistic thermodynamics. e is the energy density, n the particle number density and c the speed of the light in vacuum. In particular the non-equilibrium part Δ of p_{iijj} corresponds to the non-equilibrium part of Eq. (3.56) which is equal to $-12c^2\pi$, where π is the dynamic pressure, i.e. the non-equilibrium contribution to the pressure. Thus we have

$$\pi = -\frac{1}{12c^2}\Delta, \tag{3.57}$$

and it follows that the dynamic pressure in a gas is a relativistically small quantity.

4. EXTENDED THERMODYNAMICS OF MOLECULAR IDEAL GASES

The objective of extended thermodynamics of molecular gases is the determination of 17 scalar fields of

$$\begin{array}{ll} \text{density} & \varrho, \\ \text{velocity} & v_i, \\ \text{pressure tensor} & p_{ij}, \\ \text{translational heat flux} & q_i^T, \\ \text{intrinsic energy density} & \varrho g \ \text{and} \\ \text{intrinsic heat flux} & g_i. \end{array} \tag{4.1}$$

The translational energy density $\varrho\varepsilon^T$ is related to the pressure p by $3p = 2\varrho\varepsilon^T$. It is therefore implicit included in this list of fields. The intrinsic energy density represents the rotational or vibrational energy of the molecules.

The necessary field equations are based upon balance laws, and constitutive equations are formulated to close the system of equations. The constitutive equations are restricted by *the principle of material frame indifference* and by *the entropy principle*.

It turns out that all but 6 out of 34 constitutive coefficients can be related to the thermal and caloric equations of state by a careful and systematic exploitation of these principles. The remaining coefficients can either be related to the shear viscosity and thermal conductivity, or they follow from the coefficient of self-diffusion or the absorption and dispersion of sound waves. Alternatively these coefficients may be found by comparison with kinetic theory of gases, notably with the Eucken formula.

It is shown that in non-equilibrium the temperature is discontinuous at thermometric walls. The dynamic pressure and the volume viscosity, are discussed and it is shown how extended thermodynamics and ordinary thermodynamics are related.

This section is based on the works of (Kremer 1985 and 1989a).

4.1 Thermodynamic Processes

The necessary field equations for the determination of the 17 fields (4.1) are based on balance equations which, in an inertial frame free of external supplies, have the form

$$\dot{\varrho} + \varrho \frac{\partial v_i}{\partial x_i} = 0, \tag{4.2}$$

$$\varrho \dot{v}_i + \frac{\partial p_{ij}}{\partial x_j} = 0, \tag{4.3}$$

$$\dot{p}_{ij} + p_{ij}\frac{\partial v_k}{\partial x_k} + \frac{\partial p_{\langle ijk \rangle}}{\partial x_k} + \frac{4}{5}\frac{\partial q^T_{(i}}{\partial x_{j)}} + \frac{2}{5}\frac{\partial q^T_k}{\partial x_k}\delta_{ij} + 2p_{k(i}\frac{\partial v_{j)}}{\partial x_k} = P_{ij}, \tag{4.4}$$

$$\dot{q}^T_i + \frac{7}{5}q^T_i\frac{\partial v_k}{\partial x_k} + \frac{1}{2}\frac{\partial p_{ijjk}}{\partial x_k} + p_{\langle ijk \rangle}\frac{\partial v_j}{\partial x_k} + \frac{2}{5}q^T_j\frac{\partial v_j}{\partial x_i} + \frac{7}{5}q^T_j\frac{\partial v_i}{\partial x_j}$$
$$- \frac{p_{ij}}{\varrho}\frac{\partial p_{jk}}{\partial x_k} - \frac{p_{jj}}{2\varrho}\frac{\partial p_{ik}}{\partial x_k} = P^T_i, \tag{4.5}$$

$$\varrho \dot{g} + \frac{\partial g_i}{\partial x_i} = -\frac{1}{2}P_{ii}, \tag{4.6}$$

$$\dot{g}_i + g_i\frac{\partial v_k}{\partial x_k} + \frac{\partial g_{ij}}{\partial x_j} + g_j\frac{\partial v_i}{\partial x_j} - g\frac{\partial p_{ij}}{\partial x_j} = P^I_i. \tag{4.7}$$

Eqs. (4.2)-(4.5) are formally the usual 13 balance equations of extended thermodynamics of monatomic gases with vanishing productions of mass and momentum. The trace of Eq. (4.4) is twice the balance of translational energy density $\varrho\varepsilon^T$ and $\frac{1}{2}P_{ii}$ is its production. Thus the balance of total energy with density $\varrho(g + \varepsilon^T + \frac{1}{2}v^2)$ is a conservative law. All tensors are symmetric.

The system of balance equations (4.2)-(4.7) is closed by considering

$$p_{\langle ijk \rangle},\ p_{ijjk},\ g_{ij},\ P_{ii},\ P_{\langle ij \rangle},\ P^T_i \text{ and } P^I_i \tag{4.8}$$

as constitutive quantities that depend upon the fields given in (4.1), or equivalently, upon the 17 fields

$$\varrho,\ v_i,\ p_{ij},\ q^T_i,\ \Delta \quad \text{and} \quad g_i, \tag{4.9}$$

where

$$\Delta = g - \varepsilon^I(\varrho, p). \tag{4.10}$$

$\varepsilon^I(\varrho, p)$ is the equilibrium value of the intrinsic energy g while Δ is its deviation from equilibrium.

If the explicit form of the constitutive functions is known, one can obtain an explicit set of field equations from Eqs. (4.2)-(4.7). Every solution of this system is called a *thermodynamic process.*

4.2 Constitutive Theory

The constitutive functions are restricted by the entropy principle and by the principle of material frame indifference.For more details one is refered to Section 2.

Here I will write down only the representations of the constitutive functions and limit the attention to processes in the neighbourhood of equilibrium i.e. where $p_{\langle ij\rangle}$, q_i^T, g_i and Δ vanish. Therefore I write

$$p_{\langle ijk\rangle} = 0, \tag{4.11}$$

$$p_{ijjk} = (\beta_o + \beta_1\Delta)\delta_{ik} + \gamma p_{\langle ik\rangle}, \tag{4.12}$$

$$g_{ik} = (\nu_o + \nu_1\Delta)\delta_{ik} + \mu p_{\langle ik\rangle}, \tag{4.13}$$

$$P_{ii} = \xi\Delta, \tag{4.14}$$

$$P_{\langle ij\rangle} = \sigma p_{\langle ij\rangle} \tag{4.15}$$

$$P_i^T = \alpha_1 q_i^T + \alpha_2\, g_i, \tag{4.16}$$

$$P_i^I = \chi_1 q_i^T + \chi_2\, g_i, \tag{4.17}$$

$$\phi_k = (\varphi_1 + \varphi_2\Delta)q_k^T + (\varphi_3 + \varphi_4\Delta)\, g_k + \varphi_5 p_{\langle kr\rangle} q_r^T + \varphi_6 p_{\langle kr\rangle}\, g_r, \tag{4.18}$$

$$\begin{aligned} h = \varrho s + h_1\Delta + h_2\Delta^2 + h_3\Delta^3 + (h_4 + h_5\Delta)p_{\langle rs\rangle}p_{\langle rs\rangle} \\ +(h_6 + h_7\Delta)q_r^T q_r^T + (h_8 + h_9\Delta)\, g_r\, g_r + (h_{10} + h_{11}\Delta)q_r^T\, g_r \\ +h_{12}p_{\langle rs\rangle}p_{\langle sk\rangle}p_{\langle kr\rangle} + h_{13}p_{\langle rs\rangle}q_r^T q_s^T + h_{14}p_{\langle rs\rangle}q_r^T\, g_s + h_{15}p_{\langle rs\rangle}\, g_r\, g_s, \end{aligned} \tag{4.19}$$

All coefficients β_o through h_{15} may depend on ϱ and p.

4.3 Results of the Constitutive Theory

From the exploitation of the entropy inequality with the Lagrange multipliers it follows the equation

$$ds = \frac{\partial s}{\partial \varepsilon}\Big|_E \left(d\varepsilon\Big|_E - \frac{p}{\varrho^2}d\varrho\right), \tag{4.20}$$

where $\varepsilon\,|_E = \varepsilon^T + \varepsilon^I$ is the specific internal energy in equilibrium.

Thus by comparision with the Gibbs equation $\frac{\partial s}{\partial \varepsilon}\,|_E$ is identified as $\frac{1}{T}$ where T is the absolute temperature.

Once the absolute temperature is introduced in this way, it is now possible to replace the pair (ϱ, p) of equilibrium variables by the more familiar pair (ϱ, T) and I shall do this in the sequel.

There is one immediate consequence: The Gibbs equation implies the integrability condition

$$\frac{\partial \varepsilon\,|_E}{\partial \varrho} = \frac{1}{\varrho^2}\left[p - T\frac{\partial p}{\partial T}\right]. \tag{4.21}$$

Here we are dealing with an ideal gas where

$$p = \varrho\frac{k}{m}T \qquad \text{and} \qquad \varepsilon^T = \frac{3}{2}\frac{k}{m}T \tag{4.22}$$

hold. Therefore (4.21) implies

$$\frac{\partial \varepsilon^I}{\partial \varrho} = 0, \tag{4.23}$$

i.e. the specific intrinsic energy is independent of the density, it may only depend on the absolute temperature.

Moreover, the constitutive functions (4.11)-(4.13), (4.18) and (4.19) obtain the following explicit forms as a result of the entropy principle

$$p_{\langle ijk\rangle} = 0, \tag{4.24}$$

$$p_{ijjk} = \left[\frac{5p^2}{\varrho}\left(1 + \frac{a_1T^{5/2}}{7p}\right) + a_2\right]\delta_{ik} + \frac{7p}{\varrho}\left(1 + \frac{a_1T^{5/2}}{7p}\right)p_{\langle ik\rangle}, \tag{4.25}$$

$$g_{ik} = (p\varepsilon^I + p\Delta + a_3)\delta_{ik} + \varepsilon^I p_{\langle ik\rangle}, \tag{4.26}$$

$$\phi_k = \frac{1}{T}(q_k^T + g_k) - \frac{1}{T^2\varepsilon_T^I}\Delta\, g_k - \frac{2}{5pT}p_{\langle kr\rangle}q_r^T, \tag{4.27}$$

$$\begin{aligned} h = {} & \varrho\left(\frac{k}{m}\ln\frac{T^{3/2}}{\varrho} + \int\frac{\varepsilon_T^I}{T}dT + a_4\right) + \frac{\varrho}{T}\Delta - \frac{\varrho}{2T^2\varepsilon_T^I}\Delta^2 + \frac{\varrho}{3T^3(\varepsilon_T^I)^2}\left(1 + \frac{T}{2\varepsilon_T^I}\frac{d\varepsilon_T^I}{dT}\right)\Delta^3 \\ & - \frac{1}{4pT}p_{\langle rs\rangle}p_{\langle rs\rangle} - \frac{\varrho}{5p^2T}\left(1 + \frac{a_1T^{5/2}}{2p}\right)^{-1}q_r^Tq_r^T - \frac{1}{2pT^2\varepsilon_T^I}g_r\,g_r \\ & + \frac{1}{2T^2p(\varepsilon_T^I)^2}\left(\frac{2}{T} + \frac{1}{\varepsilon_T^I}\frac{d\varepsilon_T^I}{dT}\right)\Delta\,g_r\,g_r + \frac{1}{6p^2T}p_{\langle rs\rangle}p_{\langle sk\rangle}p_{\langle kr\rangle} \\ & + \frac{9\varrho}{25p^3T}\left(1 + \frac{a_1T^{5/2}}{2p}\right)^{-1}p_{\langle rs\rangle}q_r^Tq_s^T - \frac{1}{2p^2T^2\varepsilon_T^I}p_{\langle rs\rangle}g_rg_s. \end{aligned} \tag{4.28}$$

ε_T^I stands for $\frac{d\varepsilon^I}{dT}$ and a_1 through a_4 are constants of integration.

The amazing feature of these results is there specific nature : All coefficients in (4.11)-(4.13) and (4.18), (4.19) (some, or most of them coefficients of non-equilibrium terms) are specifically related to the thermal and caloric equation of state $p(\varrho, T)$ and $\varepsilon^I(T)$ except for the constants a. This specific nature of the results is a recurring feature of extended thermodynamics, first observed by (Liu and Müller 1983).

However, not everything is quite so specific. Indeed the 6 constitutive coefficients of the productions

$$P_i^T = \alpha_1 q_i^T + \alpha_2 g_i, \qquad P_{\langle ij \rangle} = \sigma p_{\langle ij \rangle}, \tag{4.29}$$

$$P_i^I = \chi_1 q_i^T + \chi_2 g_i, \qquad P_{ii} = \xi \Delta, \tag{4.30}$$

are only restricted by some inequalities, viz.

$$\chi_2 \leq 0, \qquad \alpha_1 \left(1 + \frac{a_1 T^{5/2}}{2p}\right)^{-1} \leq 0, \qquad \sigma \leq 0, \qquad \xi \leq 0, \tag{4.31}$$

$$\alpha_1 \chi_2 - \frac{m\varepsilon_T^I}{10k}\left(\alpha_2 + \frac{5k}{2m\varepsilon_T^I}\chi_1\right)^2 \geq 0. \tag{4.32}$$

In the next section I shall discuss ways on how to identify the remaining unknown coefficients α_1, α_2, χ_1, χ_2, σ and ξ. These are called the transport coefficients.

If we take the intrinsic energy to be the vibrational energy of the molecules, it is possible to show that the above results of extended thermodynamics are identical to those of statistical mechanics, if only the constants a_1, a_2 and a_3 are set equal to zero (Kremer 1989b). Therefore in the further development of this Section I shall ignore these constants. In fact, several contributions from a_1 were already ignored in writing (4.32) in order to make that inequality less cumbersome.

In conclusion I list the field equations, i.e. the balance laws combined with the constitutive Eqs. (4.24)-(4.26), (4.29) and (4.30).

$$\dot{\varrho} + \varrho \frac{\partial v_i}{\partial x_i} = 0, \tag{4.33}$$

$$\varrho \dot{v}_i + \frac{\partial p_{ij}}{\partial x_j} = 0, \tag{4.34}$$

$$\varrho \dot{\varepsilon} + \frac{\partial q_i^T}{\partial x_i} + \frac{\partial g_i}{\partial x_i} + p_{ij}\frac{\partial v_i}{\partial x_j} = 0, \tag{4.35}$$

$$\dot{p}_{\langle ij \rangle} + p_{\langle ij \rangle}\frac{\partial v_k}{\partial x_k} + \frac{4}{5}\frac{\partial q^T_{\langle i}}{\partial x_{j\rangle}} + 2p_{k\langle i}\frac{\partial v_{j\rangle}}{\partial x_k} = \sigma p_{\langle ij \rangle}, \tag{4.36}$$

$$\dot{q}_i^T + \frac{7}{5} q_i^T \frac{\partial v_k}{\partial x_k} + \frac{4}{5} q_j^T \frac{\partial v_{(i}}{\partial x_{j)}} + q_j^T \frac{\partial v_i}{\partial x_j} + \frac{5p^2}{2\varrho T}\frac{\partial T}{\partial x_i} - \frac{p_{\langle ij \rangle}}{\varrho}\frac{\partial p_{jk}}{\partial x_k}$$

$$+\frac{7p}{2\varrho T}p_{\langle ik\rangle}\frac{\partial T}{\partial x_k}+\frac{p}{\varrho}\frac{\partial p_{\langle ik\rangle}}{\partial x_k}=\alpha_1 q_i^T+\alpha_2\, g_i, \tag{4.37}$$

$$\varrho\Big(1-\frac{\varepsilon_T^I}{\varepsilon_T}\Big)\dot{\Delta}+\Big(1-\frac{\varepsilon_T^I}{\varepsilon_T}\Big)\frac{\partial g_i}{\partial x_i}-\frac{\varepsilon_T^I}{\varepsilon_T}\frac{\partial q_i^T}{\partial x_i}-\frac{\varepsilon_T^I}{\varepsilon_T}p_{ij}\frac{\partial v_i}{\partial x_j}=-\frac{1}{2}\xi\Delta, \tag{4.38}$$

$$\dot{g}_i+g_i\frac{\partial v_k}{\partial x_k}+g_j\frac{\partial v_i}{\partial x_j}+p\varepsilon_T^I\frac{\partial T}{\partial x_i}+p\frac{\partial \Delta}{\partial x_i}-\Delta\frac{\partial p_{\langle ij\rangle}}{\partial x_j}$$

$$+\varepsilon_T^I p_{\langle ik\rangle}\frac{\partial T}{\partial x_k}=\chi_1 q_i^T+\chi_2 g_i. \tag{4.39}$$

I have reordered these equations somewhat compared to (4.2)-(4.7) with respect to the equations of balance of energy. Thus (4.35) is the balance of internal energy, viz. the sum of the trace of (4.4) and of (4.6). Also I have used (4.35) to eliminate $\dot{T}$ from the balance (4.6) of intrinsic energy. This was done in order to facilitate a subsequent argument. ε_T in (4.38) stands for the derivative of $\varepsilon\,|_E=\frac{3p}{2\varrho}+\varepsilon^I(T)$ with respect to temperature.

Note that the left-hand sides of Eqs. (4.33)-(4.39) are explicit in the variables

$\varrho,\ v_i,\ T,\ p_{\langle ij\rangle},\ q_i^T,\ \Delta,\ g_i,$

provided we know $\varepsilon^I=\varepsilon^I(T)$. Of course, p is always equal to $\varrho\frac{k}{m}T$.

4.4 Identification of transport coefficients

4.4.1 *The limit of ordinary thermodynamics.* It is a most natural question, of course, to ask how to get back from extended to ordinary thermodynamics of a molecular gas. In ordinary thermodynamics, instead of the set (4.1) we should have the basic fields

$\varrho,\ v_i,\ p$ (or T)

and there would be conservation laws of mass, momentum and energy, or the balance of internal energy

$$\dot{\varrho}+\varrho\frac{\partial v_i}{\partial x_i}=0, \tag{4.40}$$

$$\varrho\dot{v}_i+\frac{\partial p_{ij}}{\partial x_j}=0, \tag{4.41}$$

$$\varrho\dot{\varepsilon}+\frac{\partial q_i}{\partial x_i}+p_{ij}\frac{\partial v_i}{\partial x_j}=0. \tag{4.42}$$

To close the system of equations, in ordinary thermodynamics we must have constitutive equations for

the pressure deviator $p_{\langle ij\rangle}$

the heat flux $q_i=q_i^T+g_i$

the internal energy $\varepsilon=\dfrac{3p}{2\varrho}+\varepsilon^I(T)+\Delta.$

Adapting an iterative scheme from the kinetic theory (the so-called Maxwellian iteration), we derive the desired constitutive equations from the Eqs. (4.36)-(4.39) in an approximation manner as follows: In the first step we introduce equilibrium values for $p_{\langle ij\rangle}$ q_i^T, Δ and g_i on the left-hand sides of (4.36)-(4.39) and calculate first iterates.

$$p_{\langle ij\rangle} = 2\frac{p}{\sigma}\frac{\partial v_{\langle i}}{\partial x_{j\rangle}}, \tag{4.43}$$

$$\alpha_1 q_i^T + \alpha_2\, g_i = \frac{5p^2}{2\varrho T}\frac{\partial T}{\partial x_i}, \tag{4.44}$$

$$\Delta = \frac{2p\varepsilon_T^I}{\xi\varepsilon_T}\frac{\partial v_k}{\partial x_k}, \quad \text{i.e.} \quad \varepsilon = \frac{3p}{2\varrho} + \varepsilon^I + \frac{2p\varepsilon_T^I}{\xi\varepsilon_T}\frac{\partial v_k}{\partial x_k}, \tag{4.45}$$

$$\chi_1 q_i^T + \chi_2 g_i = p\varepsilon_T^I\frac{\partial T}{\partial x_i}. \tag{4.46}$$

The iteration proceeds by putting these first iterates into the left-hand sides of (4.36)-(4.39) and calculating second iterates, etc. However, I shall not go beyond the first step.

From (4.44) and (4.46) one gets for $q_i = q_i^T + g_i$

$$q_i = -p\frac{k}{m}\frac{\frac{m}{k}\varepsilon_T^I(\alpha_1 - \alpha_2) - \frac{5}{2}(\chi_1 - \chi_2)}{\alpha_2\chi_1 - \alpha_1\chi_2}\frac{\partial T}{\partial x_i}. \tag{4.47}$$

Thus with (4.43),(4.45) and (4.47) we have the desired constitutive equations which can be used to close the system (4.40)-(4.42).

Eqs. (4.43) and (4.47) are recognized as the constitutive relations of Navier-Stokes and Fourier, respectively, and we conclude that $-\frac{p}{\sigma}$ is the coefficient of shear viscosity μ while the factor of $-\frac{\partial T}{\partial x_i}$ in (4.47) is the coefficient of of thermal conductivity κ

$$\mu = -\frac{p}{\sigma}, \qquad \kappa = p\frac{k}{m}\frac{\frac{m}{k}\varepsilon_T^I(\alpha_1 - \alpha_2) - \frac{5}{2}(\chi_1 - \chi_2)}{\alpha_2\chi_1 - \alpha_1\chi_2}. \tag{4.48}$$

Since μ and κ are measurable quantities, these equations permit the determination of σ and provide a restriction on the transport coefficients α_1, α_2, χ_1 and χ_2. Thus there remain four unknows in the theory at this stage.

4.4.2 *Comparision with the kinetic theory. Eucken formulae.* In the literature there is considerable interest in the ratio $\frac{\kappa}{\mu}$ which indicates the relative efficiency of heat and momentum transfer. According to the elementary kinetic theory we should have

$$\frac{\kappa}{\mu} = \frac{5}{2}\varepsilon_T^T + \varepsilon_T^I \tag{4.49}$$

where ε_T^T stands for derivative of the translational internal energy ε^T with respect to the temperature. This reflects the fact that momentum exchange in a gas and the exchange of intrinsic energy are less efective than the exchange of translational energy by a factor $\frac{2}{5}$, e.g. (Sommerfeld 1977). Equation (4.49) is called the Eucken formula and it has been revised several times as the kinetic theory has become more and more sophisticated. Thus there is a proposition by (Chapman and Cowling 1970) which reads

$$\frac{\kappa}{\mu} = \frac{5}{2}\varepsilon_T^T + \frac{\varrho D}{\mu}\varepsilon_T^I. \tag{4.50}$$

Here D is the coefficient of self-diffusion. There is also a more complex formula, due to (Wang Chang, Uhlenbeck and de Boer 1964) which reads

$$\frac{\kappa}{\mu} = A\left\{1 - \frac{4}{5}\frac{m}{k}B\varepsilon_T^I + \frac{4}{25}\left(\frac{m}{k}\right)^2 C(\varepsilon_T^I)^2\right\}, \tag{4.51}$$

in which the coefficients are definite expressions that follow from Chapman-Enskog solution of the Boltzmann equation for molecular gases.

With μ and κ given by (4.48) we obtain in the present theory

$$\frac{\kappa}{\mu} = \frac{\sigma}{\alpha_1\chi_2 - \alpha_2\chi_1}\left[\frac{5k}{2m}(\chi_2 - \chi_1) + \varepsilon_T^I(\alpha_1 - \alpha_2)\right] \tag{4.52}$$

and, if we abbreviate the positive expression on the left-hand side of (4.32) by δ we can recast (4.51) in the form

$$\frac{\kappa}{\mu} = a\left\{1 - \frac{4}{5}\frac{m}{k}b\varepsilon_T^I + \frac{4}{25}\left(\frac{m}{k}\right)^2 c(\varepsilon_T^I)^2\right\}, \tag{4.53}$$

where a, b and c stand for

$$a = \frac{5k\sigma(2\chi_2 - \chi_1)^2}{8m\chi_2(\alpha_1\chi_2 - \alpha_2\chi_1)}, \quad b = \frac{\alpha_2}{2\chi_2 - \chi_1} - \frac{2\delta\chi_2^2}{(2\chi_2 - \chi_1)^2}, \quad c = \frac{\alpha_2^2}{(2\chi_2 - \chi_1)^2}. \tag{4.54}$$

Note that (4.53) has the same form as the Eucken formula in the Wang Chang-Uhlenbeck-de Boer generalization. A comparison between (4.51) and (4.53) would allow us to have three restrictions on the values of the transport coefficients α_1, α_2, χ_1 and χ_2 so that now only one transport coefficient remains unknown in the theory, say ξ.

There is a tempting special case to be considered, namely the case

$$\alpha_2 = 0, \quad \text{and} \quad \chi_1 = 0 \tag{4.55}$$

so that the two equations (4.44) and (4.46) uncouple. We call this case the case of negligible coupling between translational and intrinsic energy. The ratio $\frac{\kappa}{\mu}$ in (4.51) becomes

$$\frac{\kappa}{\mu} = \frac{-\frac{5p^2}{2\varrho T\alpha_1}}{\mu} - \frac{p}{\mu\chi_2}\varepsilon_T^I. \tag{4.56}$$

By comparision with (4.44) the numerator of the first term is seen to be equal to the "translational thermal conductivity". If we accept, again from the kinetic theory (Chapman and Cowling 1970) that this thermal conductivity is very close to $\frac{15k}{4m}\mu$, we may write

$$\frac{\kappa}{\mu} = \frac{5}{2}\varepsilon_T^T - \frac{p}{\mu\chi_2}\varepsilon_T^I. \tag{4.57}$$

This formula compares well with the Eucken formula (4.50) in the Chapman and Cowling generalization and we conclude that there is a relation between χ_2 and the measurable coefficient of self-diffusion, viz.

$$-\chi_2 = \frac{p}{\varrho}D. \tag{4.58}$$

Note that, here again, with the assumptions (4.55) and the identification (4.58) of χ_2, we now have only one remaining unknown transport coefficient, viz. ξ. One way of determining that coefficient is the measurement of the attenuation of sound waves. I proceed to show this.

4.4.3 *Plane harmonic waves of small amplitude.* From the investigation of solutions of the system (4.33)-(4.39) in the form of plane harmonic waves of small amplitude one may obtain the absorption coefficient α^*. It is a measurable quantity and turns out to be a function of ξ. For the details one is refered to (Kremer 1989a). Inversely one may write

$$\xi = \frac{\frac{2\varrho^2 T}{\mu}\left(\frac{k}{m}\right)^2 \frac{\varepsilon_T^I}{\varepsilon_T^2}}{\frac{2V_o^3\varrho\alpha^*}{\omega^2\mu} - \frac{15k}{4m}\frac{1}{\varepsilon_T(m\varepsilon_T/k+1)} - \frac{\varepsilon_T^I}{\varepsilon_T(m\varepsilon_T/k+1)}\frac{\varrho D}{\mu} - \frac{4}{3}}. \tag{4.59}$$

In Eq. (4.59) V_o is the adiabatic speed of sound, i.e. $V_o = \left[\left(1 + \frac{k}{m\varepsilon_T}\right)\frac{kT}{m}\right]^{1/2}$, and ω the frequency of the wave.

In order to get Eq. (4.59) I have considered the special case (4.55) of negligible coupling. If I had not restricted the attention to this case in the analysis of waves, the absorption and dispersion of sound waves should have provided information about other transport coefficients as well.

4.5 Final Remarks

4.5.1 *The role of temperature.* The reason why temperature is such an appropriate variable in thermodynamics is that it is considered to be easily measurable and the reason for that is its continuity at a thermometric wall. Let us consider how this stands in non-equilibrium.

A thermometric wall does not actively participate in the processes on either side, in particular it does not produce or store entropy so that the entropy flux coming in on one side is going out on the other side. Thus, if e_i is the unit normal to the wall, we have

$$[\phi_i]e_i = 0. \tag{4.60}$$

The brackets denote the difference of the bracketed quantity on the two sides.

The heat flux $q_i = q_i^T + g_i$ (or rather its normal component) is also continuous on a thermometric wall, since such a wall does not store energy either. Therefore insertion of (4.27) into (4.60) gives

$$\left[\frac{1}{T}\right] q_i e_i = \left[\frac{1}{T^2 \varepsilon_T^I} \Delta g_i\right] e_i + \left[\frac{2}{5pT} p_{\langle ir \rangle} q_r^T\right] e_i. \tag{4.61}$$

If we neglect non-linear terms, the right-hand side is zero and (4.61) implies that indeed the temperature is continuous. That result, however, is only as good as the linear approximation, because (according to (4.61)) there is a jump of T, if on either or both sides of the wall the bracketed expressions cannot be neglected. In such a case, the temperature looses much of its value as a suitable variable.

4.5.2 *The dynamic pressure.* We recall (4.45) where $\varepsilon^T = \frac{3p}{2\varrho} = \frac{3k}{2m}T$ and $\varepsilon^I = \varepsilon^I(T)$ holds and write

$$\varepsilon = \frac{3k}{2m}T + \varepsilon^I(T) + 2\frac{p}{\xi}\frac{\varepsilon_T^I}{\varepsilon_T}\frac{\partial v_i}{\partial x_i}. \tag{4.62}$$

Thus ε contains a term proportional to $\frac{\partial v_i}{\partial x_i}$. In the kinetic theory of polyatomic gases, e.g. the works of (Wang Chang, Uhlenbeck and de Boer 1964) or of (Waldmann 1956) this is not done. Rather these authors define ε as

$$\varepsilon = \varepsilon_T^T + \varepsilon_T^I, \quad \text{hence} \quad \varepsilon_T^T = \frac{3p}{2\varrho} = \varepsilon - \varepsilon^I. \tag{4.63}$$

Elimination of $(\varepsilon - \varepsilon^I)$ from (4.62) and (4.63) leads to

$$p = \varrho\frac{k}{m}T - \varsigma\frac{\partial v_i}{\partial x_i} \quad \text{where} \quad \varsigma = -\frac{4}{3}\frac{k}{m}\frac{\varrho^2 T}{\xi}\frac{\varepsilon_T^I}{\varepsilon_T}, \tag{4.64}$$

if non-linear terms in $\frac{\partial v_i}{\partial x_i}$ are neglected.

It is customary to call the term $-\varsigma \frac{\partial v_i}{\partial x_i}$ the *dynamic pressure* and ς is the *volume viscosity*. By (4.31) ς is non-negative.

Thus it is possible to have either ε or p depending on $\frac{\partial v_i}{\partial x_i}$. Both versions are possible and equivalent as was already remarked by (Kogan 1973).

5. EXTENDED THERMODYNAMICS OF MIXTURE OF IDEAL GASES

The objective of extended thermodynamics of mixtures of ν ideal monatomic gases is the determination of $13\,\nu$ scalar fields of

$$\begin{array}{lll} \text{partial densities} & \varrho_\alpha, & \\ \text{partial velocities} & v_i^\alpha, & \\ \text{partial pressure tensors} & p_{ij}^\alpha & \text{and} \\ \text{partial heat fluxes} & q_i^\alpha, & (\alpha = 1, 2, ..., \nu). \end{array} \tag{5.1}$$

In this theory the exploitation of the requirement of material frame indifference and of the entropy principle leads to linear constitutive equations which, apart from unknown constants, are the same as those of the 13ν moment equations of the kinetic theory of gases.

The transition from extended thermodynamics to ordinary thermodynamics leads to constitutive equations for diffusion flux, heat flux and pressure tensor corresponding to the laws of Fick, Fourier and Navier-Stokes, respectively.

It is shown, without the use of statistical arguments, how the Onsager reciprocity relations in the presence of axial fields come out of the phenomenological extended thermodynamics of a rotating mixture of ideal gases.

This section is based on the works of (Kremer 1985 and 1987a).

5.1 Thermodynamic Processes

The necessary field equations for the determination of the 13ν fields (5.1) are based on the following balance equations

$$\frac{\partial \varrho_\alpha}{\partial t} + \frac{\partial \varrho_\alpha v_i^\alpha}{\partial x_i} = 0, \tag{5.2}$$

$$\varrho_\alpha\left(\frac{\partial v_i^\alpha}{\partial t}+\frac{\partial v_i^\alpha}{\partial x_j}v_j^\alpha\right)+\frac{\partial p_{ij}^\alpha}{\partial x_j}=P_i^\alpha+\varrho_\alpha(f_i^\alpha+I_i)+2W_{ij}\varrho_\alpha v_j^\alpha, \tag{5.3}$$

$$\frac{\partial p_{ij}^\alpha}{\partial t}+\frac{\partial p_{ij}^\alpha v_k^\alpha}{\partial x_k}+\frac{\partial p_{\langle ijk\rangle}^\alpha}{\partial x_k}+\frac{4}{5}\frac{\partial q_{(i}^\alpha}{\partial x_{j)}}+\frac{2}{5}\frac{\partial q_k^\alpha}{\partial x_k}\delta_{ij}+2p_{k(i}^\alpha\frac{\partial v_{j)}^\alpha}{\partial x_k}=P_{ij}^\alpha+4p_{k(i}^\alpha W_{j)k}, \tag{5.4}$$

$$\frac{\partial q_i^\alpha}{\partial t}+\frac{\partial q_i^\alpha v_j^\alpha}{\partial x_j}+\frac{1}{2}\frac{\partial p_{ijjk}^\alpha}{\partial x_k}+p_{\langle ijk\rangle}^\alpha\frac{\partial v_j^\alpha}{\partial x_k}+\frac{2}{5}q_j^\alpha\frac{\partial v_j^\alpha}{\partial x_i}+\frac{7}{5}q_j^\alpha\frac{\partial v_i^\alpha}{\partial x_j}$$
$$+\frac{2}{5}q_i^\alpha\frac{\partial v_k^\alpha}{\partial x_k}-\frac{p_{ij}^\alpha}{\varrho_\alpha}\frac{\partial p_{jk}^\alpha}{\partial x_k}-\frac{p_{rr}^\alpha}{2\varrho_\alpha}\frac{\partial p_{ik}^\alpha}{\partial x_k}=Q_i^\alpha-\frac{3}{2}\frac{p_{(ij}^\alpha}{\varrho_\alpha}P_{j)}^\alpha+2W_{ij}q_j^\alpha. \tag{5.5}$$

Eqs. (5.2)-(5.5) refer to non-inertial systems. f_i^α denotes the partial specific external body force and $I_i+2W_{ij}v_j^\alpha$ the inertial force which consists of the following components

$$I_i+2W_{ij}v_j^\alpha=\begin{cases}2W_{ij}(v_j^\alpha-\dot b_j) & \text{Coriolis force}\\ -W_{ij}^2(x_j-b_j) & \text{centrifugal force}\\ +\dot W_{ij}(x_j-b_j) & \text{Euler force}\\ +\ddot b_i & \text{force of relative translation}\end{cases} \tag{5.6}$$

In (5.6) $W_{ij}=-W_{ji}$ is the matrix of angular velocity of the frame with respect to an inertial frame while b_i the radius vector between the origins of the two frames.

The production term P_i^α is contrained by the requirement

$$\sum_{\alpha=1}^{\nu}P_i^\alpha=0, \tag{5.7}$$

which express the conservation of linear momentum of the mixture.

The system of balance equations (5.2)-(5.5) is closed by considering

$$p_{\langle ijk\rangle}^\alpha,\ p_{ijjk}^\alpha,\ P_i^\alpha,\ P_{ij}^\alpha,\quad\text{and}\quad Q_i^\alpha \tag{5.8}$$

as constitutive quantities that depend upon the fields given in (5.1).

If the explicit form of the constitutive functions is known, one can obtain an explicit set of field equations from Eqs. (5.2)-(5.5). Every solution of this system is called a ***thermodynamic process***.

5.2 Constitutive Theory.

The constitutive functions are restricted by the entropy principle and by the principle of material frame indifference.

From the exploitation of the principle of material frame indifference it follows that the constitutive functions cannot depend on the partial velocities v_i^β, but only on the relative velocities

$$V_i^\beta = v_i^\beta - v_i^\nu. \tag{5.9}$$

Moreover, the constitutive functions are isotropic functions of all remaining variables.

Since I am interested in a linear theory, I suppose that the constitutive equations for $\{p^\alpha_{\langle ijk\rangle}, p^\alpha_{ijjk}, P^\alpha_i, P^\alpha_{ij}, Q^\alpha_i\}$ are linear in the variables $p^\alpha_{\langle ij\rangle}$ (partial pressure deviators), q^α_i (partial heat fluxes) and V_i^α (relative velocities). For h and Φ_k and I will write down representations up to the third order in these variables.

According to the kinetic theory of mixtures of ideal gases the material quantities $p^\alpha_{\langle ijk\rangle}$ and p^α_{ijjk} depend only on the variables of the constituent α. Therefore I assume that

$$p^\alpha_{\langle ijk\rangle} = \hat{p}^\alpha_{\langle ijk\rangle}(\varrho_\alpha, V_i^\alpha, p^\alpha_{ij}, q^\alpha_i) \qquad \text{and} \qquad p^\alpha_{ijjk} = \hat{p}^\alpha_{ijjk}(\varrho_\alpha, V_i^\alpha, p^\alpha_{ij}, q^\alpha_i), \tag{5.10}$$

and write the representations for the constitutive quantities as follows

$$p^\alpha_{\langle ijk\rangle} = 0, \tag{5.11}$$

$$p^\alpha_{ijjk} = \chi^0_\alpha(\varrho_\alpha, p_\alpha)\delta_{ik} + \chi^1_\alpha(\varrho_\alpha, p_\alpha)p^\alpha_{\langle ik\rangle}, \tag{5.12}$$

$$P^\alpha_i = \sum_{\beta=1}^{\nu} M^q_{\alpha\beta} q^\beta_i + \sum_{\beta=1}^{\nu-1} M^V_{\alpha\beta} V^\beta_i, \tag{5.13}$$

$$P^\alpha_{ii} = R_\alpha, \tag{5.14}$$

$$P^\alpha_{\langle ij\rangle} = \sum_{\beta=1}^{\nu} \sigma_{\alpha\beta} p^\beta_{\langle ij\rangle}, \tag{5.15}$$

$$Q^\alpha_i = \sum_{\beta=1}^{\nu} Q^q_{\alpha\beta} q^\beta_i + \sum_{\beta=1}^{\nu-1} Q^V_{\alpha\beta} V^\beta_i, \tag{5.16}$$

$$h = h^0 + \frac{1}{2}\sum_{\beta=1}^{\nu}\sum_{\gamma=1}^{\nu}\{h^1_{\beta\gamma} p^\beta_{\langle rs\rangle} p^\gamma_{\langle rs\rangle} + h^2_{\beta\gamma} q^\beta_r q^\gamma_r\}$$
$$+\frac{1}{6}\sum_{\alpha=1}^{\nu}\sum_{\beta=1}^{\nu}\sum_{\gamma=1}^{\nu} h^3_{\alpha\beta\gamma} p^\alpha_{\langle rs\rangle} p^\beta_{\langle sk\rangle} p^\gamma_{\langle kr\rangle} + \frac{1}{2}\sum_{\alpha=1}^{\nu}\sum_{\beta=1}^{\nu}\sum_{\gamma=1}^{\nu} h^4_{\alpha\beta\gamma} p^\alpha_{\langle rs\rangle} q^\beta_r q^\gamma_s, \tag{5.17}$$

$$\Phi_k = \sum_{\alpha=1}^{\nu} \phi^1_\alpha q^\alpha_k + \sum_{\alpha=1}^{\nu-1} \phi^2_\alpha V^\alpha_k + \sum_{\alpha=1}^{\nu}\sum_{\beta=1}^{\nu} \phi^3_{\alpha\beta} p^\alpha_{\langle kr\rangle} q^\beta_r + \sum_{\alpha=1}^{\nu}\sum_{\beta=1}^{\nu-1} \phi^4_{\alpha\beta} p^\alpha_{\langle kr\rangle} V^\beta_r. \tag{5.18}$$

The coefficients $M^q_{\alpha\beta}$ through $\phi^4_{\alpha\beta}$ depend on the partial mass densities ϱ_γ and on the partial pressures $p_\gamma = p^\gamma_{rr}/3$. Moreover we have

$$\sum_{\alpha=1}^{\nu} M^V_{\alpha\beta} = \sum_{\alpha=1}^{\nu} M^q_{\alpha\beta} = 0 \qquad \text{because} \qquad \sum_{\alpha=1}^{\nu} P^\alpha_i = 0, \tag{5.19}$$

$$h^1_{\alpha\beta} = h^1_{\beta\alpha}, \quad h^2_{\alpha\beta} = h^2_{\beta\alpha}, \quad h^4_{\alpha\beta\gamma} = h^4_{\alpha\gamma\beta}, \quad h^3_{\alpha\beta\gamma} \quad \text{symmetric in } \alpha, \beta \text{ and } \gamma. \tag{5.20}$$

5.3 Results of the Constitutive Theory

From the exploitation of the entropy inequality with the Lagrange multipliers it follows the equation

$$d\eta^0_\alpha = d\left(\frac{h^0_\alpha}{\varrho_\alpha}\right) = \frac{2}{3}\frac{\partial h^0_\alpha}{\partial p_\alpha}\left(d\varepsilon_\alpha - \frac{p_\alpha}{\varrho^2_\alpha}d\varrho_\alpha\right), \tag{5.21}$$

where $\varepsilon_\alpha = \frac{3p_\alpha}{2\varrho_\alpha}$ is the specific internal energy of the constituent α in the mixture. Thus by comparison with the Gibbs equation $\frac{2}{3}\frac{\partial h^0_\alpha}{\partial p_\alpha}$ is identified as $\frac{1}{T_\alpha}$ where T_α is the absolute temperature of the constituent α in the mixture.

Once the absolute temperature is introduced, it is possible to replace the pair $(\varrho_\alpha, p_\alpha)$ of equilibrium variables by the more familiar pair $(\varrho_\alpha, T_\alpha)$ and I shall do this in the sequel.

The Gibbs equation implies the integrability condition

$$\varrho_\alpha\frac{\partial p_\alpha}{\partial \varrho_\alpha} + \frac{2}{3}T_\alpha\frac{\partial p_\alpha}{\partial T_\alpha} - \frac{5}{3}p_\alpha = 0, \tag{5.22}$$

whose general solution reads

$$p_\alpha = T^{5/2}_\alpha F_\alpha(z_\alpha) \qquad \text{with} \qquad z_\alpha = \frac{\varrho_\alpha}{T^{3/2}_\alpha}. \tag{5.23}$$

F_α are arbitrary functions. Eq. (5.23) represents the general form of the thermal equation of state for ideal gases.

Moreover, the constitutive equations (5.12), (5.17) and (5.18) have explicit forms in terms of the functions $F_\alpha(z_\alpha)$, viz.

$$p^\alpha_{ijjk} = \left\{5T^{7/2}_\alpha\left(\int\frac{F_\alpha F'_\alpha}{z_\alpha}dz_\alpha + a^1_\alpha\right) + a^2_\alpha\right\}\delta_{ik} + 7\frac{T_\alpha}{F_\alpha}\left(\int\frac{F_\alpha F'_\alpha}{z_\alpha}dz_\alpha + a^1_\alpha\right)p^\alpha_{\langle ik\rangle}, \tag{5.24}$$

$$h = \sum_{\alpha=1}^{\nu}\left\{\varrho_\alpha\left[\frac{3}{2}\int\left(\frac{F'_\alpha}{z_\alpha} - \frac{5}{3}\frac{F_\alpha}{z^2_\alpha}\right)dz_\alpha + a^3_\alpha\right] - \frac{p^\alpha_{\langle rs\rangle}p^\alpha_{\langle rs\rangle}}{4p_\alpha T_\alpha}\right.$$
$$\left. - \frac{2q^\alpha_r q^\alpha_r}{5p_\alpha T_\alpha\left[7\frac{T_\alpha}{F_\alpha}\left(\int\frac{F_\alpha F'_\alpha}{z_\alpha}dz_\alpha + a^1_\alpha\right) - 5\frac{p_\alpha}{\varrho_\alpha}\right]}\right\}, \tag{5.25}$$

$$\Phi_k = \sum_{\alpha=1}^{\nu}\left\{\frac{q^\alpha_k}{T_\alpha} + \varrho_\alpha\left[\frac{3}{2}\int\left(\frac{F'_\alpha}{z_\alpha} - \frac{5}{3}\frac{F_\alpha}{z^2_\alpha}\right)dz_\alpha + a^3_\alpha\right]u^\alpha_k - \frac{2p^\alpha_{\langle kr\rangle}q^\alpha_r}{5p_\alpha T_\alpha}\right\}, \tag{5.26}$$

where a^1_α, a^2_α, a^3_α are constants.

Since the thermal equation of state of a classical ideal gas is given by

$$p_\alpha = \varrho_\alpha\frac{k}{m_\alpha}T_\alpha \qquad \text{i.e.} \qquad F_\alpha(z_\alpha) = \frac{k}{m_\alpha}z_\alpha, \tag{5.27}$$

insertion of Eq. (5.27) into Eqs. (5.24)-(5.26) gives the constitutive equations of p^α_{ijjk}, h and Φ_k for a mixture of classical ideal gases which read

$$p^\alpha_{ijjk} = \Big(5\frac{p_\alpha^2}{\varrho_\alpha} + 5T_\alpha^{7/2}a_\alpha^1 + a_\alpha^2\Big)\delta_{ik} + \Big(7\frac{p_\alpha}{\varrho_\alpha} + 7\frac{T_\alpha^{7/2}}{p_\alpha}a_\alpha^1\Big)p^\alpha_{\langle ik\rangle} \tag{5.28}$$

$$h = \sum_{\alpha=1}^{\nu}\Big\{\frac{k}{m_\alpha}\varrho_\alpha\Big(ln\frac{T_\alpha^{3/2}}{\varrho_\alpha} + a_\alpha^3\Big) - \frac{p^\alpha_{\langle rs\rangle}p^\alpha_{\langle rs\rangle}}{4p_\alpha T_\alpha} - \Big[\frac{5p_\alpha^2 T_\alpha}{\varrho_\alpha}\Big(1 + \frac{7T_\alpha^{7/2}\varrho_\alpha a_\alpha^1}{2p_\alpha^2}\Big)\Big]^{-1} q_r^\alpha q_r^\alpha\Big\}, \tag{5.29}$$

$$\Phi_k = \sum_{\alpha=1}^{\nu}\Big\{\frac{q_k^\alpha}{T_\alpha} + \frac{k}{m_\alpha}\varrho_\alpha\Big[ln\frac{T_\alpha^{3/2}}{\varrho_\alpha} + a_\alpha^3\Big]u_k^\alpha - \frac{2p^\alpha_{\langle kr\rangle}q_r^\alpha}{5p_\alpha T_\alpha}\Big\}. \tag{5.30}$$

k is the Boltzmann constant and m_α is the molecular mass of constituent α. There is a complete agreement of this theory with Grad's method of $13\,\nu$ moments in the kinetic theory of gases providing the constants a_α^1 and a_α^2 in the above equation are choosen to be zero.

5.4 Rational Thermodynamcs of Mixtures of 5ν Fields

5.4.1 *Constitutive equations in rational thermodynamics.* The objective of rational thermodynamics of mixtures of ν ideal gases is the determination of the 5ν fields of (Truesdell 1984, Müller 1985)

$$\begin{array}{lll} \text{partial mass densities} & \varrho_\alpha, & \\ \text{partial velocities} & v_i^\alpha & \text{and} \\ \text{partial temperatures} & T_\alpha, & (\alpha = 1, 2, ..., \nu). \end{array} \tag{5.31}$$

To reach this objective the balance equations (5.2) and (5.3) and the trace of the balance equation (5.4) are used.

$$\frac{\partial \varrho_\alpha}{\partial t} + \frac{\partial \varrho_\alpha v_i^\alpha}{\partial x_i} = 0, \tag{5.32}$$

$$\varrho_\alpha\Big(\frac{\partial v_i^\alpha}{\partial t} + \frac{\partial v_i^\alpha}{\partial x_j}v_j^\alpha\Big) + \frac{\partial p_{ij}^\alpha}{\partial x_j} = P_i^\alpha + \varrho_\alpha(f_i^\alpha + I_i) + 2W_{ij}\varrho_\alpha v_j^\alpha, \tag{5.33}$$

$$\varrho_\alpha\Big(\frac{\partial \varepsilon_\alpha}{\partial t} + \frac{\partial \varepsilon_\alpha}{\partial x_i}v_i^\alpha\Big) + \frac{\partial q_i^\alpha}{\partial x_i} + p_{ki}^\alpha\frac{\partial v_i^\alpha}{\partial x_k} = \frac{1}{2}P_{ii}^\alpha. \tag{5.34}$$

The constitutive quantities of rational thermodynamics of mixtures of ideal gases are the partial pressure deviators $p^\alpha_{\langle ij\rangle}$, the partial heat fluxes q_i^α, and the production terms P_i^α and P_{ii}^α. For a viscous simple mixtures the constitutive equations depend on $(\varrho_\alpha, T_\alpha, V_i^\alpha, \frac{\partial T_\alpha}{\partial x_i}, \frac{\partial v^\alpha_{(i}}{\partial x_{j)}})$. In order to get such constitutive equations I shall use the traceless part of Eq. (5.4) and Eq. (5.5). Hence, the substitution of Eqs. (5.11), (5.15), (5.16) and (5.24) into those equations leads to

$$\left\{\delta_{ri}\delta_{sj}\frac{d}{dt}+\delta_{ri}\left(\frac{\partial v_{[j}}{\partial x_{s]}}-2W_{js}\right)+\delta_{rj}\left(\frac{\partial v_{[i}}{\partial x_{s]}}-2W_{is}\right)\right\}p^{\alpha}_{\langle rs\rangle}+\frac{\partial p^{\alpha}_{\langle ij\rangle}}{\partial x_k}u^{\alpha}_k$$
$$+\frac{5}{3}p^{\alpha}_{\langle ij\rangle}\frac{\partial v_r}{\partial x_r}+\frac{4}{5}\frac{\partial q^{\alpha}_{\langle i}}{\partial x_{j\rangle}}+p^{\alpha}_{\langle ki\rangle}\left(\frac{\partial v^{\alpha}_{\langle j}}{\partial x_{k\rangle}}+\frac{\partial u^{\alpha}_{[j}}{\partial x_{k]}}\right)+p^{\alpha}_{\langle kj\rangle}\left(\frac{\partial v^{\alpha}_{\langle i}}{\partial x_{k\rangle}}+\frac{\partial u^{\alpha}_{[i}}{\partial x_{k]}}\right)$$
$$-\frac{2}{3}p^{\alpha}_{\langle kr\rangle}\frac{\partial v^{\alpha}_{\langle r}}{\partial x_{k\rangle}}\delta_{ij}+2p_{\alpha}\frac{\partial v^{\alpha}_{\langle i}}{\partial x_{j\rangle}}+\frac{5}{3}p^{\alpha}_{\langle ij\rangle}\frac{\partial u^{\alpha}_r}{\partial x_r}=\sum_{\beta=1}^{\nu}\sigma_{\alpha\beta}p^{\beta}_{\langle ij\rangle}, \tag{5.35}$$

$$\left(\delta_{ij}\frac{d}{dt}+\frac{\partial v_{[i}}{\partial x_{j]}}-2W_{ij}\right)q^{\alpha}_j+\frac{\partial q^{\alpha}_i}{\partial x_j}u^{\alpha}_j+2q^{\alpha}_i\frac{\partial u^{\alpha}_k}{\partial x_k}+2q^{\alpha}_i\frac{\partial v_k}{\partial x_k}$$
$$+\frac{\partial u^{\alpha}_{[i}}{\partial x_{k]}}q^{\alpha}_k+\frac{9}{5}q^{\alpha}_j\frac{\partial v^{\alpha}_{\langle i}}{\partial x_{j\rangle}}+\frac{5p_{\alpha}}{4T_{\alpha}}\left(\chi^1_{\alpha}-\frac{5p_{\alpha}}{\varrho_{\alpha}}\right)\frac{\partial T_{\alpha}}{\partial x_i}-\frac{p^{\alpha}_{\langle ij\rangle}}{\varrho_{\alpha}}\frac{\partial p^{\alpha}_{\langle jk\rangle}}{\partial x_k}$$
$$+\left(\chi^1_{\alpha}-\frac{5p_{\alpha}}{\varrho_{\alpha}}\right)\left(\frac{7}{4T_{\alpha}}\frac{\partial T_{\alpha}}{\partial x_k}-\frac{1}{2p_{\alpha}}\frac{\partial p_{\alpha}}{\partial x_k}\right)p^{\alpha}_{\langle ik\rangle}+\frac{1}{2}\left(\chi^1_{\alpha}-\frac{5p_{\alpha}}{\varrho_{\alpha}}\right)\frac{\partial p^{\alpha}_{\langle ik\rangle}}{\partial x_k}$$
$$=\sum_{\beta=1}^{\nu}\left(Q^q_{\alpha\beta}-\frac{5p_{\alpha}}{2\varrho_{\alpha}}M^q_{\alpha\beta}\right)q^{\beta}_i+\sum_{\beta=1}^{\nu-1}\left(Q^V_{\alpha\beta}-\frac{5p_{\alpha}}{2\varrho_{\alpha}}M^V_{\alpha\beta}\right)V^{\beta}_i, \tag{5.36}$$

where χ^1_{α} is given by

$$\chi^1_{\alpha}=\frac{7T_{\alpha}}{F_{\alpha}}\left(\int\frac{F_{\alpha}F'_{\alpha}}{z_{\alpha}}dz_{\alpha}+a^1_{\alpha}\right). \tag{5.37}$$

In Eqs. (5.35) and (5.36) v_i and u^{α}_i are the velocity of the mixture and the velocity of diffusion of the constituent α, respectively. They are defined by

$$v_i=\frac{1}{\varrho}\sum_{\alpha=1}^{\nu}\varrho_{\alpha}v^{\alpha}_i,\qquad u^{\alpha}_i=v^{\alpha}_i-v_i. \tag{5.38}$$

$\varrho=\sum_{\alpha=1}^{\nu}\varrho_{\alpha}$ denotes the mass density of the mixture.

Now the constitutive equations for $p^{\alpha}_{\langle ij\rangle}$ and q^{α}_i are obtained from Eqs. (5.35) and (5.36) by using a method akin to the Maxwellian procedure (Ikenberry and Truesdell 1956). For the first iteration step the equilibrium values $p^{\alpha(0)}_{\langle ij\rangle}=0$ and $q^{\alpha(0)}_i=0$ are inserted on the left-hand side of Eqs. (5.35) and (5.36) and, on the right-hand side, it follows the first iterated values $p^{\alpha(1)}_{\langle ij\rangle}$ and $q^{\alpha(1)}_i$, viz.

$$p^{\alpha(1)}_{\langle ij\rangle}=-2\sum_{\beta=1}^{\nu}\mu_{\alpha\beta}\frac{\partial v^{\beta}_{\langle i}}{\partial x_{j\rangle}}, \tag{5.39}$$

$$q^{\alpha(1)}_i=-\sum_{\beta=1}^{\nu}\kappa_{\alpha\beta}\frac{\partial T_{\beta}}{\partial x_i}-\sum_{\beta=1}^{\nu-1}D_{\alpha\beta}J^{\beta}_i, \tag{5.40}$$

where

$$\mu_{\alpha\beta} = -\sigma^{-1}_{\alpha\beta}p_\beta, \qquad D_{\alpha\beta} = [B^{-1}EF]_{\alpha\beta}, \qquad \kappa_{\alpha\beta} = -\frac{5}{4}B^{-1}_{\alpha\beta}\frac{p_\beta}{T_\beta}\left(\chi^1_\beta - \frac{5p_\beta}{\varrho_\beta}\right), \tag{5.41}$$

$$B_{\alpha\beta} = \left(Q^q_{\alpha\beta} - \frac{5p_\alpha}{\varrho_\alpha}M^q_{\alpha\beta}\right), \qquad E_{\alpha\beta} = \left(Q^V_{\alpha\beta} - \frac{5p_\alpha}{\varrho_\alpha}M^V_{\alpha\beta}\right), \qquad F_{\beta\alpha} = \left(\frac{\delta_{\alpha\beta}}{\varrho_\alpha} + \frac{1}{\varrho_\nu}\right). \tag{5.42}$$

In Eq. (5.40) I have introduced the ***diffusion flux*** J^α_i of the constituent α in the mixture which is defined by $J^\alpha_i = \varrho_\alpha u^\alpha_i$.

For the second iteration step we insert the first iterated values $p^{\alpha(1)}_{\langle ij\rangle}$ and $q^{\alpha(1)}_i$ on the left-hand side of Eqs. (5.35) and (5.36) and get the second iterated values $p^{\alpha(2)}_{\langle ij\rangle}$ and $q^{\alpha(2)}_i$

$$\begin{aligned}
p^{\alpha(2)}_{\langle ij\rangle} = {} & -2\sum_{\beta=1}^{\nu}\mu_{\alpha\beta}\frac{\partial v^\beta_{\langle i}}{\partial x_{j\rangle}} - \frac{4}{5}\sum_{\gamma=1}^{\nu}\sigma^{-1}_{\alpha\gamma}\frac{\partial}{\partial x_{\langle j}}\left[\sum_{\beta=1}^{\nu}\kappa_{\gamma\beta}\frac{\partial T_\beta}{\partial x_{i\rangle}} + \sum_{\beta=1}^{\nu-1}D_{\gamma\beta}J^\beta_{i\rangle}\right] \\
& -\frac{10}{3}\sum_{\beta=1}^{\nu}\sum_{\gamma=1}^{\nu}\sigma^{-1}_{\alpha\gamma}\mu_{\gamma\beta}\frac{\partial v^\beta_{\langle i}}{\partial x_{j\rangle}}\frac{\partial v_r}{\partial x_r} - \underline{2\sum_{\beta=1}^{\nu}\sum_{\gamma=1}^{\nu}\sigma^{-1}_{\alpha\gamma}\left\{\delta_{ri}\delta_{sj}\frac{d}{dt}\right.} \\
& \underline{\left. +\delta_{ri}\left(\frac{\partial v_{[j}}{\partial x_{s]}} - 2W_{js}\right) + \delta_{rj}\left(\frac{\partial v_{[i}}{\partial x_{s]}} - 2W_{is}\right)\right\}\left(\mu_{\gamma\beta}\frac{\partial v^\beta_{\langle r}}{\partial x_{s\rangle}}\right)},
\end{aligned} \tag{5.43}$$

$$\begin{aligned}
q^{\alpha(2)}_i = {} & -\sum_{\beta=1}^{\nu}\kappa_{\alpha\beta}\frac{\partial T_\beta}{\partial x_i} - \sum_{\beta=1}^{\nu-1}D_{\alpha\beta}J^\beta_i + \frac{4}{5}\sum_{\beta=1}^{\nu}\sum_{\gamma=1}^{\nu}\kappa_{\alpha\beta}\frac{T_\gamma}{p_\gamma}\frac{\partial}{\partial x_j}\left(\mu_{\gamma\beta}\frac{\partial v^\beta_{\langle i}}{\partial x_{j\rangle}}\right) \\
& -2\sum_{\gamma=1}^{\nu}B^{-1}_{\alpha\gamma}\left(\sum_{\beta=1}^{\nu}\kappa_{\gamma\beta}\frac{\partial T_\beta}{\partial x_i} + \sum_{\beta=1}^{\nu-1}D_{\gamma\beta}J^\beta_i\right)\frac{\partial v_r}{\partial x_r} - \frac{4}{5}\sum_{\beta=1}^{\nu}\sum_{\gamma=1}^{\nu}\kappa_{\alpha\gamma}\mu_{\gamma\beta}\frac{T_\gamma}{p^2_\gamma}\frac{\partial p_\gamma}{\partial x_j}\frac{\partial v^\beta_{\langle i}}{\partial x_{j\rangle}} \\
& \underline{-\sum_{\gamma=1}^{\nu}B^{-1}_{\alpha\gamma}\left(\delta_{ij}\frac{d}{dt} + \frac{\partial v_{[i}}{\partial x_{j]}} - 2W_{ij}\right)\left(\sum_{\beta=1}^{\nu}\kappa_{\gamma\beta}\frac{\partial T_\beta}{\partial x_j} + \sum_{\beta=1}^{\nu-1}D_{\gamma\beta}J^\beta_j\right)}.
\end{aligned} \tag{5.44}$$

I have not considered non-linear terms in Eqs. (5.43) and (5.44) since I have assumed linear representations for the constitutive quantities.

5.4.2 *The laws of Navier-Stokes and Fourier. The dependence on the frame of the constitutive equations.* The first iterated values of the partial pressure deviators $p^{\alpha(1)}_{\langle ij\rangle}$ and of the partial heat fluxes $q^{\alpha(1)}_i$ (Eqs. (5.39) and (5.40)) represent, respectively, the laws of Navier-Stokes and Fourier appropriated to a mixture of ν ideal gases in the 5ν-field theory.

The second iterated values are more complex since they depend on second derivatives of the basic fields. Moreover, the constitutive equations (5.43) and (5.44) depend on the frame because the underlined terms contain the matrix of angular velocity W_{ij} of the frame with respect to an inertial frame. The dependence on the frame contradicts the principle of material frame indifference and a discussion and interpretation of these dependence can be found in (Müller 1972, Heck and Müller 1983 and Müller 1985).

5.5 Linear Irreversible Thermodynamics of Mixtures of $(\nu + 4)$ fields.

5.5.1 *Constitutive equations in linear irreversible thermodynamics.I.* The objective of linear irreversible thermodynamics is the determination of the $(\nu + 4)$ fields of (Meixner and Reik 1959 and de Groot and Mazur 1962)

density $\quad \varrho = \sum_{\alpha=1}^{\nu} \varrho_\alpha,$

concentrations $\quad c_\alpha = \frac{\varrho_\alpha}{\varrho} \qquad (\alpha = 1, 2, ..., \nu - 1),$ (5.45)

velocity $\quad v_i = \sum_{\alpha=1}^{\nu} \frac{\varrho_\alpha}{\varrho} v_i^\alpha \qquad$ and

absolute temperature $\quad T.$

From now on I assume that the constituents of the mixture have the same temperature T.

To achieve this objective the following field equations, which result from Eqs. (5.32)-(5.34) through addition and rearrangement, are used

$$\dot{\varrho} + \varrho \frac{\partial v_i}{\partial x_i} = 0, \tag{5.46}$$

$$\dot{c}_\alpha + \frac{1}{\varrho}\frac{\partial J_i^\alpha}{\partial x_i} = 0, \qquad (\alpha = 1, 2, ..., \nu - 1) \tag{5.47}$$

$$\varrho \dot{v}_i + \frac{\partial p_{ij}}{\partial x_j} = \varrho(f_i + I_i) + 2W_{ij}\varrho v_j, \tag{5.48}$$

$$\varrho \dot{\varepsilon} + \frac{\partial q_i}{\partial x_i} + p_{ij}\frac{\partial v_i}{\partial x_j} = 0. \tag{5.49}$$

In the above equations I have introduced the pressure tensor of the mixture p_{ij}, the specific internal energy of the mixture ε and the heat flux of the mixture q_i which are defined by

$$p_{ij} = \sum_{\alpha=1}^{\nu} (p_{ij}^\alpha + \varrho_\alpha u_i^\alpha u_j^\alpha), \tag{5.50}$$

$$\varepsilon = \frac{3p}{2\varrho} = \frac{3}{2\varrho}\sum_{\alpha=1}^{\nu} p_\alpha = \frac{3}{2\varrho}\sum_{\alpha=1}^{\nu} T^{5/2} F_\alpha\left(\frac{\varrho_\alpha}{T^{3/2}}\right), \tag{5.51}$$

$$q_i = \sum_{\alpha=1}^{\nu}\left(q_i^\alpha + p_{(ij)}^\alpha u_j^\alpha + \frac{5}{2}p_\alpha u_i^\alpha + \frac{1}{2}\varrho_\alpha u_\alpha^2 u_i^\alpha\right). \tag{5.52}$$

The constitutive quantities of linear irreversible thermodynamics of mixtures of ideal gases are the pressure deviator $p_{\langle ij\rangle}$, the heat flux q_i and the diffusion fluxes J_i^α.

In rational thermodynamics of mixtures we know how $p_{\langle ij\rangle}$ and q_i depend on (ϱ_α, T_α, $\frac{\partial T_\alpha}{\partial x_i}$, J_i^α, $\frac{\partial v^\alpha_{\langle i}}{\partial x_{j\rangle}}$). Below I write the first iterated value of q_i^α for a mixture whose constituents are at the same temperature

$$q_i^{\alpha(1)} = -\kappa_\alpha \frac{\partial T}{\partial x_i} - \sum_{\beta=1}^{\nu-1} D_{\alpha\beta} J_i^{\beta(1)}, \tag{5.53}$$

where

$$\kappa_\alpha = -\frac{5}{4} \sum_{\beta=1}^{\nu} B_{\alpha\beta}^{-1} \frac{p_\beta}{T} \Big(\chi_\beta^1 - \frac{5p_\beta}{\varrho_\beta}\Big). \tag{5.54}$$

In order to get the diffusion fluxes J_i^α the equations

$$\begin{aligned}
&\Big(\delta_{ij}\frac{d}{dt} + \frac{\partial v_{[i}}{\partial x_{j]}} - 2W_{ij}\Big)\Big(\sum_{\beta=1}^{\nu-1} F_{\alpha\beta} J_j^\beta\Big) - \frac{1}{\varrho_\alpha}\frac{\partial p_\alpha}{\partial x_i} - \frac{1}{\varrho_\nu}\frac{\partial p_\nu}{\partial x_i} + \frac{1}{\varrho_\alpha}\frac{\partial p^\alpha_{\langle ij\rangle}}{\partial x_j} - \frac{1}{\varrho_\nu}\frac{\partial p^\nu_{\langle ij\rangle}}{\partial x_j} \\
&+\frac{\partial u^\alpha_{\langle i}}{\partial x_{j\rangle}} u_j^\alpha - \frac{\partial u^\nu_{\langle i}}{\partial x_{j\rangle}} u_j^\nu + \frac{1}{3}\frac{\partial u_r^\alpha}{\partial x_r} u_i^\alpha - \frac{1}{3}\frac{\partial u_r^\nu}{\partial x_r} u_i^\nu + \frac{\partial v_{\langle i}}{\partial x_{j\rangle}} \sum_{\beta=1}^{\nu-1} F_{\alpha\beta} J_j^\beta + \frac{1}{3}\frac{\partial v_r}{\partial x_r} \sum_{\beta=1}^{\nu-1} F_{\alpha\beta} J_i^\beta \\
&+\frac{\partial u^\alpha_{[i}}{\partial x_{j]}} u_j^\alpha - \frac{\partial u^\nu_{[i}}{\partial x_{j]}} u_j^\nu = \sum_{\beta=1}^{\nu} [FM^q]_{\alpha\beta} q_i^\beta + \sum_{\beta=1}^{\nu-1} [FM^V F]_{\alpha\beta} J_i^\beta, \ (\alpha = 1, 2, ..., \nu-1)
\end{aligned} \tag{5.55}$$

are used. They follow from Eq. (5.32) by subtraction of the νth equation from the αth equation. Moreover, I have performed some rearrangements, inserted the constitutive equation of P_i^α (5.13) and assumed that the partial specific body forces f_i^α are the same for all $\alpha = 1, 2, ..., \nu$; i.e. $f_i^\alpha = f_i$.

For the first iteration step the equilibrium values of J_i^α and $p^\alpha_{\langle ij\rangle}$, namely $J_i^{\alpha(0)} = 0$ and $p^{\alpha(0)}_{\langle ij\rangle} = 0$, are inserted on the left-hand side of (5.55) and it follows

$$\begin{aligned}
J_i^{\alpha(1)} = &-\Big[T^2 \sum_{\gamma=1}^{\nu} [G^{-1} FM^q]_{\alpha\gamma} \kappa_\gamma + \frac{5}{2} T \sum_{\gamma=1}^{\nu-1} G^{-1}_{\alpha\gamma} \Big(\frac{p_\gamma}{\varrho_\gamma} - \frac{p_\nu}{\varrho_\nu}\Big)\Big] \frac{\partial}{\partial x_i}\Big(\frac{1}{T}\Big) \\
&-\sum_{\beta=1}^{\nu-1} T G^{-1}_{\alpha\beta} \frac{\partial}{\partial x_i}\Big(-\frac{\mu_\beta - \mu_\nu}{T}\Big),
\end{aligned} \tag{5.56}$$

where μ_α is the chemical potential of the constituent α in the mixture and

$$G_{\alpha\beta} = [FM^V F]_{\alpha\beta} - [FM^q D]_{\alpha\beta}. \tag{5.57}$$

The first iterated value of the heat flux of the mixture is obtained by insertion of Eqs. (5.39), (5.53) and (5.56) into Eq. (5.52) and by linearization

$$q_i^{(1)} = \sum_{\alpha=1}^{\nu}\Big\{T^2\kappa_\alpha - \sum_{\beta=1}^{\nu-1}\Big[\frac{5}{2}\Big(\frac{p_\alpha}{\varrho_\alpha} - \frac{p_\nu}{\varrho_\nu}\Big)\delta_{\alpha\beta} - D_{\alpha\beta}\Big]\Big[T^2\sum_{\gamma=1}^{\nu}[G^{-1}FM^q]_{\beta\gamma}\kappa_\gamma$$
$$+\frac{5}{2}T\sum_{\gamma=1}^{\nu-1} G^{-1}_{\beta\gamma}\Big(\frac{p_\gamma}{\varrho_\gamma} - \frac{p_\nu}{\varrho_\nu}\Big)\Big]\Big\}\frac{\partial}{\partial x_i}\Big(\frac{1}{T}\Big) - \sum_{\beta=1}^{\nu-1} T\Big\{\sum_{\gamma=1}^{\nu-1}\frac{5}{2}\Big(\frac{p_\gamma}{\varrho_\gamma} - \frac{p_\nu}{\varrho_\nu}\Big)G^{-1}_{\gamma\beta}$$
$$-\sum_{\alpha=1}^{\nu}[DG^{-1}]_{\alpha\beta}\Big\}\frac{\partial}{\partial x_i}\Big(-\frac{\mu_\beta - \mu_\nu}{T}\Big). \tag{5.58}$$

Moreover, the first iterated value of the pressure deviator of the mixture is obtained by insertion of Eqs. (5.39) and (5.56) into Eq. (5.50) and by linearization

$$p^{(1)}_{\langle ij\rangle} = -2\sum_{\alpha=1}^{\nu}\sum_{\beta=1}^{\nu}\mu_{\alpha\beta}\frac{\partial v^\beta_{\langle i}}{\partial x_{j\rangle}}. \tag{5.59}$$

5.5.2 *Laws of Fick, Navier-Stokes and Fourier. The Onsager relations.* First I write Eqs. (5.56) and (5.58) as

$$J_i^{\alpha(1)} = \sum_{\beta=1}^{\nu-1} L_{\alpha\beta}\frac{\partial}{\partial x_i}\Big(-\frac{\mu_\beta - \mu_\nu}{T}\Big) + \tilde{L}_\alpha\frac{\partial}{\partial x_i}\Big(\frac{1}{T}\Big), \tag{5.60}$$

$$q_i^{(1)} = T^2\kappa\frac{\partial}{\partial x_i}\Big(\frac{1}{T}\Big) + \sum_{\alpha=1}^{\nu-1} L_\alpha\frac{\partial}{\partial x_i}\Big(-\frac{\mu_\alpha - \mu_\nu}{T}\Big), \tag{5.61}$$

by introducing

$$L_{\alpha\beta} = -TG^{-1}_{\alpha\beta}, \tag{5.62}$$

$$\tilde{L}_\alpha = -\Big\{T^2\sum_{\gamma=1}^{\nu}[G^{-1}FM^q]_{\alpha\gamma}\kappa_\gamma + \frac{5}{2}T\sum_{\gamma=1}^{\nu-1} G^{-1}_{\alpha\gamma}\Big(\frac{p_\gamma}{\varrho_\gamma} - \frac{p_\nu}{\varrho_\nu}\Big)\Big\}, \tag{5.63}$$

$$L_\alpha = -T\Big\{\sum_{\gamma=1}^{\nu-1}\frac{5}{2}\Big(\frac{p_\gamma}{\varrho_\gamma} - \frac{p_\nu}{\varrho_\nu}\Big)G^{-1}_{\gamma\alpha} - \sum_{\gamma=1}^{\nu}[DG^{-1}]_{\gamma\alpha}\Big\}, \tag{5.64}$$

$$\kappa = \sum_{\alpha=1}^{\nu}\Big\{\kappa_\alpha - \sum_{\beta=1}^{\nu-1}\Big[\frac{5}{2}\Big(\frac{p_\alpha}{\varrho_\alpha} - \frac{p_\nu}{\varrho_\nu}\Big)\delta_{\alpha\beta} - D_{\alpha\beta}\Big]\Big[\sum_{\gamma=1}^{\nu}[G^{-1}FM^q]_{\beta\gamma}\kappa_\gamma$$
$$+\frac{5}{2T}\sum_{\gamma=1}^{\nu-1} G^{-1}_{\beta\gamma}\Big(\frac{p_\gamma}{\varrho_\gamma} - \frac{p_\nu}{\varrho_\nu}\Big)\Big]\Big\}. \tag{5.65}$$

Eqs. (5.59), (5.60) and (5.61) represent, respectively, the laws of Navier-Stokes, Fick and Fourier for a mixture and we identify the coefficients as

$\mu_{\alpha\beta}$ – coefficients of shear viscosity,

$L_{\alpha\beta}$ – diffusion coefficients ,

$\tilde{L}_\alpha$ – coefficients of thermal-diffusion ,

L_α – diffusion-thermo coefficients ,

κ – coefficient of thermal conductivity .

In linear irreversible thermodynamics the following symmetric relations are postulated (de Groot and Mazur 1962)

$$L_{\alpha\beta} = L_{\beta\alpha}, \qquad L_\alpha = \tilde{L}_\alpha, \tag{5.66}$$

which are known as the Onsager reciprocity relations.

In this case, one can read off from Eqs. (5.62)-(5.64) that the Onsager reciprocity relations (5.66) are valid if

$$G_{\alpha\beta} \text{ is symmetric and } \quad D_{\alpha\beta} = 0, \qquad M^q_{\alpha\beta} = 0. \tag{5.67}$$

The condition $D_{\alpha\beta} = 0$ holds when the partial heat fluxes do not depend on the diffusion fluxes, while the condition $M^q_{\alpha\beta} = 0$ holds when the productions of momentum densities do not depend on the partial heat fluxes. Both assumptions are almost the same as those of (Müller 1973).

On the other hand, insertion of Eq. (5.67) into Eq. (5.57) leads to

$$G_{\alpha\beta} = [FM^V F]_{\alpha\beta}. \tag{5.68}$$

We read off from Eq. (5.68) that $G_{\alpha\beta}$ is symmetric if $M^V_{\alpha\beta}$ is. But $M^V_{\alpha\beta}$ is symmetric if one assumes that

(*i*) $M^V_{\alpha\beta}$ is independent of the density ϱ_γ of constituent γ for all $\gamma \neq \alpha, \beta$;

(*ii*) $M^V_{\alpha\beta}$ tends to zero if ϱ_α and/or ϱ_β tends to zero.

The first assumption means that there are only binary drags, while the second means that the binary drags are zero if one constituent is absent. These assumption are due to (Truesdell 1962).

Hence the expressions for the coefficients given in Eqs. (5.62)-(5.65) which confirm the symmetry relations (5.66) are given by

$$L_{\alpha\beta} = -T[F^{-1}(M^V)^{-1}F^{-1}]_{\alpha\beta}, \qquad L_{\alpha\beta} = L_{\beta\alpha}, \tag{5.69}$$

$$L_\alpha = \tilde{L}_\alpha = -\frac{5T}{2}\sum_{\gamma=1}^{\nu-1}[F^{-1}(M^V)^{-1}F^{-1}]_{\alpha\gamma}\left(\frac{p_\gamma}{\varrho_\gamma} - \frac{p_\nu}{\varrho_\nu}\right), \tag{5.70}$$

$$\kappa = \sum_{\alpha=1}^{\nu}\kappa_\alpha - \frac{25}{4T}\sum_{\beta=1}^{\nu-1}\sum_{\gamma=1}^{\nu-1}\left(\frac{p_\beta}{\varrho_\beta} - \frac{p_\nu}{\varrho_\nu}\right)[F^{-1}(M^V)^{-1}F^{-1}]_{\beta\gamma}\left(\frac{p_\gamma}{\varrho_\gamma} - \frac{p_\nu}{\varrho_\nu}\right). \tag{5.71}$$

5.5.3 *Constitutive equations in linear irreversible thermodynamics.II.* Here I shall make use of the assumptions of Section 5.5.2 which have led to the confirmation of the Onsager reciprocity relations. Moreover I am interested only on constitutive equations for the diffusion fluxes J_i^α and for the heat flux of the mixture q_i without the terms $\frac{\partial v_r}{\partial x_r}$ and $\frac{\partial v_{(i}^\alpha}{\partial x_{j)}}$. This is the reason why I shall not consider both in the second iteration step.

In order to get the second iterated value of the diffusion fluxes J_i^α the first iterated values are inserted on the left-hand side of Eq. (5.55) and by linearization and rearrangement it follows

$$J_i^{\alpha(2)} = \sum_{\alpha=1}^{\nu-1} L_{ij}^{\alpha\beta}\frac{\partial}{\partial x_j}\Big(-\frac{\mu_\beta-\mu_\nu}{T}\Big) + \tilde{L}_{ij}^{\alpha}\frac{\partial}{\partial x_j}\Big(\frac{1}{T}\Big), \tag{5.72}$$

where

$$L_{ij}^{\alpha\beta} = -T\sum_{\gamma=1}^{\nu-1}[F^{-1}(M^V)^{-1}F^{-1}]_{\alpha\gamma}(\delta_{ij}\delta_{\gamma\beta} + A_{ij}[(M^V)^{-1}F^{-1}]_{\gamma\beta}), \tag{5.73}$$

$$A_{ij} = \delta_{ij}\frac{d}{dt} + \frac{\partial v_{[i}}{\partial x_{j]}} - 2W_{ij}, \tag{5.74}$$

$$\tilde{L}_{ij}^{\alpha} = \frac{5}{2}\sum_{\beta=1}^{\nu-1} L_{ij}^{\alpha\beta}\Big(\frac{p_\beta}{\varrho_\beta} - \frac{p_\nu}{\varrho_\nu}\Big). \tag{5.75}$$

$L_{ij}^{\alpha\beta}$ is the tensor of the diffusion coefficients and $\tilde{L}_{ij}^{\alpha}$ is the tensor of the coefficients of thermal- diffusion.

On the other hand, the second iterated value of the heat flux of the mixture q_i follows by insertion of Eqs. (5.92) and (5.44) into Eq. (5.52) and by linearization

$$q_i^{(2)} = T^2\kappa_{ij}\frac{\partial}{\partial x_j}\Big(\frac{1}{T}\Big) + \sum_{\alpha=1}^{\nu-1} L_{ij}^{\alpha}\frac{\partial}{\partial x_j}\Big(-\frac{\mu_\alpha-\mu_\nu}{T}\Big), \tag{5.76}$$

where

$$\kappa_{ij} = \frac{5}{2T^2}\sum_{\alpha=1}^{\nu-1}\tilde{L}_{ij}^{\alpha}\Big(\frac{p_\alpha}{\varrho_\alpha} - \frac{p_\nu}{\varrho_\nu}\Big) + \sum_{\alpha=1}^{\nu}\sum_{\beta=1}^{\nu}(\delta_{ij}\delta_{\alpha\beta} + (Q_{\alpha\beta}^q)^{-1}A_{ij})\kappa_\beta, \quad L_{ij}^{\alpha} = \tilde{L}_{ij}^{\alpha}. \tag{5.77}$$

κ_{ij} is the tensor of thermal conductivity of the mixture and L_{ij}^{α} is the tensor of the diffusion-thermo coefficients.

5.5.4 *The dependence on the frame of the constitutive equations. The Onsager reciprocity relations in the presence of axial fields.* From Eqs. (5.72)-(5.77) one can conclude that the constitutive equations of the diffusion fluxes J_i^α and of the heat flux of the mixture

q_i contain the matrix of angular velocity W_{ij}. This means that the constitutive equations depend on the frame and this dependence contradicts the principle of material frame indifference. Further, it is easy to see from these equations that

$$L_{ij}^{\alpha\beta}(\Omega_s,\omega_s) = L_{ji}^{\beta\alpha}(-\Omega_s,-\omega_s), \tag{5.78}$$

$$\tilde{L}_{ij}^{\alpha}(\Omega_s,\omega_s) = \tilde{L}_{ji}^{\alpha}(-\Omega_s,-\omega_s) = L_{ij}^{\alpha}(\Omega_s,\omega_s) = L_{ji}^{\alpha}(-\Omega_s,-\omega_s), \tag{5.79}$$

$$\kappa_{ij}(\Omega_s,\omega_s) = \kappa_{ji}(-\Omega_s,-\omega_s). \tag{5.80}$$

Eqs. (5.78)-(5.80) are the Onsager reciprocity relations in the presence of the axial fields Ω_s and ω_s which are defined by

$$\Omega_s = \frac{1}{2}\varepsilon_{ijs}\frac{\partial v_{[i}}{\partial x_{j]}}, \qquad \omega_s = \frac{1}{2}\varepsilon_{ijs}W_{ij} \tag{5.81}$$

Finally, I call attention to the fact that the Onsager reciprocity relations (5.66) and (5.78)-(5.80) came out in this theory without the use of statistical arguments.

ACKNOWLEDGMENT

The support by the CNPq of Brazil is gratefully acknowledged.

REFERENCES

Chapman,S. and T.G.Cowling. 1970. *The mathematical theory of non- uniform gases.* Cambridge: Cambridge University Press.

de Groot, S.R. and P. Mazur. 1962. *Non-equilibrium thermodynamics.* Amsterdam: North-Holland.

Dreyer, W. and W. Weiss. 1986. The classical limit of relativistic extended thermodynamics. *Ann. Inst. Henri Poincaré 45*: 401-418.

Heck, M. and I.Müller. 1983. Frame dependence, entropy, entropy flux and wave speeds in mixtures of gases. *Acta Mech 50*: 71-95.

Huang, K. 1963. *Statistical mechanics.* New York: John Wiley.

Ikenberry, E. and C. Truesdell. 1956. On the pressures and the flux of energy in a gas according to Maxwell's kinetic theory I. *J. Ratl. Mech. Anal. 5*: 1-54.

Kogan, M.N. 1973. Molecular gas dynamics. *Ann. Rev. Fluid Mech. 5*: 383-404.

Kremer, G.M. and C.Beevers. 1984. Extended thermodynamics of dense gases. In *Recent developments in non-equilibrium thermodynamics, Lecture Notes in Physics 199*, eds. J. Casas-Vázquez, D. Jou and G. Lebon, 429-436. Berlin: Springer.

Kremer, G.M. 1985. *Zur erweiterten Thermodynamik idealer und dichter Gase.* Dissertation. Technische Univesität Berlin.

Kremer, G.M. 1986. Extended thermodynamics of ideal gases with 14 fields. *Ann. Inst.*

Henri Poincaré 45: 419-440.
Kremer, G.M. 1987a. Extended thermodynamics of mixtures of ideal gases. *Int. J. Engn. Sci. 25*: 95-115.
Kremer, G.M. 1987b. Extended thermodynamics of non-ideal gases. *Physica 144A*: 156-178.
Kremer, G.M. and I. Müller. 1987. Extended thermodynamics of monatomic real gases. In *Kinetic theory and extended thermodynamics*, eds. I. Müller and T. Ruggeri, 181-195. Bologna: Pitagora Editrice.
Kremer, G.M. 1989a. Extended thermodynamics of molecular ideal gases. *Continuum Mech. Thermodyn. 1*: 21-45.
Kremer, G.M. 1989b. Extended thermodynamics and statistical mechanics of a polyatomic ideal gas. *J. Non-Equilib. Thermodyn. 14*: 363-373.
Liu, I-Shih. 1985. Extended thermodynamcis of fluids and virial equations of state. *Arch. Ration. Mech. Anal. 88*: 1-23.
Liu, I-Shih and G.M. Kremer. 1989. Hyperbolic system of field equations for viscous fluids. *Submited for publication.*
Liu, I-Shih and I. Müller. 1983. Extended thermodynamics of classical and degenerate ideal gases. *Arch. Ration. Mech. Anal. 83*: 285-332.
Liu, I-Shih, I. Müller und T. Ruggeri. 1986. Relativistic thermodynamics of gases. *Ann. of Physics 169*: 191-219.
Meixner, J. and H.G. Reik. 1959. Thermodynamik der irreversiblen Prozesse. In *Handbuch der Physik, Bd. III/2*, ed. S. Flügge, 413-523. Berlin. Springer.
Müller, I. 1966. *Zur Ausbreitungsgeschwindigkeit von Störungen in kontinuierlichen Medien.* Dissertation. Technische Hochschule Aachen.
Müller, I. 1967. Zum Paradoxon der Wärmeleitungstheorie. *Z.Physik 198*: 329-344.
Müller, I. 1972. On the frame dependence of stress and heat flux. *Arch. Ration. Mech. Anal. 45*: 241-250.
Müller, I. 1973. A new approach to thermodynamics of simple mixtures. *Z. Naturforsch. 28a*: 1801-1813.
Müller, I. 1985. *Thermodynamics.* London: Pitman.
Sommerfeld, A. 1977. *Thermodynamik und statistik.* Thun: Harri Deutsch.
Truesdell, C. 1962. Mechanical basis of diffusion. *J. Chem Phys. 37*: 2336-2344.
Truesdell, C. 1984. *Rational thermodynamics.* New York: Springer.
Waldman, L. 1956. Transporterscheinungen in Gasen von mittlerem Druck. In *Handbuch der Physik, Bd. XII*, ed. S. Flügge, 295-514. Berlin: Springer.
Wang Chang, C.S., G.E. Uhlenbeck and J. de Boer. 1964. The heat conductivity and viscosity of polyatomic gases. In *Studies in statistical mechanics, Vol. II*, eds. J. de Boer and G.E. Uhlenbeck, 241- 268. Amsterdam: North-Holland.

Kinetic Theory and Irreversible Thermodynamics of Multicomponent Fluids

Byung Chan Eu
Department of Chemistry and Department of Physics
McGill University, Montreal, Quebec, Canada

ABSTRACT

An axiomatic formulation of extended irreversible thermodynamics is reviewed for a mixture of simple fluids. The axiomatic theory is then given kinetic theory foundations by means of the Boltzmann equation for a mixture of dilute momatomic gases and the generalized Boltzmann equation for a mixture of dense simple fluids, both being subject to an external field (e.g., an electric field). Applications of the theory so formulated are discussed in connection with rheology and nonequilibrium thermodynamic functions for fluids subject to shearing in various flow configurations. Shear–induced melting point depression is also discussed.

I. INTRODUCTION

In linear irreversible thermodynamics the classical theory of thermodynamics of Kelvin(1851), Clausius(1850) and Gibbs(1876) is extended into the realm of nonequilibrium, when nonequilibrium processes occur near equilibrium. The theory of linear irreversible thermodynamics is founded on a pair of fundamental assumptions that the entropy change accompanying irreversible processes in an elementary volume around a point in the fluid obeys the equilibrium Gibbs relation(Kirkwood and Oppenheim, 1961) and that there hold linear constitutive relations between fluxes and thermodynamic forces (gradients) that drive the processes. The first assumption is called the local equilibrium hypothesis and the constitutive relations in the second assumption are simply called linear force–flux relations. These relations obey two symmetry principles: the Curie principle (de Groot and Mazur, 1962) and the Onsager reciprocal theorem(Onsager, 1931). The former is a symmetry principle pertaining to spatial transformations of vectors and tensors, and the latter originates from the time reversal invariance of microscopic events, i.e., microscopic reversibility. The pair of aforementioned assumptions supplemented with constitutive relations subject to the two symmetry principles forms the basis of the theory of linear irreversible thermodynamics, and it allows descriptions of irreversible processes occurring near equilibrium. The literature in linear irreversible thermodynamics is extensive(e.g., de Groot and Mazur, 1962; Prigogine, 1967; Gyarmati, 1970).

From the viewpoint of fluid dynamics linear irreversible thermodynamics corresponds to the Navier–Stokes–Fourier theory of hydrodynamics(Landau et al., 1959; Batchelor, 1967) since

---

This work is supported in part by the Natural Sciences and Engineering Research Council of Canada and Fonds FCAR, Province of Quebec.

the constitutive relations, namely, linear thermodynamic force–flux relations, give rise to the Navier–Stokes equation and the Fourier equation when the fluxes given by the constitutive equations are substituted into the momentum and energy balance equations. In the case of a mixture mass diffusion equations arise when diffusion (mass) fluxes from the linear constitutive equations are substituted into the concentration balance equations. The constitutive relations for fluxes also involve various transport coefficients such as shear viscosity, thermal conductivity, diffusion coefficients, etc. which are determined by measuring the response of the substance to the applied thermodynamic forces of various kinds. In experiments dealing with processes occurring near equilibrium the magnitudes of thermodynamic forces are sufficiently small. One can perform experiments, or observe natural phenomena, in which thermodynamic forces driving the process of interest are not so small. In the case of such phenomena, if the transport coefficients are calculated from the measured fluxes and applied forces by using the conventional linear force–flux relations in linear irreversible thermodynamics, then the transport coefficients thus calculated turn out to depend on thermodynamic forces applied on the system. Such situations often arise in rheology(Eirich, 1956; Bird et al., 1977), and one speaks of non–Newtonian viscosities vis–a–vis Newtonian viscosities, and the former sensitively depend on the shear rate applied. Such a thermodynamic force dependence of transport coefficients implies in effect that the constitutive relations are no longer linear with respect to thermodynamic forces and fluxes. Therefore, for proper description the transport processes involved require nonlinear constitutive relations for fluxes or thermodynamic forces. We will term such processes *nonlinear transport processes.* In the sense that linear transport processes occur near equilibrium the nonlinear transport processes are generally associated with a large magnitude of thermodynamic forces and for this reason they are said to occur far from equilibrium. The terminology "far from equilibrium" as qualified above is admittedly still loose, but it will be used in connection with nonlinear transport processes only.

If the processes of interest are nonlinear with respect to fluxes or thermodynamic forces, then the theory required is already out of the realm of linear irreversible thermodynamics. It is then natural to seek an extension of linear irreversible thermodynamics into the nonlinear regime. Such a quest logically leads to the question of whether the local equilibrium hypothesis underlying the linear theory would still hold valid for nonlinear processes. Under the local equilibrium hypothesis one seeks for the entropy density of a fluid mixture as a function of conserved variables: namely, the internal energy, density (or specific volume) and concentrations. Since there are no other conserved variables for simple fluids, any revision of the local equilibrium hypothesis cannot avoid an addition of nonconserved variables to the set of variables chosen for entropy density. In extended irreversible thermodynamics(Lebon et al., 1980; Eu, 1980, 1982, 1987; Casas–Vazquez et al., 1981) the local equilibrium hypothesis is so extended as to include nonconserved variables in the set of independent variables for entropy in such a way that the second law of thermodynamics is rigorously satisfied.

A phenomenological theory of extended irreversible thermodynamics can be formulated axiomatically, and there are a number of versions available in the literature(Eu, 1982; Lebon, 1982; Müller, 1984). In most of the versions of extended irreversible thermodynamics available in the literature the constitutive equations are still linear. This, however, is a severe limitation that leaves out most of fascinating nonlinear phenomena from consideration and thereby renders the theory thus extended inapplicable to nonlinear phenomena far from equilibrium. Here only the nonlinear version(Eu, 1980, 1982, 1987, 1989c) is reviewed. Despite its importance for understanding nonlinear irreversible processes, it is not easy to formulate nonlinear constitutive equations that can be experimentally adequate and thermodynamically appropriate because of the mathematical and physical complexity associated with nonlinear phenomena. To these ends we rely on statistical mechanical theories of matter in nonequilibrium. It must be emphasized that molecular theories are, in fact, indispensible for constitutive equations. In this article I would like to describe theoretical results achieved by making an extension of linear irreversible thermodynamics to nonlinear regimes of processes in mixtures of simple fluids, dilute or dense, in both molecular (statistical mechanical) and continuum mechanical approaches.

In Section II an axiomatic theory of irreversible thermodynamics is reviewed which extends the existing theory of linear irreversible thermodynamics. In the kinetic theory (statistical mechanical) part (Sections III and IV) of this review the theory will be discussed couched, first, in the language of dilute gas kinetic theory and then an extension of the kinetic theory to dense simple fluid mixtures will be discussed. In Sections V, VI and VII applications of the theories reviewed will be discussed. Section VIII is for concluding remarks.

II. AXIOMATIC THEORY OF IRREVERSIBLE PROCESSES IN MIXTURES

Irreversible thermodynamics of nonlinear transport processes may be axiomatically formulated in a general form for mixtures of fluids(Eu, 1982). The axiomatic theory so formulated then can be supported by kinetic theories of gases and dense fluids so that various phenomenological coefficients appearing in the axiomatic theory can be calculated in terms of molecular parameters. There are other versions(Lebon et al., 1982; Müller, 1984) of phenomenolgical theories formulated on the basis of postulates in the line reminiscent of rational thermodynamics(Truesdell, 1969), but such theories invariably involve linear constitutive equations which are inadequate to account for experimental data for nonlinear transport processes. The axiomatic theory reviewed below can be applied to nonlinear transport processes in a mixture of fluids or solids.

Let us assume that the system consists of r chemically inert components contained in a volume V. The mole numbers of components are denoted by n_a, $a = 1, 2, \ldots, r$. The itialic subscripts denote species in what follows. In nonequilibrium situations n_a may be regarded as dependent on position $\mathbf{r}$ and time t. We also introduce local mass densities $\rho_a(\mathbf{r},t)$ and $\rho(\mathbf{r},t)$ defined by

$$\rho_a(\mathbf{r},t) = m_a n_a(\mathbf{r},t) \tag{2.1}$$

and

$$\rho(\mathbf{r},t) = \sum_{a=1}^{r} \rho_a(\mathbf{r},t). \tag{2.2}$$

Mass fractions are defined by

$$c_a(\mathbf{r},t) = \rho_a(\mathbf{r},t)/\rho(\mathbf{r},t) \tag{2.3}$$

and the specific volume per unit mass by

$$v = 1/\rho. \tag{2.4}$$

The local internal energy density $\mathcal{E}(\mathbf{r},t)$ is defined such that the total internal energy E per unit volume is given as

$$E \equiv \rho(\mathbf{r},t)\,\mathscr{E}(\mathbf{r},t) = \sum_{a=1}^{r} \rho_a(\mathbf{r},t)\,\mathscr{E}_a(\mathbf{r},t) \tag{2.5}$$

where $\mathscr{E}_a(\mathbf{r},t)$ is the internal energy density of species a. The fluid velocity will be denoted by $\mathbf{u}$. It is a mean velocity of the fluid defined by

$$\mathbf{u} = \sum_{a=1}^{r} \rho_a(\mathbf{r},t)\mathbf{u}_a(\mathbf{r},t)/\rho \tag{2.6}$$

where $\mathbf{u}_a$ is the velocity of the fluid species a in the r component mixture. Various flows of mechanical quantities such as mass, energy and momentum and diffusions are denoted by the symbol $\Phi_a^{(\alpha)}$ and their densities $\hat{\Phi}_a^{(\alpha)} = \Phi_a^{(\alpha)}/\rho$. For example,

$$\begin{aligned}
&\Phi_a^{(1)} \equiv [\mathbf{P}_a]^{(2)} \equiv \Pi_a = \text{traceless symmetric part of stress tensor } \mathbf{P}_a \text{ for species } a,\\
&\Phi_a^{(2)} \equiv \tfrac{1}{3}\mathrm{Tr}\mathbf{P}_a - p_a \equiv \Delta_a = \text{trace part of } \mathbf{P}_a \text{ in excess of the hydrostatic pressure } p_a,\\
&\Phi_a^{(3)} \equiv \mathbf{Q}_a - \hat{h}_a\mathbf{J}_a \equiv \mathbf{Q}'_a = \text{heat flow},\\
&\Phi_a^{(4)} \equiv \mathbf{J}_a = \text{mass flux of species } a \text{ due to diffusion.}
\end{aligned} \tag{2.7}$$

Here $\hat{h}_a$ stands for the enthalpy density of species a defined by

$$\hat{h}_a = \mathscr{E}_a + p_a v_a, \qquad (v_a = 1/\rho_a),$$

and $\mathbf{Q}_a$ is the energy flux for species a. These quantities listed above are of mechanical origin since they are computable by a mechanical means once an appropriate statistical averaging method is taken as will be shown in Sections III and IV below. For instance, they may be computed by taking the time average of their corresponding microscopic definitions as is commonly done in molecular dynamics.

In the axiomatic theory formulated by the present author(1982) the theory is phrased in a set of axioms and definitions which are presented below. It is assumed that the nonequilibrium (thermodynamic) state of a system is specified by a set of macroscopic variables of mechanical origin. The following postulate is made to that effect.

Postulate I. *There exists a set of macroscopic variables by which the nonequilibrium state of a macroscopic system is specified: They are* $\{\mathscr{E}, v, \mathbf{u}, c_a, \hat{\Phi}_a^{(\alpha)}: a = 1, 2,\ldots, r;\ \alpha = 1,2,\ldots, \ell\}$ *where* ℓ *depends on the system and the macroscopic processes in hand. These are in general dependent on the position* $\mathbf{r}$ *of the fluid and time.*

These variables are collectively denoted by

$$\mathscr{Z} \equiv \{ \mathscr{Z}_a : a = 1,2,\ldots,m\}$$

$$\equiv \{ \mathscr{E}, v, \mathbf{u}, c_a, \hat{\Phi}_a^{(\alpha)} : a = 1, 2,\ldots, r;\ \alpha = 1,2,\ldots, \ell\}. \tag{2.8}$$

Historically, since Clausius(1865) was able to define entropy for reversible processes only, a controversy was raised by some authors(Meixner, 1970; Kestin, 1975) regarding the validity of the entropy concept for nonequilibrium. However, it has been shown(Eu, 1988a) that a global entropy exists for nonequilibrium processes. This was possible to show by regarding the *compensated* and *uncompensated heats*(Clausius, 1865) as independent quantities that must be elucidated of their basic character. Thus, the concept of entropy originally put forward by Clausius(1865) not only extends into nonequilibrium, but also is essentially a nonequilibrium attribute of the system. Nevertheless, local entropy density is another matter and we postulate the following:

Postulate II. *There exists a piecewise differentiable function called local entropy density which depends on the variables* $\{ \mathscr{Z}_a : a = 1,2,\ldots,m\}$ *and which reaches a maximum if the isolated system is brought to equilibrium. The entropy density* $\mathscr{S}$ *is additive in the following sense:*

$$\mathscr{S} = \sum_{a=1}^{r} c_a \mathscr{S}_a . \tag{2.9}$$

The system is said to be in equilibrium if $\mathscr{E}$, v and c_a are independent of $\mathbf{r}$ and t over the volume and if $\mathbf{u} = 0$ and $\hat{\Phi}_a^{(\alpha)} = 0$ for all a and α. Therefore, there is no flow of energy and matter in the system at equilibrium. The global entropy is given by the integral

$$S(t) = \int_V d\mathbf{r}\, \rho(\mathbf{r},t)\, \mathscr{S}(\mathbf{r},t), \tag{2.10}$$

where the integration is over the whole volume. Since the entropy density is a function of $\mathscr{Z}$:

$$\mathscr{S}(\mathbf{r},t) = \mathscr{S}(\mathscr{Z}), \tag{2.11}$$

the entropy density may be regarded as a surface in the (m+1)–dimensional space spanned by $\mathscr{S}$ and $\mathscr{Z}$, which we call the nonequilibrium Gibbs space or simply the Gibbs space.

Since a surface may be regarded as an envelope of tangents to the surface, the tangents to the entropy surface are introduced and given the following physical significance.

Definition I

temperature:
$$T^{-1} = (\partial \mathscr{S} / \partial \mathscr{E})_{\mathfrak{z}}, \tag{2.12}$$

pressure:
$$p/T = (\partial \mathscr{S} / \partial v)_{\mathfrak{z}}, \tag{2.13}$$

chemical potential:
$$\hat{\mu}_a/T = -(\partial \mathscr{S} / \partial c_a)_{\mathfrak{z}}, \tag{2.14}$$

generalized potential: $$X_a^{(\alpha)}/T = (\partial \mathcal{S}/\partial \hat{\Phi}_a^{(\alpha)})_{\mathcal{Z}}, \tag{2.15}$$

where the subscript $\mathcal{Z}$ means keeping other Gibbs variables constant. These derivatives are quantities to be measured or expressed in terms of the measurables in phenomenological theory. The first three derivatives look similar to those in thermostatics except for the additional conditions of constant fluxes $\hat{\Phi}_a^{(\alpha)}$. Postulates I and II and Definition I naturally give rise to the entropy change in time interval dt in the substantial frame as follows:

$$T\frac{d\mathcal{S}}{dt} = \frac{d\mathcal{E}}{dt} + p\frac{dv}{dt} - \sum_{a=1}^{r} \hat{\mu}_a \frac{d}{dt}c_a + \sum_{\alpha=1}^{\ell} \sum_{a=1}^{r} X_a^{(\alpha)} \odot \frac{d}{dt}\hat{\Phi}_a^{(\alpha)} \tag{2.16}$$

where $\odot$ stands for an appropriate scalar product of the tensors involved. This is an *extended Gibbs relation for nonequilibrium.* It obviously shows that the evolution of entropy is determined by those of the variables chosen $\{ \mathcal{Z}_a : a = 1,2,...,m\}$ in complete accordance with the postulates made. Eq. (2.16) clearly reduces to the equilibrium Gibbs relation if the fluxes or the generalized potentials vanish.

Postulate IIIa. *The variables* $\mathcal{Z}_a$, a = 1,2,...,m, *obey the balance equations and the evolution equations*

$$\frac{d\rho}{dt} = -\rho\nabla\cdot\mathbf{u}, \tag{2.17}$$

$$\rho\frac{d}{dt}c_a = -\nabla\cdot\mathbf{J}_a \tag{2.18}$$

$$\rho\frac{d\mathbf{u}}{dt} = -\nabla\cdot\mathbf{P}, \tag{2.19}$$

$$\rho\frac{d\mathcal{E}}{dt} = -\nabla\cdot\mathbf{Q} - \mathbf{P}:\nabla\mathbf{u}, \tag{2.20}$$

$$\mathbf{P}_{abc} = \mathbf{P}_{acb}, \tag{2.21}$$

$$\rho\frac{d}{dt}\hat{\Phi}_a^{(\alpha)} = Z_a^{(\alpha)} + \Lambda_a^{(\alpha)}, \tag{2.22}$$

where $\mathbf{P}_{abc}$ *denotes the* (bc) *component of stress tensor* $\mathbf{P}_a$ *and* $Z_a^{(\alpha)}$ *and* $\Lambda_a^{(\alpha)}$ *are functions of* $\{\mathcal{Z}_a\}$ *as well as* T, p, $\hat{\mu}_a$ *and* $X_a^{(\alpha)}$ *in general.*

Here the total stress tensor **P** and the total heat flux **Q** are respectively given by

$$\mathbf{P} = \sum_{a=1}^{r} \mathbf{P}_a, \qquad \mathbf{Q} = \sum_{a=1}^{r} \mathbf{Q}_a \tag{2.23}$$

in terms of respective species components. Eqs. (2.17)–(2.20) hold for any fluids free from an external force, but (2.21) holds only for isotropic fluids and represents the angular momentum conservation law. The equation (2.22) collectively stands for constitutive equations which differ from system to system. These constitutive equations can be highly nonlinear for system removed far from equilibrium. Therefore, the forms for $Z_a^{(\alpha)}$ and $\Lambda_a^{(\alpha)}$ must be assumed specifically for each substance considered. Eq. (2.22) replaces the constitutive assumptions for fluxes in linear irreversible thermodynamics.

If the set is to describe properly thermodynamic processes in a macroscopic system, the Gibbs variables and the constitutive equations cannot be arbitrary, but must be subject to additional restrictions demanded by the thermodynamic laws. The quantities $Z_a^{(\alpha)}$, $\Lambda_a^{(\alpha)}$ and $X_a^{(\alpha)}$ are tensors of the same rank as $\Phi_a^{(\alpha)}$, and $\Lambda_a^{(\alpha)}$ is said to be positive relative to $X_a^{(\alpha)}$ if the bilinear form

$$(X,\Lambda) \equiv \sum_a^r \sum_\alpha^\ell X_a^{(\alpha)} \odot \Lambda_a^{(\alpha)} \tag{2.24}$$

is positive and equal to zero only if fluxes are equal to zero or $X_a^{(\alpha)}$ vanish. To make the whole formalism consistent with the second law of thermodynamics it is necessary to postulate the following:

Postulate IIIb. *The dissipative contributions* $\Lambda_a^{(\alpha)}$ *are positive relative to* $X_a^{(\alpha)}$.

Since it is known that there occur various fluxes such as heat flux, mass fluxes in response to thermodynamic forces linear thermodynamic forces are defined by spatial gradients of T, **u** and $\hat{\mu}_a$ as follows:

Definition II.

$$
\begin{aligned}
&\text{shear:} && \chi_a^{(1)} = -[\nabla \mathbf{u}]^{(2)}, \\
&\text{compression or dilatation:} && \chi_a^{(2)} = -\nabla \cdot \mathbf{u}, \\
&\text{heat flow:} && \chi_a^{(3)} = -\nabla \ln T, \\
&\text{diffusion:} && \chi_a^{(4)} = -(\nabla \hat{\mu}_a)_T + v\nabla p,
\end{aligned}
\tag{2.25}
$$

where the subscript T means that the differentiation is performed with T fixed and $[\mathbf{A}]^{(2)}$ means the traceless symmetric part of tensor **A**. This notation is used for the traceless symmetric part of tensor **A** throughout this article. In linear irreversible thermodynamics these thermodynamic forces are in linear relationship to the fluxes subject to the Curie principle and the Onsager reciprocal theorem(de Groot et al., 1962; Onsager, 1931). The following abbreviations can be useful:

$$\bar{\chi}_a^{(\alpha)} = \chi_a^{(\alpha)}/T, \tag{2.26}$$

$$\bar{X}_a^{(\alpha)} = X_a^{(\alpha)}/T. \tag{2.27}$$

The last postulate is:

Postulate IV. *There exists a quantity called the entropy flux* $\mathbf{J}_s$ *which is associated with transmissions of heat, energy, and matter and also with their intricate interplays within the system. It is defined by*

$$\mathbf{J}_s = (\mathbf{Q} - \sum_{a=1}^{r} \hat{\mu}_a \mathbf{J}_a)/T + \sum_{a=1}^{r} \sum_{\alpha=1}^{\ell} \mathscr{J}_a^{(\alpha)} \tag{2.28}$$

where $\mathscr{J}_a^{(\alpha)}$ *are determined by the equations*

$$\nabla \cdot \mathscr{J}_a^{(\alpha)} + Z_a^{(\alpha)} \odot \bar{X}_a^{(\alpha)} + \boldsymbol{\Phi}_a^{(\alpha)} \odot \bar{\chi}_a^{(\alpha)} = 0 \tag{2.29}$$

for all α *and* a.

The second term in the right hand side of (2.28) is a nonclassical contribution to the entropy flux. A form equivalent to (2.29) was originally introduced by I. Müller(1967), but (2.29) is not the same as his in detail. We will see its kinetic theory counterpart in the subsequent sections. To determine $\mathscr{J}_a^{(\alpha)}$ it is necessary to solve the partial differential equations subject to suitable boundary conditions.

Eq. (2.16) must be supplied with constitutive relations for the generalized specific heat(caloric equation of state), pressure, chemical potentials and generalized potentials:

$$C_v = T(\partial \mathscr{E} / \partial T)_{\mathscr{J}} = C_v(\mathscr{E}, v, \mathbf{u}, c_a, \hat{\boldsymbol{\Phi}}_a^{(\alpha)}, \mathscr{S}), \tag{2.30}$$

$$p = p(\mathscr{E}, v, \mathbf{u}, c_a, \hat{\boldsymbol{\Phi}}_a^{(\alpha)}, \mathscr{S}), \tag{2.31}$$

$$\hat{\mu}_a = \hat{\mu}_a(\mathscr{E}, v, \mathbf{u}, c_a, \hat{\boldsymbol{\Phi}}_a^{(\alpha)}, \mathscr{S}), \tag{2.32}$$

$$X_a^{(\alpha)} = X_a^{(\alpha)}(\mathscr{E}, v, \mathbf{u}, c_a, \hat{\boldsymbol{\Phi}}_a^{(\alpha)}, \mathscr{S}). \tag{2.33}$$

By using (2.16) and postulates III and IV it is shown(Eu, 1982) that the entropy density follows a balance equation. This result can be summarized by the following theorem.

Theorem. *The entropy density satisfies the balance equation*

$$\rho \frac{d\mathscr{S}}{dt} + \nabla \cdot \mathbf{J}_s = \sigma, \tag{2.34}$$

where σ is the entropy production and is given by the formula

$$\sigma = (\bar{X},\Lambda) = \sum_{a}^{r}\sum_{\alpha}^{\ell} \bar{X}_a^{(\alpha)} \odot \Lambda_a^{(\alpha)} \geq 0 \tag{2.35}$$

where the equality holds at equilibrium.

The inequality (2.35) follows from postulate IIIb concerning the construction of the evolution equations and is a mathematical statement of the second law of thermodynamics. It gives a clear physical meaning to the notion of entropy production and reduces to the well known expression for entropy production in linear irreversible thermodynamics if $\Lambda_a^{(\alpha)}$ and $X_a^{(\alpha)}$ are linear with respect to fluxes and linear force–flux relations hold valid. Eq. (2.35) is a nonlinear generaliztion of the Rayleigh–Onsager dissipation function(Rayleigh, 1945; Onsager, 1931) appearing in the linear theory.

It is important to note that the entropy production is intimately related to the dissipative terms $\Lambda_a^{(\alpha)}$ in the evolution equations for fluxes. This is a feature not present in the theories of Coleman and Noll(1963), and Bataille and Kestin(1979) who simply assume its positivity— the Clausius–Duhem inequality. It must be noted that in the present theory the entropy production σ as a mathematical representation for the second law of thermodynamics serves as a constraint on the constitutive equations since it is not possible to take any forms for them but those satisfying the inequality (2.35).

The axiomatic theory of irreversible thermodynamics presented in this section can be supported by kinetic theories of dilute and dense simple fluids(Eu, 1979a, 1980, 1981, 1987). Results of kinetic theories will be reviewed in the subsequent sections. In fact, kinetic theories can give much more than the phenomenological theory demands. Especially, in the case of mixtures formulations of constitutive equations for the species components are not generally possible without kinetic theories, namely, molecular theories since it is not experimentally possible to decompose all observables into their species components and their composition dependence is not possible to obtain without the help of a molecular theory.

III. KINETIC THEORY: DILUTE GASES

Molecular, statistical mechanical foundations of the axiomatic theory of irreversible processes can be given if a suitable kinetic equation is used, provided that the kinetic equation is irreversible and there exists a statistical model for entropy. There is an obvious kinetic equation to use for this purpose. That is the Boltzmann equation(1872, 1964) for dilute gases. Let us assume that the system consists of r gaseous components which may be treated as structureless particles. They are assumed to be chemically inert so that there are no chemical reactions. If the system is assumed to be dilute, then its kinetic evolution can be described by Boltzmann equations for singlet distribution functions $f_a(\mathbf{r},\mathbf{v}_a;t)$ where $\mathbf{r}$ and $\mathbf{v}_a$ are the position and velocity of a molecule of species $a \in r$. In the standard notation the Boltzmann equations are

$$(\partial_t + \mathbf{v}_a\cdot\nabla + \mathbf{F}_a\cdot\nabla_{va})f_a(\mathbf{r},\mathbf{v}_a;t) = \sum_{b=1}^{r} C(f_a,f_b),\ a = 1,2,\ldots,r \tag{3.1}$$

where $\partial_t = \partial/\partial t$, $\nabla_{va} = \partial/\partial\mathbf{v}_a$ and the Boltzmann collision integral $C(f_a,f_b)$ is defined by

$$C(f_a,f_b) = \int d\mathbf{v}_b \int_0^\infty dbb \int_0^{2\pi} d\phi\, g_{ab}[f_a^*(\mathbf{r},\mathbf{v}_a;t)f_b^*(\mathbf{r},\mathbf{v}_b;t) - f_a(\mathbf{r},\mathbf{v}_a;t)f_b(\mathbf{r},\mathbf{v}_b;t)]. \tag{3.2}$$

Here the asterisk denotes the post–collision value. Since the Boltzmann equations are irreversible and there holds the H theorem, there exists a kinetic theory model for entropy for the system, and it is possible to investigate the statistical mechanical foundation of the axiomatic theory of irreversible thermodynamics discussed in Sec. II, albeit for dilute gas mixtures.

It is well known that the Chapmann–Enskog method(Chapman et al., 1970) of solution for the Boltzmann equations, when limited to first order in thermodynamic gradients, yields a theory of linear irreversible thermodynamics at the thermodynamic level and the Navier–Stokes, Fourier and Fick's diffusion equations at the hydrodynamic level. When the entropy change due to applied thermodynamic forces is calculated, the local equilibrium Gibbs relation is recovered for it(Prigogine, 1949). As is well known, since the second or higher order Chapman–Enskog solutions do not lead to a structure of theory of irreversible processes compatible with thermodynamic laws, it is clearly necessary to find a way to achieve the desired goal.

It is shown in a series of papers by the present author(1980, 1986, 1987, 1989a) that the modified moment method can adequately meet the need. Before proceeding further, we note that the entropy density $\mathcal{S}$, entropy flux $\mathbf{J}_s$ and entropy production σ for a mixture of gases are formally given by the statistical mechanical formulas(Chapman et al., 1970)

$$\rho\mathcal{S} = -k_B \sum_{a=1}^{r} \int d\mathbf{v}_a f_a(\mathbf{r},\mathbf{v}_a;t)(\ln f_a - 1), \tag{3.3}$$

$$\mathbf{J}_s = -k_B \sum_{a=1}^{r} \int d\mathbf{v}_a(\mathbf{v}_a - \mathbf{u}) f_a(\mathbf{r},\mathbf{v}_a;t)(\ln f_a - 1), \tag{3.4}$$

$$\sigma = -k_B \sum_{a=1}^{r} \sum_{b=1}^{r} \int d\mathbf{v}_a \ln f_a C(f_a,f_b). \tag{3.5}$$

These three quantities form the entropy balance equation

$$\rho(d\mathcal{S}/dt) = -\nabla\cdot\mathbf{J}_s + \sigma \tag{3.6}$$

where d/dt is the substantial derivative. Note that this equation is in the same form as the entropy balance equation (2.34) in the the axiomatic theory. Thus, it is a first formal result provided by the Boltzmann kinetic theory in support of the axiomatic theory although there still remains the question of correspondence between the two equations since $\mathcal{S}$ must be shown to possess all the attributes of entropy. It is the principal aim of this section to show their correspondence. The density ρ and the fluid velocity $\mathbf{u}$ are defined by the statistical formulas

$$\rho = \sum_{a=1}^{r} \langle m_a f_a \rangle, \tag{3.7}$$

$$\rho\mathbf{u} = \sum_{a=1}^{r} \langle m_a \mathbf{v}_a f_a \rangle, \tag{3.8}$$

where the angular brackets mean integration over $\mathbf{v}_a$ space. Since the statistical formulas (3.3)–(3.5) for entropy, etc., however, are too formal to indicate their significance to irreversible thermodynamics or even equilibrium thermodynamics as they stand, they are not useful for formulating a theory of irreversible thermodynamics unless their macroscopic variable dependence must be made explicit.

The H theorem may be expressed equivalently as $\sigma \geq 0$, the equality holding at equilibrium. Since this positivity of σ is a statistical mechanical expression for the second law of thermodynamics if $\mathcal{S}$ is taken equal to the entropy density in thermodynamics, any thermodynmaically consistent representations for $\mathcal{S}$, $\mathbf{J}_s$ and σ in terms of macroscopic observables must be such that the second law of thermodynamics, i.e., $\sigma \geq 0$, is rigorously satisfied. Of course, a consistency with the second law is not enough for a theory of irreversible processes to be correct and useful; it also must be able to account for observed natural and experimental phenomena within the experimental accuracy. And there lies the challenge posed by nonlinear irreversible phenomena, if a formulation of theory is aimed at in such a way consistent with thermodynamic laws.

To obtain the desired representations for $\mathcal{S}$, $\mathbf{J}_s$ and σ it is necessary to solve the Boltzmann equations, but since the latter are not analytically solvable, approximation methods must be employed to obtain the distribution functions. Since approximate solutions may not necessarily guarantee that the second law will be satisfied by them, it is essential to develop approximate solution methods that are rigorously consistent with the second law. *This requirement that the second law of thermodynamics be rigorously satisfied by the approximate solution taken for the Boltzmann equations is the tenet under which the modified moment method reviewed here is developed*(Eu, 1980, 1986, 1987, 1989c). The evolution equations for macroscopic variables derived under the tenet, therefore, are fully consistent with the second law.

The balance equations for mass, concentrations (mass fractions), momentum and internal energy listed in (2.17)–(2.20) can be easily derived from the Boltzmann equations (3.1) by using appropriate statistical definitions for mass density, momentum, and internal energy. The angular momentum conservation law expressed by (2.21) can be easily proven with (3.1) and the statistical definition for the stress tensor of species a which must be a symmetric tensor if the fluid is simple. The evolution equations (2.22) for nonconserved variables can be derived similarly to the ones for conserved variables (2.17)–(2.20) and the dissipative terms $\Lambda_a^{(\alpha)}$ is found to be in the form

$$\Lambda_a^{(\alpha)} = \sum_{b=1}^{r} \langle h_a^{(\alpha)} C(f_a, f_b) \rangle \tag{3.9}$$

where $h_a^{(\alpha)}$, $a = 1,2,\ldots,r$; $\alpha = 1,2,\ldots,\ell$, are:

$$h_a^{(1)} = m_a[\mathcal{C}_a \mathcal{C}_a]^{(2)}; \qquad h_a^{(2)} = \tfrac{1}{3} m_a \mathcal{C}_a^2 - m_a p_a/\rho_a;$$

$$h_a^{(3)} = \frac{1}{2}m_a \mathscr{C}_a^2 \mathscr{C}_a - m_a \hat{h}_a \mathscr{C}_a; \qquad h_a^{(4)} = m_a \mathscr{C}_a; \text{ etc.} \tag{3.10}$$

with $\mathscr{C}_a$ denoting the peculiar velocity $\mathscr{C}_a = \mathbf{v}_a - \mathbf{u}$. Note that $\Phi_a^{(\alpha)} = \langle h_a^{(\alpha)} f_a \rangle$. Other notations in (3.10) are as follows: p_a is the partial hydrostatic pressure of species a; $\hat{h}_a^{(\alpha)}$ its enthalpy density per unit mass; ρ_a its mass density.

The convective term $Z_a^{(\alpha)}$ in (2.22) are given by the statistical formula

$$Z_a^{(\alpha)} = -\nabla \cdot \langle \mathscr{C}_a h_a^{(\alpha)} f_a \rangle + \langle \mathbf{v}_a \cdot (\nabla h_a^{(\alpha)}) f_a \rangle. \tag{3.11}$$

More explicit forms are listed in Table 1. The evolution equations for $\Phi_a^{(\alpha)}$ derived from the Boltzmann equations (3.1) form an open hierarchy of differential equations because they are sequentially coupled in the sense that the lower order (moment) equations contain higher order moments. This open hierarchy must be suitably closed. There can be different theories about closure. One classical method(Ikenberry et al., 1956) is to use an idea similar to the Chapman–Enskog method by which nonconserved variables (moments) are systematically expanded in series of a nonuniformity parameter. The results obtained this way agree with the Chapman–Enskog solutions in the leading terms, but, as in the latter case, the higher order results are not consistent with the second law. The generally accepted, mathematically systematic expansion schemes yield results that are consistent with thermodynamic laws only at the first order leading to the existing theory of linear irreversible themodynamics. This defect makes it necessary to look for a closure that yields a formalism consistent with the thermodynamic laws, perhaps, at the cost of mathematically systematic schemes which in any case cannot guarantee convergence of the expansions made thereby and consequently make the mathematical systematism a mere symbolic gesture. Note that there is no proof of convergence, for example, for the Chapman–Enskog expansion. Unfortunately, at present there is no universally accepted theory of closure and a closure of the open set must be guided by the physical nature of the problem in hand and the efficacy of the formalism thus obtained by the closure.

Table 1. $Z_a^{(\alpha)}$ in the Flux Evolution Equations for Dilute Gases†.

$$Z_a^{(1)} = -2[(d_t\mathbf{u})\mathbf{J}_a]^{(2)} - 2[\mathbf{P}_a \cdot \nabla\mathbf{u}]^{(2)}$$

$$Z_a^{(2)} = -\frac{2}{3}(d_t\mathbf{u}) \cdot \mathbf{J}_a - \frac{2}{3}\mathbf{P}_a : \nabla\mathbf{u} - p_a d_t \ln(p_a v^{5/3})$$

$$Z_a^{(3)} = -d_t\mathbf{u} \cdot (\mathbf{P}_a - p_a\mathbf{U}) - \mathbf{Q}'_a \cdot \nabla\mathbf{u} - \mathbf{J}_a d_t \hat{h}_a - \mathbf{P}_a \cdot \nabla\mathbf{u}$$

$$Z_a^{(4)} = -\nabla \cdot (\mathbf{P}_a - c_a\mathbf{P}) - \mathbf{P} \cdot \nabla c_a - \mathbf{J}_a \cdot \nabla\mathbf{u}$$

† In these convective terms $Z_a^{(\alpha)}$ the moments not belonging to the Gibbs variable set are

discarded according to the discussion given in the text. The external force is assumed to be zero for the convective terms presented. See Eu(1985b, 1987) for the case of an external force.

One important consideration to make in this connection is that the moment evolution equations, especially for the nonconserved moments, are constitutive equations for the system and there are only a finite number of them required for description of the material properties of the substance of interest. Moreover, there is little physical sense to include the moments that are not experimentally accessible. The experimentally accessible moments are in fact those appearing in the conservation equations, namely, mass fluxes, stress tensor, and heat fluxes in the case of simple fluids; other moments are not measured experimentally. Therefore, it is sensible to include in the Gibbs variable set only those macroscopic variables that are experimentally accessible, such as mass density, mass fractions, stress tensor, heat fluxes and mass fluxes, and then look for the distribution functions in terms of them. In fact, the distribution functions are regarded as functionals of the Gibbs variables chosen, whose temporal and spatial evolution determines the evolution of the distribution functions. Thus *we may simply drop the moments other than those corresponding to the Gibbs variables*, and Table 1 contains the convective terms $Z_a^{(\alpha)}$ obtained in that approximation (a method of closure). This truncation of an open set of nonconserved moments represents a closure in the modified moment method. This way of closing an open set is commonly practiced in moment methods(Grad, 1958); we would like to make it explicit here. It must be noted that in the usual moment methods the higher order moments are computed in terms of lower order moments included in the chosen variable set(Theodosopulu et al., 1972).

There still remains the question of determining the distribution functions in terms of the Gibbs variables. In the modified moment method(Eu, 1980, 1986) they are determined such that the entropy production is positive—the second law of thermodynamics. This aim and a deeper knowledge of entropy production can be simultaneously achieved if the entropy production is looked for as a linear combination of dissipative terms $\Lambda_a^{(\alpha)}$ in the constitutive equations

$$\sigma = T^{-1} \sum_{a=1}^{r} \sum_{\alpha=1}^{\ell} X_a^{(\alpha)} \odot \Lambda_a^{(\alpha)} \tag{3.12}$$

where $\Lambda_a^{(\alpha)}$ are given by (3.9) and $X_a^{(\alpha)}$ are as yet to be determined such that σ as given in (3.12) remains positive so that the second law is satisfied for all $X_a^{(\alpha)}$ at the approximation made. Comparison of (3.12) with the statistical formula (3.5) for σ suggests that

$$f_a = \exp[-\beta(H_a + H_a^{(1)} - \mu_a)] \tag{3.13}$$

where

$$H_a = \tfrac{1}{2} m_a \mathscr{C}_a^2 + \Psi_a, \tag{3.14}$$

$$H_a^{(1)} = \sum_{\alpha=1}^{\ell} X_a^{(\alpha)} \odot h_a^{(\alpha)}, \tag{3.15}$$

and μ_a is the normalization factor such that

$$\exp(-\beta\mu_a) = n_a^{-1}<\exp[-\beta(H_a + H_a^{(1)})]>. \tag{3.16}$$

Here Ψ_a is the potential of external force $\mathbf{F}_a$: $\mathbf{F}_a = -\nabla\Psi_a$; $\beta = 1/k_BT$; and T is the kinetic theory temperature which is defined by

$$\frac{3}{2}nk_BT = \sum_a <\frac{1}{2}m_a \mathscr{C}_a^2 f_a> = \sum_a <\frac{1}{2}m_a \mathscr{C}_a^2 f_a^{(0)}> \tag{3.17}$$

where

$$f_a^{(0)} = n_a \exp[-\beta H_a]/<\exp[-\beta H_a]>. \tag{3.18}$$

It will turn out that T is also the thermodynamic temperature. In kinetic theory the pressure must be defined along with temperature. It is defined by

$$p_a = <\frac{1}{3}m_a \mathscr{C}_a^2 f_a^{(0)}>. \tag{3.19}$$

It must be emphasized that p_a is defined with $f_a^{(0)}$ for which T is defined by (3.17). In kinetic theory the pressure is usually defined as one third the trace of stress tensor. In the case of dilute gases this definition agrees with that given in (3.19) since by virtue of (3.17)

$$p_a = <\frac{1}{3}m_a \mathscr{C}_a^2 f_a^{(0)}> = <\frac{1}{3}m_a \mathscr{C}_a^2 f_a> = \frac{1}{3}\mathrm{Tr}\mathbf{P}_a = n_a k_B T \tag{3.20}$$

where $\mathbf{P}_a$ is the stress tensor of species a. Therefore, it seems pointless to define p_a as in (3.19), but this difference becomes important in the case of dense fluids since then (3.17) cannot be used to calculate p_a so simply as in (3.20) because of the fact that the statistical definition of stress tensor for dense fluids now contain contributions from intermolecular forces and the latter do not appear in the definition of temperature (3.17) which must remain the same for both dilute gases and dense fluids. The definition (3.19) for p_a has an implication for bulk viscosity. In the case of dilute gases, because of (3.20) it can be deduced that there is no bulk viscosity. On the other hand, since (3.20) is no longer true in the case of dense fluids it is expected that the excess trace part Δ_a of $\mathbf{P}_a$, which is related to the bulk viscosity of the fluid, does not vanish and therefore there should be a nonvanishing bulk viscosity as generally observed for dense fluids(Rice et al., 1968). This experimental fact is one of the reasons that the pressure is not defined by p_a = $\mathrm{Tr}\mathbf{P}_a/3$ as is often done in kinetic theory(Chapman et al., 1970; Ferziger et al., 1972); such a definition gives rise to a difficulty in the case of compressible dense fluids for which there should be a bulk viscosity as discussed earlier.

Now, by substituting (3.12) and the statistical expression for $\mathbf{J}_s$ computed with (3.13) into the entropy balance equation and making use of various balance equations and the constitutive equations for $\hat{\Phi}_a^{(\alpha)}$, we obtain the extended Gibbs relation for entropy change

$$T\frac{d\mathscr{S}}{dt} = \frac{d\mathscr{E}}{dt} + p\frac{dv}{dt} - \sum_{a=1}^{r}\hat{\mu}_a\frac{d}{dt}c_a + \sum_{\alpha=1}^{\ell}\sum_{a=1}^{r} X_a^{(\alpha)} \odot \frac{d}{dt}\hat{\Phi}_a^{(\alpha)}, \tag{3.21}$$

provided the following consistency condition holds for $X_a^{(\alpha)}$:

$$\sum_{a=1}^{r}\sum_{\alpha}[\nabla\cdot(\psi_a^{(\alpha)}\odot X_a^{(\alpha)}/T) + T^{-1}(Z_a^{(\alpha)}\odot X_a^{(\alpha)} + \Phi_a^{(\alpha)}\odot\chi_a^{(\alpha)}] + \sum_{a=1}^{r}\nabla\cdot(k_B J_a/m_a) = 0 \tag{3.22}$$

where

$$X_a^{(\alpha)} = T^{-1}(\partial\mathscr{S}/\partial\hat{\Phi}_a^{(\alpha)}). \tag{3.23}$$

This means that the consistency condition is a partial differential equation for $\mathscr{S}$

$$\sum_{i=1}^{r}\sum_{\alpha}\{\nabla\cdot[\psi_a^{(\alpha)}\odot(\partial\mathscr{S}/\partial\hat{\Phi}_a^{(\alpha)})] + Z_a^{(\alpha)}\odot(\partial\mathscr{S}/\partial\hat{\Phi}_a^{(\alpha)}) + \Phi_a^{(\alpha)}\odot\chi_a^{(\alpha)}/T + \sum_{a=1}^{r}\nabla.(k_B J_a/m_a) = 0. \tag{3.24}$$

This equation is coupled to the evolution equations for the Gibbs variables chosen: (ρ, c_i, $\mathbf{u}$, $\hat{\Phi}_a^{(\alpha)}$); namely, (2.17)–(2.22). Thus, when the macroscopic equations (2.17)–(2.22) and (3.24) are solved together for variables (ρ, c_a, $\mathbf{u}$, $\hat{\Phi}_a^{(\alpha)}$, $\mathscr{S}$), the distribution functions are determined through (3.23) for $X_a^{(\alpha)}$.

The consistency condition (3.22) is similar to the one used by Müller in his work where an extended Gibbs relation is derived for $d\mathscr{S}$. In Müller's form for extended Gibbs relation $X_a^{(\alpha)}$ are linear with respect to $\hat{\Phi}_a^{(\alpha)}$. Since $X_a^{(\alpha)}$ are generally nonlinear in $\hat{\Phi}_a^{(\alpha)}$ the extended Gibbs relation (3.21) is more general, holding for any nonlinear processes. The consistency condition as in (3.24) is not present in Müller's(1967) work. Since the set of moments is limited to $\hat{\Phi}_a^{(\alpha)}$ for α = 1,2,3,4 and the moments $\psi_a^{(\alpha)}$ in (3.24) that are not in the chosen variable set must be dropped, there is in fact only one term remaining; namely, $\psi_a^{(4)}$ which is equal to $\mathbf{P}_a$; all others are set equal to zero.

Although the consistency condition (3.24) must be solved, in principle, together with other hydrodynamic equations, it is in practice convenient to solve it separately from the hydrodynamic equations; this is an approximation in which the conserved variables appearing in (3.24) are kept constant. In this kind of approximation (3.24) may be solved by a perturbation method in which the spatial derivatives of fluxes are treated as a perturbation. In such a treatment the leading order approximate solutions give

$$X_a^{(\alpha)} = -g_a^{(\alpha)} \Phi_a^{(\alpha)} \tag{3.25}$$

where

$$g_a^{(1)} = 1/2p_a;\ g_a^{(2)} = 5/3p_a;\ g_a^{(3)} = 1/p_a\hat{h}_a;\ g_a^{(4)} = 1/c_a. \tag{3.26}$$

There now remains the problem of calculating the dissipative term $\Lambda_a^{(\alpha)}$. It can be calculated by means of a cumulant expansion(Eu, 1980, 1985b) for the entropy production. To first order cumulant approximation the entropy production is given by the formula

$$\sigma = k_B g^{-1} \kappa(X) \sinh\kappa(X), \tag{3.27}$$

$$g = (m_r/2k_B T)^{1/2}/n^2 d^2 \tag{3.28}$$

where $\kappa(X)$ is related to the Rayleigh–Onsager dissipation function when the approximation (3.25) is made for $X_a^{(\alpha)}$ as is clear from the following formula

$$\kappa(X) = (\sum_{ab}^{r} \sum_{\alpha\beta}^{\ell} R_{ab}^{(\alpha\gamma)} X_a^{(\alpha)} \odot X_b^{(\gamma)})^{1/2} \tag{3.29}$$

where $R_{ab}^{(\alpha\gamma)}$ are collision bracket integrals appearing in the first order Chapman–Enskog theory(Chapman et al., 1970). Its details appear in the paper by Eu(1985b). Then, comparison of (3.27) with (3.12) yields the statistical formulas for the dissipative terms $\Lambda_a^{(\alpha)}$:

$$\Lambda_a^{(\alpha)} = -g^{-1} \sum_{b\gamma}^{r\ell}\sum R_{ab}^{(\alpha\gamma)} X_b^{(\gamma)} \sinh\kappa(X)/\kappa(X). \tag{3.30}$$

It must be emphasized that the constitutive equations with $\Lambda_a^{(\alpha)}$ given by (3.30) are fully consistent with the thermodynamic second law since $\sigma \geq 0$ according to (3.27). If the approximation (3.25) is used for $X_a^{(\alpha)}$, we then obtain

$$\Lambda_a^{(\alpha)} = -g^{-1} \sum_{b\gamma}^{r\ell}\sum R_{ab}^{(\alpha\gamma)} g_b^{(\gamma)} \Phi_b^{(\gamma)} \sinh\kappa(X)/\kappa(X). \tag{3.31}$$

Note that in the limit of small $\Phi_b^{(\gamma)}$ (3.31) reduces to a linear form if $\sinh\kappa/\kappa$ may be put equal to unity when κ is sufficiently small:

$$\Lambda_a^{(\alpha)} = -g^{-1} \sum_{b\gamma}\sum R_{ab}^{(\alpha\gamma)} g_b^{(\gamma)} \Phi_b^{(\gamma)}. \tag{3.32}$$

In this case the entropy production becomes a quadratic function of fluxes $\Phi_b^{(\gamma)}$, namely, the Rayleigh–Onsager dissipation function. It must be pointed out that the entropy production as given in (3.27) is always positive no matter what approximation is made to $X_a^{(\alpha)}$ and the bracket integrals $R_{ab}^{(\alpha\gamma)}$ satisfy the Onsager reciprocal relations.

When (3.31) is substituted for $\Lambda_a^{(\alpha)}$ in the flux evolutions for $\Phi_a^{(\alpha)}$ in (2.22), the hydrodynamic equations (2.17)–(2.22) become complete, and their solutions subject to suitable initial and boundary conditions provide all the necessary information for the behavior of a fluid—a dilute gas mixture. The set of hydrodynamic equations (2.17)–(2.22), which are called *generalized hydrodynamic equations*, has been successfully applied to a number of problems in gas dynamics(Khayat and Eu, 1988, 1989), plasmas(Clause et al., 1982, Eu, 1985a), rheology(Eu, 1983; Ohr and Eu, 1984; Bhattacharya and Eu, 1986, 1987), flow problems of non–Newtonian fluids(Eu, 1988b, 1989a, 1989b; Khayat and Eu, 1990), etc. in good agreement with experiments.

The distribution functions f_a takes a very suggestive form if (3.21) and (3.24) are used to replace T and $X_a^{(\alpha)}$ with entropy derivatives in the exponent of f_a in (3.13):

$$f_a = \exp\{-k_B^{-1}[(\partial \mathcal{S}/\partial \mathcal{E})H_a + \sum_\alpha(\partial \mathcal{S}/\partial\hat{\Phi}_a^{(\alpha)})\odot h_a^{(\alpha)} + m_a(\partial \mathcal{S}/\partial c_a)]\} \tag{3.33}$$

and this form of f_a, when substituted into the statistical formula of entropy (3.3), gives the entropy density in the form

$$\mathcal{S} = (\partial \mathcal{S}/\partial \mathcal{E})\mathcal{E} + (\partial \mathcal{S}/\partial v)v + \sum_a^r c_a(\partial \mathcal{S}/\partial c_a) + \sum_{a\alpha}^{r\ell}\sum(\partial \mathcal{S}/\partial\hat{\Phi}_a^{(\alpha)})\odot\hat{\Phi}_a^{(\alpha)}. \tag{3.34}$$

This form suggests that $\mathcal{S}$ is a first–degree homogeneous function of Euler in the variables $\mathcal{E}$, v, c_a and $\hat{\Phi}_a^{(\alpha)}$. Furthermore, the slopes of the entropy surface in the Gibbs space determines the distribution function at variables (H_a, $h_a^{(\alpha)}$, m_a). Therefore, according to this result, it is possible to determine the distribution function if the thermodynamic information on the entropy surface is obtained by transport and thermodynamic experiments. Since the thermodynamic information provides f_a, the distribution function as in (3.33) is called the thermodynamic branch of solution for the Boltzmann equations (3.1). Thus, such solutions for the Boltzmann equations are obtained in strict conformation to the thermodynamic laws as enunciated earlier by the guiding tenet for solving the kinetic equation. The form (3.33) for distribution function f_a is somewhat reminiscent of the famous relation for distribution function used by Einstein(1910) who expressed the distribution function in terms of entropy and its derivatives in the case of microcanonical ensemble. It must be noted in this connection that f_a in (3.33) is a canonical distribution function.

The bracket integrals in (3.31) or (3.30) can be related to transport coefficients that are measured in experiments on systems near equilibrium. This connection is easily made by linearizing the dissipative terms and the constitutive equations in which the substantial time derivatives of fluxes are also set equal to zero when the fluxes are steady since then the constitutive equations become the well known Newtonian law of viscosity, Fourier's law of thermal conductivity and Fick's law of diffusion. From these linearized constitutive relations it is possible to identify the bracket integrals $R_{ab}^{(\alpha\gamma)}$ with various linear transport coefficients such as shear viscosity, thermal conductivity and diffusion coefficients. Since the technical details of such connections are available in the literature(Eu, 1985b), we simply quote the results here: for example, for α, $\gamma = 1$

$$(\boldsymbol{\eta}^{-1})_{ab} = g^{-1} g_a^{(1)} R_{ab}^{(11)} g_b^{(1)} \equiv \bar{R}_{ab}^{(11)} \tag{3.35}$$

where $\boldsymbol{\eta}$ is the viscosity matrix which (3.35) defines; $\boldsymbol{\eta}$ is calculated by inverting the matrix $\bar{R}^{(11)} \equiv (\bar{R}_{ab}^{(11)})$. This viscosity matrix gives the component viscosities in the constitutive relations for the traceless symmetric parts Π_a of the stress tensors $\mathbf{P}_a$ for various species:

$$\Pi_a = -\sum_{b=1}^{r} 2\eta_{ab}\gamma, \tag{3.36}$$

$$\Pi = \sum_{a=1}^{r} \Pi_a = -\sum_{a=1}^{r}\sum_{b=1}^{r} 2\eta_{ab}\gamma \equiv -2\eta_0\gamma, \quad (\gamma = [\nabla \mathbf{u}]^{(2)}). \tag{3.37}$$

Other bracket integrals can be identified with phenomenological transport coefficients. For example, $R_{ab}^{(34)}$, $R_{ab}^{(43)}$, $R_{ab}^{(33)}$ and $R_{ab}^{(44)}$ can be identified if the phenomenological relations are first written for $\mathbf{Q}'_a$ and $\mathbf{J}_a$ as follows:

$$\mathbf{Q}'_a = \sum_b \lambda_{ab}\chi_b^{(3)} + \sum_b D_{ab}^{(qd)}\chi_b^{(4)}, \tag{3.38}$$

$$\mathbf{J}_a = \sum_b D_{ab}^{(dq)}\chi_b^{(3)} + \sum_b D_{ab}\chi_b^{(4)}, \tag{3.39}$$

where λ_{ab}, D_{ab}, $D_{ab}^{(qd)}$ and $D_{ab}^{(dq)}$ are, respectively, the thermal conductivities, diffusion coefficients, and thermal diffusion coefficients. If the linear relations (3.38) and (3.39) are solved for thermodynamic forces $\chi_a^{(3)}$ and $\chi_a^{(4)}$ and the results are compared with the corresponding relations obtained from linearized steady constitutive equations for $\hat{\Phi}_a^{(\alpha)}$ in the case of $\alpha = 3$ and 4, the bracket integrals in question are obtained in terms of phenomenological coefficients. Since (3.38) and (3.39) may be written in matrix form as

$$\begin{bmatrix} \mathbf{Q}'_1 \\ \vdots \\ \mathbf{J}_1 \\ \vdots \end{bmatrix} = \begin{bmatrix} \lambda & D^{(qd)} \\ D^{(dq)} & D \end{bmatrix} \begin{bmatrix} \chi_1^{(3)} \\ \vdots \\ \chi_1^{(4)} \\ \vdots \end{bmatrix}$$

$$\equiv \mathbf{M} \begin{bmatrix} \chi_1^{(3)} \\ \vdots \\ \chi_1^{(4)} \\ \vdots \end{bmatrix}, \tag{3.40}$$

inversion of this relation yields the linear relations

$$\chi_a^{(3)} = \sum_b (\mathbf{M}^{-1})_{ab}\mathbf{Q}'_b + \sum_b (\mathbf{M}^{-1})_{a,b+r}\mathbf{J}_b, \tag{3.41a}$$

$$\chi_a^{(4)} = \sum_b (\mathbf{M}^{-1})_{a+r,b}\mathbf{Q}'_b + \sum_b (\mathbf{M}^{-1})_{a+r,b+r}\mathbf{J}_b, \tag{3.41b}$$

where $\mathbf{M}$ is a $(2r-1)\times(2r-1)$ matrix: note that we take $(r-1)$ linearly independent diffusion fluxes since $\Sigma_{a=1}^{r}\mathbf{J}_a = 0$ by virtue of the definition of diffusion fluxes. Comparison of (3.41a,b) with the equations

$$p_a \hat{h}_a \chi_a^{(3)} = g^{-1}\sum_b [R_{ab}^{(33)} g_b^{(3)} \mathbf{Q}'_b + R_{ab}^{(34)} g_b^{(4)} \mathbf{J}_b], \tag{3.42a}$$

$$c_a \chi_a^{(4)} = g^{-1}\sum_b [R_{ab}^{(43)} g_b^{(3)} \mathbf{Q}'_b + R_{ab}^{(44)} g_b^{(4)} \mathbf{J}_b], \tag{3.42b}$$

which are linearized constitutive equations, gives relations between the phenomenological coefficients and the collision bracket integrals:

$$(\mathbf{M}^{-1})_{ab} = g^{-1} g_a^{(3)} R_{ab}^{(33)} g_b^{(3)}, \quad 1 \le a, b \le r, \tag{3.43a}$$

$$(\mathbf{M}^{-1})_{a,b+r} = g^{-1} g_a^{(3)} R_{ab}^{(34)} g_b^{(4)}, \quad 1 \le a \le r,\ 1 \le b \le r, \tag{3.43b}$$

$$(\mathbf{M}^{-1})_{a+r,b} = g^{-1} g_a^{(4)} R_{ab}^{(43)} g_b^{(3)}, \quad 1 \le a \le r-1,\ 1 \le b \le r, \tag{3.43c}$$

$$(\mathbf{M}^{-1})_{a+r,b+r} = g^{-1} g_a^{(4)} R_{ab}^{(44)} g_b^{(4)}, \quad 1 \le a, b \le r-1. \tag{3.43d}$$

Note that if $\mathbf{D}^{(qd)}$ and $\mathbf{D}^{(dq)}$ are neglected in $\mathbf{M}$, then the block diagonal matrices of $\mathbf{M}^{-1}$ gives, respectively, the inverse of thermal conductivity matrix $\boldsymbol{\lambda}$ and the inverse of diffusion coefficient matrix $\mathbf{D}$.

Now, if the bracket integrals thus identified are substituted into the nonlinear dissipative terms $\Lambda_a^{(\alpha)}$ as in (3.31) and the Rayleigh–Onsager dissipation function $\kappa^2(\hat{\Phi})$, the constitutive equations are expressed in terms of linear transport coefficients. This form of dissipative terms contains only the bracket integrals appearing in the first order Chapman–Enskog solution and yet the dissipative terms are highly nonlinear by virtue of the nonlinearity vested in the nonlinear factor $q_e \equiv \sinh\kappa(\hat{\Phi})/\kappa(\hat{\Phi})$.

The question is then, how good are the nonlinear constitutive equations when their predictions are compared with experiments? Answering this question basically establishes the utility of the formalism obtained by means of the modified moment method which is reviewed in this section. This question will be answered in Section V after a parallel formalism is presented for dense simple fluids.

IV. KINETIC THEORY: DENSE FLUIDS

Since the formalism presented for gaseous mixtures in Section III has been deduced from the Boltzmann equations for dilute gases, one can question its utility for liquid mixtures or dense gas mixtures. There is no question that the conservation equations must remain the same since mass, momentum and energy must be conserved regardless of whether a fluid is a dilute gas or a liquid. It is expected that the constitutive equations should be different for liquids and gases as is the case for the equation of state. However, as we will see, the constitutive equations for dense fluids are structurally almost the same as for those for gases except for the material functions such as transport coefficients and some terms related to intermolecular interaction forces. This structural similarity of constitutive equations between gases and liquids has been shown by means of a kinetic equation for dense simple fluids(Eu, 1979a, 1981b, 1987) and even for dense polyatomic fluids(Eu, 1989c).

To develop a theory of irreversible thermodynamics for dense fluids on the basis of a statistical model it is essential that the kinetic equation for the distribution function must be irreversible in the sense that it is of broken time reversal symmetry and there exists an H function for the fluid. From this viewpoint the Liouville equation is obviously not useful since the Liouville equation predicts that the entropy of the system in question remains invariant in time. The Liouville equation has a fundamentally different structure from an irreversible kinetic equation, a typical example of which is the Boltzmann equation, since the former lacks a dissipation term that accounts for contraction of information as time progresses. Phrased in other terms, the Liouville equation is time–reversal invariant whereas the Boltzmann equation breaks the time–reversal symmetry because of the presence of the irreversible collision operator in it. As far as irreversibility is concerned, it is in fact a minor point that the Boltzmann equation is a kinetic equation for a singlet distribution function while the Liouville equation is an N–body distribution function. From the consideration made above, it is clear that a suitable kinetic equation must be, first of all, of broken time–reversal symmetry. Based on this clue, it is possible to formulate *a similarity principle by which one can construct a set of irreversible kinetic equations that are not only similar in structure to the Boltzmann equation and of broken time–reversal symmetry but also include the Boltzmann equation as the low density limit element of the set.* Such a set of kinetic equations for a dense mixture of charged particles in a solvent (an r–component mixture) is found to be(Eu, 1979a, 1981b, 1987, 1989c)

$$[\partial_t + L_\alpha^{(s)}]f_\alpha^{(s)}(x^{(s)};t) = \int dx^{(\mathcal{N}-s)}(-i)T_{s_1,s_2,\dots,s_N}(z)\prod_{\beta=1}^{N} f_\beta^{(s)}(x^{(s)};t) \qquad (4.1)$$

where s denotes the number of particles in N subsystems making up the whole system of $\mathcal{N}$ particles ($\mathcal{N} = sN$) contained in volume $\mathcal{V} = NV$ and

$$s = \sum_{a=1}^{r} s_a,$$

where s_a denotes the number of particles of species a. The symbol $x^{(s)}$ stands collectively for the phases of s particles; $L_\alpha^{(s)}$ for the Liouville operator of the αth subsystem consisting of s particles and $T_{s_1,s_2,\ldots,s_N}(z)$ for the collision operator describing the collisional evolution of N subsystems of s particles. This collision operator obeys the classical Lippmann–Schwinger equation(Ernst et al., 1969, Miles et al., 1970; Eu, 1971a, 1971b)

$$T_c = \mathscr{L}_c' - \mathscr{L}_c' R_c^{(0)} T_c. \tag{4.2}$$

Here the subscript c stands for clusters $(s_1,s_2,\ldots,s_N)$ into which the whole system is decomposed, $\mathscr{L}_c' = -iL_c'$ with L_c' denoting the intercluster interaction Liouville operator and $R_c^{(0)}$ is the resolvent operator defined by

$$R_c^{(0)} = (\mathscr{L}_o - z)^{-1} \quad (z = i\epsilon,\ \epsilon > 0) \tag{4.3}$$

with $\mathscr{L}_o = -iL_o$ where L_o is the sum of Liouville operators for N disconnected subsystems of s particles each. The kinetic equation (4.1) reduces to the Boltzmann equation if s = 1 and the low density limit of the N–body collision operator $T_{s_1,s_2,\ldots,s_N}$ for $(s_1,s_2,\ldots,s_N) = (1,1,\ldots,1)$ is taken, namely, if it is approximated by a binary collision operator T_{12}. Note that in that case the Liouville operator on the left side of (4.1) becomes $L_\alpha^{(1)} = \mathbf{v}_1 \cdot \nabla$ which is the streaming term in the Boltzmann equation. It must be noted that the $\epsilon \to 0$ limit must be understood in the kinetic equation (4.1).

It can be shown that for a class of functions belonging to the 0–class of functions(Eu, 1979a) for which the eigenvalue of L_o is zero the H theorem holds for kinetic equation (4.1) and therefore there exists an entropy for the fluid. Consequently, it is possible to formulate a theory of irreversible thermodynamics for dense fluids in a completely parallel manner to that of a dilute gas mixture presented in Sec. III. One can derive the conservation equations and constitutive equations for various fluxes by using the kinetic equation (4.1). Since their derivations are rather lengthy but available in the literature(Eu, 1981b, 1987, 1989c), the salient results of the theory are listed here:

(1)The extended Gibbs relation holds as in (3.21) for a dilute gas mixture, provided the following consistency condition holds:

$$\sum_\alpha \sum_a \{\nabla\cdot(T^{-1} X_a^{(\alpha)} \odot \psi_a^{(\alpha)}) + T^{-1}(\Phi_a^{(\alpha)} \odot \chi_a^{(\alpha)} + X_a^{(\alpha)} \odot Z_a^{(\alpha)})\} + \sum_a [(p\hat{v}_a + \Psi_a) J_a + \mathscr{F}_a] = 0 \tag{4.4}$$

where Ψ_a is the potential of the external force on species a, $\hat{v}_a = 1/\rho_a$, and

$$\mathscr{F}_a = -\frac{1}{2}\sum_{ja=1}^{\mathscr{N}_a}\sum_{kb=1}^{\mathscr{N}_b} <(\partial V_{jakb}/\partial \mathbf{r}_{jakb})\mathbf{r}_{jakb}\cdot \mathscr{C}_{ja}\delta(\mathbf{r}_{ja}-\mathbf{r})> \tag{4.5}$$

with V_{jakb} denoting the potential energy between particles ja and kb and $\mathbf{r}_{jakb}$ their relative distance vector. Here $\mathscr{N}_a$ denotes the number of particles of species a in the whole system consisting of $\mathscr{N}$ particles. Other symbols have the same meanings as before. This consistency condition is somewhat different from the corresponding condition for dilute gas mixtures because of intermolecular interactions present in the dense fluid mixture.

(2) The streaming terms $Z_a^{(\alpha)}$, with the moments other than various fluxes (i.e., stress tensors, heat fluxes and diffusion fluxes) neglected in accordance with the closure discussed before, are as follows:

$$Z_a^{(1)} = -2[(d_t\mathbf{u})\mathbf{J}_a]^{(2)} - 2[\mathbf{P}_a\cdot\nabla\mathbf{u}]^{(2)} + 2(z_a - z)[\mathbf{E}_a\mathbf{J}_a]^{(2)}, \tag{4.6}$$

$$Z_a^{(2)} = -\tfrac{2}{3}(d_t\mathbf{u})\cdot\mathbf{J}_a - \tfrac{2}{3}\mathbf{P}_a:\nabla\mathbf{u} - p_a d_t \ln(p_a v^{5/3}) + \tfrac{2}{3}(z_a - z)\mathbf{E}_a\cdot\mathbf{J}_a, \tag{4.7}$$

$$Z_a^{(3)} = -d_t\mathbf{u}\cdot(\mathbf{P}_a - p_a\mathbf{U}) - \mathbf{Q}'_a\cdot\nabla\mathbf{u} - \mathbf{J}_a d_t \hat{h}_a - \mathbf{P}_a\cdot\nabla\mathbf{u} + (z_a - z)\mathbf{E}_a\cdot(\mathbf{P}_a - p_a\mathbf{U}), \tag{4.8}$$

$$Z_a^{(4)} = -\nabla\cdot(\mathbf{P}_a - c_a\mathbf{P}) - \mathbf{P}\cdot\nabla c_a - \mathbf{J}_a\cdot\nabla\mathbf{u} + \rho_a(z_a - z)\mathbf{E}_a, \tag{4.9}$$

where $\mathbf{E}_a$ is the external electric field $\mathbf{E}_a = -\nabla\Psi_a$ and z_a is the charge number of species a and $z = \sum_a c_a z_a$. We have abbreviated the substantial derivative by $d_t = d/dt$. It is useful to compare these equations with those in Table 1 which are obtained from the Boltzmann equation for gases. With the external force put equal to zero, they are exactly the same.

(3) The dissipative terms $\Lambda_a^{(\alpha)}$ are given by the formula

$$\Lambda_a^{(\alpha)} = \sum_{ja}^{\mathscr{N}_a} <h_a^{(\alpha)}\delta(\mathbf{r}_{ja} - \mathbf{r})C[\prod_{\beta=1}^{N} f_\beta^{(s)}]> \tag{4.10}$$

where

$$C[\prod_{\beta=1}^{N} f_\beta^{(s)}] = \int dx^{(\mathscr{N}-s)}(-i)T_{s_1,s_2,\ldots,s_N}(z)\prod_{\beta=1}^{N} f_\beta^{(s)}(x^{(s)};t) \tag{4.11}$$

and $h_a^{(\alpha)}$ are the molecular formulas for nonconserved moments corresponding to the dilute gas counterparts; when averaged over the phase space, $h_a^{(\alpha)}$ gives $\Phi_a^{(\alpha)}$. The $h_a^{(\alpha)}$ for dense fluid mixtures contain contributions from the interaction forces between the particles in the system whereas the dilute gas counterparts do not have such constributions. We list them below:

$$h_{ja}^{(1)} = [m_a \mathscr{C}_{ja}\mathscr{C}_{ja} + \frac{1}{2}\sum_b \sum_{k\,b=1}^{\mathscr{N}_a} F_{jakb}\mathbf{r}_{jakb}\mathbf{r}_{jakb}]^{(2)}, \tag{4.12}$$

$$h_{ja}^{(2)} = \frac{1}{3}m_a \mathscr{C}_{ja}^2 + \frac{1}{6}\sum_b \sum_{k\,b=1}^{\mathscr{N}_a} F_{jakb}\mathbf{r}_{jakb}\cdot\mathbf{r}_{jakb} - m_a p_a/\rho_a, \tag{4.13}$$

$$h_{ja}^{(3)} = [\frac{1}{2}m_a \mathscr{C}_{ja}^2 \mathbf{U} + \frac{1}{2}\sum_b \sum_{k\,b=1}^{\mathscr{N}_a} (V_{jakb}\mathbf{U} + F_{jakb}\mathbf{r}_{jakb}\mathbf{r}_{jakb})]\cdot \mathscr{C}_{ja} - \hat{h}_a m_a \mathscr{C}_{ja}, \tag{4.14}$$

$$h_{ja}^{(4)} = m_a \mathscr{C}_{ja}, \tag{4.15}$$

where

$$F_{jakb} = -(\partial V_{jakb}/\partial r_{jakb})/r_{jakb}. \tag{4.16}$$

If $\Lambda_a^{(\alpha)}$ are calculated to first order cumulant approximation as was done for the Boltzmann collision integral for dilute gases, it can be shown that $\Lambda_a^{(\alpha)}$ is given by the same form as in (3.30) except that the collision bracket integrals $R_{ab}^{(\alpha\gamma)}$ this time are defined with the collision operator $C[\prod_{\beta=1}^{N} f_\beta^{(s)}]$ in (4.11) and $h_a^{(\alpha)}$ are given by (4.12)–(4.15). Because of these differences in the definitions of the collision bracket integrals in the dilute gas and dense fluid versions, the material functions (transport coefficients) appearing in the constitutive equations for stress tensors, heat fluxes and mass fluxes have different density and temperature dependences, etc. and thereby give rise to different flow properties.

(4) The temperature is still defined by the formula

$$\frac{3}{2}nk_B T = \sum_{ja} \langle \frac{1}{2}m_a \mathscr{C}_{ja}^2 \delta(\mathbf{r}_{ja} - \mathbf{r})\rangle. \tag{4.17}$$

(5) The pressure p_a is defined by the formula

$$p_a = \mathscr{V}^{-\mathscr{N}}\int dx^{(\mathscr{N})}\frac{1}{3}\sum_{ja} (h_{ja}^{(1)}:\mathbf{U})\delta(\mathbf{r}_{ja} - \mathbf{r})f_{loc}^{(\mathscr{N})}(x^{(\mathscr{N})};\beta) \tag{4.18}$$

where

$$f_{loc}^{(\mathscr{N})} = \exp\{-\beta[H^{(\mathscr{N})}(x^{(\mathscr{N})},\mathbf{u}) - \mathscr{A}_e^{(\mathscr{N})}]\}, \tag{4.19}$$

$$\exp(-\beta\mathscr{A}_e^{(\mathscr{N})}) = \langle \exp[-\beta H^{(\mathscr{N})}(x^{(\mathscr{N})},\mathbf{u})]\rangle, \tag{4.20}$$

$$H^{(\mathcal{N})}(x^{(\mathcal{N})},\mathbf{u}) = \sum_{a=1}^{r}\sum_{j\,a}^{\mathcal{N}_a}\left[\tfrac{1}{2}m_a\mathcal{C}_{ja}^2 + \tfrac{1}{2}\sum_{b}\sum_{k\,b=1}^{\mathcal{N}_b}V_{jakb} + \Psi_a(\mathbf{r}_{ja})\right]. \tag{4.21}$$

It must be noted that the pressure defined here reduces to that for a dilute gas mixture when the contribution from the intermolecular forces are neglected and that in such limit the equation of state is $p_a = n_a k_B T$ by virtue of (4.17). The pressure p_a defined in (4.18) obviously is not equal to $\mathrm{Tr}P_a/3$ in contrast to the case of dilute gases. Since $p_a = \mathrm{Tr}P_a/3$ for dilute monatomic gases and as a consequence the excess trace part Δ_a of the stress tensor vanishes, they do not have a bulk viscosity. If this definition of pressure were taken in the case of dense fluids, then the bulk viscosity would have been equal to zero, and this is against the experimental fact. Therefore, the definition (4.18) for p_a is consistent with the experimental observation; it seems also sensible in view of the notion commonly accepted for hydrostatic pressure. We note that there is another line of attempt to resolve the tricky question of bulk viscosity by van Beijren et al.(1988) who introduce more than one temperature scale. The present author does not believe that such a scheme is a solution to the question.

The thermodynamic formalism for irreversible processes in dense fluids is the same as for those in dilute gaseous mixtures, provided that the allowances are made as to the items enumerated in the previous paragraph. The reason for the identical formalisms is simple: since irreversible thermodynamics is a macroscopic, phenomenological theory that holds for continuum matter regardless of the constituents, there is no molecular picture present in it. However, when the phenomenological theory is supported by a statistical mechanical theory based on, for example, the kinetic equation (4.1), then the picture gets sharpened and we are able to distinguish fluids of different molecularities through transport coefficients that depend on density, temperature and the structure of the molecule. As the formalism reviewed here shows, statistical mechanical foundations are now given, to a satisfactory degree, to the axiomatic formalism presented in Section II for irreversible thermodynamics of fluid mixtures, dilute or dense.

V. TEST OF CONSTITUTIVE EQUATIONS

The constitutive equations—the evolution equations for fluxes—obtained for dilute gases and liquids have been tested for some special cases, especially, in connection with rheology(Ohr et al., 1984; Eu et al., 1984; Eu, 1986) of polymer solutions and with molecular dynamics simulations for argon(Eu, 1983; Bhattacharya et al., 1987). If the fluids are assumed to be of incompressible, non–heat–conducting and of a single component, the problem becomes simplified and if we assume a simple flow configuration such as plane Couette flow, then the constitutive equations can be sometimes solved analytically. The most simplifying feature of plane Couette flow is that the velocity profile is linear for both Newtonian and non–Newtonian fluids, and this feature makes it possible to solve the constitutive equations separately from the momentum balance equation. Based on such a flow configuration, the constitutive equation for the stress tensor of a non–Newtonian fluid can be well represented by the stress tensor evolution equation

$$\rho\frac{d}{dt}\hat{\Pi} = -2p[\nabla\mathbf{u}]^{(2)} - 2\rho[\hat{\Pi}\cdot\nabla\mathbf{u}]^{(2)} - 2\rho[\hat{\Pi}\cdot\omega - \omega\cdot\hat{\Pi}] - \frac{p\rho\hat{\Pi}}{\eta_0}(\sinh\kappa/\kappa) \tag{5.1}$$

in a familiar notation where $\hat{\Pi} = \Pi/\rho$ with Π denoting shear stress and ω the vorticity tensor. Here, κ^2 is the Rayleigh–Onsager dissipation function already introduced. The steady state solution of this equation predicts a non–Newtonian viscosity for the fluid whose shear rate dependence has been shown(Eu, 1983; Ohr et al. 1984; Bhattacharya et al. 1986, 1987) to be in excellent agreement with experiments on polymer solutions(Huppler et al., 1967; Christiansen et al., 1974) and with the results(Ashurst et al., 1975) by molecular dynamics simulation methods. When the normal stresses are negligible, the non–Newtonian viscosity of a fluid is given by the formula

$$\eta = \eta_0 \sinh^{-1}(\tau\gamma)/\tau\gamma \qquad (5.2)$$

where τ is a parameter of dimension time and γ is the shear rate. This parameter τ may be expressible in terms of temperature, density and η_0 itself as will be indicated later.

The constitutive equation has been used to study viscoelastic properties of polymer solutions and good agreement(Eu, 1986) is obtained when compared with experimental data and it is also a basis to understand the Cox–Merz rule(Cox et al., 1958) for frequency dependent viscosity.

The constitutive equation can be solved together with the momentum balance equation to study the evolution of flow properties for polymer melts. This kind of study is rather relevant to polymer processings such as blow molding, film processing, etc. A study in bubble growth (cavitation) in polymer melts(Khayat et al., 1990) shows that the constitutive equation gives rise to correct results for the rate of bubble growth for various industrially important polymer melts.

VI. NONEQUILIBRIUM THERMODYNAMIC FUNCTIONS

As classical thermodynamics furnishes theoretical and conceptual foundations on which various theories of measurement for equilibrium macroscopic properties are based, the theory of extended irreversible thermodynamics should also provide such foundations for measuring nonequilibrium properties of matter near or far from equilibrium. However, development of the theory in this direction is not extensive, but there are some efforts reported in the literature which will be reviewed here. There is a major difficulty in relating nonequilibrium properties of matter (e.g., fluids) to the material functions and experimental (boundary) conditions to which the fluid of interest is subject since to achieve the desired goal it is generally necessary to solve a set of hydrodynamic equations which are not only partial differential equations but also generally nonlinear. Since analytic solutions of such equations are, for most of cases, virtually impossible, it is rather difficult to put forth a theory in terms of molecular parameters and experimentally accessible observables, which are often the boundary and initial conditions, that experiments can readily test for the validity. Nevertheless, it is possible to make some progress on the basis of known analytic solutions for some relatively simple flow problems and thereby gain valuable insights into the questions of more general nature. Here the current status of the development in the subject is reviewed.

We have established that the entropy change accompanying the displacement of a nonequilibrium state to another can be given by the extended Gibbs relation which we now write in a differential form

$$T d\mathcal{S} = d\mathcal{E} + p dv - \sum_{a=1}^{r} \hat{\mu}_a dc_a + \sum_{a=1}^{r} \sum_{\alpha=1}^{\ell} X_a^{(\alpha)} \odot d\hat{\Phi}_a^{(\alpha)}. \qquad (6.1)$$

It holds for an r component fluid where ℓ fluxes $\hat{\Phi}_a^{(\alpha)}$ occur. Instead of treating $\mathcal{S}$ as the dependent variable in the Gibbs space made up with the variables ($\mathscr{S}$, $\mathscr{E}$, v, c_a, $\hat{\Phi}_a^{(\alpha)}$: $a = 1,2,\ldots, r$; $\alpha = 1,2,\ldots, \ell$) we may treat $\mathscr{E}$ as the dependent variable. Then the extended Gibbs relation (6.1) may be cast into the form

$$d\mathscr{E} = Td\mathscr{S} - pdv + \sum_{a=1}^{r} \hat{\mu}_a dc_a - \sum_{a=1}^{r} \sum_{\alpha=1}^{\ell} X_a^{(\alpha)} \odot d\hat{\Phi}_a^{(\alpha)}. \tag{6.2}$$

In this mode of representation we regard ($\mathscr{S}$, v, c_a, $\hat{\Phi}_a^{(\alpha)}$) as the Gibbs variables. It is useful to introduce the following thermodynamic functions in analogy to the equilibrium theory:

enthalpy: $$h = \mathscr{E} + pv, \tag{6.3a}$$

work function: $$\mathscr{a} = \mathscr{E} - T\mathscr{S}, \tag{6.3b}$$

Gibbs free energy: $$\mathscr{g} = h - T\mathscr{S}. \tag{6.3c}$$

Then the differential forms for h, $\mathscr{a}$ and $\mathscr{g}$ are:

$$dh = Td\mathscr{S} + vdp + \sum_{a=1}^{r} \hat{\mu}_a dc_a - \sum_{a=1}^{r} \sum_{\alpha=1}^{\ell} X_a^{(\alpha)} \odot d\hat{\Phi}_a^{(\alpha)}, \tag{6.4}$$

$$d\mathscr{a} = -\mathscr{S}dT - pdv + \sum_{a=1}^{r} \hat{\mu}_a dc_a - \sum_{a=1}^{r} \sum_{\alpha=1}^{\ell} X_a^{(\alpha)} \odot d\hat{\Phi}_a^{(\alpha)}, \tag{6.5}$$

$$d\mathscr{g} = -\mathscr{S}dT + vdp + \sum_{a=1}^{r} \hat{\mu}_a dc_a - \sum_{a=1}^{r} \sum_{\alpha=1}^{\ell} X_a^{(\alpha)} \odot d\hat{\Phi}_a^{(\alpha)}. \tag{6.6}$$

It is helpful to note that the extended Gibbs relations (6.1)–(6.6) take the forms reminiscent of those for an open system in thermostatics since in irreversible thermodynamics the theory is of necessity formulated in a local form, and in a local description the mass is not conserved in a small elementary volume at a position in the fluid because of the exchange of matter between the elementary volume and its surroundings; that is, the elementary volume is thermodynamically open.

Chemical potentials $\hat{\mu}_a$ may be defined by

$$\hat{\mu}_a = (\partial\mathscr{S}/\partial c_a)_{\mathscr{E},v,c',\hat{\Phi}}, \tag{6.7a}$$

but equivalently as a partial molar Gibbs free energy

$$\hat{\mu}_a = (\partial\mathscr{g}/\partial c_a)_{T,p,c',\hat{\Phi}} \tag{6.7b}$$

where the subscript c' means the exclusion of c_a from the set (c_1, c_2,..., c_r). Since there appear in (6.7b) the conditions that $\hat{\Phi}_a^{(\alpha)}$ be kept constant, the chemical potentials $\hat{\mu}_a$ are functions of fluxes in addition to T, p and c_a on which equilibrium chemical potentials also depend. Eq. (6.7b) suggests that $\mathscr{G}$ is a homogeneous function of c_a of degree one. Therefore

$$\mathscr{G} = \sum_{a=1}^{r} \hat{\mu}_a c_a. \tag{6.8}$$

Thus, substitution of (6.8) into the extended Gibbs relation (6.6) yields

$$\sum_{a=1}^{r} [c_a d\hat{\mu}_a + \mathscr{S}_a dT - \bar{v}_a dp + \sum_{\alpha=1}^{\ell} X_a^{(\alpha)} \odot d\hat{\Phi}_a^{(\alpha)}] = 0 \tag{6.9}$$

where

$$\mathscr{S}_a = (\partial \mathscr{S} / \partial c_a)_{T,p,c',\hat{\Phi}}, \qquad \bar{v}_a = (\partial v / \partial c_a)_{T,p,c',\hat{\Phi}}. \tag{6.10}$$

These are partial molar quantities, but for nonequilibrium. Eq. (6.9) is an extension of the Gibbs–Duhem equation since it reduces to the equilibrium Gibbs–Duhem equation when $\hat{\Phi}_a^{(\alpha)} = 0$ or $X_a^{(\alpha)} = 0$ for a and α. The extended Gibbs–Duhem equation gives methods to determine nonequilibrium effects for intensive variables $\hat{\mu}_a$, T and p. The change in chemical potential $\hat{\mu}_a$ may be expressed by the relation

$$d\hat{\mu}_a = -\mathscr{S}_a dT + \bar{v}_a dp + \sum_{b=1}^{r} \hat{\mu}_{ab} dc_b + \sum_{\alpha=1}^{\ell} \sum_{b=1}^{r} \hat{\mu}_{ab}^{(\alpha)} \odot d\hat{\Phi}_b^{(\alpha)} \tag{6.11}$$

where

$$\hat{\mu}_{ab} = (\partial \hat{\mu}_a / \partial c_b)_{\xi'}, \quad \hat{\mu}_{ab}^{(\alpha)} = (\partial \hat{\mu}_a / \partial \hat{\Phi}_b^{(\alpha)})_{\xi'}$$

with ξ' meaning that T, p, c_a and $\hat{\Phi}_b^{(\alpha)}$, $a \neq b$, are kept fixed. The differential forms presented above, when appropriately integrated, can in principle give the desired thermodynamic functions, provided that constitutive relations are supplied for the coefficients to various differentials in the differential forms. As was indicated in Sec. II, one of important constitutive relations is the caloric equation of state given by one of specific heats which are defined by

$$C_v = (\partial \mathscr{E} / \partial T)_{v,c,\hat{\Phi}}, \tag{6.12a}$$

$$C_p = (\partial h / \partial T)_{p,c,\hat{\Phi}}. \tag{6.12b}$$

These specific heats pertain to a nonequilibrium system since $\hat{\Phi} \neq 0$, and may be decomposed into species components in the case of a mixture. Because of the presence of $\hat{\Phi}$ among the variables kept constant, the specific heats depend on fluxes $\hat{\Phi}_a^{(\alpha)}$. Their flux dependence may be calculated, once the associated flow problem is solved. The equilibrium specific heats are then defined by the vanishing flux limits of C_p and C_v:

$$C_v^o = \lim_{\hat{\Phi}\to 0} (\partial \mathscr{E}/\partial T)_{v,c,\hat{\Phi}} , \tag{6.13a}$$

$$C_p^o = \lim_{\hat{\Phi}\to 0} (\partial h/\partial T)_{p,c,\hat{\Phi}} . \tag{6.13b}$$

It is quite obvious from the various nonequilibrium thermodynamic quantities presented up to this point that they can be determined in terms of material functions and the experimental (namely, boundary) conditions, only if the evolution of fluxes are explicitly given in terms of the latter for the processes of interest. This means that the associated fluid dynamical problem must be solved in order for such relations to be determined.

Before solutions of fluid dynamic equations are obtained for particular problems, it is necessary to develop thermodynamic methods of calculating nonequilibrium thermodynamic functions. Since the same methods apply to both single–component fluids and mixtures and the formalism is simpler for single–component fluids, we will couch the discussion in a single–component formalism. It is easy to extend the results obtained to the case of mixtures.

It is convenient to decompose nonequilibrium thermodynamic functions into equilibrium and nonequilibirum contributions. To achieve this aim the path of an irreversible process Γ is decomposed into two parts Γ_{p1} and Γ_{p2} such that in Γ_{p1} the process is of equilibrium and in Γ_{p2} the system is taken isothermally and isobarically to a nonequilibrium state from equilibrium state at the end of path Γ_{p1}: that is,

$$(T_o, p_o, 0) \to (T, p, 0) \qquad \text{for } \Gamma_{p1},$$
$$(T, p, 0) \to (T, p, \hat{\Phi}^{(\alpha)}) \qquad \text{for } \Gamma_{p2}.$$

Thus the path Γ_{p1} is that part embedded in the reversible part of the Gibbs space. Then it is shown(Eu, 1989a) that the enthalpy can be given in the form

$$h(T, p, \hat{\Phi}) = h_e(T,p) + h_{ne}(T,p,\hat{\Phi}) \tag{6.14}$$

where

$$h_e(T,p) = h(T_o,p_o,0) + \int_{T_o\,(\Gamma_{p1})}^{T} dT\, C_p^o(T,p_o) + \int_{p_o\,(\Gamma_{p1})}^{p} dp[v - T(\partial v/\partial T)_{p,0}], \tag{6.15a}$$

$$h_{ne}(T,p,\hat{\Phi}) = \sum_{\alpha} \int_{0\ (\Gamma_{p2})}^{\hat{\Phi}^{(\alpha)}} d\hat{\Phi}^{(\alpha)} \odot [T(\partial X^{(\alpha)}/\partial T)_{p,\hat{\Phi}} - X^{(\alpha)}]. \tag{6.15b}$$

Eqs. (6.14)–(6.15b) can be obtained by integrating dh along the paths Γ_{p1} and Γ_{p2} and making use of Maxwell's relations(Eu, 1981a) for the extended Gibbs relations (6.2) and (6.4)–(6.6). Eq. (6.15b) clearly suggests that as mentioned before, the nonequilibrium part h_{ne} cannot be given in terms of material functions, boundary conditions on $\hat{\Phi}^{(\alpha)}$, etc. unless the constitutive equations for $X^{(\alpha)}$ are given and $\hat{\Phi}^{(\alpha)}$ are obtained from the solutions of the relevant hydrodynamic equations, and it is now clear what was meant by the assertion.

It is easy to calculate C_p from (6.14) and (6.12b):

$$C_p(T,p,\hat{\Phi}) = C_p^o(T,p) + C_p^{(n)}(T,p,\hat{\Phi}) \tag{6.16}$$

where

$$C_p^{(n)}(T,p,\hat{\Phi}) = \sum_{\alpha=1}^{\ell} \int_{0\ (\Gamma_{p2})}^{\hat{\Phi}^{(\alpha)}} d\hat{\Phi}^{(\alpha)} \odot T(\partial^2 X^{(\alpha)}/\partial T^2)_{p,\hat{\Phi}} \tag{6.17}$$

is the nonequilibrium contribution to the specific heat C_p. Clearly, it vanishes as $\hat{\Phi}^{(\alpha)}$ vanish.

The entropy density can be calculated similarly:

$$\mathcal{S}(T,p,\hat{\Phi}) = \mathcal{S}_e(T,p) + \mathcal{S}_{ne}(T,p,\hat{\Phi}) \tag{6.18}$$

where the equilibrium part $\mathcal{S}_e$ is given by

$$\mathcal{S}_e(T,p) = \mathcal{S}(T_o,p_o,0) + \int_{T_o\ (\Gamma_{p1})}^{T} dTC_p^o(T,p_o)/T - R\ln(p/p_o) + \int_{p_o\ (\Gamma_{p1})}^{p} dp[R/p - (\partial v/\partial T)_{p,0}], \tag{6.19a}$$

and the nonequilibrium part by

$$\mathcal{S}_{ne}(T,p,\hat{\Phi}) = \sum_{\alpha} \int_{0\ (\Gamma_{p2})}^{\hat{\Phi}^{(\alpha)}} d\hat{\Phi}^{(\alpha)} \odot (\partial X^{(\alpha)}/\partial T)_{p,\hat{\Phi}} . \tag{6.19b}$$

The Gibbs free energy is then easily computed with (6.15), (6.18) and (6.19):

$$\mathcal{G}(T,p,\hat{\Phi}) = \mathcal{G}_e(T,p) + \mathcal{G}_{ne}(T,p,\hat{\Phi}) \tag{6.20}$$

where the equilibrium part $\mathcal{G}_e$ and the nonequilibrium part $\mathcal{G}_{ne}$ are respectively given by

$$\mathcal{G}_e(T,p) = \mathcal{G}_e^*(T) + RT\ln p + \underset{(\Gamma_{p1})}{\int_{p_0}^{p}} dp[v - RT/p) \tag{6.21a}$$

and

$$\mathcal{G}_{ne}(T,p,\hat{\Phi}) = -\sum_\alpha \underset{(\Gamma_{p2})}{\int_0^{\hat{\Phi}^{(\alpha)}}} d\hat{\Phi}^{(\alpha)} \odot X^{(\alpha)} \tag{6.21b}$$

with the definitions

$$\mathcal{G}_e^* = h_e^* - T\mathcal{S}_e^*, \tag{6.22}$$

$$h_e^*(T) = \lim_{p_0 \to 0} \{h(T_0,p_0,0) + \underset{(\Gamma_{p1})}{\int_{T_0}^{T}} dTC_p^0(T,p_0) - \underset{(\Gamma_{p1})}{\int_0^{p_0}} dp[v - T(\partial v/\partial T)_{p,0}]\}, \tag{6.23a}$$

$$\mathcal{S}_e^*(T) = \lim_{p_0 \to 0} \{\mathcal{S}(T_0,p_0,0) + \underset{(\Gamma_{p1})}{\int_{T_0}^{T}} dTC_p^0(T,p_0)/T + R\ln p_0 - \underset{(\Gamma_{p1})}{\int_0^{p_0}} dp[R/p - (\partial v/\partial T)_{p,0}]\}. \tag{6.23b}$$

The Gibbs free energy in (6.20) may be written in a compact form in terms of nonequilibrium fugacity in analogy to the equilibrium fugacity:

$$\mathcal{G}(T,p,\hat{\Phi}) = \mathcal{G}_e^*(T) + RT\ln(pf_n) \tag{6.24}$$

where

$$f_n = \exp\{(RT)^{-1}[\underset{(\Gamma_{p1})}{\int_{p_0}^{p}} dp[v - RT/p) - \sum_\alpha^{\ell} \underset{(\Gamma_{p2})}{\int_0^{\hat{\Phi}^{(\alpha)}}} d\hat{\Phi}^{(\alpha)} \odot X^{(\alpha)}]\}. \tag{6.25}$$

Therefore, the nonequilibrium fugacity is a measure of nonideality as well as displacement from equilibrium.

Other thermodynamic functions can be obtained in a similar manner. It is rather straightforward to obtain nonequilibrium thermodynamic functions for a mixture of fluids by using the same method as for those presented earlier. The results presented in this section indicate that irreversible thermodynamics and fluid dynamics are intimately connected to each other, and the connection is now made clear although such an intimation was present at the early stage in the development of irreversible theormodynamic theory reviewed in this article. In this regard it must be emphasized that it is not sufficient to know thermodynamic functions in terms of fluxes since the latter do not give the experimental input to the system.

VII. APPLICATIONS

Solutions of flow problems are required for calculating nonequilibrium thermodynamic functions if the formalism presented in Section V is used. Since flow problems are not easily solved for arbitrary flow configurations, only special cases that render analytic solutions have been so far studied. Phenomenological treatments of macroscopic phenomena are a powerful approach and merit studies in depth. Such studies may be done by means of extended irreversible thermodynamic methods. If use of the extended irreversible thermodynamics formalism is to be made as wide spread as equilibrium thermodynamics, more extensive studies will have to be made in a number of directions. Here we review the status of the subject. To make the presentation as brief as possible, the review will be confined to the results valid at the asymptotic longtime limit.

VII.1. Plane Couette Flow

First, let us assume that the fluid is Newtonian, incompressible and non–heat–conducting. It is also assumed that the fluid is confined in an infinitely long rectangular channel and one of the plates bounding the channel are moving in the x direction at velocity u_0. Then there is a velocity gradient in the y direction, if we take the z direction as that of the neutral direction. This is the plane Couette flow configuration and the flow profile is well known in this case: it is linear in y.

The shear stress Π for the flow configuration is then given by

$$\Pi = -2\eta_0\gamma \qquad (7.1)$$

where η_0 is the shear viscosity and

$$\gamma = (\partial u_x/\partial y) \qquad (7.2)$$

is the shear rate; u_x is the x component of the fluid velocity. The velocity profile is well known

$$u_x(y) = u_0 y/D \qquad (7.3)$$

where D is the distance between the two plates. The corresponding shear stress then is easily calculated from (7.1)–(7.3):

$$\Pi(y) = -\eta_0 u_0/D. \qquad (7.4)$$

By the assumption made earlier regarding the nature of the fluid there is only one term for $X^{(\alpha)}$

corresponding to shear. If we assign $\alpha = 1$ to shear, then we may take

$$X^{(1)} = -\Pi/2p \tag{7.5}$$

where p is the pressure. Thus, substitution of (7.4) into (7.5) yields

$$X^{(1)} = \eta_0 u_0/2Dp. \tag{7.6}$$

By using these results the nonequilibrium part of the entropy is obtained:

$$\mathscr{S}_{ne} = -(\eta_0/\rho p)(\partial\eta_0/\partial T)_p(u_0/D)^2 \tag{7.7}$$

and thus its global value S_{ne}, which is a volume integral of $\rho\mathscr{S}_{ne}$, is given by the formula

$$S_{ne} = -V(\eta_0/p)(\partial\eta_0/\partial T)_p(u_0/D)^2. \tag{7.8}$$

The nonequilibrium part of the Gibbs free energy is similarly calculated:

$$\mathscr{G}_{ne} = (\eta_0^2/\rho p)(u_0/D)^2. \tag{7.9}$$

The corresponding enthalpy h_{ne} and specific heat $C_p^{(n)}$ are

$$h_{ne} = (\eta_0^2/\rho p)(u_0/D)^2[1 - (\partial \ln\eta_0/\partial \ln T)_p] \tag{7.10}$$

and

$$C_p^{(n)} = -(\eta_0^2/\rho p)(u_0/D)^2[\frac{\partial}{\partial T}(\partial \ln\eta_0/\partial \ln T)]_p. \tag{7.11}$$

Eq. (7.11) for the nonequilibrium part of the specific heat shows that shearing a fluid affects specific heat measurement, and its effect will be determined by the temperature dependence of shear viscosity.

VII.2. Plane Couette Flow of a Non–Newtonian Fluid

If a non–Newtonian fluid is subject to a shear in the same experimental configuration as in the plane Couette flow described in Section VII.1, the situation becomes rather different. The non–Newtonian fluid considered obeys the steady–state constitutive equation

$$\Pi(\sinh\kappa/\kappa) = -\eta_0(\partial u_x/\partial y). \tag{7.12}$$

Here

$$\kappa = (\tau_p/\sqrt{2}\eta_0)(\Pi^2)^{1/2},$$

$$\tau_p = [2\eta_0(m_r k_B T/2)^{1/2}]^{1/2}/ndk_B T$$

with n denoting the number density, m_r the reduced mass and d the size parameter of the molecule. The solution of the steady hydrodynamic equation for plane Couette flow with Π defined by (7.12) yields

$$u_x = (u_o/D)y, \tag{7.13}$$

$$\Pi = -\eta_o q(\gamma)\gamma \tag{7.14}$$

with γ now defined by $\gamma = u_o/D$ and $q(\gamma) = \sinh^{-1}(\tau\gamma)/\tau\gamma$, $\tau = \tau_p/\sqrt{2}$. These results give rise to the following results for thermodynamic functions in the steady state(Eu, 1989a)

$$\mathcal{J}_{ne}(\infty) = (\eta_o^2/\rho p\tau^2)\ln^2\{\tau\gamma + [1 + (\tau\gamma)^2]^{1/2}\}, \tag{7.15}$$

$$\mathcal{S}_{ne}(\infty) = -(\eta_o^2/\rho p)(q\gamma)^2[\partial\ln(\eta_o q)/\partial T]_{p,\hat{\Pi}}. \tag{7.16}$$

Since $q\gamma \approx \ln(\tau\gamma)$ for large $\tau\gamma$, $\mathcal{J}_{ne}$ and $\mathcal{S}_{ne}$ do not grow quadratically with the shear rate γ as is the case for a Newtonian fluid.

VII.3. Tube Flow of Non–Newtonian Fluids

Laminar flow in a circular tube is also another example for which analytical solutions are possible for both Newtonian fluids and non–Newtonian fluids obeying the constitutive equation (7.12). If the axis of the tube is assumed to be parallel to the z axis of the coordinate system, then the velocity u_z is found to be in the form

$$u_z(r) = (2\eta_o L/\Delta\mathcal{P}\tau^2)[\cosh a - \cosh(ar/R)] \tag{7.17}$$

where

$$a = \tau\Delta\mathcal{P}R/2\eta_o L, \quad \Delta\mathcal{P} = \Delta p + \rho gL$$

with R and L denoting the radius and length of the tube, Δp the pressure difference between the entrance and exit of the tube, ρ the density of the fluid and g the gravitational constant. The shear stress is easily found to be in the form

$$\Pi = \Delta\mathcal{P}r/2L \tag{7.18}$$

and

$$p = p_i - (\Delta p/L)z$$

where p_i is the pressure at the entrance of the tube. Eq. (7.18) and the expression for $\mathcal{J}_{ne}$ yield

$$\mathcal{J}_{ne}(\infty) = [(\Delta\mathcal{P})^2/8\rho p](r/R)^2. \tag{7.19}$$

Since this local quantity depends on position, a mean value is more useful. We define it as follows:

$$\bar{\mathcal{F}}_{ne}(\infty) = (\pi R^2 L)^{-1} \int_0^L dz \int_0^R dr\, 2\pi r\, \mathcal{F}_{ne}(r). \tag{7.20}$$

Substitution of (7.19) gives

$$\bar{\mathcal{F}}_{ne}(\infty) = \frac{(\Delta \mathcal{P})^2 R^2}{16\rho \Delta p L^2} \ln(1 + \Delta p/p_f) \tag{7.21}$$

where p_f is the pressure at the exit of the tube. The volume flow rate Q can be easily calculated with the velocity profile (7.17):

$$Q = 2\pi \int_0^R dr\, r u_z(r) = \frac{\pi R^4 \Delta \mathcal{P}}{8L\eta_o}(1 + \Delta Q) \tag{7.22}$$

where

$$\Delta Q = 8a^{-2}[\tfrac{1}{2}\cosh a + a^{-2}(\cosh a - 1) - a^{-1}\sinh a] - 1. \tag{7.23}$$

It is easy to show that ΔQ vanishes as $a \to 0$ which is equivalent to $\Delta \mathcal{P} \to 0$. In the limit of $\Delta Q \to 0$, (7.22) reduces to the well known Hagen–Poiseuille volume flow rate formula(Landau et al., 1959; Reiner, 1960). Since the volume flow rate Q is directly accessible to measurement it is convenient to express $\bar{\mathcal{F}}_{ne}$ in terms of Q:

$$\bar{\mathcal{F}}_{ne}(\infty) = \frac{4\eta_o^2}{\pi^2 \rho R^6 \Delta p} \ln(1 + \Delta p/p_f) Q^2 (1 + \Delta Q)^{-2}, \tag{7.24}$$

which, in the case of a Newtonian fluid, becomes

$$\bar{\mathcal{F}}_{ne}(\infty) = \frac{4\eta_o^2}{\pi^2 \rho R^6 \Delta p} \ln(1 + \Delta p/p_f) Q^2. \tag{7.25}$$

These expressions suggest that the mean nonequilibrium part of the Gibbs free energy can be measured by determining the volume flow rate of the fluid passing through the tube.

VII.4. Shear–Induced Melting Point Depression

It is known that condensed matter undergoes a phase transition induced by shear. There are many examples: such as melting and transition to an amorphous structure from a crystalline structure of a colloidal suspension(Clark et al., 1980, 1981); shear–induced shift in critical consolute temperature of cyclohexane–aniline mixtures(Pines et al., 1987); shear–induced phase changes observed in numerical simulations(Evans, 1982). All of these are nonequilibrium phenomena and fall into the category of studies in fluid dynamics. However, fluid dynamic calculations of such phenomena can be rather complicated and difficult. There are some renormalization–group calculations available in the literature(Satten et al., 1986), but compared to the amount of calculations required, the theoretical returns in terms of flow properties are rather limited. Here we review an alternative irreversible thermodynamic theory based on a simplified model(Eu, 1989b). Because of the nature of the model the results do not exactly correspond to the experimental situation, but the model indicates a general method of attack for the kinds of problems mentioned earlier.

Suppose an incompressible liquid is confined between two solid plates which are the solidified liquid at temperature T_m. The two phases of the solid and liquid are in thermal equilibrium in the absence of shearing. We will assume that the substance is non–heat–conducting for the simplicity of the formalism the assumption produces. One of the plates is then made to move at a steady velocity u_o in a direction and the system is let reach the steady state. Therefore, the flow geometry is that of plane Couette flow and the flow is assumed to be laminar. Under these assumptions the irreversible thermodynamic results presented in this section and Section V are applicable. In particular, the chemical potential for the single–component fluid in hand is given by

$$\mu(T,p,\Pi) = \mu_o^{\ell}(T,p) + RT\ln\Delta_n \tag{7.26}$$

where

$$\mu_o^{\ell}(T,p) = \mu_e(t,p) + RT\ln(pf_e), \tag{7.27}$$

$$\Delta_n = \exp[(\eta^2/\rho pRT)(u_o/2D)^2, \tag{7.28}$$

with $\mu_e(T,p)$ denoting the equilibrium reference chemical potential and f_e the equilibrium fugacity coefficient. The viscosity coefficient η is as follows:

$$\eta = \begin{cases} \eta_o & \text{for Newtonian liquids,} \\ \eta_o \sinh^{-1}(\tau u_o/2D)/(\tau u_o/2D) & \text{for non-Newtonian liquids.} \end{cases}$$

Here η_o is the Newtonian (zero shear) viscosity and the parameter τ and other symbols were defined previously.

Let us denote the chemical potential of the solid phase by $\mu_e^s(T_m,p)$. Then the equilibrium condition before shearing is

$$\mu_o^{\ell}(T_m,p) = \mu_e^s(T_m,p). \tag{7.29}$$

When there is a steady shear, this condition does not hold. To see the effect of shear, we consider the following:

$$[\mu(T,p,\hat{\Pi}] - \mu_e^s(T,p)]/T.$$

By differentiating it with T at constant p and $\hat{\Pi}$, we obtain

$$[\frac{\partial}{\partial T}(\mu - \mu_e^s)/T]_{p,\Pi} = -(\Delta h_m + h_n)/T^2 \qquad (7.30)$$

where

$$\Delta h_m = h_e^{\ell}(T,p) - h_e^s(T,p) \qquad (7.31)$$

and h_n is the nonequilibrium part of the enthalpy. Note that $\Delta h_m > 0$ and $h_n \geq 0$. Integration of (7.27) over T and use of (7.26) and (7.29) yield the following equation

$$R\ln\Delta_n = -\int_{T_m}^{T} dT(\Delta h_m + h_n)/T^2. \qquad (7.32)$$

The left member of (7.32) is positive since $\Delta_n \geq 1$ for $\gamma = u_o/2D \geq 0$. Since the integrand in (7.32) is positive, it is possible to conclude

$$\Delta T = T_m - T \geq 0 \qquad (7.33)$$

for $\gamma \geq 0$. That is, a steady shear induces a lowering of the melting point of the solid. This is a rigorous irreversible thermodynamic result demanded ultimately by the second law of thermodynamics.

By using suitable temperature dependences for Δh_m, h_n and the viscosity it is possible to estimate ΔT. It is shown(Eu, 1989b) that

$$\Delta T/T_m \sim R_e^2 \qquad (7.34))$$

where R_e is the Reynolds number defined by

$$R_e = \rho u_o D/\eta^* \qquad (7.35)$$

with D denoting the gap between the plates and η^* is such that(Frenkel, 1946)

$$\eta_o = \eta^* \exp(A/RT), \qquad (7.36)$$

namely, the viscosity at $T = \infty$. The estimate (7.34) for ΔT is in qualitiative agreement with experiment on randomly stirred critical mixtures(Pines et al., 1987).

VIII. CONCLUDING REMARKS

In this article both the kinetic theory and the axiomatic theory of irreversible processes are reviewed along the lines studied by this author. Attempts have been made to show that it is indispensible to have molecular, statistical mechanical theories for theory of irreversible processes for systems far from equilibrium since nonlinear macroscopic equations are generally involved and they are not easy to obtain correctly by simply extending the linear theory versions for lack of a coherent strategy for extension. The theories reviewed also strongly indicate that fluid dynamics is intimately tied up with irreversible thermodynamics. This aspect, however, has not been apparent in the linear theory of irreversible processes as is evident from the conventional investigations in linear irreversible processes which are usually done detached from fluid dynamics. It is, however, an inevitable conclusion of the series of studies reviewed in this article that *the subjects of irreversible thermodynamics and fluid dynamics form a coherent whole that must be treated as such to the benefits of both subjects.* Some examples discussed in this connection are merely indicators for what one might be able to do in the phenomenological approach to study of macroscopic phenomena; things one can investigate with the formalisms reviewed are numerous. On the kinetic theory side the collision bracket integrals, especially, for dense fluids have not been calculated as yet except for one special case of shear viscosity for hard spheres(Eu, 1979b), but they would be very useful for calculating flow properties of real fluids in terms of molecular parameters and boundary conditions only.

In the Ginzburg–Landau theory(1950), for example, the free energy of a nonequilibrium system is simply taken from the equilibrium theory and then extended into the nonequilibrium domain without establishing the validity of such an extension. The theory of extended irreversible thermodynamics and nonequilibrium thermodynamic functions calculated therewith may be relevant to the question just mentioned.

Eqs. (2.17)–(2.22) reduce to the equations in classical hydrodynamics when the constitutive equations (2.22) are approximated by steady linear forms which give rise to, for example, (3.37)–(3.39). Therefore, (2.17)–(2.22) are generalizations of classical hydrodynamic equations and hydrodynamics based on the former is termed *generalized hydrodynamics.* The flow problems studied so far by means of generalized hydrodynamics indicate its superior capability beyond the classical hydrodynamics predictions when the systems are far from equilibrium. Further extensive studies would be useful for making generalized hydrodynamics and extended irreversible thermodynamic valuable theoretical tools for macroscopic phenomena.

REFERENCES

Ashurst, W.T. and W.G. Hoover. 1975. Dense–fluid shear viscosity via nonequilibrium molecular dynamics. Phys. Rev. A11, 658.

Bampi, F. and A. Morro. 1984. Nonequilibrium thermodynamics: a hidden variable approach. In Recent Developments in Nonequilibrium Thermodynamics, eds. J. Casas–Vazquez, D. Jou and G. Lebon, p. 211. Berlin: Springer.

Bataille, J. and J. Kestin. 1979. Irreversible processes and physical interpretation of rational thermodynamics. J. Nonequil. Thermodyn. **4**, 229.

Batchelor, G.K. 1967. An Introduction to Fluid Dynamics. Cambridge: Cambridge University Press.

Bhattacharya, D.K. and B.C. Eu. 1986. Theory of dynamic viscosities and normal stress coefficients of dense fluids. Mol. Phys. **59**, 1145.

Bhattacharya and B.C. Eu. 1987. Viscoelastic properties of dense simple fluids: two–dimensional viscosities. Phys. Rev. A**35**, 4850.

Bird, R.B., R.C. Armstrong and O. Hassager. 1977. Dynamics of Polymeric Liquids. New York: Wiley.

Boltzmann, L. 1872. Weitere Studien ueber das Waermegleichgewicht unter Gasmolekulen. Wien Ber. **66**, 275; Wien Anz. **9**, 23; Abh. **1**, 316.

1964. Lectures on Gas Theory. Translated by S. G. Brush. Berkeley: University of California Press.

Casas–Vazquez, J. and D. Jou. 1981. On a phenomenological non–equilibrium entropy for a class of rigid heat conductors. J. Phys. A**14**, 1225.
Chapman, S. and T.G. Cowling. 1970. Mathematical Theory of Nonuniform Gases. Cambridge: Cambridge University Press.
Christiansen, E.B. and W.R. Leppard. 1974. Steady–state and oscillatory flow properties of polymer solutions. Trans. Soc. Rheol. **18**, 65.
Clark, N.A. and B.J. Ackerson. 1980. Observation of the coupling of concentration fluctuations to steady state shear flow. Phys. Rev. Lett. **44**, 1005; 1981. Shear–induced melting. ibid. **46**, 123.
Clause, J.P. and R. Balescu. 1982. A nonlinear approach to the kinetic theory of heat conductivity in a plasma. Plasma Phys. **24**, 1492.
Clausius, R. 1850. Ueber die bewegende Kraft der Waerme und die Gesetze, welche sich daraus fuer die Waermelehre selbst ableiten lassen. Ann. Phys. (Leipzig) **79**, 368.
1865. Ueber verschiedene fuer die Anwendung bequeme Formen der Hauptgleichungen der mechanischen Waermetheorie. Ann. Phys. (Leipzig) **125**, 353.
Coleman, B. D. and W. Noll. 1963. The thermodynamics of elastic materials with heat conduction and viscosity. Arch. Rat. Mech. Anal. **13**, 167.
Cox, W.P. and E. H. Merz. 1958 Correlation of dynamic and steady flow viscosities. J. Polym. Sci. **28**, 619
de Groot, S.R. and P. Mazur. 1962. Nonequilibirum Thermodynamics. Amsterdam: North–Holland.
Einstein, A. 1910. Theorie der Opaleszens von homogenen Fluessigkeiten und Fluessigkeitsgemischen in der Naehe des kritischen Zustandes. Ann. Phys. **33**, 1275.
Eirich, F.R., ed. 1956. Rheology. New York: Academic Press.
Ernst, M. H., L. K. Haines and J. R. Dorfman. 1969. Theory of transport coefficients for moderately dense gases. Rev. Mod. Phys. **41**, 296.
Eu, B.C. 1971a. Classical theory of scattering: three–body problems. J. Chem. Phys. **54**, 559.
Eu, B.C. 1971b. Cluster expansion and autocorrelation functions: self–diffusion coefficient. J. Chem. Phys. **55**, 4613.
Eu, B.C. 1979a. Kinetic theory of dense fluids. I. Ann. Phys.(N.Y.) **118**, 187.
Eu, B.C. 1979b. Kinetic theory of dense fluids. III. Density dependence of transport coefficients. Ann. Phys.(N.Y.) **120**, 423.
Eu, B.C. 1980. A modified moment method and irreversible thermodynamics. J. Chem. Phys. **73**, 2858.
Eu, B.C. 1981a. On the extended Gibbs relations and nonlinear irreversible thermodynamics. J. Chem. Phys. **74**, 2998.
Eu, B.C. 1981b. The modified moment method, irreversible thermodynamics and the nonlinear viscosity of a dense fluid. J. Chem. Phys. **74**, 6362.
Eu, B.C. 1982. Irreversible thermodynamics of fluids. Ann. Phys.(N.Y.) **140**, 341.
Eu, B.C. 1983. Shear rate dependence of non–Newtonian viscosity of fluids. J. Chem. Phys. **79**, 2315.
Eu, B.C. and Y.G. Ohr. 1984. Theory of non–Newtonian viscosity and normal stress coefficients. J. Chem. Phys. **81**, 2756.
Eu, B.C. 1985a. Nonlinear heat conductivity in plasmas. Phys. Fluids. **28**, 222.
Eu, B.C. 1985b. Theory of viscoelasticity of fluids. J. Chem. Phys. **82**, 4283.
Eu, B.C. 1986. On the modified moment method and irreversible thermodynamics. J. Chem. Phys. **85**, 1592.
Eu, B.C. 1987. Kinetic theory of dense fluids subject to an external field. J. Chem. Phys. **87**, 1220.
Eu, B.C. 1988a. Entropy for irreversible processes. Chem. Phys. Lett. **143**, 65.
Eu, B.C. 1988b. Velocity profiles in two–dimensional channel flow of a non–Newtonian fluid under pressure. Phys. Rev. A**37**, 4504.
Eu, B.C. 1989a. Nonequilibrium thermodynamic functions for sheared fluids. Physica A**160**, 53.

Eu, B.C. 1989b. Irreversible thermodynamic theory of shear–induced melting point depression. Physica **A160**, 87.
Eu, B.C. 1989c. Integral equation for nonequilibrium chemical potential and the Kirkwood diffusion equation: derivation from the generalized Boltzmann equation. J. de Phys. France **50**, 3497.
Evans, D.J. 1982. Shear–induced melting of soft–sphere crystals. Phys. Rev. **A25**, 2788.
Ferziger, J.H. and H. G. Kaper. 1972. Mathematical Theory of Transport processes in Gases. Amsterdam: North–Holland.
Frenkel, J. 1946. Kinetic Theory of Liquids. Cambridge: Cambridge University Press.
Gibbs, J.W. 1876. On the heterogeneous equilibria. Trans. Conn. Acad. **III**, 108; The Scientific Papers of J. W. Gibbs. New York: Dover.
Ginzburg, V.L. and L. D. Landau. 1950. On the theory of superconductivity. Zh. Exsp. Teor. Fiz. **20**, 1064.
Grad, H. 1958. Principles of Kinetic Theory of Gases in Handbuch der Physik, ed. S. Flügge, Vol. 12, p. 205. Berlin: Springer.
Gyarmati, I. 1970. Nonequilibrium Thermodynamics. Berlin: Springer.
Huppler, J.D., E. Ashare and L.A. Holms. 1967. Rheological properties of three solutions. Part 1. non–Newtonian viscosity, normal stresses, and complex viscosity. Trans. Soc. Rheol. **11**, 159.
Ikenberry, E. and C. Truesdell. 1956. On the pressures and the flux of energy in a gas according to Maxwell's kinetic theory. I. J. Rat. Mech. Anal. **5**, 1.
Jou, D., J. Casas–Vazquez and G. Lebon. 1988. Extended irreversible thermodynamics. Repts. Prog. Phys. **51**, 1105.
Lord Kelvin 1851. On the dynamical theory of heat, with numerical results deduced from Mr. Joule's equivalent of a thermal unit, and M. Regnault's observation on steam. Trans. Roy. Soc. Edinburgh **20**, 261.
Kestin, J., ed. 1976. The Second Law of Thermodynamics. Stroudsburg, Pa.: Dowden,Hutchinson & Ross.
Khayat,R.E. and B.C. Eu. 1988. Nonlinear transport processes and fluid dynamics: cylindrical Couette flow of Lennard–Jones fluids. Phys. Rev. **A38**, 2492.
Khayat, R.E. and B.C. Eu. 1989. Density profiles and entropy production in cylindrical Couette flow: comparison of generalized hydrodynamics and Monte Carlo results. Prog. Astronaut. Aeronaut. **118**, 396.
Khayat, R.E. and B.C. Eu. 1990. Generalized hydrodynamics and dynamics of polmer bubble growth. To be submitted.
Kirkwood, J.G. and I. Oppenheim. 1961. Chemical Thermodynamics. New York: McGraw–Hill.
Landau, L.D.and E.M. Lifshitz. 1959. Fluid Mechanics. London: Pergamon.
Lebon, D. Jou and J. Casas–Vazquez. 1980. An extension of the local equilibrium hypothesis. J. Phys. **A13**, 275.
Lebon, G. and D. Jou. 1982. A nonclassical thermodynamic description of heat conducting viscous fluids. J. Chem. Phys. **77**, 970.
Meixner, J.. 1970. On the foundation of thermodynamics of processes. In A Critical Review of Thermodynamics, eds. E.B. Stuart, B. Gal–Or and A. J. Brainard, p. 544. Baltimore: Mono Book Corp.
Miles, J.R.N. and J. S. Dahler. 1970. Classical scattering theory: elastic collisions. J. Chem. Phys. **52**, 616.
Müller, I. 1967. Zum Paradoxon der Waermeleitungstheorie. Z. Phys. **198**, 329.
Müller, I. 1984. Extended thermodynamics—past, present, future. In Recent Developments in Nonequilibrium Thermodynamics, eds. J. Casas–Vazquez, D. Jou and G. Lebon, p. 32. Berlin: Springer.
Nettleton, R. 1984. Early applications of extended irreversible thermodynamics. In Recent Developments in Nonequilibrium Thermodynamics, eds. J. Casas–Vazquez, D. Jou and

G. Lebon, p. 1. Berlin: Springer.
Ohr, Y.G. and B.C. Eu. 1984. Non–Newtonian shear viscosity, normal stress coefficients and corresponding states in rheology. Phys. Lett. **A101**, 338.
Onsager, L. 1931a. Reciprocal relations in irreversible processes. I. Phys. Rev. **37**, 405.
Onsager, L. l931b. Reciprocal processes in irreversible processes. II. Phys. Rev. **38**, 2265.
Pines, D.J., N. Easwar, J.V. Maher and W.I. Goldburg. 1987. Turbulent suppression of spinodal decomposition. Phys. Rev. A**35**, 1756.
Prigogine, I. 1967. Thermodynamics of Irreversible Processes. Third edition. New York: Interscience.
Prigogine, I. 1949. Le domaine de validite de la thermodynamique des phenomenes irreversibles. Physica **15**, 271.
Lord Rayleigh, 1945. Theory of Sound. New York: Dover.
Reiner, M. 1960. Deformation and Flow. London: Lewis.
Rice, S.A., J.P. Boon and H.T. Davis. 1968. On the experimental and theoretical study of transport phenomena in simple fluids. In Simple Dense Fluids, eds. H. L. Frisch and Z. W. Salsburg, p. 251. New York: Academic Press.
Satten, G. and D. Ronis. 1986. Critical phenomena in randomly stirred fluids: correlation functions, equation of motion, and crossover behavior. Phys. Rev. A**33**, 3415(1986).
Theodosopulu, M and J.S. Dahler. 1972. Theory of second order stress in a dense gas. Phys. Fluids. **15**, 10.
Truesdell, C. 1969. Rational Thermodynamics. New York: McGraw–Hill.
van Beijren, H., J. Karkheck and J.V. Sengers. 1988. Nonequilibrium temperature and bulk viscosity for a dense fluid of square–well molecules. Phys. Rev. A**37**, 2274.

Reciprocity in Extended Thermodynamics

R. E. Nettleton
Department of Physics
University of the Witwatersrand
Johannesburg, South Africa

ABSTRACT

The extended non–equilibrium thermodynamics provides additional reciprocity relations resulting from introduction of heat, matter, and momentum fluxes as state variables. These reciprocity relations are useful in calculating transport coefficients in liquids. Early applications were made to generalizations of the Cattaneo–Vernotte and Maxwell stress–relaxation equations to include coupled flows of heat and matter and viscoelasticity in presence of several creep mechanisms. Reciprocity reduces the number of coefficients which must be estimated from simple molecular models or from molecular dynamics. More recent work has utilized reciprocity in the estimation of non–linear effects in heat flow and diffusion. If **J** is the flow of heat or particles, one can estimate at high density the $0(\mathbf{J}^2)$ contributions to the thermal conductivity or binary diffusion coefficient. To do this, one expresses the driving term in the Cattaneo–Vernotte equation in terms of the thermodynamic pressure and applies an anti–reciprocity relation leading to calculation of the force associated with **J**. One can also calculate the $0(\mathbf{J}^2)$ term in $T - \theta$, the difference between thermodynamic and local equilibrium temperatures. The non–linear effects appear negligibly small in dense fluids. Non–linear reciprocity has received a statistical foundation via the Fokker–Planck equation derived for the distribution g of values of operators whose averages are variables such as **J**. The equation for g is obtained by applying Zwanzig or Grabert projection operators to the Liouville equation. The maximum entropy formalism provides an ansatz for the solution of this equation which gives a finite number of moments exactly in the spirit of Grad. By using this ansatz and taking moments of the equation for g, one derives kinetic equations exhibiting reciprocity in the non–linear regime.

1. INTRODUCTION

The original applications of extended thermodynamics (Nettleton 1959a,b, 1960a,b, 1961, 1963, 1964a,b) were designed to utilize reciprocity in the calculation of transport coefficients in condensed phases, principally liquids. For example, the heat flow in a pure liquid can be divided into two parts, of which one, $\mathbf{J}_1$, is carried by hypersound ($10^9 - 10^{12}$ Hz) waves and the other, $\mathbf{J}_2$, by self–diffusing particles. If coupled equations are given relating the rates of change, $d\mathbf{J}_1/dt$ and $d\mathbf{J}_2/dt$, to $\mathbf{J}_1$, $\mathbf{J}_2$, and the gradients of temperature and concentration, reciprocity can assist in determining the coefficients which couple $d\mathbf{J}_2/dt$ to $\mathbf{J}_1$. These are not easily accessible to calculation by simple molecular models. They can be obtained by computer simulation, but then we should like to have a simple means of making estimates to check the latter. When the rate equations are specialized to the steady state by equating time derivatives to zero, one can solve for $\mathbf{J}_1$ and $\mathbf{J}_2$ in terms of ∇T.

Substituting the result into the total heat flux, we get

$$\mathbf{J} = \mathbf{J}_1 + \mathbf{J}_2 = -\lambda \, \nabla T \tag{1.1}$$

and, in the process, a theoretical expression for the thermal conductivity, λ (Nettleton 1960a, 1961, 1963).

To apply reciprocity to the coupled kinetic equations for $\mathbf{J}_1$ and $\mathbf{J}_2$, one needs to suppose that number density, n temperature, T, plus $\mathbf{J}_1$ and $\mathbf{J}_2$ are state variables describing a non–equilibrium state characterized by an entropy, S, internal energy, E, and free energy $F \equiv E - TS$, which are functions of these variables. The idea that this could be done came from the Cattaneo–Vernotte equation (Cattaneo 1958, Vernotte 1958),

$$\partial \mathbf{J}/\partial t = -(\mathbf{J}/\tau) - (\lambda/\tau)\nabla T \tag{1.2}$$

and the Maxwell stress–relaxation equation (Maxwell 1967),

$$\partial P^{\dagger}_{ij}/\partial t = -(P^{\dagger}_{ij}/\tau_1) - \lambda_2 B^{\dagger}_{ij} \tag{1.3}$$

where $P^{\dagger}_{ij}$ is the traceless pressure tensor and $B^{\dagger}_{ij}$ the traceless total rate of strain. λ_2 is the high–frequency shear modulus. Equation (1.2) was obtained (Vernotte 1958) by adding the inertial term $\partial \mathbf{J}/\partial t$ to Fourier's law of heat conduction. This addition removes the paradox of propagation of thermal effects with infinite velocity which characterizes the theory based on substitution of the Fourier law into the hydrodynamic equation of energy conservation. A statistical mechanical derivation of a transport law with inertial term included was given by Bearman and Kirkwood (1958, 1959). Examples of more recent derivations are given below, based on taking moments of kinetic equations for condensed phases derived from the classical Liouville equation.

From (1.2) and (1.3), we see that during the short times τ and τ_1 required for approach to a steady state, $\mathbf{J}$ and $P^{\dagger}_{ij}$ are independent variables. Since the regression of these is an irreversible process, there should exist an entropy $S(n, T, \mathbf{J}, P^{\dagger}_{ij})$ which increases. This implies a Gibbs equation,

$$T\, dS = dE + P\, dV + \mathbf{\Phi} \cdot d\mathbf{J} + \sum_{i,j} \Psi^{\dagger}_{ij}\, dP^{\dagger}_{ij} \tag{1.4}$$

as an extension of the second law which reduces to the latter in equilibrium. P is the thermodynamic pressure, usually taken to be one–third the pressure trace, although we point out in Section 7 below that P may have a different significance in some formulations of extended thermodynamics. $\mathbf{\Phi}$ and the traceless tensor $\Psi^{\dagger}_{ij}$ are thermodynamic forces respectively conjugate to $d\mathbf{J}/dt$ and $dP^{\dagger}_{ij}/dt$. Equation (1.4) in turn implies that

$$dF = -S\, dT - P\, dV - \mathbf{\Phi} \cdot d\mathbf{J} - \sum_{i,j} \Psi^{\dagger}_{ij}\, dP^{\dagger}_{ij} \; . \tag{1.5}$$

The Gibbs equation in the form (1.5) has integrability conditions:

$$\partial S/\partial \mathbf{J} = \partial \mathbf{\Phi}/\partial T \tag{1.6a}$$

$$\partial P/\partial J = \partial \Phi/\partial V \qquad (1.6b)$$

and similar conditions involving $P^{\dagger}_{ij}$ and $\Psi^{\dagger}_{ij}$. From (1.6a,b) one can calculate the J– and $P^{\dagger}_{ij}$–dependence of S and P if $\mathbf{\Phi}$ and $\Psi^{\dagger}_{ij}$ are known. To identify conjugate fluxes and forces, one can use (1.4) to calculate

$$T\,dS/dt - V\,\nabla\cdot\mathbf{J}_s = -VT^{-1}\mathbf{J}\cdot\nabla T - \sum_{i,j} P^{\dagger}_{ij}B^{\dagger}_{ij}V + \mathbf{\Phi}\cdot d\mathbf{J}/dt + \sum_{i,j} \Psi^{\dagger}_{ij}dP^{\dagger}_{ij}/dt \qquad (1.7)$$

To obtain (1.7), one needs the hydrodynamic equations of mass and energy conservation for dE/dt and dn/dt, $n \equiv N/V$. $\mathbf{J}_s$ is the entropy flux which, classically, is

$$\mathbf{J}_s = T^{-1}\mathbf{J} \qquad (1.8)$$

If the right–hand member of (1.7) is to be positive definite when we substitute for d**J**/dt from (1.2), a necessary condition is the relation

$$\mathbf{\Phi} = a\mathbf{J} \qquad (1.9a)$$

$$\lambda T/\tau V = -a^{-1} \qquad (1.9b)$$

This is an anti–reciprocity relation. To see this, observe that if Equation (1.2) is cast in terms of the forces $\mathbf{\Phi}$ and $-T^{-1}\nabla T$, and we add the trivial relation,

$$\mathbf{J} = \mathbf{J} = a^{-1}\mathbf{\Phi} \quad , \qquad (1.10)$$

then (1.9b) asserts that the coefficient of $-VT^{-1}\nabla T$ in d**J**/dt is minus the coefficient of $\mathbf{\Phi}$ in (1.10). If a theoretical result is available for λ from e.g. molecular dynamics, we can use (1.9b) to find the coefficient a in (1.9b) and from it the **J**–dependence of S and P via (1.6a,b).

An objection to the general applicability of reciprocity in extended thermodynamics can be based on (1.8). In the kinetic theory of dilute gases, the Boltzmann H–theorem yields an expression of the form (Jou et al. 1988, p.1115)

$$J_{si} = T^{-1}J_i + \alpha_s \sum_j P^{\dagger}_{ij}J_j + \ldots \qquad (1.11)$$

provided one chooses as ansatz for the single–particle momentum distribution a function which is an expansion in moments of the distribution. Equation (1.11), with the bilinear term in $P^{\dagger}_{ij}$ and **J**, has been a basic postulate of many recent formulations (Jou et al. 1988, p.1114ff) of extended thermodynamics. If (1.11) is necessary, it calls in question applications of reciprocity based on the use of (1.8) in (1.7). Early applications (Nettleton 1959a,b, 1960a,b, 1961, 1963, 1964a,b) did not consider problems involving non–zero **J** and $P^{\dagger}_{ij}$ in the same problem, and so (1.8) could be assumed. It is pointed out in Section 7 below that (1.8) and (1.11) probably belong to equivalent formulations having thermodynamic potentials differing in higher–order terms.

When coupled fluxes, e.g. $\mathbf{J}_1$ and $\mathbf{J}_2$, enter the formalism, we have associated forces,

$$\Phi_1 = a_{11}J_1 + a_{12}J_2 = -\partial F/\partial J_1 \tag{1.12a}$$
$$\Phi_2 = a_{21}J_1 + a_{22}J_2 = -\partial F/\partial J_2 \tag{1.12b}$$

which depend on both these extended variables. The integrability conditions imply

$$a_{12} = a_{21} \tag{1.13}$$

which reduces the number of independent coefficients to be calculated (Nettleton 1961, 1963). In Section 2, we use (1.12a,b) to cast the kinetic equations for dJ_1/dt and dJ_2/dt in canonical form, linear in Φ_1 and Φ_2. In the process, J_1 and J_2 are expressed in terms of the Φ_i (i = 1,2). Symmetric reciprocity then relates the coefficient of Φ_2 in dJ_1/dt to the coefficient of Φ_1 in dJ_2/dt. Anti–reciprocity obtains between the coefficients of the gradient terms in dJ_1/dt and dJ_2/dt and the coefficients of Φ_1 and Φ_2 in J_1 and J_2, as in (1.9b).

These examples illustrate the fact that, in a given problem, reciprocity and the integrability conditions for F reduce the number of independent coefficients in the rate equations for e.g. dJ_1/dt and dJ_2/dt which must be calculated from a model. The quantities which are known, vary from problem to problem, and so there is no way of laying down a recipe for the use of reciprocity in all cases. In any given problem of liquid transport, one must list what is calculable from structural, statistical, or kinetic models and then determine whether reciprocity provides the missing elements. If not, molecular dynamic or other calculations may be required before one can apply the principles of phenomenological irreversible thermodynamics. The resulting estimates can be used in conjunction with computer simulations as an independent test of the magnitudes and analytic form of results extracted from the latter. An example (Nettleton 1987c) is discussed in Section 3 where a prediction is made of nonanalyticities seen in computer studies of liquids at high steady rates of shear.

The last–mentioned example shows that, whereas the earlier work (Nettleton 1959–64) emphasized linear extended thermodynamics and applications to liquid transport, the more recent work has centred on non–linear $0(J^2)$ contributions to P, E, and F. These contributions are shown to be small, except perhaps in dilute gases, and so it is only with computers that reliable estimates can be made with which to compare extended thermodynamic predictions. Much of this work (Nettleton 1988c, Nettleton and Freidkin 1989) involves the development of statistical entropy analogues for calculating S and F. The dependence of E and P on extended variables such as **J** is then obtained from (1.6a,b). Two exceptions (Nettleton 1987a, 1988a) involve the use of the $0(J^2)$ terms in P to calculate non–linear contributions to the thermal conductivity and diffusion coefficients. The latter are shown to be approximately proportional to P.

In all the aforementioned applications, the objective is to use mechanical, structural, and statistical models to make quantitative estimates for comparison with experimental or computer results. These objectives can be contrasted with other work in this field which has sought to derive in most general possible form a set of constitutive equations from basic principles including objectivity, symmetry, and equipresence. The original work on *rational thermodynamics* (Truesdell 1985) gave rise to the work of Müller (1967, 1985) and modifications by Lebon (Jou et al. 1988) and others. The coefficients in the constitutive equations obtained via these mathematical theories are fitted to experiment and not, in general, calculated from models. The entropy is undefined. In a few cases, the results are compared with the kinetic theory Boltzmann equation (Lebon 1978, García–Colín and Fuentes y Martinez 1982, Lopez de Haro et al. 1984), although this does not provide a foundation for applications to condensed phases. Reciprocity is not used, since it cannot be derived without statistical entropy analogues and microscopic mechanical models, which are extraneous to the macroscopic hypotheses on which rational thermodynamics and its derivative theories are built. In the early versions of the work reviewed here (Nettleton, 1959a,b, 1960a,b, 1961,, 1963, 1964a,b) which were linear, reciprocity was introduced following de Groot (1951) who erected reciprocal relations as independent hypotheses of non–equilibrium thermodynamics and justified them using microscopic reversibility and fluctuation theory. Such arguments are limited to the linear case and depend on the assumption that thermodynamic potentials are applicable to the extended theory. The latter idea has found acceptance only recently because of its usefulness in describing states far from equilibrium. In Section 7, general statistical and mechanical derivations are described which extend reciprocity to the non–linear regime.

These derivations show that while an information–theoretic statistical model for entropy can be made the basis for a non–linear extended thermodynamics that includes reciprocity, it is not easy to include in such approaches all the usual assumptions. e.g. Equation (1.8) and $P = (1/3)\mathrm{Tr}\ (P_{ij})$. Statistical entropy analogues possess a non–uniqueness requiring caution in the application of any particular one to non–linear problems.

In addition to statistical derivations of reciprocity, we should mention the Lagrangian approach of Landau and Lifshitz (1958). These authors set $F = K + U$ where K is quadratic in $\{\dot{Q}_i\}$, the set $\{Q_i\}$ being state variables. U is the part of F which does not depend on $\dot{Q}$. Defining a Lagrangian $L = K - U$, they cast the kinetic equations for the $\dot{Q}_i$, including second–order time–derivatives, in Lagrangian form and show that it is equivalent to an Onsager–Casimir formalism with $-\partial F/\partial Q_i$ and $-\partial F/\partial \dot{Q}_i$ as thermodynamic forces. The coefficients exhibit reciprocity. The Lagrangian approach was designed to lead to constitutive equations containing inertial terms. It was applied by Sannikov (1962a,b) to derive equations of motion of magneto–elastic and thermo–elastic solids. Landau and Lifshitz (1960) also proposed an extended free energy F containing terms quadratic in a set $\{J_a\}$ of electric currents. It is not clear whether these authors proposed to found reciprocity on the Lagrangian approach or vice versa. They should, in any case, be credited with setting up the earliest forms of extended thermodynamics which led to practical applications. The Lagrangian approach is generalized to the non–linear regime in Section 6.

In the following sections, we shall give details of the early applications (Nettleton 1959–64) of reciprocity to linear transport in liquids and of more recent applications (Nettleton 1985–89) to the non–linear regime. Section 2 deals with linear vector processes, including coupled flows of heat and matter. Section 3 discusses viscoelasticity, including steady states far from equilibrium and associated non–analyticity. Section 4 reviews scalar processes such as ultrasonic absorption and domain–wall motion in a ferroelectric. These scalar applications are formal and of lesser interest, since reciprocity affords fewer opportunities in the scalar case to predict measurable values not given by models. Section 5 discusses non–linear effects in chemical reactions, where a computer can simulate a steady state maintained by absorption of ultra–violet radiation in a system of dissociating dimers. Section 6 reviews some theorems relevant to reciprocity, including the extension of the Lagrangian formalism to general non–linear phenomena. Section 7 summarizes the statistical foundations of non–linear reciprocity, based on generalizations of work by Zwanzig (1960, 1961) and Grad (1949). A summary and conclusions are presented in Section 8.

2. HEAT CONDUCTION AND DIFFUSION

In this section we discuss in more detail the vector processes described in connection with Equations (1.12a,b) and (1.13). This involves first–of–all the linear coupling of heat flow and self– or binary–diffusion and the use of reciprocity to find some of the coupling coefficients not given by the simple molecular models used. Following this, we shall consider more recent work on non–linear effects in the description of heat and matter flows.

Linear Treatment of Heat Flow and Diffusion

Here and in most of the later discussions of transport, we take the system to be a cube of side $\ell \sim 10^{-6}$ m immersed in a large, homogeneous, non–uniform fluid phase. The system is macroscopically small, but large enough for the application of statistical mechanics. Kinetic equations are written for the relaxation of the extended variables, i.e. the fluxes of heat and matter, to values proportional to the gradients of temperature and concentration. These processes are so fast that, in first approximation, they may be treated as adiabatic. Subsequent evolution of the system following relaxation of the fast variables is assumed describable via hydrodynamics. Arbitrary non–equilibrium states cannot be described by using only a few *fast* variables, e.g. $\mathbf{J}_1$ and *slow* variables, e.g. number density n. To predict dispersion equations for sound to all orders in the propagation vector, $\mathbf{k}$, we should have to augment the set of state variables by including, ∇n, $\nabla \mathbf{J}$, and higher–order space–derivatives. However, we are interested in steady states, including those far from equilibrium, and we introduce simplified kinetic equations applicable to the neighbourhood of a steady state. Emphasis is on description of the latter rather than on the approach to it.

One spatial derivative which enters the description of steady heat conduction under large ∇T is ∇n. A large temperature gradient sets up a mass flow which continues until pressure P is uniform across the system and $\nabla n = -(\partial P/\partial T)/(\partial P/\partial n)\nabla T$. As we approach a steady state, the rate $(d/dt)\nabla n$ is a sum of terms in $\mathbf{J}_1$ and $\mathbf{J}_2$, respectively the hypersound and self–diffusion components of the total heat flux, $\mathbf{J}$. To express this circumstance (Nettleton 1961), we define a variable

$$\boldsymbol{\rho} \equiv \nabla n - \chi \nabla \theta \tag{2.1}$$

where θ is the local equilibrium temperature, $\cong$ T in a linear calculation, χ is introduced because $\boldsymbol{\rho} \to 0$ as frequency $\omega \to \infty$, and this makes ∇n proportional to $\nabla \theta$ with a proportionality constant calculable from the hypersound limit of the hydrodynamic equation of energy conservation.

The rate equation (2.6 below) for $d\boldsymbol{\rho}/dt$ is coupled to those for $\mathbf{J}_1$ and $\mathbf{J}_2$:

$$\dot{\mathbf{J}}_1 = -(\tau_1^{-1} + \kappa_1)\mathbf{J}_1 + \kappa_2\mathbf{J}_2 - A_1^{(1)}\nabla T - A_2^{(1)}\nabla n \tag{2.2a}$$

$$\dot{\mathbf{J}}_2 = -(\tau_2^{-1} + \kappa_2)\mathbf{J}_2 + \kappa_1\mathbf{J}_1 - A_1^{(2)}\nabla T - A_2^{(2)}\nabla n \tag{2.2b}$$

Models can be found (Nettleton 1961) for all the coefficients in (2.2a,b) except κ_1, the coefficient which provides for excitation of a self–diffusion flow, carrying with it the heat flow component $\mathbf{J}_2$, by the presence of a hypersound energy flow $\mathbf{J}_1$. A reciprocity relation is available for κ_1, however. We introduce thermodynamic forces, $\boldsymbol{\Phi}_1$ and $\boldsymbol{\Phi}_2$, obeying

$$\bar{\omega}_1\boldsymbol{\Phi}_1 = C_1^{(1)}\mathbf{J}_1 + C_2^{(1)}\mathbf{J}_2 \tag{2.3a}$$

$$\bar{\omega}_2\boldsymbol{\Phi}_2 = C_1^{(2)}\mathbf{J}_1 + C_2^{(2)}\mathbf{J}_2 \tag{2.3b}$$

$$C_2^{(1)}/\bar{\omega}_1 = C_2^{(2)}/\bar{\omega}_2 \tag{2.3c}$$

Equations (2.3a–c) are the same as (1.12a,b) and (1.13). Solving (2.3a,b) for $\mathbf{J}_1$ and $\mathbf{J}_2$ in terms of $\mathbf{\Phi}_1$ and $\mathbf{\Phi}_2$, we rewrite (2.2a,b) in the form

$$\dot{\mathbf{J}}_1 = \bar{\omega}_1\mathbf{\Phi}_1 + \bar{\omega}_{12}\mathbf{\Phi}_2 - \Upsilon_1\nabla T + \Lambda_1\mathbf{\Theta} \tag{2.4a}$$

$$\dot{\mathbf{J}}_2 = \bar{\omega}_2\mathbf{\Phi}_2 + \bar{\omega}_{21}\mathbf{\Phi}_1 - \Upsilon_2\nabla T + \Lambda_2\mathbf{\Theta} \tag{2.4b}$$

$$\Upsilon_i \equiv A_1^{(i)} + \chi A_2^{(i)} \qquad (i = 1,2) \tag{2.4c}$$

$$\Lambda_i \equiv A_2^{(i)} v \qquad (i = 1,2) \tag{2.4d}$$

$$\mathbf{\Theta} \equiv -\boldsymbol{\rho}/v \tag{2.4e}$$

$\mathbf{\Theta}$ is the thermodynamic force conjugate to $\dot{\boldsymbol{\rho}}$. There will also be an anti–reciprocal relation analogous to (1.9b). The reciprocity and integrability conditions give (2.3c) and

$$\bar{\omega}_{12} = \bar{\omega}_{21} \tag{2.5a}$$

$$n\Upsilon_1 T(C_1^{(1)}/\bar{\omega}_1) + n\Upsilon_2 T(C_1^{(2)}/\bar{\omega}_2) = -1 = n\Upsilon_1 T(C_2^{(1)}/\bar{\omega}_1) + n\Upsilon_2 T(C_2^{(2)}/\bar{\omega}_2) \tag{2.5b}$$

Equation (2.5a) is useful in determining κ_1. Various molecular models are adduced (Nettleton 1961) for making rough estimates of the remaining coefficients.

Since the equations for $\dot{\mathbf{J}}_1$ and $\dot{\mathbf{J}}_2$ depend on $\mathbf{\Theta}$, reciprocity implies that $\dot{\boldsymbol{\rho}}$ depends on Φ_1 and $\mathbf{\Phi}_2$ according to the relation:

$$\dot{\boldsymbol{\rho}} = -\Lambda_1\Phi_1 - \Lambda_2\Phi_2 \tag{2.6}$$

This equation shows why, in a realistic description of the approach to the steady state, $\mathbf{J}$ must have at least two independent components. Otherwise, we could not have $\dot{\boldsymbol{\rho}} = 0$ in the presence of a steady heat flow.

Since $\Phi_i = -\partial F/\partial J_i$ $(i = 1,2)$ and $\mathbf{\Theta} = -\partial F/\partial \rho$, we can integrate to obtain

$$F = F_0(N,V,T) - \tfrac{1}{2}(C^{(1)}/\bar{\omega}_1)J_1^2 - \tfrac{1}{2}(C_2^{(2)}/\bar{\omega}_2)J_2^2 - (C_2^{(1)}/\bar{\omega}_1)J_1\cdot J_2 + \tfrac{1}{2}v^{-1}\rho^2 \tag{2.7}$$

When ∇n is maintained non–zero by external forces with $\nabla\theta = 0$, we can solve Equations (2.2a,b) for $\mathbf{J}_1$ and $\mathbf{J}_2$ in terms of ∇n and write

$$F = F_0(N,V,T) + F_\rho(\nabla n)^2 + \ldots \; (\omega \to 0, \nabla\theta = 0)\ . \tag{2.8}$$

From the estimates made of the coefficients in (2.7), we can estimate F_ρ (Nettleton 1961). An expression of the form (2.8) is useful in calculating the density profile across a liquid–vapour phase boundary. This is done by minimizing the integral of the free energy density taken over the boundary region (Cahn and Hilliard 1958).

The foregoing formalism can be applied with minor changes to the case of a binary fluid mixture of molecules of mass m_1 and m_2 and respective mass densities ρ_1 and ρ_2. There is still a heat flux component $\mathbf{J}_1$ carried by hypersound, but the self– and binary–diffusion components are lumped into a single contribution $\mathbf{J}_2$ roughly proportional to the flow of component 2 relative to the centre–of–mass velocity (Nettleton 1963). The kinetic equations for $d\mathbf{J}_1/dt$ and $d\mathbf{J}_2/dt$ have the same form as in (2.2a,b), with ∇n replaced by $-T\nabla[T^{-1}(\mu_2 - \mu_1)]$ with $\mu_i(i = 1,2)$ the chemical potential of molecular species i. This gradient is the thermodynamic force conjugate to the diffusion flow, $\mathbf{J}_d$.

In the binary diffusion case $\mathbf{\Phi}_1$ and $\mathbf{\Phi}_2$ can be defined as in (2.3a–c) and the kinetic equations expressed in terms of these. The symmetric reciprocity relation analogous to (2.5a) can be used in calculating the thermal diffusion coefficient when coefficients other than κ_1 can be estimated from simple molecular models. Since the rate equations for $d\mathbf{J}_1/dt$ and $d\mathbf{J}_2/dt$ depend on $-T^{-1}\nabla T$ and on $-T\nabla[T^{-1}(\mu_2 - \mu_1)]$, there exist anti–reciprocal relations which assert that

$$\mathbf{J}_1 = -A_1^{(1)}TV^{-1}\mathbf{\Phi}_1 - A_1^{(2)}TV^{-1}\mathbf{\Phi}_2 \tag{2.9a}$$

$$\mathbf{J}_2 = -A_2^{(1)}TV^{-1}\mathbf{\Phi}_1 - A_2^{(2)}TV^{-1}\mathbf{\Phi}_2 \tag{2.9b}$$

Consistency and integrability conditions permit evaluation of the $C_j^{(i)}$ and of the $A_1^{(2)}$ coefficient. The latter is required, along with κ_1, for calculation of the thermal diffusion coefficient.

The free energy F in the binary diffusion case has an expansion in powers of $\mathbf{J}_1$ and $\mathbf{J}_2$ analogous to (2.7). In a steady state with $\nabla T = 0$, $\mathbf{J}_1$ and $\mathbf{J}_2$ are both proportional to $\nabla\rho_2$, so that F has a term proportional to the square of $(\nabla\rho_2)$. A formal expression has been derived (Nettleton 1963) for the coefficient in this term, but no numerical estimates are made, in contrast to the corresponding self–diffusion case (Nettleton 1961). $\nabla\rho_2$ is not treated as an independent relaxing variable. Such a variable would arise if we wanted to introduce additional independent components of $\mathbf{J}$ associated with self–diffusion and couple the relaxation of $(d/dt)(\nabla\rho_2)$ to these components, as in Equation (2.6).

<u>Non–linear Heat and Matter Transport</u>

We consider here the possibility that the thermal conductivity, λ, can have the form

$$\lambda = \lambda_0(n,T) + \tilde{\lambda}_2 J^2 + \ldots \tag{2.10}$$

while the binary diffusion coefficient, D_T, is

$$D_T = D_0 + D_2 J_d^2 + \ldots \tag{2.11}$$

where J_d is the diffusion flow, proportional to J_2 in (2.9b). In the case of heat conduction, which we discuss in connection with (2.10), the fluid is taken to be a dense pure liquid, and coupling of heat and diffusion fluxes is not considered. We are concerned therefore with the simple Cattaneo–Vernotte equation (1.2) which we write in the form

$$\partial \mathbf{J}/\partial t = a\mathbf{J} - \gamma \nabla T \tag{2.12a}$$

$$\gamma = \gamma_0(n,T) + \gamma_2 J^2 + \ldots \tag{2.12b}$$

J_d satisfies a similar equation [(2.16a) below] which generalizes Fick's law through introduction of an inertial term, $\partial J_d/\partial t$.

We have shown in (1.6b) and (1.9a,b) how we can calculate P if we can evaluate γ and how λ_2 can be calculated from γ_2 using reciprocity. We desire, therefore, to make a model calculation of γ. To derive (2.12a) with the driving term in ∇T, one can follow Zwanzig (1961) in using the phase–space distribution, f(x), to define

$$g(\mathbf{v}) = \int f(x)\delta(\hat{\mathbf{J}} - \mathbf{v})\, dx \tag{2.13}$$

$g(\mathbf{v})\, d\mathbf{v}$ is the probability that the heat flux, $\hat{\mathbf{J}}(x)$, has a value between $\mathbf{v}$ and $\mathbf{v} + d\mathbf{v}$. The explicit form of the heat flux operator is

$$\hat{\mathbf{J}} = \sum_i (p_i^2/2m)(\mathbf{p}_i/m) + \tfrac{1}{2}\sum_{i\neq j}(\phi_{ij}\boldsymbol{\delta} + \mathbf{r}_{ij}\mathbf{F}_{ij})\cdot(\mathbf{p}_i/m) - h\sum_i(\mathbf{p}_i/m) \tag{2.14}$$

with ϕ_{ij} the pair potential for the interaction between particles at $\mathbf{r}_i$ and $\mathbf{r}_j$ and h the enthalpy per particle. $\mathbf{F}_{ij}$ is the force on i produced by its interaction with j, while $\boldsymbol{\delta}$ is the Kronecker delta. An equation for $\partial g/\partial t$ is found by multiplying the Liouville equation by $\delta(\hat{\mathbf{J}} - \mathbf{v})$ and integrating over phase space. The contribution to the latter equation linear in ∇T comes from the interaction of the system with the surroundings. This interaction contributes to the equation for $\partial g/\partial t$ a term (Nettleton 1984a, 1987a), assuming ∇T to be in the x–direction,

$$\gamma(\partial g/\partial v_x)\partial T/\partial x \tag{2.15a}$$

$$\gamma = m^{-1}P(\breve{C}_p + P_0 n^{-1}\alpha_p) \tag{2.15b}$$

where P is the thermodynamic pressure; $\breve{C}_p \equiv N^{-1}(\partial E/\partial T)_p$; α_p is the coefficient of volume expansion. When the equation for $\partial g/\partial t$ is multiplied by $\mathbf{v}$ and integrated over $\mathbf{v}$–space, we obtain (2.12a). The conclusion that γ is proportional to P should be valid at high densities on the assumption that the fluid is contained by its surroundings during the short period τ required for $\mathbf{J}$ to relax to a value proportional to ∇T. The interaction with the surroundings is then proportional to P by reasoning (Nettleton 1987a) similar to the arguments used in deriving the surface contributions to the virial in a discussion of the virial equation–of–state for a fluid.

In the case of hard spheres, the equilibrium equation–of–state, $P_0(n,T)$, has been obtained by fitting a Padé approximant to computer simulations (Ree and Hoover 1964). One can find γ_0 by introducing P_0 into (2.15b). This result yields P_2 via (1.6b) and

(1.9a,b). From P_2 we find γ_2 by using (2.15b). Repeating the process, one can find the $0(J^2\mathbf{J})$ term in $\mathbf{\Phi}$ and the $0(J^4)$ term in F. Numerical estimates (Nettleton 1987a) for a hard–sphere model of Ar at 87 K at a density not far below the fluid–solid transition yield $\tilde{\lambda}_2/\lambda_0 = 3.44 \times 10^{-18}$ kg m^6 J^{-3}. One should have $|\mathbf{J}| \sim 10^9$ Wm^{-2} before non–linear effects in heat conduction become observable. Thus, at densities where diffusion across the boundaries is very small during time $\tau = -a_0^{-1}$, non–linear effects should be negligible at realizable conditions. If a fluid is subjected to large $|\nabla T|$, the linear Fourier law will still hold with the λ appropriate to small $|\mathbf{J}|$. This conclusion, viz. that $0(J^4)$ contributions to P are negligibly small, will be found to hold generally for other processes. The only important non–linear effects so far found in the material reviewed here arise in non–linear viscoelasticity discussed in Section 3.

The methods developed here for calculating $\tilde{\lambda}_2$ resemble closely an analogous procedure for estimating D_2 in a dense fluid. In the case of binary diffusion (Nettleton 1988a) we have, in place of (2.14) the flux

$$\hat{\mathbf{J}}_d = V^{-1} \sum_{i \in 1} \mathbf{p}_i \tag{2.16}$$

of one of the two components. The generalized Fick's law for $\mathbf{J}_d = \langle \hat{\mathbf{J}}_d \rangle$ is

$$\dot{\mathbf{J}}_d = (L/V)\mathbf{\Phi} - \gamma\nabla_T(\mu_1 - \mu_2) - \lambda T^{-1}\nabla T \tag{2.17a}$$

$$\gamma = \gamma_0 + \gamma_2 J_d^2 + 0(J_d^4) \tag{2.17b}$$

$$\mathbf{\Phi} = -\nu_2\mathbf{J}_d + \nu_4 J_d^2 \mathbf{J}_d + \ldots \tag{2.17c}$$

where ∇_T signifies a gradient taken at constant T. Writing the trivial equation $\mathbf{J}_d = \mathbf{J}_d$ as,

$$\mathbf{J}_d = -\gamma V^{-1}\mathbf{\Phi} \tag{2.18}$$

which exhibits anti–reciprocity, we have for consistency

$$\nu_2\gamma_0 V^{-1} = 1 \tag{2.19a}$$

$$\nu_2\gamma_2 = \nu_4\gamma_0 \tag{2.19b}$$

Calculation of the coefficients ν_2 and ν_4 in the $\mathbf{J}_d$–expansion of $\mathbf{\Phi}$ and F requires a model for estimation of γ. The same term in the Liouville equation used above in deriving the Cattaneo–Vernotte equation gives in this case a coupling term

$$-P_{xx}\partial c_1/\partial x \tag{2.20}$$

Again we assume that interactions across the boundary can, over a short time, be represented by the pressure. We assume, also, that the concentration gradient, ∇c_1, is in the x–direction. Using a simple model for the momentum transport by diffusing particles, we relate the diffusion and momentum fluxes and set

$$P_{xx} = P_0 + J_{dx}J_{dx}(\rho_1^{-1} + \rho_2^{-1}) \tag{2.21}$$

where ρ_i is the partial mass density of component i. to evaluate γ, one must cast (2.20) in the form of the first gradient term in (2.17a). This can be done (Nettleton 1988a) with the aid of the Gibbs–Duhem equation and the mechanical equilibrium condition $\partial P_{xx}/\partial x = 0$. One obtains

$$\gamma_0 = P_0\rho_1\rho_2(\kappa T\rho n) \tag{2.22}$$

and an elaborate expression for γ_2.

Numerical estimates of γ_0, γ_2, P_0, P_2, and ν_4/ν_2 are made (Nettleton 1988a) for a hard–sphere mixture at 87 K having component masses and diameters appropriate to ^{36}Ar and ^{40}Ar. It is found that $\nu_4/\nu_2 = -5.49 \times 10^{-12} m^6 J^{-1} kg^{-1}$; $D_2/D_0 = 3.17 \times 10^{-11} m^6 J^{-1} kg^{-1}$. The non–linear effects are again found to be negligibly small in realizable states.

3. VISCOELASTICITY IN LIQUIDS AND ISOTROPIC SOLIDS

The extended thermodynamics of viscoelasticity has been inspired by the Maxwell equation (1.3). This provides for a period τ_1 during which the traceless pressure $P^{\dagger}_{ij}$ relaxes to the Stokes form:

$$P^{\dagger}_{ij} = -\eta B^{\dagger}_{ij} \tag{3.1a}$$

$$\eta = \lambda_2\tau_1 \tag{3.1b}$$

with η the shear viscosity. A superscript dagger denotes a traceless tensor throughout. Equation (1.3) was originally proposed for a gas, where the use of $P^{\dagger}_{ij}$ as independent variable provides the most mathematically elegant treatment. Most recent treatments of the thermodynamics of viscoelasticity (Lebon et al. 1988) use $P^{\dagger}_{ij}$. In solids and highly viscous liquids, however, one can define an elastic strain ϵ_{ij} by

$$P^{\dagger}_{ij} = -\lambda_2\epsilon^{\dagger}_{ij} \tag{3.2}$$

$\epsilon^{\dagger}_{ij}$ is the part of the total strain which would disappear if the stress were instantaneously removed. It is measurable in highly viscous materials.

If $\epsilon^{\dagger}_{ij}$ is used as a state variable, we have an associated thermodynamic force,

$$\Psi^{\dagger}_{ij} = -\partial F/\partial\epsilon^{\dagger}_{ij} = \mu\epsilon^{\dagger}_{ij} \tag{3.3}$$

with μ a phenomenological force constant. Equation (1.3) becomes (Nettleton 1959b)

$$\dot{\epsilon}^{\dagger}_{ij} = -(1/\tau_1)\epsilon^{\dagger}_{ij} + B^{\dagger}_{ij}$$

$$= -(1/\mu\tau_1)\Psi^{\dagger}_{ij} + B^{\dagger}_{ij} \tag{3.4}$$

Since $\dot{\epsilon}^{\dagger}_{ij}$ depends on $-B_{ij}$, the force canonically conjugate to $P^{\dagger}_{ij}$, we expect by reciprocity that

$$P^{\dagger}_{ij} = V^{-1}\Psi^{\dagger}_{ij} \tag{3.5a}$$

$$\mu = -\lambda_2 V \tag{3.5b}$$

The V–factors in (3.5a,b) arise when $\Psi^{\dagger}$ is the derivative of F rather than of the free energy per unit volume.

A formulation of this kind (Nettleton 1959b) should be adequate for simple liquids. If a stress is applied to the liquid, a fraction of particles will be in regions where the local expansion is sufficient to permit a particle to diffuse, and a non–zero $\epsilon^{\dagger}_{ij}$ produces an instantaneous response, $\dot{\epsilon}^{\dagger}_{ij}$, predicted by (3.4). In a solid or very viscous liquid, there will be a delay in response until a local expansion can occur or until dislocations begin to move. Application of a stress in the latter case causes instantaneously an acceleration of creep rather than a non–zero creep–rate. To provide for this inertial effect, a new set of independent variables, the inelastic strain–rate components, need to be introduced. These are extended variables like those whose time–rate character provides for an inertial effect in the kinetic equations for e.g. scalar thermal relaxation processes.

Viscoelasticity with Creep–rate as an Independent Variable

To provide for acceleration of creep, rather than a non–zero creep–rate, as the instantaneous response to imposition of a shear stress in a polycrystal, we introduce (Nettleton 1964b) a set of independent components $\{C^{(k)\dagger}_{ij}\}$ of the total inelastic strain–ratc. These can be associated with various dislocation and vacancy diffusion mechanisms. The total strain–rate is

$$B^{\dagger}_{ij} = \sum_{k} C^{(k)\dagger}_{ij} + \dot{\epsilon}^{\dagger}_{ij} \tag{3.6}$$

The creep–rate components are taken to obey acceleration–rate equations:

$$\dot{C}^{(k)\dagger}_{ij} = \sum_{m} \kappa_{km} C^{(m)\dagger}_{ij} + \gamma_k \epsilon^{\dagger}_{ij} \qquad (k = 1,\ldots,r). \tag{3.7}$$

The time–derivatives $\dot{C}^{(k)\dagger}_{ij}$ provide for a proper inertial effect, since the inelastic strain–rate rises from zero to a value proportional to $\epsilon^{\dagger}_{ij}$ which in turn is proportional to stress. Associated with the new independent variables are thermodynamic forces

$$K_{ij}^{(k)\dagger} = \sum_{m=1}^{r} \alpha_{km} C_{ij}^{(m)\dagger} \qquad (k = 1,...,r) \tag{3.8}$$

in addition to $\Psi_{ij}^{\dagger}$ defined in (3.3).

If (3.7) is expressed in canonical form in terms of the forces, we have:

$$\dot{C}_{ij}^{(k)\dagger} = \sum_{m} \Lambda_{km} K_{ij}^{(m)\dagger} + \Lambda_{kT} \Psi_{ij}^{\dagger} \tag{3.9}$$

Similarly, (3.6) become

$$\dot{\epsilon}_{ij}^{\dagger} = -\sum_{m} \Lambda_{mT} K_{ij}^{(m)\dagger} + B_{ij}^{\dagger} \tag{3.10}$$

The Λ–coefficients in (3.9) and (3.10) are determined by

$$\sum_{j} \Lambda_{jT} \alpha_{jm} = 1 \; (m = 1,...,r) \tag{3.11a}$$

$$\Lambda_{iT} = \gamma_i / \mu \tag{3.11b}$$

$$\Lambda_{ij} = \sum_{m} \kappa_{im} (\alpha^{-1})_{mj} \tag{3.11c}$$

which make (3.9) agree identically with (3.7). In addition, the conditions of integrability and reciprocity, respectively, imply

$$\alpha_{ij} = \alpha_{ji} \tag{3.12a}$$

$$\Lambda_{ij} = \Lambda_{ji} \tag{3.12b}$$

These equations and (3.5b) permit us to calculate the constants provided the model yields all the parameters in (3.7). When this has been done, we can use the integrability conditions for F in the form:

$$\partial P / \partial C_{ij}^{(k)\dagger} = \partial K_{ij}^{(k)\dagger} / \partial V \tag{3.13a}$$

$$\partial P / \partial \epsilon_{ij}^{\dagger} = \partial \Psi_{ij}^{\dagger} / \partial V \tag{3.13b}$$

to find the $C^{\dagger}$– and $\epsilon^{\dagger}$–dependence of P. Numerical estimates were made (Nettleton (1964b) for a particular model involving the slip and climb of dislocations generated by Frank–Read sources. These estimates indicate that the $C^{\dagger}$–dependence of P might make an observable contribution to the effective third–order elastic constants of a steadily–creeping specimen. These effective contributions to the elastic constants arise because P depends on $(C^{\dagger})^2$, and, in a state of steady creep, where $C^{\dagger}$ is proportional to $\epsilon^{\dagger}$, this yields an effective dependence on $(\epsilon^{\dagger})^2$.

The work described in (3.1–3.5) (Nettleton 1959b) was nearly contemporaneous with a similar work of Kluitenberg (1962a,b,c) which used the inelastic strain rather than $\epsilon^{\dagger}_{ij}$ as a state variable. The formulation given here has been preferred because one need not keep track of the initial state.

Non–linear Viscoelastic Effects in Simple Liquids

One can extend the linear viscoelastic theory to states far from equilibrium where the thermodynamic forces are non–linear in the elastic strains and creep–rates. Assuming a single creep–rate component, $C^{\dagger}_{ij}$, we can expand the coefficients which, in the linear theory just described were independent of the extended variables, in powers of these. Thus (Nettleton 1987c) we take

$$K^{\dagger}_{ij} = a_c C^{\dagger}_{ij} \tag{3.14a}$$

$$a_c = a_1 + a_2 \sum_{i,j} C^{\dagger}_{ij} C^{\dagger}_{ji} + a_3 \sum_{i,j} \epsilon^{\dagger}_{ij} \epsilon^{\dagger}_{ji} + a_4(\zeta - \zeta_0) \tag{3.14b}$$

$$\mu = b_1 + b_2 \sum_{i,j} C^{\dagger}_{ij} C^{\dagger}_{ji} + b_3 \sum_{i,j} \epsilon^{\dagger}_{ij} \epsilon^{\dagger}_{ji} + b_4(\zeta - \zeta_0). \tag{3.14b}$$

ζ is a scalar structural parameter which, in the particular model used (Nettleton 1987c) was taken to be the volume fraction of localized spherical regions, each with a radius of the order of the intermolecular separation, in which a local expansion had occurred sufficient to permit a particle to diffuse out of the cage of surrounding neighbours. ζ_0 is the equilibrium value of ζ. This scalar obeys a rate equation:

$$\dot{\zeta} = \alpha Z \tag{3.15a}$$

$$Z = d_1(\zeta-\zeta_0)+d_2(\zeta-\zeta_0)^2+d_3(\zeta-\zeta_0)^3+d_4\sum_{i,j} \epsilon^{\dagger}_{ij}\epsilon^{\dagger}_{ji}+d_5\sum_{i,j} C^{\dagger}_{ij}C^{\dagger}_{ji} \tag{3.15b}$$

In a particular experiment, $\epsilon^{\dagger}_{ij}$ has a single non–zero component, ϵ, and $C^{\dagger}_{ij}$ a corresponding component, C. These components could be e.g. the xz components when it is the xz plane that is sheared. We neglect d_5, in line with the general conclusion that higher–order terms in the time–derivative variables are small and solve the equation Z = 0 in a steady state at large strain–rate in the form of an asymptotic expansion,

$$\zeta - \zeta_0 = D_0 + D_1\epsilon + D_2\epsilon^2 + D_{-1}\epsilon^{-1} + \ldots \tag{3.16}$$

There are two such solutions which are free energy minima separated by one which is a maximum. One of these solutions is substituted into the equation obtained by setting $\dot{C}^{\dagger}_{ij} = 0$ in (3.7). The result yields an asymptotic solution for ϵ as a sum of positive and negative power of $C^{\frac{1}{2}}$. These solutions in powers of $C^{\frac{1}{2}}$ for $\zeta - \zeta_0$ and ϵ are substituted into

$$P = (1/3)(\partial d_2/\partial V)(\zeta - \zeta_0)^3 + (\partial b_4/\partial V)(\zeta - \zeta_0)\epsilon^2 \tag{3.17}$$

to show that P has a leading contribution at high steady strain–rate which is proportional to $C^{3/2}$. The same type of term appears in the internal energy, E.

The $C^{3/2}$ –dependence of P and E at high shear–rate has been observed in a molecular dynamics study of a Lennard–Jones model of liquid Ar. Hanley and Evans (1982) set up boundary conditions at the top and bottom of a box which maintained a steady gradient, in the vertical direction, of the velocity of flow. The temperature was kept constant by periodic rescaling of the particle velocities. The magnitudes and signs of the non–analytic terms in E and P agreed with those obtained via (3.16) and (3.17). We can also explain the shear thinning effect, according to which

$$\eta = \eta_0 - \eta_1 C^{1/2} \tag{3.18}$$

which arises from the $(\zeta - \zeta_0)$–dependence of the viscosity. This has also been observed in molecular dynamics studies (Holian and Evans 1983).

While the magnitudes of non–analytic effects predicted by the thermodynamic theory described above are comparable to the results of molecular dynamics, this agreement is somewhat fortuitous. The value used for ζ_0 is anomalously large, based on a model developed to describe structural relaxation contributions to sound absorption (Nettleton 1959a). This model treated oscillations of the separation between two nearest neighbours via an independent oscillator picture, like the Einstein model of a crystal, which is not realistic.

It should be noted that the non–analyticities predicted above result from coupling of the relaxation of the scalar, ζ, and of the elastic shear component, ϵ. Whatever may be said of specific molecular models used to evaluate the b– and d–coefficients, this coupling appears to yield the simplest route to non–analytic behaviour. As remarked above, however, the models can stand improvement. Non–linear viscoelastic effects are not concerned with inertial effects. We neglect a_2 and b_2 in (3.14b,c). Thus we should not expect, on the basis of results in the present section, to obtain non–analytic behaviour in heat conduction or diffusion where the non–linear terms in P and the thermodynamic potentials are too small to become important in observable states. The coupling of relaxation of **J** and $\mathbf{J}_d$ to that of a scalar structural parameter has not been studied. Such a coupling will contribute to higher–order terms because both self– and binary–diffusion depend on the local liquid structure.

4. SCALAR RATE PROCESSES

Two cases will be considered of attempts to cast the equations for rates–of–change of scalar parameters in Onsager–Casimir canonical form, with fluxes expressed as sums of thermodynamic forces. This permits the application of reciprocity. The first case concerns theories of ultrasonic absorption in liquids. The scalar parameter there can be the fraction of molecules excited in a higher vibrational–rotational level whole relaxation is responsible for the observed acoustic absorption. Alternatively, one can take the fraction of molecules which are in a localized region whose structure permits them to diffuse out of the cage formed by their neighbours. In the other type of model considered in the present section, the scalar variables are numbers of steps of reversed polarization nucleated on the surface of 180° ferroelectric domain wall. This application of extended thermodynamics to switching, together with the discussion of polycrystals in Section 3, are the only departures from the subject of fluids.

Ultrasonic Absorption

In non–associated liquids such as CS_2 or $CC\ell_4$, sound absorption has been attributed to excitation and de–excitation of internal molecular vibrational–rotational states in collisions (Herzfeld and Litovitz 1959,, Chapter 11). If ζ_i denotes the fraction of molecules excited in a given vibrational–rotational state, and $\eta_i \equiv \dot{\zeta}_i$, then (Nettleton 1960b) we write a generalized master equation:

$$\dot{\eta}_i = -(1/\tau_i')\eta_i + \sum_j (L_{ij}/\tau_i')(\zeta_j - \zeta_{jo}) \tag{4.1}$$

Off–diagonal terms relating to $\dot{\eta}_i$ to η_j with $j \neq i$ are here neglected. The time τ_i' is of the order of a collision time, so that $\dot{\eta}_i \to 0$ when angular frequency $\omega > (\tau_i')^{-1}$. L_{ij} $(j \neq i)$ is the product of the frequency of collisions between particles of species i and j multiplied by the probability of a $j \to i$ transition. The equilibrium values ζ_{io} can be estimated from

$$\zeta_{io} = Z^{-1} \exp(-\beta\epsilon_i) \tag{4.2}$$

where ϵ_i is the energy of excitation of molecular level i and Z the molecular partition function.

To cast (4.1) in canonical form, we introduce thermodynamic forces assumed to have the analytical form,

$$\Phi_i = \sum_j \alpha_{ij}(\zeta_j - \zeta_{jo}) \tag{4.3a}$$

$$\Psi_i = \sum_j \alpha'_{ij}\, \eta_j \tag{4.3b}$$

With terms linking $\dot{\eta}_i$ to Ψ_j $(j \neq i)$ again neglected, (4.1) becomes

$$\dot{\eta}_i = \gamma_i\Psi_i + \sum_j \bar{\omega}_{ij}(n,T)\Phi_j \tag{4.4a}$$

$$\sum_k \bar{\omega}_{ik}\alpha_{kj} = L_{ij}/\tau_i' \tag{4.4b}$$

We have

$$\dot{\zeta}_i = \eta_i = \sum_j (\alpha'^{-1})_{ij}\Psi_j \tag{4.5}$$

Anti–reciprocity requires that

$$(\alpha'^{-1})_{ij} = -\,\bar{\omega}_{ji} \tag{4.6}$$

This permits us to calculate the forces Ψ_i if we know the α_{ij} and the coefficients in (4.1). The α–coefficients can be estimated by supposing that equilibrium statistical mechanics provides the leading ζ–dependence of F, i.e. the ζ–dependence in the part of F which involves no η–variables. Using (4.4b) and (4.6), we can find $\Psi_i = -\partial F/\partial\eta_i$ and thus the $0(\eta^2)$ terms in F via integration. From these we find by differentiation the $0(\eta^2)$ contributions to thermodynamic functions derivable from F.

There are, in ultrasonics, no measurements to compare with the predictions of reciprocity just cited. Furthermore, since we expect that all coefficients in (4.1) will be given by the model, reciprocity does not, as it did in the case of coupled heat and diffusion flows, provide a means of calculating some unknown coefficients in the kinetic equations in terms of others. An interesting consequence (Nettleton 1960b) can be found, however, of (4.4a), (4.5), and (4.6) if we compute $\dot{S}$ from the Gibbs equation

$$T\,dS = dE + P\,dV + \sum_i \Phi_i\,d\zeta_i + \sum_i \Psi_i\,d\eta_i \tag{4.7}$$

This implies that the irreversible rate of entropy production is

$$T\dot{S}_{irrev} = \sum_i \gamma_i \Psi_i^2 \tag{4.8}$$

as a consequence of anti–reciprocity. The steady non–equilibrium state is not a minimum of the expression $\dot{S}_{irrev}$.

In associated liquids (Nettleton 1957) a molecule can have, according to the simplest models, two structural states: H–bonded and non–H–bonded. If ζ is the fraction of non–bonded molecules, one can write a master equation similar to (4.1) for its time dependence. In some models (Hall 1948) the kinetic coefficients are obtained from the theory of absolute reaction–rates (Glasstone et al. 1941). Reciprocity has provided little that is new in this case. It might be useful if one considered coupled structural parameters, since in structural models the models might not provide all the coupling coefficients. There are few examples of these, however.

Motion of a Ferroelectric Domain Wall

Here we consider an application of extended thermodynamics with reciprocity which has no direct relation to liquids. The 180° domain wall in polarized tetragonal $BaTiO_3$ represents an abrupt reversal of the direction of spontaneous electric polarization as we move in a direction normal to the wall from one unit cell to the next. Under an applied electric field along the polarization direction, the domain polarized along the field grows at the expense of the adjacent oppositely–polarized domain, and the wall displaces in a direction normal to itself.

When the domain wall moves under a low field E, it does not move as a unit. The estimated activation energy for such a process (Kinase and Takahasi 1957) is too high at low fields. Accordingly (Nettleton 1967b) it is assumed, following Miller and Weinreich (1960) that steps of reversed polarization are nucleated at the surface of the wall and, if sufficiently large, grow and cover the wall. Steps below a critical size tend to disappear rather than grow because they have a high ratio of surface to bulk energy. Each nucleation of a step of super–critical size causes the wall to advance by one lattice constant if growth is fast and coalescence of nucleated steps negligible.

To parameterize the problem, let ν_λ be the number of nucleated steps of width k_λ having a height nearly equal to that of the wall. Smaller steps tend to disappear and are neglected. Suppose also that there are ν_A fully–nucleated steps of width k_A having a height equalling that of the wall. These parameters obey coupled kinetic equations,

$$\ddot{\nu}_\lambda = -(2\pi)^{-1}\bar{\omega}_0\, \dot{\nu}_\lambda + (2\pi)^{-1}\bar{\omega}_0 \sum_\rho \Gamma_{\lambda\rho}\nu_\rho + (2\pi)^{-1}\bar{\omega}_0 \sum_A \Gamma_{\lambda A}\nu_A + (2\pi)^{-1}\bar{\omega}_0\gamma_A E \tag{4.9a}$$

$$\ddot{\nu}_A = -(2\pi)^{-1}\bar{\omega}_0\, \dot{\nu}_A + (2\pi)^{-1}\bar{\omega}_0 \sum_\lambda \Gamma_{A\lambda}\nu_\lambda + (2\pi)^{-1}\bar{\omega}_0 \sum_B \Gamma_{AB}\nu_B + (2\pi)^{-1}\bar{\omega}_0\gamma_A E \tag{4.9b}$$

$\bar{\omega}_0$ is the frequency of the long–wavelength ferroelectric transverse optic mode. It appears in the terms on the right in (4.9a,b) because the cell polarization fluctuates with this frequency. The cell will not switch in a time less than one period of this mode. The Γ–coefficients provide for the growth from larger to smaller steps.

In matrix notation, we can introduce thermodynamic forces

$$\Phi = M\nu \tag{4.10a}$$

$$\Psi = L\eta \tag{4.10b}$$

where ν and η are vectors whose components obey $\eta_i = \dot{\nu}_i$. The kinetic equations become:

$$\dot{\eta} = (\bar{\omega}_0/2\pi)[-L^{-1}\Psi + \Gamma M^{-1}\Phi + \gamma E] \tag{4.11a}$$

$$\dot{\nu} = \eta = L^{-1}\Psi \tag{4.11b}$$

Casimir anti–reciprocity implies

$$L^{-1} = -(\bar{\omega}_0/2\pi)\Gamma M^{-1} \tag{4.12}$$

Since $\dot{\eta}$ depends on E, the polarization current, $\dot{P}$, depends on the forces Ψ_j. The model gives

$$\dot{P} = 2P_s \sum_{i,j} k_i r_i b^2 c V^{-1} (-)^{s(i)} L^{-1}_{ij}\, \Psi_j \tag{4.13}$$

where P_s is the magnitude of the spontaneous polarization; s(i) = +1 if the step causes the wall to advance forward in the direction encouraged by E and –1 otherwise; b is the lattice constant normal to the wall and c > b the lattice constant in the direction of P_s; r_i is the height of a step of type i. From anti–reciprocity

$$(\bar{\omega}_0/2\pi)\gamma_i = -2P_s \sum_j k_j r_j b^2 c (-)^{s(j)} L^{-1}_{ji} \tag{4.14}$$

Equations (4.12) and (4.14) determine L and M if we have an estimate of the ν–dependence of S (Nettleton 1967b). The latter is found from the degeneracy of the state characterized by the ν–variables. Degeneracy is estimated from the number of ways of stacking steps of specified sizes.

In a steady state, the wall velocity $v = b\dot{\nu}^{\dagger}$, where $\dot{\nu}^{\dagger}$ is the number of completely–nucleated steps which cover the entire wall. We use the above formalism to calculate the coefficient of $(\dot{\nu}^{\dagger})^2$ in F. Multiplying this by $2b^{-2}$, we get the effective mass of the domain wall. A formal expression for the effective mass is given (Nettleton 1967b).

5. EXTENDED THERMODYNAMICS OF NONLINEAR REACTION KINETICS

The thermodynamic treatment of thermal relaxation in the preceding section was inspired by the circumstance that such a process can be regarded as a unimolecular reaction with Equations (4.1) for the various species i as the reaction–kinetic equations. An ad hoc thermodynamics, postulating an F, S, and P which depend on the set $\{\zeta_i\}$ of populations, has long existed (Markham et al. 1951). This was used to calculate the absorption and dispersion of sound by adding the linearized reaction–kinetic equations to the hydrodynamic equations of mass, momentum, and energy conservation. The complex propagation constant of the sound was calculated from the secular determinantal condition that the four linearized equations should have a non–trivial sinusoidally–oscillating solution. Section 4 casts this ad hoc thermodynamics in canonical form by introducing thermodynamic forces in a manner inspired by non–equilibrium chemical thermodynamics (Prigogine and Defay 1954).

In the present section, we do not follow Prigogine and Defay in using the degree of advancement of the reaction as a state variable. If $(dN_i)_c$ represents the change in component i resulting from the reaction, the total change dN_i being the sum of this plus an increment stemming from diffusion across the boundaries, then the degree of advancement, ξ, and reaction–rate, J, are given by

$$(dN_i)_c = \nu_i \, d\xi \qquad (5.1)$$

$$J = \dot{\xi} \qquad (5.2)$$

Here ν_i is a stoichiometric coefficient. The problem with the use of ξ as a state variable arises from the need to keep track of the initial state. To avoid this, we specify (Nettleton 1988b) the state by T, J, and the set $\{N_i\}$ of particle numbers.

The use of J as an extended variable was suggested by García–Colín et al. (1983, 1985, 1986).

Consistent with this choice of independent state variables, we propose a Gibbs equation (Nettleton 1988b).

$$T\,dS = dE + P\,dV + \Phi\,dJ - \sum_i \mu_i\,dN_i \qquad (5.3a)$$

$$dF = -S\,dT - P\,dV - \Phi\,dJ + \sum_i \mu_i\,dN_i \qquad (5.3b)$$

where $\Phi = -\partial F/\partial J$ is the force conjugate to $\dot{J}$. From (5.3a) one obtains for the irreversible entropy production rate,

$$\dot{S}_{irrev} = AJ + \Phi\dot{J} + ... \qquad (5.4a)$$

$$A = -\sum_i \mu_i \nu_i \qquad (5.4b)$$

A is minus the chemical affinity. Equation (5.4a) demonstrates that J and A are conjugate flux and force as are $\dot{J}$ and Φ. In the general, non–linear case, we expand

$$\Phi = -\nu_{ro} J + \nu_{r2} J^3 + 0(J^5) \qquad (5.5)$$

which involves only odd powers of J to satisfy the requirement that J and Φ be odd under time–reversal. The kinetic equation in canonical form is

$$\dot{J} = L\Phi + \gamma_r A \qquad (5.6)$$

where, in the non–linear case, L and γ_r can depend on J. By anti–reciprocity, the trivial equation J = J should be written

$$J = -\gamma_r \Phi \qquad (5.7)$$

The coefficients in the expansion

$$\gamma_r = \gamma_{ro} + \gamma_{r2} J^2 + 0(J^4) \qquad (5.8)$$

are calculable by solving (5.5) for J as an expansion in powers of Φ and comparing the result with (5.7) and (5.8).

Since the force $\Phi = -\partial F/\partial J$ is a sum of powers of J, so are functions derivable from F. Thus we have

$$P = P_0 + P_2 J^2 + 0(J^4) \qquad (5.9a)$$

$$P_2 = -\tfrac{1}{2}\partial\nu_{ro}/\partial V = \tfrac{1}{2}(\gamma_{ro})^{-2}\partial\gamma_{ro}/\partial V \qquad (5.9b)$$

$$U = U_o + U_2 J^2 + 0(J^4) \qquad (5.9c)$$

$$2U_2 = -T(\partial\nu_{ro}/\partial T + \nu_{ro}) = \gamma_{ro}^{-1}(T\gamma_{ro}^{-1}\partial\gamma_{ro}/\partial T + 1) \qquad (5.9d)$$

(5.9b,d) follow from the integrability conditions for dF in (5.3b). We can find the small coefficients P_2 and U_2 if we can determine the J–dependence of γ_r. When U_2 has been obtained, we can calculate the coefficient of J^2 in

$$T - \theta = -U_2 J^2/C_v, \qquad C_v \equiv \partial U_o/\partial\theta \qquad (5.10)$$

where θ is the local equilibrium temperature. There is also an integrability condition,

$$\partial\mu_i/\partial J = -\partial\Phi/\partial N_i \tag{5.11}$$

which can be used to calculate the coefficients in

$$\mu_i = \mu_{io} + \mu_{i2}J^2 + 0(J^4) \tag{5.12a}$$

$$\mu_{i2} = \tfrac{1}{2}\partial\nu_{ro}/\partial N_i \tag{5.12b}$$

Further steps to evaluate γ_r invoke a specific model (Nettleton 1988b). Let the system consist of N_2 homonuclear diatomic molecules plus N_1 monomers formed by dissociation of the dimer species. We set $N = N_1 + N_2$ and $c_i = N_i/N$ (i = 1,2). In equilibrium, in a fluid like liquid nitrogen, the fraction of monomers is negligibly small because the dissociation energy is high. However, the monomer fraction can be maintained at a high steady value by absorption of ultra violet radiation. This can be simulated on a computer which removes the heat and some of the dimers produced by recombination, yielding a steady state at constant T.

From the theory of absolute reaction rates (Glasstone et al. 1941), the steady–state reaction–rate in the liquid phase is

$$J = (\kappa T/h)(N_1^2/N)\exp\left[-\Delta F_{b\ell}/\kappa T\right](\exp[z] - 1) \tag{5.13}$$

$$z \equiv A_o/\kappa T \tag{5.14}$$

where A_o is the J–dependent part of A. $\Delta F_{b\ell}$ is the energy barrier to monomer recombination. A_o appears in (5.13) because, in states with large J, we want the right–hand member of (5.6) to be linear in J, and we determine the J–dependence of L to bring this about. Otherwise, the inertial term $\dot{J}$ would not dominate at high ω and would not, in fact, provide the inertial effect. Thus, we take L to have the linear form $L\Phi = -\tau_c^{-1} J$, so that (5.6) in the steady state is

$$J = \gamma_{ro}\tau_c\kappa Tz \quad \text{(steady state)} \tag{5.15}$$

Comparison of (5.13) and (5.15) yields

$$\gamma_{ro} = h^{-2}\,\kappa T(N_1^2/N)\exp\left[-\Delta F_{b\ell}/\kappa T\right] z^{-1}(\exp[z] - 1). \tag{5.16}$$

We can now calculate P_2 and U_2 from (5.9b,c) if we can find z and A_o. These are estimated (Nettleton 1988b) from the ideal mixing approximation which gives reasonable orders–of–magnitude for non–electrolytes (Rowlinson 1959). In this way, we find

$$A_o = \mu_{2o} - 2\mu_{1o} \tag{5.17a}$$

$$\mu_{io} = \kappa T\,\ell n\left[(P_o/\kappa T)(h^2/2\pi m_i\,\kappa T)^{3/2}\right] + \kappa T\,\ell n\,c_i - \epsilon_i \quad (i = 1,2) \tag{5.17b}$$

where $\epsilon_1 = 0$ and $\epsilon_2 =$ dissociation energy. Equation (5.17b) sets the equilibrium chemical potential of a pure liquid component i at pressure P_o equal to the chemical potential of a gas at the same pressure.

Numerical estimates have been made (Nettleton 1988b) for liquid nitrogen at T = 77.4 K, $P_o = 1$ atom. From (5.9b) and (5.16), with the assumption

$$c_1 \exp[-\tfrac{1}{2}\Delta F_{b\ell}/\kappa T] = 10^{-3} \quad ,$$

it is concluded that $P_2 J^2/P_o \gtrsim 0.4$. This can become observable in a computer simulation in which c_1 is kept constant by removing dimers while T is held constant by removing the energy added through the replacement of dimers by monomers. From (5.10) and an estimate of U_2, one finds

$$(T - \theta)/T \sim -1.6 \times 10^{-5} \tag{5.18}$$

Thus, although the $0(J^2)$ contribution to P may be seen in a computer simulation of a steady state far from equilibrium, the difference between T and θ is still negligibly small at a monomer concentration where P_2 first becomes observable. A corresponding conclusion has been found in other studies (Nettleton 1987a, 1988a) of the dependence of $T - \theta$ on the extended variables, e.g. heat and diffusion flows, at liquid densities.

Aside from the possibility of observing the J–dependence of P in a computer study of a state in which c_1 is kept high, the J–dependence of P and μ_i opens up the possibility of relating the reaction–rates in two phases in contact. If a gas in which a reaction is progressing at rate J_g is in mechanical equilibrium with a liquid in which the same reaction is occurring at rate J, the pressures must be equal, i.e. $P = P_g$. This implies that

$$P_{og} = P_o + P_2 J^2 - P_{2g} J_g^2 \tag{5.19}$$

If the two phases are in chemical equilibrium so that there is no net flow of either component from one phase to the other, we have

$$\mu_i = \mu_i^g \tag{5.20}$$

Into (5.20), we substitute the expansions (5.12a) appropriate to each phase, with μ_{io}^g given by an expression of the form (5.17b) containing P_{og} in place of P_o and the gas phase concentrations, y_i, in place of c_i. Equation (5.20), under these circumstances, becomes:

$$\ell n\, c_i = [(P_2/P_o) - (\mu_{i2}/\kappa T)]J^2 - [(P_{2g}/P_o) - (\mu_{i2}^g/\kappa T)]J_g^2 + \ell n\, y_i \tag{5.21a}$$

$$c_i = y_i + 0(J^2) \tag{5.21b}$$

Application of (5.21a) of the condition

$$\sum_i c_i = 1 = \sum_i y_i \tag{5.22}$$

yields

$$(P_2/P_o - \sum_i c_i\mu_{i2}/\kappa T)J^2 = (P_{2g}/P_o - \sum_i c_i\mu^g_{i2}/\kappa T)J^2_g \,. \tag{5.23}$$

The reaction–rates in the two phases are proportional.

In a steady state, to $0(J^2)$,

$$J = \tau_c\gamma_{ro}A \tag{5.24a}$$

$$J_g = \tau_g\gamma^g_{ro}A \tag{5.24b}$$

where we invoke equality of the chemical potentials in the two phases to set $A = A_g$. τ_c and τ_g are the effective collision times in each phase. Putting (5.24a,b) into (5.23), we obtain a relation between c_1 and c_2. Thus, the condition of relative mechanical and chemical equilibrium fixes the concentrations, whereas in a single phase, chemical equilibrium does not fix c_1. This raises the possibility that if a simulation could be made of a liquid and gas phase in relative equilibrium, analytic approximate expressions could be fitted to the observed values of $c_1(n)$ and $c_2(n)$. Equations (5.21a) and (5.24a,b) then provide a way of obtaining an expression in analytical form for γ_{ro} if γ^g_{ro} is known. Alternatively, if a computer estimate is made of J, equations (5.23) and (5.24a,b) can provide an estimate for τ_c. The latter is hard to estimate from liquid structural models, where each molecule is in continuous interaction with its neighbours, making the concept of a collision hard to define.

In the application described above of extended thermodynamics to non–linear chemical kinetics, we have made much more use of the integrability conditions for F than we have of reciprocity. For example, Equations (5.11) provide a way of calculating μ_{i2} and (5.9b) of calculating P_2. The principal example of reciprocity cited is (5.7) which enables us to calculate Φ if we know γ_r. Reciprocity would come more prominently into play if we were considering an example in which there are coupled reactions.

6. PHENOMENOLOGICAL THEOREMS

In this section, we collect a number of mathematical results which can be demonstrated from the basic structure, including reciprocity, of the kinetic equations of extended thermodynamics. These results require only macroscopic reasoning, in contrast to the results of the next section where we invoke statistical models for the entropy and introduce microscopic mechanics through the Liouville equation to derive expressions exhibiting reciprocity for the phenomenological coefficients. Here we assume reciprocity without deriving it and consider various different, but equivalent, ways of formulating the phenomenological equations which exhibit it.

Transformations Which Violate Reciprocity

A statement called "Meixner's Theorem" by de Groot (1951) was proposed by him as a foundation for the application of reciprocity in non–equilibrium thermodynamics. This statement postulated that

$$\dot{S}_{irrev} = \sum_{a=1}^{n} J_a X_a \tag{6.1}$$

where the set $\{X_a\}$ are linearly–independent functions of the thermodynamic state variables. It was asserted that, if (6.1) obtains together with

$$J_a = \sum_{b=1}^{n} L_{ab} X_b \tag{6.2}$$

then

$$L_{ab} = \pm L_{ba} \tag{6.3}$$

where the upper sign in (6.3) obtains when $X_a(t)$ and $X_b(t)$ are both even or both odd under $t \to -t$, and the lower sign applies in all other cases.

This statement by de Groot goes beyond anything ever proved or asserted by Meixner himself, and indeed it has been easy to find counter examples. Coleman and Truesdell (1960) pointed out that if we substitute into (6.1) the transformation,

$$J^*_a = J_a + \sum_b W_{ab} X_b \tag{6.4a}$$

$$W_{ab} = -W_{ba} \tag{6.4b}$$

the effect is to leave (6.1) invariant in form with $J_a \to J^*_a$ and to replace the coefficient matrix in (6.2) with

$$L^*_{ab} = L_{ab} + W_{ab} \tag{6.5}$$

which does not exhibit Onsager symmetry.

One can eliminate the type of transformation put forward by Coleman and Truesdell by requiring that J_a, as well as the set $\{X_a\}$ have a definite time–reversal symmetry. Then if J_a is odd, we cannot use in (6.4a) any forces which are even. Under these circumstances, the Onsager symmetry of L_{ab} is not changed by (6.5). Similar reasoning holds if J_a is even. However, we can show (Nettleton 1962) that requiring all the fluxes and forces to have definite time–reversal symmetry is not sufficient, to assure that (6.1) and (6.2) will imply (6.3).

To assure that Onsager symmetry is preserved under transformations which preserve (6.1) as a sum of products of conjugate fluxes and forces, we must divide the fluxes and forces into two classes. In one class, all the J_a are time–derivatives of state variables and all the forces are free energy derivatives, e.g. $-\partial F/\partial \alpha$. The first class is the one considered in derivations of Onsager reciprocal relations from fluctuation theory and microscopic reversibility (de Groot 1951). In a linear formulation, Onsager symmetry should obtain with respect to the coefficients coupling fluxes and forces entirely within this class. In the other class the forces are gradients like $-T^{-1}\nabla T$ and $-\nabla T^{-1}(\mu_2 - \mu_1)$ conjugate to heat and binary diffusion flows, respectively. The latter fluxes are not time derivatives of state variables. To preserve Onsager symmetry, one must eschew transformations which mix fluxes or forces belonging to one class with those belonging to the other, as well as transformations leading to fluxes or forces which do not have definite time–reversal symmetry.

Evaluation of Phenomenological Coefficients by Comparison with a Model

Suppose we have a set of n state variables $\{\alpha_j\}$ which are even under time–reversal and an associated extended set $\eta_j \equiv \dot{\alpha}_j$ which are odd (j = 1,...,n). Furthermore, we imagine that there exists a model yielding the kinetic equations,

$$\dot{\eta}_i = \sum_j a_{ij}\eta_j + \Gamma_i(\alpha) \tag{6.6}$$

where the a_{ij} and Γ_i are functions given by the model of density, temperature, and the $\{\alpha_i\}$. We consider only equations linear in the set $\{\eta_i\}$. A discussion of the non–linear case, which applies to microscopic models, is given in Section 7.

We ask (Nettleton 1967a) whether it is possible to find a unique set of coefficients $L_{ij}^{(k)}$ and functions $F_i(\{\alpha_i\})$ and $F_i^*(\{\alpha_i\},\{\eta_i\})$ such that (6.6) is equivalent to:

$$\dot{\alpha}_i = \sum_{j=1}^{n} L_{ij}^{(2)}(\{\alpha_i\})F_j^* \tag{6.7a}$$

$$\dot{\eta}_i = \sum_{j=1}^{n} L_{ij}^{(3)}F_j + \sum_{j=1}^{n} L_{ij}^{(4)}F_j^* \tag{6.7b}$$

The functions $\{F_i\}$ are η–independent to $0(\eta^2)$ while the $\{F_j^*\}$ are linear in η and odd under time–reversal. We require that the L–coefficients exhibit Onsager symmetry, viz. that

$$L_{ij}^{(2)} = -L_{ji}^{(3)} \tag{6.8a}$$

$$L_{ij}^{(4)} = L_{ji}^{(4)} \tag{6.8b}$$

In the next section, we shall discuss microscopic derivations of (6.7a,b) and (6.8a,b) without the restriction the F_j be independent of η and F_j^* linear. There the L–coefficients can also depend on η.

To find the F–functions which make (6.7b) agree identically with the model equation (6.6), expand

$$F_i = \sum_j b_{ij}(\alpha_j - \alpha_{jo}) + \sum_{j,k} b_{ijk}(\alpha_j - \alpha_{jo})(\alpha_k - \alpha_{ko}) + \dots \tag{6.9}$$

The expansion can be inverted, since $F_i = \partial S/\partial \alpha_i$, so that b_{ij} is the matrix of a negative definite quadratic form. By such an inversion, we can expand the set $\{\alpha_i - \alpha_{io}\}$ and the L–functions in powers of F. Thus, we have

$$L_{ij}^{(3)} = L_{ij}^{(3,0)} + \sum_k L_{ijk}^{(3)} F_k + \sum_{k,m} L_{ijkm}^{(3)} F_k F_m + \dots \quad . \tag{6.10}$$

We shall assume that, to zero order in η, the forces $\{F_i\}$ can be found from the α–derivatives of the free energy calculated from equilibrium statistical mechanics. We want now to determine $L_{ij}^{(3)}$ so that

$$\sum_j L_{ij}^{(3)} F_j = \Gamma_i \tag{6.11}$$

The solution can be obtained uniquely (Nettleton 1967a).

Now if (6.7a) is to hold identically, we need

$$F_i^* = \sum_k (L^{(2)})_{ik}^{-1} \eta_k = - \sum_k (L^{(3)T})_{ik} \eta_k \tag{6.12}$$

enabling us to calculate uniquely the forces $\{F_i^*\}$. The coefficients $L_{ij}^{(4)}$ follow from

$$\sum_{j,k} a_{ik} L_{kj}^{(2)} F_j^* = \sum_j L_{ij}^{(4)} F_j^* \quad . \tag{6.13}$$

The symmetry conditions (6.8b) impose a restriction on the a_{ij} matrix elements. In Section 4, we had in (4.1) the relation $a_{ij} = -\tau_c'^{-1} \delta_{ij}$. Under these circumstances, the condition (6.8b) is satisfied if $L_{ij}^{(2)}$ is symmetric under permutation of its two indices.

From the foregoing arguments, it is evident that, neglecting $0(\eta^2)$, we can find all the F, F^*, and L functions uniquely, even when they are non–linear to arbitrary order in $\alpha - \alpha_o$. We shall find in Section 7 that this is no longer true when we consider higher orders in η. Sufficient conditions are given above to determine only the term in F^* which is linear in η. Different formulations of non–linear extended thermodynamics can be found which are apparently equivalent and which differ in the higher–order η–dependence of F and F^*.

Lagrangian Formulation of Non–linear Extended Thermodynamics

In Section 1, a brief reference is made to the work of Landau and Lifshitz (1958) who showed that phenomenological equations having the form (6.7a,b) can be cast in Lagrangian form. These authors proposed a Lagrangian L = K – V, where K is a bilinear form representing the leading η–dependence of the free energy, and V depends only on α. The Lagrangian equations are equivalent to (6,7a,b), with reciprocity included.

We propose to show here that, for a suitable choice of L and the dissipation function, (6.7a,b) can still be put in Lagrangian form when we assume an arbitrary η–dependence of $L_{ij}^{(k)}$ and the sets $\{F_i\}$ and $\{F_i^*\}$. The demonstration that equations of the form (6.7a,b) can be derived to all orders in η for a general mechanical model is discussed in Section 7. We find that the function which, in the general case, plays the role of K in the Lagrangian is not, beyond $0(\eta^2)$, the part of the free energy F which includes its entire η–dependence. Also, this function cannot be identified with kinetic energy in general.

To define the Lagrangian, suppose (Nettleton 1986) that we can expand:

$$F = F_o(\{\alpha_i\}) + \sum_{n\geq 1} \sum_{\mu(1)...\mu(2n)} F^{(2n)}_{\mu(1)...\mu(2n)} \eta_{\mu(1)} ... \eta_{\mu(2n)} \quad . \tag{6.14}$$

Next, define $\tilde{F}_{ij}(\alpha,\eta)$ so that

$$F_i^* \equiv \sum_j \tilde{F}_{ij}\eta_j \quad . \tag{6.15}$$

The Lagrangian equations are:

$$(\partial/\partial t)(\partial L/\partial \dot{\eta}_i) - \partial L/\partial \alpha_i = R_i = -\partial R/\partial \dot{\eta}_i \tag{6.16}$$

where R is a dissipation function. Setting L ≡ 2T – F in this equation, we find

$$\sum_j (2\partial^2 T/\partial \eta_i \partial \eta_j - \partial^2 F/\partial \eta_i \partial \eta_j)\dot{\eta}_j + \sum_j (2\partial^2 T/\partial \eta_i \partial \alpha_j - \partial^2 F/\partial \eta_i \partial \alpha_j)\eta_j$$

$$- 2\partial T/\partial \alpha_i + \partial F/\partial \alpha_i = R_i \quad . \tag{6.17}$$

The function T, which plays the role of K in the Landau–Lifshitz version, is chosen to satisfy

$$2\partial^2 T/\partial \eta_i \partial \eta_j - \partial^2 F/\partial \eta_i \partial \eta_j = -\tilde{F}_{ij} \quad . \tag{6.18}$$

The solution of this equation is

$$T^{(2n)}_{\mu(1)...\mu(2n)} = \tfrac{1}{2}[2n/(2n-1)]F^{(2n)}_{\mu(1)...\mu(2n)} . \tag{6.19}$$

Now suppose that R_i, which up to this point is undefined, obeys

$$R_i = -\sum_{j,m} \tilde{F}_{ij} L^{(4)}_{jm} F^*_m + \sum_j (2\partial^2 T/\partial\eta_i \partial\alpha_j - \partial^2 F/\partial\eta_i \partial\alpha_j)\eta_j - 2\partial T/\partial\alpha_i \,. \tag{6.20}$$

Substituting this definition into (6.17), we obtain

$$-\sum_i \tilde{F}_{ij}\dot{\eta}_j + \sum_{j,m} \tilde{F}_{ij} L^{(4)}_{jm} F^*_m - F_i = 0 \tag{6.21}$$

Matrix multiplying (6.21) by $\tilde{F}^{-1}_{ij}$, we put this equation in the form:

$$\dot{\eta}_i = \sum_j L^{(4)}_{ij} F^*_j - \sum_j \tilde{F}^{-1}_{ij} F_j \tag{6.22}$$

This result exhibits the symmetric reciprocity relation (6.8b) if $L^{(4)}_{ij}$ is symmetric. It also exhibits anti–reciprocity because

$$\dot{\alpha}_i = \eta_i = \sum_j \tilde{F}^{-1}_{ij} F^*_j \tag{6.23a}$$

$$L^{(2)}_{ij} = \tilde{F}^{-1}_{ij} = L^{(2)}_{ji} \tag{6.23b}$$

so that (6.8a) holds. We can thus determine R_i so that the Lagrangian equations (6.11) are equivalent to the kinetic equations, whatever the model for $L^{(4)}_{ij}$.

The Lagrangian formulation is equivalent to a Hamiltonian formalism (Nettleton 1986). Defining

$$p_i \equiv \partial L(\alpha,\eta)/\partial\eta_i \tag{6.24a}$$

$$H \equiv \sum_i p_i \eta_i - L = F \quad , \tag{6.24b}$$

we find

$$\partial H(\alpha,p)/\partial p_i = \eta_i = \dot{\alpha}_i \tag{6.25}$$

Also,

$$\sum_i p_i \eta_i = 2T \tag{6.26}$$

The generalized Lagrangian formalism has not been used, as was the original work of Landau and Lifshitz, as a means of setting up constitutive equations non–linear in η. The ansatz which must be introduced for R is no longer, in the general case, a simple quadratic form. Anti–symmetries appear in the higher–order expansion coefficients for R in powers of η.

Alternative Extended Thermodynamics

In the notation of (6.7a,b), in the general case where all coefficients and functions can be nonlinear in η, define

$$\tilde{F} \equiv F + \sum_i \overset{*}{F}_i \eta_i \tag{6.27}$$

$\tilde{F}$ should not be confused with the matrix defined in (6.15). Equation (6.27) is a Legendre transformation to a thermodynamic potential depending on N, V, T, $\{\alpha_i\}$, and $\{\overset{*}{F}_i\}$ as independent state variables, We also introduce a pseudo–entropy, $\tilde{S}$, obeying

$$\tilde{F} \equiv E - T\tilde{S} \tag{6.28}$$

Invoking the Gibbs equation

$$T\dot{\tilde{S}} = \dot{E} + P\dot{V} + \sum_i (F_i \dot{\alpha}_i + \overset{*}{F}_i \dot{\eta}_i) \quad , \tag{6.29}$$

we find that, in an adiabatic process where $\dot{E} = -P\dot{V}$,

$$T\, d\tilde{S}/dt = (S - \tilde{S})\dot{T} - \sum_i \eta_i \dot{\overset{*}{F}}_i + \sum_i F_i \dot{\alpha}_i \tag{6.30}$$

Ignoring the term in $\dot{T}$, we see that if fluxes and forces are picked from (6.30) as they are from the irreversible rate of entropy production in usual formulations of non–equilibrium thermodynamics, then the roles of $-\eta_i$ and $\overset{*}{F}_i$ are interchanged. This suggests that there should exist a formulation of extended thermodynamics in which $\overset{*}{F}_i$ is the extended variable and $-\eta_i$ the associated thermodynamic force. We examine whether such a formulation can be set up, preserving reciprocity (Nettleton 1989a).

In the notation of (6.15), Equations 6.7a,b) become:

$$\dot{\eta}_i = \sum_{j,k} L^{(4)}_{ij} \tilde{F}_{jk} \eta_k + \sum_j L^{(3)}_{ij} F_j \tag{6.31a}$$

$$\dot{\alpha}_i = -\sum_{j,k} L^{(3)}_{ij} \tilde{F}_{jk} \eta_k \tag{6.31b}$$

Taking the matrix product of (6.31b) by $\tilde{F}_{ij}$, we obtain

$$\dot{\overset{*}{F}}_i - \sum_j \dot{\tilde{F}}_{ij} \eta_j = \sum_{j,r,s} \tilde{F}_{ij} L^{(4)}_{jr} \tilde{F}_{rs} \eta_s + \sum_{j,r} \tilde{F}_{ij} L^{(3)}_{jr} F_r \tag{6.32}$$

Now define

$$\tilde{L}^{(4)}_{ij} \equiv \dot{\tilde{F}}_{ij} + \sum_{k,r} \tilde{F}_{ik} L^{(4)}_{ir} \tilde{F}_{rj} = \tilde{L}^{(4)}_{ji} \quad . \tag{6.33}$$

From (6.23b), we have

$$L^{(3)}_{ij} = -L^{(2)}_{ij} = -\tilde{F}^{-1}_{ij} \quad . \tag{6.34}$$

Equation (6.32) becomes

$$\dot{F}^*_i = \sum_j \tilde{L}^{(4)}_{ij} \eta_j - F_i \quad . \tag{6.35}$$

If we compare this with (6.33), we see that the coefficient of F_i in (6.35) equals the coefficient of $-\eta_i$ in the equation,

$$\dot{\alpha}_i = -(-\eta_i) \tag{6.36}$$

An Onsager–type coupling has been shown still to exist, but it is purely symmetric. The alternative formulation offers few advantages if we can calculate F^* from a mechanical model. An example where this is difficult is a chemical reaction. In that case, setting up an equation of the form (6.31a), which is made linear in η to satisfy an inertial condition, we can infer the unknown coefficients by invoking integrability conditions (Nettleton 1989a).

7. MICROSCOPIC STATISTICAL APPROACHES

We found in the preceding section that all the coefficients in extended thermodynamic kinetic equations could be found by fitting them to the corresponding rate constants derived from a model except the $0(\eta^4)$ terms in the free energy, F. These terms have been found to be negligibly small in cases such as heat conduction (Nettleton 1987a) and diffusion (Nettleton 1988a) where quantitative estimates could be made. However, no general approach for evaluating these free energy terms has been proposed in the work discussed in earlier sections. We proceed to introduce here statistical models for the entropy which, together with projection operator techniques applied to the Liouville equation, yield formal expressions for all the coefficients in $L^{(k)}_{ij}$ and F. It is found that there is still a non–uniqueness and that several formulations of extended thermodynamics are possible which differ in their $0(\eta^4)$ contributions to F.

Zwanzig Projection Operator and Reciprocity in Nonlinear Regime

The first statistical derivation of Equations (6.7a,b) and (6.8a,b), in the general case where all coefficients are nonlinear in η, which we shall examine here is based on a projection operator introduced by Zwanzig (1961). The kinetics of the problem is based on the classical Liouville equation,

$$i\partial f/\partial t = \hat{L}f \tag{7.1a}$$

$$-i\hat{L}f \equiv \{f,\hat{H}\} \tag{7.1b}$$

for a large, closed system of N particles. f is the distribution in Γ–space, and $\{f,\hat{H}\}$ is the classical Poisson bracket. We introduce state variables α and η as in (6.7a,b). To define these assume that there is a set $\{A_i(x)\}$ of functions of the phase co–ordinates which are even under reversal of the particle momenta. Defining $\dot{A}_i \equiv i\hat{L}A_i$, which are odd under momentum reversal, we have for i = 1,...,n:

$$\alpha_i = \int fA_i(x)\, dx \tag{7.2a}$$

$$\eta_i = \int f\dot{A}_i(x)\, dx \quad . \tag{7.2b}$$

The averages in (7.2a,b) can be calculated via a distribution function

$$g(a,v) = \int f\delta(a - a)\delta(\dot{A} - v)\, dx \tag{7.3}$$

where $\delta(A - a)$ denotes a product of factors $\delta(A_i - a_i)$. $g(a,v)\, da\, dv$ is the probability that A_i has a value between a_i and $a_i + da_i$ and $\dot{A}_i$ between v_i and $v_i + dv_i$, etc. An equation of Fokker–Planck type can be derived for g if we multiply (7.1a) by $\delta(A - a)\delta(\dot{A} - v)$ and integrate over Γ–space, assuming the initial condition,

$$g(a,v,0) = \delta(a - a_o)\delta(v - v_o) \quad . \tag{7.4}$$

We obtain in this way (Nettleton 1964a):

$$\partial g(a,v,t)/\partial t = - \sum_{j=1}^{n} \{(\partial/\partial a_j)[v_j g] + (\partial/\partial v_j)[q_j g]\}$$

$$+ \sum_{j=1}^{n} \sum_{k=1}^{n} \int_o^t ds \int da'\, dv'\, (\partial/\partial v_j)[W(a,v)K_{jk}(a,v,a',v',s)](\partial/\partial v_k')$$

$$\cdot\, [g(a',v',t - s)W(a',v')^{-1}] \tag{7.5a}$$

$$W(a,v) \equiv \int \delta(A - a)\delta(\dot{A} - v)\, dx \tag{7.5b}$$

$$A_i^\dagger \equiv (i\hat{L})^2 A_i \qquad (i = 1,...,n) \tag{7.5c}$$

$$q_i \equiv W(a,v)^{-1} \int A_i^\dagger \delta(A - a)\delta(\dot{A} - v)\, dx \equiv \langle A_i;a\rangle \tag{7.5d}$$

$$K_{jk} \equiv \langle A_j^\dagger \{\exp[-is(1 - P)\hat{L}]\}(1 - P)A_k^\dagger (A - a');a\rangle \tag{7.5e}$$

$$P\bar{f}(x) \equiv W^{-1} \int \bar{f}\, \delta(A(x) - A(x'))\delta(\dot{A}(x) - \dot{A}(x'))\, dx'. \tag{7.5f}$$

P is a projection operator introduced by Zwanzig (1960, 1961) so that Pf is the "relevant" part of f, i.e. the part which gives $\langle A\rangle$ and $\langle\dot{A}\rangle$ correctly, but not the ensemble average of an arbitrary function. W(a,v) is the structure function. In general, W is not analytic, and so to use it here we have to assume that we can replace it with an analytic approximation. In a closed system, the phase space distribution is microcanonical, and so all points in Γ–space consistent with $\hat{H} = E$, $A = \alpha$, and $\dot{A} = \eta$ are given equal weight. Then the information–theoretic entropy appropriate to this distribution is

$$S = \kappa \, \ell n \, W(a,b) \tag{7.6}$$

This model S must be approximated by an analytic function to provide a statistical basis for phenomenological theories in which S is analytic. This latter approximation gives rise to a possible inconsistency.

Since the system is closed, we can invoke fluctuation theory to obtain an approximate g which is the probability of a fluctuation in the values of A and $\dot{A}$ from their ensemble averages α and η, respectively. If ΔS represents the quadratic and higher terms in the Taylor expansion of $S(a,b)$ about $S(\alpha,\eta)$, we suppose that

$$\begin{aligned} g \cong g^{(o)} &= C(\alpha,\eta) \exp\,(\Delta S/\kappa) \\ &= C(\alpha.\eta) \exp\,\{S'(a,v) - S'(\alpha,\eta) - \sum_i [(a_i - \alpha_i)(\partial S'/\partial \alpha_i) \\ &\qquad + (v_i - \eta_i)(\partial S'/\partial \eta_i)]\} \end{aligned} \tag{7.7}$$

where $S' \equiv S/\kappa$. This reduces to the Einstein fluctuation formula in equilibrium and is sharply–peaked about the ensemble average values α and η in a general non– equilibrium state.

To derive kinetic equations having the form (6.7a,b), we substitute (7.7) into (7.5a) and multiply first by a_i and then by v_i, integrating in each case over all a,v–space. We get the trivial result $\dot{\alpha}_i = \eta_i$ plus

$$\begin{aligned} \dot{\eta}_i = \int q_i g^{(o)}\, da\, dv - \int_o^t ds \int da\, dv\, da'\, dv' \sum_{r=1}^{n} K_{ir}(\partial/\partial v'_r) \\ \cdot [g^{(o)}(a',v',t-s) W(a',v')^{-1}] \qquad (i = 1,\dots,n) \end{aligned} \tag{7.8}$$

Next we make a Markovian approximation by supposing that K_{ir} decays rapidly enough with s so that we can replace $g^{(o)}(t-s) \to g^{(o)}(t)$. Finally, one expands in powers of $\partial S'/\partial \eta_i = F^*_i/\kappa$ and regroups the terms so that (7.8) assumes the form (6.7b) with $F_i = \kappa\, \partial S'/\partial \alpha_i$.

This process yields explicit expressions for $L^{(k)}_{ij}$ which are found to exhibit reciprocity expressed by (6.8a,b). We thus have an argument that reciprocity should hold to all orders in η. This generalizes an earlier approach (Nettleton 1964a) which kept only the terms in (7.8) which were linear in η and assumed a Gaussian for (7.7). However, we have had to neglect possible corrections to $g^{(o)}$. Also, although explicit expressions are obtained for $L_{ij}^{(k)}$, these are not, in general, the analytic functions assumed in phenomenological treatments, and they are difficult to evaluate. We shall shortly, therefore, proceed to a more general approach which postulates a system in contact with a heat bath and reduces to the canonical rather than the microcanonical ensemble in equilibrium.

Uniqueness Demonstrable only to Terms $0(\eta^2)$ in F

We propose to discuss now (Nettleton 1987b) the question of whether the $L_{ij}^{(k)}$ and the coefficients in the η–expansion of F can be found by introducing expansions in powers of η and $\{\alpha_i - \alpha_{io}\}$ for all functions in (6.7a,b) and comparing the result term–by–term with a model equation such as (1.6). This discussion invokes the microscopic calculation just described in connection with (7.8) to generalize the discussion put forward in a purely phenomenological way in the preceding section. We assume that there exist expansions

$$F_i = \tilde{F}_i + \sum_{j,k} f_{i,jk}\eta_j\eta_k + 0(\eta^4) \qquad (i = 1,...,n) \tag{7.9a}$$

$$F_i^* = -\sum_j \mu_{ij}\eta_j + \sum_{j,k,m} \mu_{ijkm}\eta_j\eta_k\eta_m + 0(\eta^5) \tag{7.9b}$$

$$L_{ij}^{(r)} = \lambda_{ij}^{(r)} + \sum_k \lambda_{ijk}^{(r)}\eta_k + \sum_{k,m} \lambda_{ijkm}^{(r)}\eta_k\eta_m + ... \tag{7.9c}$$

The equation $\dot{\alpha}_i = \eta_i$ follows from (6.7a) if

$$\lambda_{ij}^{(2)} = -(\mu^{-1})_{ij} \tag{7.10}$$

From anti–reciprocity,

$$\sum_j \lambda_{ij}^{(3)}\tilde{F}_j = -\sum_j \lambda_{ji}^{(2)}\tilde{F}_j \equiv \Psi_i \tag{7.11}$$

from which we can find $\lambda_{ij}^{(2)}$. Comparing terms linear in η in (6.7b) and (6.6), we have

$$\lambda_{ij}^{(4)} = -\sum_m a_{im}(\mu^{-1})_{mj} \tag{7.12}$$

Equations (7.10 – 7.12) summarize what is done in the preceding section.

When it comes to finding higher order coefficients, the expressions derived via the Zwanzig projection operator technique show that

$$\lambda_{ijk}^{(3)} = 0 \neq \lambda_{ijk}^{(4)} \tag{7.13}$$

Taking into account the number of coefficients predicted (Nettleton 1987b) not to vanish in general, we find the equations for the higher–order λ–coefficients to be insufficient in number to yield a unique solution. This means that we cannot determine the free energy F uniquely unless we can do this using the statistical model for S. Equation (7.6), because of problems with its evaluation, does not in general solve this problem. We must go on, therefore, to the maximum entropy formalism which yields an analytic entropy function.

Maximum Entropy Formalism

As an entropy model appropriate to a non–equilibrium system in contact with a heat bath, we can adopt the information–theoretical expression (Jaynes 1957a,b)

$$S = -\kappa \int \bar{\rho}\, \ell n\, \bar{\rho}\, dx \qquad . \tag{7.14}$$

The maximization of the functional in (7.14) is carried out subject to subsidiary conditions obtained by putting $f = \bar{\rho}$ in (7.2a,b). The $\bar{\rho}$ which solves this variational problem is (Nettleton and Freidkin 1989):

$$\bar{\rho}(x) = Z^{-1} \exp\,[-\beta\hat{H} - \sum_i \gamma_i A_i - \sum_i \tilde{\gamma}_i \dot{A}_i] \tag{7.15}$$

where Z normalizes $\bar{\rho}$ to unity and where the sets $\{\gamma_i\}$, $\{\tilde{\gamma}_i\}$ are Lagrange multipliers. These are determined to satisfy the conditions $<A_i> = \alpha_i$ and $<\dot{A}_i> = \eta_i$. If (7.15) is substituted into (7.14) and we use the result to derive a Gibbs equation, we find $F_i = \kappa T\gamma_i$, $F_i^* = \kappa T\tilde{\gamma}_i$ play the role of thermodynamic forces.

If $\bar{\rho}$ is used as the phase space distribution, we obtain a corresponding

$$\bar{g}(a,v) = \int \bar{\rho}(x)\psi_a(x)dx = Z_c p_\beta(a,v)\exp[\beta F(\alpha,\eta) + \sum_i \beta F_i(\alpha_i - a_i) + \sum_i \beta F_i^*(\eta_i - v_i)] \tag{7.16a}$$

$$\psi_a \equiv \delta(A - a)\delta(\dot{A} - v) \tag{7.16b}$$

$$\rho_\beta \equiv Z_c^{-1} \exp\,(-\beta\hat{H}) \tag{7.16c}$$

$$p_\beta \equiv \int \rho_\beta \psi_a\, dx \tag{7.16d}$$

The expression (7.15) for $\bar{\rho}$ yields an analytic S and F when substituted into (7.14) in contrast to the Zwanzig entropy analogue (7.6). We can use $\bar{g}$ in the kinetic equation (7.5a) for g in place of (7.7). On taking moments of the kinetic equation, we obtain equations for $\dot{\alpha}_i$ and $\dot{\eta}_i$ whose terms can be regrouped and cast in the form (6.7a,b) with coefficients which obey the reciprocity relations (6.8a,b). However, the terms in $L_{ij}^{(2)}$ and $L_{ij}^{(4)}$ are not uniquely determined. The equation for $\partial g/\partial t$ still has the form (7.5a), but (Nettleton and Freidkin 1989) the projection operator $\hat{P}$ is the one introduced by Grabert (1982) to make $\bar{g}$ reduce to p_β in equilibrium in the case of systems interacting with a heat bath. The use of $\hat{P}$ instead of Zwanzig's P modifies (7.5d,e). We now have

$$q_i = \int p_\beta^{-1} \rho_\beta \psi_a (i\hat{L})^2 A_i \, dx \tag{7.17a}$$

$$K_{ij} = \int dx \, \rho_\beta \psi_{a'} A_j^\dagger (1 - \hat{P}) \exp [i\hat{L}(1 - \hat{P})(t - s)] \psi_a A_i^\dagger \, . \tag{7.17b}$$

Since conditions like $\alpha = \langle A \rangle$ determine the multipliers $\{\gamma_i\}$ and $\{\tilde{\gamma}_i\}$ in (7.15) or $\{F_i\}$ and $\{F_i^*\}$ in (7.16a), we have an unambiguous way of finding the $0(\eta^4)$ contributions to the free energy, F. This is satisfactory for a system with rigid boundaries. If we substitute (7.15) into (7.14), we can extract the Gibbs equation,

$$T\, dS = dE + P\, dV + \sum_i F_i \, d\alpha_i + \sum_i F_i^* \, d\eta_i \tag{7.18}$$

It does not follow automatically, however, that the function P in (7.18) is given by

$$P = (1/3) \sum_i \int \bar{\rho} \, \hat{P}_{ii} \, dx \tag{7.19}$$

where $\hat{P}_{ii}$ is the momentum flux tensor operator. To achieve a consistent statistical basis for a phenomenology based on (7.18) and (7.19), one can introduce additional terms and lagrange multipliers into the exponent in (7.15).

Specifically, one can add a term

$$\sum_i \bar{\gamma}_i B_i(x) \tag{7.20}$$

to those already appearing in the exponent of (7.15). The functions $\{B_i(x)\}$ can be chosen from a complete set of functions of particle momenta or be selected so that the derivation of Onsager reciprocity relations goes through as before (Nettleton and Freidkin 1989). When additional Lagrange multipliers are introduced, conditions such as the consistency requirements $\alpha = \langle A \rangle$, $\eta = \langle \dot{A} \rangle$, plus (7.19), lead to coupled equations for γ, $\tilde{\gamma}$, and $\bar{\gamma}$ and to forces F and F^* which depend on several of these multipliers. The more closely we seek to make the formalism derived from information theory agree with the usual phenomenology, including the Gibbs equation, the more complicated the problem of using this theory to calculate the free energy.

The possibility of introducing (7.20) means that one may derive several extended thermodynamic formulations which can differ among them in the higher–order terms in the free energy. The simplest of these can be used in the application of reciprocity to the kinetic equations, but one must be aware that it will not lead, in general, to consistency between (7.18) and (7.19). It cannot be used, in general to calculate the thermodynamic forces to be used in a Gibbs equation, especially if the latter is to be used in calculating the pressure P from the forces in question.

A particular example (Nettleton 1988c) has been worked out for the case where the heat flux, for which the operator is given by (2.14), is the only operator appearing in the exponent in (7.15). It was originally pointed out by Jou et al. (1984) that the force $\mathbf{\Phi} = -\partial F / \partial \mathbf{J}$ could be calculated uniquely from the exponential ansatz of the form (7.15) plus the condition

$$\mathbf{J} = \int \bar{\rho}\, \hat{\mathbf{J}}\, dx \tag{7.21}$$

The coefficients in F to $0(J^4)$ have been worked out (Nettleton 1988c) for a dense hard–sphere fluid. The $0(J^2)$ term in F can be calculated from the Wertheim–Thiele (Wertheim 1963, Thiele 1963) exact solution to the Percus–Yevick equation for the pair distribution function. The $0(J^4)$ terms in the free energy were calculated approximately by invoking Einstein fluctuation theory, since the five–particle distribution function for hard spheres, which we should need to follow the procedure used in connection with the $0(J^2)$ term, is not available. Assuming a dense hard–sphere model of Ar. at 87 K, we find the ratio of the coefficient of the $0(J^4)$ term in F to that of the $0(J^2)$ term is estimated to be $-4.58 \times 10^{-19} m^4 s^2 J^{-2}$. This is about half the value estimated by purely phenomenological reasoning (Nettleton 1987a). It verifies the general conclusion that the higher–order dependence of F on the extended variables tends to be negligibly small at liquid density and for accessible states.

This calculation, with **J** as the only extended variable, ignores the fact that, in a steady state at large $|\nabla T|$, a density gradient ∇n builds up until the mechanical equilibrium condition, $\nabla P = 0$, is satisfied. To provide for this, we consider the model in which the system is a cube of side $\ell \sim 10^{-6}$m immersed in a continuous fluid phase. One can define a density gradient operator (Nettleton 1989b),

$$\begin{aligned} \hat{A} &= \ell^{-1}[\hat{n}(x = \tfrac{1}{2}\ell) - \hat{n}(x = -\tfrac{1}{2}\ell)] \\ &= \ell^{-1}(2\pi)^{-3} \int \sum_i \exp(i\mathbf{k}\cdot\mathbf{r}_i)[\exp(-ik_x\ell/2) - \exp(ik_x\ell/2)]d\mathbf{k} \end{aligned} \tag{7.22}$$

The limits of the **k**–space integration in (7.22) are taken to be

$$-2\pi/\ell \leqq k_x \leqq 2\pi/\ell,\ -\pi/\sigma \leqq k_y \leqq \pi/\sigma,\ -\pi/\sigma \leqq k_z \leqq \pi/\sigma. \tag{7.23}$$

for hard spheres of diameter σ. These conditions make the free energy extensive and assure that $\langle\hat{A}\rangle$ involves only density variations over distances $\gtrsim \ell$. This is the case for the temperature gradient, since T is a variable which characterizes the entire system and is not so defined that it can vary from place to place in the interior of the cube.

For the case where ∇n is the variable, we have in place of (7.15) the expression

$$\bar{\rho} = Z^{-1} \exp[-\beta(\hat{H} + X\hat{A})] \tag{7.24}$$

where X is a thermodynamic force so determined that

$$\partial n/\partial x = \int \bar{\rho}\, \hat{A}\, dx \tag{7.25}$$

The determination of X from (7.25) is analogous to the determination of $\mathbf{\Phi}$ from (7.21). A rough estimate is made (Nettleton (1989b) of the term proportional to $(\partial n/\partial x)^2$ in the thermodynamic pressure. This is found to be sufficiently large in steady heat conduction to dominate the $0(J^2)$ terms predicted when ∇n is neglected. Both types of terms, however, are so small that it would be difficult to detect them.

8. SUMMARY

The genesis of the applications of extended thermodynamics discussed in Sections 2–4 was contemporary with the formulation of Landau and Lifshitz (1959) who showed that an Onsager–Casimir approach with reciprocity was equivalent to a Lagrangian representation of the macroscopic equations of motion. The latter representation was aimed at deriving constitutive equations, an application now superseded by rigorous mathematical theories devised by Müller (1985), Lebon (1978), and others. The Onsager–Casimir formalism, with reciprocity, was designed to calculate coupling coefficients, such as the one which expresses the excitation of a diffusion flow, $\mathbf{J}_d$, by a heat flow, $\mathbf{J}$, which are difficult to estimate from simple molecular models. Using reciprocity to estimate the unknown coupling coefficients, one can specialize the kinetic equations to the steady state and solve for $\mathbf{J}$ and $\mathbf{J}_d$ in terms of the gradients of temperature and the chemical potentials. One extracts in this way theoretical formulae for the transport coefficients in condensed phases.

This early work, mainly on thermal diffusion and viscoelasticity, was carried out prior to 1963. The amplitudes of fluctuations from equilibrium were assumed small, and the equations were linear in these amplitudes. By contrast, the recent trend in extended thermodynamics has been to seek applications to states far from equilibrium, where the free energy F can have terms $0(J^4)$. Much of the focus of publications reviewed here and written since 1983 has been to make quantitative estimates of these terms and show that they are negligibly small at liquid densities for most of the cases studied to date. The major example where observable non–linear effects were predicted was a Lennard–Jones liquid under a high, steady rate–of–shear (Nettleton 1987c). The effects in question did not come, however, from dependence of pressure and thermodynamic potentials on inertial variables such as the creep–rate components. Rather the effects predicted arise from non–linear coupling of shear relaxation to that of a scalar structural parameter.

A question which has arisen concerning extended thermodynamics is whether the thermodynamic potentials are unique. There appears to be no problem so long as the kinetic equations are restricted to linear dependence on the η–variables and the free energy to $0(\eta^2)$ (Nettleton 1967a, 1987b). However, if we go to terms in F^* which are $0(\eta^3)$, these can be estimated uniquely only via the maximum entropy formalism (Nettleton 1988c, 1989b, Nettleton and Freidkin 1989). In maximizing the entropy functional, one can impose auxiliary conditions such as (7.19) in addition to the consistency condition $\alpha = \langle A \rangle$. One apparently needs to impose such auxiliary conditions in order to derive a Gibbs equation in which P is one–third the pressure trace. Thus the calculation of thermodynamic potentials via the maximum entropy formalism may lead to F–functions differing from one another in fourth and higher– order terms. One must take care in using such a function to ascertain that it is the one appropriate to the use to be made of it. The simplest case (Nettleton and Freidkin 1989) involves the application of reciprocity without invoking anywhere a Gibbs equation.

One circumstance which long inhibited the development of applications of extended thermodynamics is size of predicted effects and the difficulty of producing steady states far enough from equilibrium so that they could be observed. Molecular dynamics promises to permit the study of such states, e.g. high steady rates of shear (Hanley and Evans 1982). By removing the heat generated, either by velocity rescaling or via fictitious damping forces, one can simulate steady states at constant temperature which could not be produced in the laboratory. The recent non–linear work reviewed here promises to provide simple estimates for comparison with computer simulations. Such estimates can provide assurance that the computer results are not artifacts of the boundary conditions. It is hoped that computers will verify predictions already made (Nettleton 1987a) that Fourier's law holds for large as well as small temperature gradients.

A possible new application of extended thermodynamics which is opened up but minimally exploited in the papers reviewed here stems from the dependence of pressure and thermodynamic potentials on the square of the chemical reaction–rate (Nettleton 1988b). Although these terms are very small in comparison with the leading terms in these functions, the conditions of mechanical and chemical equilibrium of two fluid phases in contact, in both of which the same reaction proceeds at a finite rate, permits us to use the terms in question to relate the reaction–rates in the two phases. This can lead, e.g., to an analytic expression for the reaction–rate in the liquid.

REFERENCES

Bearman, R.J., and Kirkwood, J.G. 1958. Statistical Mechanics of Transport Processes. XI. Equations of Transport in Multicomponent Systems. J. Chem. Phys. 28: 136–145.

Bearman, R.J., 1959. On the Linear Phenomenological Equations. II. The Linear Statistical Mechanical Theory. J. Chem. Phys. 31: 751–755.

Cahn, J.W., and Hilliard, J.E. 1958. Free Energy of a Nonuniform System. I. Interfacial Free Energy. J. Chem. Phys. 28: 258–267.

Cattaneo, C. 1958. A Form of Heat Equation which Eliminates the Paradox of Instantaneous Propagation. Compt. Rend. Acad. Sci. Paris 247: 431–433.

Coleman, B.D., and Truesdell, C. 1960. On the Reciprocal Relations of Onsager. J. Chem. Phys. 33: 28–31.

García–Colín, L.S., and Fuentes y Martinez, G.J. 1982. A Kinetic Derivation of Extended Irreversible Thermodynamics. J. Stat. Phys. 29: 387–396.

García–Colín, L.S., and de la Selva, S.M.T. 1983. Flux–Force Relation for Chemical Processes in Extended Thermodynamics. J. Non–Equilib. Thermodyn. 8: 277–292.

García–Colín, L.S., de la Selva, S.M.T., and Pina, E. 1985. The Mass Action Law in Extended Irreversible Thermodynamics. Phys. Lett. 110A: 363–365.

García–Colín, L.S., de la Selva, S.M.T., and Pina, E. 1986. Consistency of the Kinetic Mass Action Law with Thermodynamics. J. Phys. Chem. 90: 953–956.

Glasstone, S., Laidler, K.J., and Eyring, H. 1941. The Theory of Rate Processes. (McGraw–Hill, New York).

Grabert, H. 1982. Projection Operator Techniques in Nonequilibrium Statistical Mechanics, (Springer, Berlin).

Grad, H. 1949. On the Kinetic Theory of Rarefield Gases. Commun. Pure Appl. Math. 2: 331–407.

Groot, S.R.de 1951. Thermodynamics of Irreversible Processes. (North–Holland, Amsterdam).

Hall, L. 1948. The Origin of Ultrasonic Absorption in Water. Phys. Rev. 73: 775–781.

Hanley, H.J.M., and Evans, D.J. 1982. A Thermodynamics for a System under Shear. J. Chem. Phys. 76: 3225–3232.

Herzfeld, K.F., and Litovitz, T.A. 1959. Absorption and Dispersion of Ultrasonic Waves. (Academic Press, New York).

Holian, B.L., and Evans, D.J. 1983. Shear Viscosities Away from the Melting Line: A comparison of Equilibrium and Nonequilibrium Molecular Dynamics. J. Chem. Phys. 78: 5147–5150.

Jaynes, E.T. 1957a. Information Theory and Statistical Mechanics. Phys. Rev. 106: 620–630.

Jaynes, E.T. 1957b. Information Theory and Statistical Mechanics. II. Phys. Rev. 108: 171–190.

Jou, D., Pérez–García, C., and Casas–Vázquez, J. 1984. On a Non–Equilibrium Partition Function of Heat Conduction. J. Phys. A : Math. Gen. 17: 2799–2805.

Jou, D., Casas–Vázquez, J., and Lebon, G. 1988. Extended Irreversible Thermo–dynamics. Rep. Progr. Phys. 5: 1105–1179.

Kinasi, W., and Takahasi, H. 1957. On the 180^0–type Domain Wall of $BaTiO_3$ Crystal. J. Phys. Soc. Japan 12: 464–476.

Kluitenberg, G.A. 1962a. Thermodynamical Theory of Elasticity and Plasticity. Physica 28: 217–232.

Kluitenberg, G.A. 1962b. A Note on the Thermodynamics of Maxwell Bodies, Kelvin Bodies (Voigt Bodies), and Fluids. Physica 28: 561–568.

Kluitenberg, G.A. 1962c. On Rheology and thermodynamics of Irreversible Processes. Physica 28: 1173–1183.

Landau, L.D., and Lifshitz, E.M. 1958. Statistical Physics. (Pergamon, London).

Landau, L.D., and Lifshitz, E.M. 1960. Electrodynamics of Continuous Media. (Pergamon, London).

Lebon, G. 1978. Derivation of Generalized Fourier and Stokes–Newton Equations Based on Thermodynamics of Irreversible Processes. Bull. R. Soc. Belg. Class. Sci. 64: 456.

Lebon, G., Pérez–García, C., and Casas–Vázquez, J. 1988. On the Thermodynamic Foundations of Viscoelasticity. J. Chem. Phys. 88: 5068–5075.

Lopez de Haro, M., Rodriguez, R.F., and García–Colín, L.S., 1984. Effects of Spatial Variations in Extended Irreversible Thermodynamics. Physica A 128: 535–540.

Markham, J.J., Beyer, R.T., and Lindsay, R.B. 1951. Absorption of Sound in Fluids. Rev. Modern Phys. 23: 353–411.

Maxwell, J.C. 1867. On the Dynamical Theory of gases. Trans. Roy. Soc. London 157: 49–101.

Miller, R.C., and Weinreich, G. 1960. Mechanism for the Sidewise Motion of 180^0 Domain Walls in Barium titanate. Phys. Rev. 117: 1460–1466.

Müller, I. 1985. Thermodynamics. (Pitman Publishing, Ltd. London).

Nettleton, R.E. 1957. Theory of the Equation of State and Ultrasonic Absorption in Associated Liquids at Ordinary Temperatures. Phys. Rev. 106: 631–637.

Nettleton, R.E. 1959a. Compressional Relaxation in Liquids. J. Acoust. Soc. Am. 31: 557–567.

Nettleton, R.E. 1959b. Thermodynamics of Viscoelasticity in Liquids. Phys. Fluids 3: 256–263.

Nettleton, R.E. 1960a. Relaxation Theory of Thermal Conduction in Liquids. Phys. Fluids 3: 216–225.

Nettleton, R.E. 1960b. Casimir Coefficients and Minimum Entropy Production. J. Chem. Phys. 33: 237–241.

Nettleton, R.E. 1961. Density Fluctuations and Heat Conduction in a Pure Liquid. Phys. Fluids 4: 74–84.

Nettleton, R.E. 1962. Invalidity of Meixner's Theorem in Irreversible Thermodynamics. J. Res. Natl. Bur. Stds. 66B: 101–107.

Nettleton, R.E. 1963. Inertial Effects in the Phenomenological theory of Thermal Diffusion in Liquids. Il Nuovo Cimento 29: 952–969.

Nettleton, R.E. 1964a. Anti–reciprocity and Memory in the Statistical Approach to Irreversible Thermodynamics. J. Chem. Phys. 40: 112–116.

Nettleton, R.E. 1964b. Non–equilibrium Thermodynamics of Creep in Polycrystals. Physica 30: 1989–2002.

Nettleton, R.E. 1967a. Uniqueness Theorem for Nonlinear Irreversible Thermodynamics. Phys. Lett. 24A: 231–232.

Nettleton, R.E. 1967b. Effective Mass of 180^0 Domain Wall in Single Crystal Barium Titanate. J. Phys. Soc. Japan 22: 1375–1387.

Nettleton, R.E. 1984a. Heat Fluctuation Distribution from Non–equilibrium States. J. Phys. A : Math. Gen. 17: 2149–2155.

Nettleton, R.E. 1984b. Coupled Fluctuations in Temperature and Heat Flux from a Nonequilibrium State. J. Chem. Phys. 81: 2458–2462.

Nettleton, R.E. 1985a. Onsager Reciprocity in the Nonlinear Regime. Physica A 132: 143–163.

Nettleton, R.E. 1985b. Scalar Fluctuations from Extended Non–equilibrium Thermodynamic States. Z. Naturforsch. 40a: 976–985.

Nettleton, R.E. 1986. Lagrangian Formulation of non–linear Extended Irreversible Thermodynamics. J. Phys. A : Math. Gen. 19: L295–L297.

Nettleton, R.E. 1987a. Non–linear Heat Conduction in Dense Fluids. J. Phys. A : Math. Gen. 20: 4017–4025.

Nettleton, R.E. 1987b. Applications of Reciprocity to Non–linear Extended Thermodynamic Kinetic Equations. Physica A 144: 219–234.

Nettleton, R.E. 1987c. Shear–Rate nonanalyticity in Simple Liquids. J. Non–Equilib. Thermodyn. 12: 273–290.

Nettleton, R.E. 1988a. Non–linear Isotopic Diffusion in Dense Fluids. J. Phys. A : Math. Gen. 21: 1079–1090.
Nettleton, R.E. 1988b. Non–linear Extended Thermodynamics of Reaction Kinetics in Liquids. Il Nuovo Cimento 101B: 53–68.
Nettleton, R.E. 1988c. Information–Theoretic Extended Entropy for Steady Heat Conduction in Dense Fluids. J. Phys. A : Math. Gen. 21: 3939–3949.
Nettleton, R.E. 1989a. Non–linear Alternative Extended Thermodynamics. J. Phys. A: Math. Gen. 22: L353–L355.
Nettleton, R.E. 1989b. Corrections to Maximum Entropy Formalism for Steady Heat Conduction. J. Phys. A : Math. Gen. 22: 5281–5290.
Nettleton, R.E., and Freidkin, E.S. 1989. Nonlinear Reciprocity and the Maximum Entropy Formalism. Physica A 158: 672–690.
Prigogine, I., and Defay, R. 1954. Chemical Thermodynamics. (Longmans green and Co., London).
Ree, F.H., and Hoover, W.G. 1964. Fifth and Sixth Virial Coefficients for Hard Spheres and Hard Disks. J. Chem. Phys. 40: 939–950.
Rowlinson, J.S. 1959. Liquids and Liquid Mixtures. (Butterworth's, London).
Sannikov, D.G. 1962a. Dispersion in Ferroelectrics. Soviet Phys. – JETP 14: 98–101.
Sannikov, D.G. 1962b. Electromagnetic and Acoustic Waves in Ferroelectrics. Soviet Phys. – Solid State 4: 1187–1192.
Thiele, E. 1963. Equation of State for Hard Spheres. J. Chem. Phys. 39: 474–479.
Truesdell, C. 1985. Rational Thermodynamics. (McGraw–Hill, New York).
Vernotte, P. 1958. Paradoxes in the Continuous Theory of the Heat Equation. Compt. Rend. Acad. Sci. Paris 246: 3154–3155.
Wertheim, M.S. 1963. Exact Solution of the Percus–Yevick Integral Equation for Hard Spheres. Phys. Rev. Lett. 10: 321–323.
Zwanzig, R. 1960. Ensemble Method in the Theory of Irreversibility. J. Chem. Phys. 33: 1338–1341.
Zwanzig, R. 1961. Memory Effects in Irreversible thermodynamics. Phys. Rev. 124: 983–992.

Extended Irreversible Thermodynamics and Fluctuation Theory

J. Casas-Vázquez and D. Jou
Departamento de Física, Física Estadística
Universitat Autònoma de Barcelona
08193 Bellaterra, Catalonia, Spain

ABSTRACT This chapter outlines the mutual contributions between extended irreversible thermodynamics and fluctuation theory. The former one postulates a generalized nonequilibrium entropy which depends on the dissipative fluxes and gives an explicit macroscopic expression for it. Fluctuation theory allows to compute the values of the coefficients of the nonclassical terms in the entropy and in the entropy flux . One of the main aims of extended irreversible thermodynamics is to obtain evolution equations for the dissipative fluxes; this is a second connection with fluctuation theory, because nonequilibrium statistical mechanics based on fluctuation-dissipation relations attributes a central role to the time correlation function of the fluctuations of the fluxes. Finally, the generalized entropy yields explicit values for nonequilibrium corrections to the second moments of the fluctuations in steady states.

1 Introduction

The most fundamental problem in irreversible thermodynamics is the definition or even the existence of entropy in nonequilibrium situations. The classical version of irreversible thermodynamics (De Groot and Mazur 1962, Glansdorff and Prigogine 1971, Gyarmati 1970) is built on the local-equilibrium hyphothesis, namely, on the assumption that in nonequilibrium the classical equilibrium entropy keeps its validity at a local scale. As a consequence, the whole classical thermodynamic formalism (Gibbs equation, equations of state, stability conditions...) remains valid. No new insights into the entropy are gained from this point of view.

The practical effectiveness of the local-equilibrium hypothesis must not be discussed: it has produced an elegant formalism of great unifying power and has led to useful predictions concerning transport coefficients. However, from the microscopic point of view a strict local-equilibrium distribution is untenable, because it leads to vanishing dissipative transport; thus, it may only be considered as a working hypothesis without a truly fundamental basis. Though in many situations the deviations with respect to the local equilibrium are practically unobservable at the level of the entropy and of the equations of state, this is not at all a general rule. Typical nonequilibrium parameters, as dissipative fluxes, gradients of intensive quantities or several kinds of internal variables, may contribute to the entropy. When the relaxation time of these variables is much shorter or much longer than the observation time, such variables act as constant parameters. The situation is much more involved when the observation time and the relaxation time are of the same order: the dynamical behaviour of the additional variables becomes crucial in the description of the system.

Extended irreversible thermodynamics (EIT) (Casas-Vázquez et al 1984; Müller 1985; Jou et al 1988; García-Colín 1988) is a thermodynamic theory which takes the dissipative fluxes as independent variables, besides the usual conserved ones. The main aim of the theory is to find evolution equations for the dissipative fluxes satisfying the second law of thermodynamics. However, the classical version of the second law, based on the local-equilibrium entropy, is not sufficient for such endeavour, and it may even be contradictory with it. Thus, a generalized entropy depending on the dissipative fluxes is proposed, which leads to an expression for the second law compatible with their independent character. Such a generalized entropy is assumed, as the usual entropy, to be an additive and convex function, with a production rate which is definite positive and which is related to the probability of the fluctuations around equilibrium in the same way as the classical entropy (Jou et al 1988). In this way, EIT, which is not based on the local-equilibrium hypothesis, provides a more complete formulation of nonequilibrium thermodynamics able to give new insights into the meaning and definition of nonequilibrium entropy and nonequilibrium equations of state.

In the usual descriptions of hydrodynamic systems, the independent variables related to conservation laws are the mass density ρ, the baricentric velocity $\vec{v}$ or the momentum density $\rho\vec{v}$, and the specific internal energy u. In the simplest version of extended irreversible thermodynamics as applied to hydrodynamic systems, the further additional variables are the common dissipative fluxes, namely, heat flux, shear and bulk viscous pressure, diffusion fluxes and electrical flux. Of course, not all of them are always needed: in one-component and uncharged fluids, the latter fluxes are identically vanishing. In more elaborated versions of EIT, one may take into account that the dissipative fluxes may contain several additive contributions, each with their own relaxation time and their own evolution equation. Such contributions may be due, in some cases, to internal variables, as for instance internal degrees of freedom in macromolecular fluids. From this perspective, EIT is more general than the usual theories with internal variables, as it contains in a natural way many of them. The wider generality of EIT is evident, because it does not reduce to the classical theory even in the case of monoatomic ideal gases.

The independent character of the dissipative fluxes is perceptible when working at frequencies comparable with their relaxation times. For systems with internal degrees of freedom, as in polymeric fluids, such times are of macroscopic order thus leading to well-known phenomena as for instance viscoelasticity. In ideal monoatomic gases, the relaxation times are of the order of the collision time. From the experimental point of view, such situations are achieved in ultrasound propagation or in light scattering in gases, in neutron scattering in liquids or solids, in shock waves and in numerical simulations. All these fields of research received a strong impetus during the decades of the 1960's and the 1970's and fostered an important progress in several areas of nonequilibrium statistical mechanics as in kinetic theory, linear response theory and projection operator techniques. One of the central aims of these researches was the understanding and computation of generalized transport laws, expressed in terms of memory functions or of generalized frequency and wave vector-dependent transport coefficients. Whereas this progress stands outside the scope of classical irreversible thermodynamics, it is closely related to the aims and views of EIT, which provides a macroscopic thermodynamic framework to some of these microscopic developments.

Fluctuation theory plays an important role in the classical nonequilibrium thermodynamics and in EIT. In the former theory, it was at the basis of the derivation of the Onsager's reciprocity relations. In EIT, fluctuation theory is also useful in several aspects: a) it allows to compute in a simple and direct way the coefficients appearing in the generalized entropy; b) it provides relations amongst several quantities, thus reducing the number of independent parameters in the theory; c) it provides a link between EIT and nonequilibrium statistical mechanics along the fluctuation-dissipation domain; d) it enlightens the role and meaning of the generalized entropy and e) it outlines some open problems and present limitations of the theory. In this review, we will try to make evident all these points.

This chapter is organized as follows: Section 2 is an introduction to EIT in one-component hydrodynamic systems; in Section 3, we present the macroscopic results concerning fluctuation expressions around equilibrium; Section 4 is devoted to the microscopic computation of the coefficients appearing into the entropy; Section 5 deals with those appearing in the generalized entropy flux; and Section 6 is devoted to problems on fluctuations around nonequilibrium steady states.

2 Extended irreversible thermodynamics

In the classical description, the state of a one-component fluid is determined by a kinematical variable, the velocity $\vec{v}$, and two thermodynamic variables, as for instance the specific volume v (the inverse of the mass density ρ) and the internal energy per unit mass u. The evolution of these variables is governed by the well-known balance equations of momentum, mass and internal energy

$$\rho\dot{v} = \nabla \cdot \vec{v} \tag{2.1}$$

$$\rho\dot{\vec{v}} = -\nabla \cdot \mathbf{P} + \rho\vec{\mathbf{F}} \tag{2.2}$$

$$\rho\dot{u} = -\nabla \cdot \vec{q} - \mathbf{P} : \nabla\vec{\mathbf{v}} \tag{2.3}$$

An upper dot stands for the material time derivative, $\mathbf{P}$ is the pressure tensor, $\vec{q}$ is the heat flux and $\vec{F}$ the body force per unit mass. To simplify slightly the expressions to be derived below, we will assume that $\vec{F}$ vanishes. The inclusion of $\vec{F}$ is straightforward and it does not modify in any essential way the conclusions of the present chapter.

The pressure tensor $\mathbf{P}$ is usually split into an equilibrium part, the thermodynamic pressure p, and a viscous part $\mathbf{P^v}$. The latter one can still be decomposed into a scalar bulk viscous presure p^v and a traceless shear viscous tensor $\overset{\circ}{\mathbf{P}}{}^{\mathbf{v}}$

$$\mathbf{P} = \mathbf{pU} + \mathbf{p^vU} + \overset{\circ}{\mathbf{P}}{}^{\mathbf{v}} \tag{2.4}$$

with $\mathbf{U}$ the identity tensor. In absence of internal structure in the molecules, $\mathbf{P}$ is symmetric and both internal and external angular momenta are separately conserved. Similarly, the velocity gradient is split into a scalar, a traceless symmetric $\overset{\circ}{\mathbf{V}}$ and an antisymmetric part $\mathbf{W}$ as

$$\nabla\vec{v} = (1/3)(\nabla \cdot \vec{v})\mathbf{U} + \overset{\circ}{\mathbf{V}} + \mathbf{W} \tag{2.5}$$

According to the local-equilibrium hypothesis, the Gibbs equation keeps its classical form

$$ds = T^{-1}du + T^{-1}pdv \tag{2.6}$$

with T the absolute temperature. The time derivative of the entropy is thus given by

$$\rho\dot{s} = T^{-1}\rho\dot{u} + T^{-1}p\rho\dot{v} \tag{2.7}$$

Introduction of (2.1) and (2.3) into (2.7) yields

$$\rho\dot{s} = -\nabla \cdot (T^{-1}\vec{q}) - T^{-2}\vec{q} \cdot \nabla T - T^{-1}p^v\nabla \cdot \vec{v} - T^{-1}\, \overset{\circ}{\mathbf{P}}{}^{\mathbf{v}} : \overset{\circ}{\mathbf{V}} \tag{2.8}$$

Comparison of (2.8) with the most general form of the entropy balance equation

$$\rho\dot{s} + \nabla \cdot \vec{J^s} = \sigma \tag{2.9}$$

allows the following identification of the entropy flux $\vec{J^s}$ and of the entropy production σ

$$\vec{J^s} = T^{-1}\vec{q} \tag{2.10}$$

$$\sigma = -T^{-2}\vec{q} \cdot \nabla T - T^{-1}p^v\nabla \cdot \vec{v} - T^{-1}\, \overset{\circ}{\mathbf{P}}{}^{\mathbf{v}} : \overset{\circ}{\mathbf{V}} \tag{2.11}$$

The entropy production σ is a bilinear form in the thermodynamic fluxes $\vec{q}$, p^v and $\overset{\circ}{\mathbf{P}}{}^{\mathbf{v}}$ and their respective conjugate forces.

The classical version of the second law states that the entropy production is positive semidefinite. The simplest way to relate the fluxes to the forces in a way that ensures the positive character of σ is to assume a linear relation of the form

$$\vec{q} = -\mu_1 T^{-2} \nabla T \tag{2.12}$$

$$p^v = -\mu_o T^{-1} \nabla \cdot \vec{v} \tag{2.13}$$

$$\overset{\circ}{\mathbf{P}}{}^{\mathbf{v}} = -\mu_2 T^{-1} \overset{\circ}{\mathbf{V}} \tag{2.14}$$

When these relations are compared to the Fourier and Stokes-Newton equations, the positive coefficients μ_1 , μ_o and μ_2 may be related to the thermal conductivity λ , bulk viscosity ς and shear viscosity η respectively by

$$\mu_1 = \lambda T^2, \qquad \mu_o = \varsigma T, \qquad \mu_2 = 2\eta T \tag{2.15}$$

These classical transport laws have been very useful in many contexts. However, these equations are not satisfactory at high frequencies or short wavelengths, or in systems with long relaxation times, as viscoelastic liquids, heat conductors at low temperatures or superconductors. Furthermore, nonlinear terms should be taken into account, but we will not deal with this problem here.

In contrast with the classical theory just presented, EIT assumes a generalized entropy depending on the dissipative fluxes besides the classical variables. The generalized Gibbs equation corresponding to it is supposed to have the form

$$ds = T^{-1}du + T^{-1}pdv - T^{-1}v\alpha_o p^v dp^v - T^{-1}v\alpha_1 \vec{q} \cdot d\vec{q} - T^{-1}v\alpha_2 \overset{\circ}{\mathbf{P}}{}^{\mathbf{v}} : d\,\overset{\circ}{\mathbf{P}}{}^{\mathbf{v}} \tag{2.16}$$

with α_1, α_o and α_2 parameters whose physical meaning will be identified. For the moment, we neglect second-order contributions of the fluxes to the absolute temperature and to the pressure.

Furthermore, we assume that the entropy flux depends on all the dissipative fluxes and not only on the heat flux. Up to second order in the fluxes, $\vec{J}^s$ is assumed to have the form

$$\vec{J}^s = T^{-1}\vec{q} + \beta_o p^v \vec{q} + \beta_1 \overset{\circ}{\mathbf{P}}{}^{\mathbf{v}} \cdot \vec{q} \tag{2.17}$$

wherein the first term is the classical entropy flux. The physical meaning of the parameters β_o and β_1 will be identified later.

Introduction of (2.16) and (2.17) into the general form of the entropy balance (2.9) , and assuming, for simplicity, that the coefficients β are constant (an unnecessary but simplifying hypothesis) yields for the entropy production

$$\begin{aligned}\sigma = \vec{q} \cdot [\nabla T^{-1} + \beta_1 \nabla \cdot \overset{\circ}{\mathbf{P}}{}^{\mathbf{v}} + \beta_o \nabla p^v - T^{-1}\alpha_1 \dot{\vec{q}}] \\ + p^v[-T^{-1}\nabla \cdot \vec{v} - T^{-1}\alpha_o \dot{p}^v + \beta_o \nabla \cdot \vec{q}] \\ + \overset{\circ}{\mathbf{P}}{}^{\mathbf{v}} : [-T^{-1} \overset{\circ}{\mathbf{V}} - T^{-1}\alpha_2 (\dot{\overset{\circ}{\mathbf{P}}}{}^{\mathbf{v}}) + \beta_1 (\nabla \overset{\circ}{\vec{q}})^s]\end{aligned} \tag{2.18}$$

One observes, as in classical irreversible thermodynamics, that the entropy production has the structure of a bilinear form in the fluxes and their respective conjugate expressions, which contain the time derivatives of the fluxes. Since our aim is to determine the evolution equations for the fluxes, we write in the linear approximation the following expressions

$$\nabla T^{-1} - T^{-1}\alpha_1\dot{\vec{q}} + \beta_1\nabla\cdot\overset{\circ}{\mathbf{P}}{}^{\mathbf{v}} + \beta_o\nabla p^v = \mu_1^{-1}\vec{q} \tag{2.19}$$

$$- T^{-1}\nabla\cdot\vec{v} - T^{-1}\alpha_o\dot{p^v} + \beta_o\nabla\cdot\vec{q} = \mu_o^{-1}p^v \tag{2.20}$$

$$- T^{-1}\overset{\circ}{\mathbf{V}} - T^{-1}\alpha_2(\dot{\overset{\circ}{\mathbf{P}}{}^{\mathbf{v}}}) + \beta_1(\overset{\circ}{\nabla\vec{q}})^s = \mu_2^{-1}\overset{\circ}{\mathbf{P}}{}^{\mathbf{v}} \tag{2.21}$$

The positive character of σ implies that μ_o, μ_1 and μ_2 must be positive.

Several of the coefficients appearing in (2.19)-(2.21) may be identified by comparing these equations with the so-called Maxwell-Cattaneo equations (Cattaneo 1985, Jou et al 1988, Joseph and Preziosi 1989) namely

$$\tau_1\dot{\vec{q}} + \vec{q} = -\lambda\nabla T \tag{2.22}$$

$$\tau_o\dot{p^v} + p^v = -\varsigma\nabla\cdot\vec{v} \tag{2.23}$$

$$\tau_2(\dot{\overset{\circ}{\mathbf{P}}{}^{\mathbf{v}}}) + \overset{\circ}{\mathbf{P}}{}^{\mathbf{v}} = -2\eta\overset{\circ}{\mathbf{V}} \tag{2.24}$$

Here, τ_1, τ_o and τ_2 are the relaxation times of the corresponding dissipative fluxes. When the relaxational terms are neglected, these equations reduce to the usual Fourier and Newton-Stokes laws. The relaxational terms are important for high-frequency phenomena or when the relaxation times are of macroscopic order. This comparison leads to the identifications (2.15) for the coefficients μ and to the new relations

$$\alpha_1 = \tau_1(\lambda T)^{-1}, \qquad \alpha_o = \tau_o\varsigma^{-1}, \qquad \alpha_2 = \tau_2(2\eta)^{-1} \tag{2.25}$$

Because of relations (2.15) and (2.25), equations (2.19)-(2.21) may be written as

$$\tau_1\dot{\vec{q}} = -(\vec{q} + \lambda\nabla T) + \beta_1\lambda T^2\nabla\cdot\overset{\circ}{\mathbf{P}}{}^{\mathbf{v}} + \beta_o\lambda T^2\nabla p^v \tag{2.26}$$

$$\tau_o\dot{p^v} = -(p^v + \varsigma\nabla\cdot\vec{v}) + \beta_o\varsigma T\nabla\cdot\vec{q} \tag{2.27}$$

$$\tau_2(\dot{\overset{\circ}{\mathbf{P}}{}^{\mathbf{v}}}) = -(\overset{\circ}{\mathbf{P}}{}^{\mathbf{v}} + 2\eta\overset{\circ}{\mathbf{V}}) + 2\beta_1\eta T(\overset{\circ}{\nabla\vec{q}})^s \tag{2.28}$$

Recently, a wide review of the application of these equations in several areas has been written by Jou, Casas-Vázquez and Lebon (1988).

Similarly, relations (2.25) allow us to write the generalized entropy in the following explicit way

$$ds = T^{-1}du + T^{-1}pdv - (v\tau_1/\lambda T^2)\vec{q}\cdot d\vec{q} - (v\tau_o/\varsigma T)p^v dp^v - (v\tau_2/2\eta T)\,\overset{\circ}{\mathbf{P}}{}^{\mathbf{v}} : d\,\overset{\circ}{\mathbf{P}}{}^{\mathbf{v}} \tag{2.29}$$

All the terms in this equation have a concrete physical meaning, and thus define explicitly the generalized entropy.

The values of the relaxation times may be found from the propagation speeds of heat and of viscous pulses. Note, indeed, that introduction of the Maxwell-Cattaneo equation (2.22) for the heat flux into the energy balance equation, and taking into account an equation of state of the form $du = cdT$ with c a specific heat, one is led to the equation

$$\tau_1(\partial^2 T/\partial t^2) + (\partial T/\partial t) - (\lambda/\rho c)\nabla^2 T = 0 \tag{2.30}$$

In the high-frequency limit this equation leads to the following phase speed for thermal waves

$$U = (\lambda/\rho c \tau_1)^{1/2} \tag{2.31}$$

In the limit of vanishing relaxation time τ_1 , (2.30) reduces to the classical heat equation, which predicts an infinite speed of propagation in the high-frequency limit, i.e. for the speed of thermal pulses. A wide review of the bibliography on theoretical and experimental aspects of heat waves may be found in Jou et al (1988) and in Joseph and Preziosi (1989).

On the other hand, introducing (2.24) into the momentum balance equation (2.2) , one may obtain for the high-frequency limit of the phase speed of shear waves (transverse velocity waves)

$$U = (\eta/\rho\tau_2)^{1/2} \tag{2.32}$$

Therefore, the relaxation times are in principle measurable from the high-frequency limit of thermal and viscous waves. Measurements of shear-wave speeds have been reported in Joseph et al (1986).

After having obtained an expression for the nonequilibrium entropy in (2.29) , the second-order contributions of the fluxes to the temperature and the pressure may be easily evaluated from the equality of second-order mixed derivatives of s. In this way we obtain, for instance, the following expression for the generalized absolute temperature under a heat flux

$$T^{-1}(u,v,q) = T_{eq}^{-1}(u,v) - [\partial(\tau_1 v/2\lambda T^2)/\partial u]\vec{q}\cdot\vec{q} \tag{2.33}$$

When the generalized temperature including second-order corrections is used in (2.22) for the analysis of phase speed of thermal waves under a heat flux one obtains instead of (2.30) (Casas-Vázquez and Jou 1989)

$$\tau_1(\partial^2 T/\partial t^2) + \lambda T^2 q_o[\partial(v\tau_1/2\lambda T^2)/\partial u](\partial\nabla T/\partial t) - (\lambda/\rho c)\nabla^2 T + (\partial T/\partial t) = 0 \tag{2.34}$$

This equation predicts a difference of speed of heat pulses along the steady heat flux $\vec{q_o}(U_+)$ and against $\vec{q_o}(U_-)$. It turns out that

$$\Delta U = U_- - U_+ = -(\lambda T^2/2)q_o[\partial(v\tau_1/2\lambda T^2)/\partial u] \tag{2.35}$$

It may be seen (Jou, Llebot, Casas-Vázquez 1982a) that the derivative between square brackets is negative, so that the speed against $\vec{q_o}$ is higher than the speed along it, a prediction first made by Coleman, Fabrizio and Owen (1982). This is a prediction of the generalized theory which could provide a test of it. Unfortunately, no experiments in this direction have yet been performed.

At this point, one may summarize the main aspects of the formulation of EIT just presented in this Section. The main achievements are the following ones:

1) An explicit version of a nonequilibrium entropy has been found in (2.29). This expression shows that when relaxational effects are taken into account in the transport equations, the entropy becomes modified and the local-equilibrium assumption is no longer valid. Note that in the limit of vanishing relaxation times, (2.29) reduces to the classical local-equilibrium entropy and the Maxwell-Cattaneo equations (2.22)-(2.24) yield the classical transport laws.

2) As formulated in terms of the generalized entropy, the second law of thermodynamics is compatible with the Maxwell-Cattaneo laws. This is not the case with the classical entropy, as the ensuing entropy production may become negative in presence of relaxational terms. Since the Maxwell-Cattaneo equations have been used to give a successful description of thermal waves observed in dielectric solids at low temperatures,the generalized entropy is more satisfactory than the classical one in the description of fast phenomena.

3) The generalized entropy is confirmed from kinetic theory up to second order in the fluxes. In contrast, the classical entropy only coincides with the kinetic entropy up to the first order in the fluxes. Furthermore, kinetic theory corroborates some predictions of EIT, as for instance the reciprocity relation between the coefficients linking $\dot{\vec{q}}$ and $\nabla \cdot \overset{\circ}{\mathbf{P}}^{\mathrm{v}}$ and $(\dot{\overset{\circ}{\mathbf{P}}}{}^{\mathrm{v}})$ and $(\overset{\circ}{\nabla q})^s$ on the one side, and $\dot{\vec{q}}$ and ∇p^v and $\dot{p^v}$ and $\nabla \cdot \vec{q}$ on the other side. Though these relations are a particular case of Onsager reciprocity relations, they have the important feature that they can be macroscopically deduced from the formalism of EIT, whereas the usual Onsager relations between first-order transport coefficients cannot be derived from a purely thermodynamic formalism.

4) EIT focuses its interest on the evolution equations for the dissipative fluxes, rather than on the usual constitutive equations. Besides avoiding the paradox of infinite speeds of propagation, this point of view is close to the one adopted in modern nonequilibrium statistical physics, where the time correlation functions of the fluxes play a central role in the determination of memory functions.

The main limitations of the theory, as presented here, are the following ones:

1) Besides the usual three transport coefficients (λ , η , and ς) appearing in the

classical theory, EIT introduces five additional parameters, the relaxation times τ_1, τ_o and τ_2 and β_o , β_1. This is a high number of undetermined parameters in the theory.

2) The dissipative fluxes have been assumed to behave as a single physical quantity. However, they could contain several different physical contributions with different relaxation times. This is so, for instance, in nonideal gases, where the viscous pressure tensor and the heat flux contain a kinetic plus an interaction part, or in polymeric fluids, where, besides the kinetic contribution due to the motion of the center of mass, the pressure tensor contains contributions from several internal degrees of freedom.

3) In ideal gases, the relaxation times of the fluxes are of the order of the collision times. At this time scale, however, not only the usual dissipative fluxes should be taken as additional independent variables, but all the possible fast variables of the system, whose relaxation time cannot be shorter than the collision time. Thus a very high number of independent variables should be introduced .

4) The physical meaning and the physical consequences of the generalized equations of state obtained as partial derivatives of the entropy are not yet clear. In particular, their implications on the stability of nonequilibrium steady states and their relation with the second moments of nonequilibrium fluctuations should be clarified from both the conceptual and the experimental points of view. The inclusion of higher-order terms into the generalized entropy is necessary to carry out this task.

Our group in the Autonomous University of Barcelona, in Bellaterra, has addressed its attention to all these features, and to some other ones which cannot be covered here, as for instance its application to a wide diversity of physical systems, ranging from generalized hydrodynamics to phonon hydrodynamics, from biophysical to cosmological problems. In this chapter we deal mainly with points 1, 2 and 4 of the list above, with special attention to the aspects more closely related with fluctuation theory. Point 3 will be analysed separately in another chapter of this volume (see Pérez-García and Jou, this volume).

3 Fluctuation theory

Up to now we have formulated a thermodynamic theory which leads to evolution equations for dissipative fluxes. In this Section, we focus our interest on the generalized Gibbs equation (2.29) , which is the central thermodynamic result of extended irreversible thermodynamics. The generalized entropy plays a central role in the description of the fluctuations around equilibrium states not only of conserved variables but also of dissipative fluxes.

The main reasons to study the problem of fluctuations is that it provides a link between entropy and probability and that it allows to compute the nonclassical coefficients of the entropy in terms of the equilibrium distribution function. The line of thought involving fluctuations is somewhat intermediate between the macroscopic and the microscopic ones and some authors refer to it as a mesoscopic description.

In this work we start from the Einstein relation for the probability of fluctuations. For isolated systems, the probability of fluctuations of thermodynamic variables with respect to an equilibrium reference state is related to the change in entropy ΔS associated with the fluctuation

$$\Delta S = S(fluctuation) - S(equilibrium) \tag{3.1}$$

The relation between the probability and ΔS was obtained by Einstein by inverting the well-known Boltzmann-Planck formula for the entropy

$$S = k \ln W \tag{3.2}$$

with W the probability of the respective macrostate and k Boltzmann's constant. One is led to

$$W \sim \exp(\Delta S/k) \tag{3.3}$$

If the fluctuations are small enough, the quantity ΔS may be developed as

$$\Delta S = (\delta S)_{eq} + (1/2)(\delta^2 S)_{eq} \tag{3.4}$$

δS and $\delta^2 S$ being the first and second differentials of the entropy. For an isolated system the entropy is a maximum in equilibrium, so that $(\delta S)_{eq} = 0$ and $(\delta^2 S)_{eq} < 0$. Introduction of this approximation into (3.3) yields

$$W \sim \exp(\delta^2 S/2k) \tag{3.5}$$

This is the well-known Einstein formula for the probability of fluctuations. This relation is valid for moments up to second order, but it fails for higher-order moments (Callen 1960). Other forms for the probability distribution of the fluctuations may be written for nonisolated systems by taking into account the presence of the surroundings (Landau and Lifshitz 1980). Up to the second order, they may be written as (3.5), but with the second differential of S evaluated under the conditions fixed by the definite physical situation under study.

We may now proceed to use the generalized entropy of EIT to obtain information on the fluctuations around equilibrium states. The second differential of the generalized entropy in an equilibrium state may be derived directly from the Gibbs equation (2.29). For the moment, we focus our attention on the fluctuations of dissipative fluxes, because the equilibrium fluctuations of the classical variables are well known. From (2.29) and (3.5) one obtains

$$W(\delta\vec{q}, \delta p^v, \delta \overset{\circ}{\mathbf{P}}{}^{\mathbf{v}}) \sim \exp\{(M/2k)[-(v\tau_o/\varsigma T)\delta p^v \delta p^v - (v\tau_1/\lambda T^2)\delta\vec{q}\cdot\delta\vec{q} - (v\tau_2/2\eta T)\delta \overset{\circ}{\mathbf{P}}{}^{\mathbf{v}} : \delta \overset{\circ}{\mathbf{P}}{}^{\mathbf{v}}]\} \tag{3.6}$$

where M denotes the mass of the system.

To obtain the second moments of the fluctuations recall that for a multivariate Gaussian probability distribution of the form

$$W \sim \exp\left[-(1/2)E_{ij}\delta x_i \delta x_j\right] \tag{3.7}$$

the second moments are given by (Landau and Lifshitz 1980)

$$< \delta x_i \delta x_j >= (E^{-1})_{ij} \tag{3.8}$$

Thus, one obtains for the second moments of the fluctuations of the dissipative fluxes

$$< \delta q_i \delta q_j >= k\lambda T^2(\tau_1 V)^{-1}\delta_{ij} \tag{3.9}$$

$$< \delta p^v \delta p^v >= k\varsigma T(\tau_o V)^{-1} \tag{3.10}$$

$$< \delta \overset{\circ}{P^v_{ij}} \delta \overset{\circ}{P^v_{kl}} >= k\eta T(\tau_2 V)^{-1}\Delta_{ijkl} \tag{3.11}$$

with $\Delta_{ijkl} = (\delta_{ik}\delta_{jl} + \delta_{il}\delta_{jk} - (2/3)\delta_{ij}\delta k)$ and with V the volume of the system $V = Mv$ (Jou and Careta, 1982; Llebot, Jou and Casas-Vázquez, 1982).

When these expressions are combined with the relaxational evolution equations one obtains for the time correlation functions of the fluctuations

$$< \delta q_i(0)\delta q_j(t) >= k\lambda T^2(\tau_1 V)^{-1}\delta_{ij}\exp\left(-t/\tau_1\right) \tag{3.12}$$

$$< \delta p^v(0)\delta p^v(t) >= k\varsigma T(\tau_o V^{-1}\exp\left(-t/\tau_o\right) \tag{3.13}$$

$$< \delta \overset{\circ}{P^v_{ij}}(0) \delta \overset{\circ}{P^v_{kl}}(t) >= k\eta T(\tau_2 V)^{-1}\Delta_{ijkl}\exp\left(-t/\tau_2\right) \tag{3.14}$$

In the limit of vanishing τ, (3.12)-(3.14) yield the well-known delta correlated expressions for the white hydrodynamic noise as proposed by Landau and Lifshitz (1981) which stand at the basis of fluctuating hydrodynamics. Here, in difference with the Landau-Lifshitz approach, the second moments of the noise have been obtained on the same grounds as those of the classical variables (Keizer 1983, García-Colín et al 1984).

These expressions relate the dissipative coefficients λ , ς and η to the fluctuations of the fluxes and they may be interpreted in two alternative ways: either by stating that the dissipative coefficients determine the strength of the fluctuations or by saying that the fluctuations determine the dissipative coefficients. The latter point of view has been exploited in the Green-Kubo relations for transport coefficients, which are one of the central results of modern nonequilibrium statistical mechanics.

Here, we outline the connection of (3.9-11) with the Green-Kubo formulae for thermal conductivity and shear viscosity, which read respectively as

$$\lambda = (V/kT^2)\int_0^\infty < \delta q_i(0)\delta q_i(t) > dt \tag{3.15}$$

$$\eta = (V/kT) \int_0^{\infty} < \delta \overset{\circ}{P^v_{ij}} (0) \delta \overset{\circ}{P^v_{ij}} (t) > dt \tag{3.16}$$

Here, no summation is implied over repeated indices, and $i \neq j$, and angular brackets stand for equilibrium average. If the evolution of the fluctuations of the fluxes is described by the Maxwell-Cattaneo relaxational equations (2.22-24) , then (3.15-16) , when integrated, reduce to

$$\lambda = (\tau_1 V/kT^2) < \delta q_i(0) \delta q_i(0) > \tag{3.17}$$

$$\eta = (\tau_2 V/kT) < \delta \overset{\circ}{P^v_{ij}}(0) \; \delta \overset{\circ}{P^v_{ij}} (0) > \tag{3.18}$$

which are just (3.9) and (3.11) , respectively.

This derivation shows that the present theory is limited to a very particular dynamics for the fluxes, namely the exponential one, whereas the Green-Kubo relations are valid for all kinds of dynamics. The Green-Kubo relations have fostered a wide interest for the study of the evolution of dissipative fluxes in statistical mechanics. The time correlation function of the fluxes has become one of the central quantities in nonequilibrium statistical mechanics, comparable in some way to the partition function in equilibrium statistical mechanics. This point is worth to emphasize, because one of the main aims of EIT is the determination of evolution equations for the dissipative fluxes, but from a macroscopic point of view. This provides an important area of contact between the macroscopic and the microscopic approaches. From this perspective, extended irreversible thermodynamics is much closer to the microscopic theories than classical irreversible thermodynamics.

The Green-Kubo relations have been used to obtain expressions for the generalized frequency and wavevector-dependent transport coefficients appearing in generalized hydrodynamics. The simplicity of (3.15) and (3.16) is, however, only apparent, because the whole Liouvillian appears in the evolution of the dynamical variables. As a consequence, in order to compare the theory with the experimental results some modelling is needed. Thermodynamic criteria as the ones provided by EIT may be useful as restrictions on the possible modelizations of the memory functions or generalized transport coefficients.

4 Microscopic evaluation of the coefficients of the generalized entropy

The expressions (3.9)-(3.11) for the second moments of fluctuations allow one to compute the nonclassical coefficients of the Gibbs equation (2.29) from the equilibrium distribution function of the system and the microscopic expressions of the dissipative fluxes. In this section we proceed to compute them for ideal gases, nonideal gases and Rouse polymers by using the expressions for the second moments derived previously.

Ideal gases

The microscopic expressions for $\vec{q}$ and P^v for ideal gases are (Chapman and Cowling 1970)

$$\vec{q} = \int (1/2)m(\vec{c}-\vec{v})^2(\vec{c}-\vec{v})f\,d\vec{c} \tag{4.1}$$

$$\overset{\circ}{\mathbf{P}}{}^{\mathbf{v}} = \int [m(\vec{c}-\vec{v})(\vec{c}-\vec{v}) - (1/3)p\mathbf{U}]\mathbf{f}d\vec{\mathbf{c}} \tag{4.2}$$

with f the distribution function, $\vec{c}$ the velocity of a molecule, $\vec{v}$ the mean velocity and m the mass of a molecule, $\mathbf{U}$ the identity tensor and p the thermodynamic pressure.

The fluctuations of $\vec{q}$ near equilibrium are due to the fluctuations δf of the distribution function and to the fluctuations $\delta\vec{v}$ of the mean velocity. Up to first order in the fluctuations, one has

$$\delta\vec{q} = \int (1/2)m(\vec{c}-\vec{v})^2(\vec{c}-\vec{v})\delta f\,d\vec{c} - \int (1/2)m(\vec{c}-\vec{v})^2 f_{eq}d\vec{c}\delta\vec{v} - \int m(\vec{c}-\vec{v})(\vec{c}-\vec{v})f_{eq}d\vec{c}\delta\vec{v} \tag{4.3}$$

By setting $\vec{C} = \vec{c} - \vec{v}$, the second and third integrals in the right-hand side of (4.3) for an ideal gas give respectively $u\rho\delta\vec{v}$ and $p\delta\vec{v}$,and hence

$$\delta\vec{q} = \int (1/2)m\vec{C}^2\vec{C}\delta f\,d\vec{c} - \rho h\delta\vec{v} \tag{4.4}$$

where $h = u + (p/\rho)$ is the enthalpy per unit mass, which for a classical ideal gas is $h = 5kT/2m$.

Since $\delta\vec{v}$ is given by

$$\rho\delta\vec{v} = \int m\vec{C}\delta f\,d\vec{c} \tag{4.5}$$

(4.4) can be written as

$$\delta\vec{q} = \int [(1/2)mC^2 - (5/2)kT]\vec{C}\delta f\,d\vec{c} \tag{4.6}$$

Using this expression, also known as reduced heat flux, to compute the second moments of the fluctuations one gets

$$<\delta\vec{q}\delta\vec{q}> = \int d\vec{c}\int d\vec{c}'[(1/2)m\vec{C}^2 - (5/2)kT]\vec{C}[(1/2)m\vec{C}'^2 - (5/2)kT]\vec{C}' <\delta f(\vec{c})\delta f(\vec{c}')> \tag{4.7}$$

For the fluctuations of the nondiagonal components of the viscous pressure tensor, an analogous derivation leads to

$$\delta\overset{\circ}{\mathbf{P}}{}^{\mathbf{v}} = \int d\vec{c}m\vec{C}\vec{C}\delta f \tag{4.8}$$

from which

$$<\delta\overset{\circ}{P}{}^{v}_{ij}\,\delta\overset{\circ}{P}{}^{v}_{kl}> = \int d\vec{c}\int d\vec{c}'mC_iC_jmC'_kC'_l <\delta f(\vec{c})\delta f(\vec{c}')> \tag{4.9}$$

The expression for the second moments of the fluctuations of the distribution function may be obtained (Lax 1960) from the Einstein formula (3.5) when Boltzmann's formula for the entropy

$$S = -kV \int f \ln f d\vec{c} \tag{4.10}$$

is used. According to (3.5) and (4.10) ,

$$< \delta f(\vec{c}, t) \delta f(\vec{c}', t) >= (1/V) f_{eq}(\vec{c}) \delta(\vec{c} - \vec{c}') \tag{4.11}$$

As a consequence, (4.7) and (4.9) reduce to

$$< \delta \vec{q} \delta \vec{q} >= \int d\vec{c} [(1/2)mC^2 - (5/2)kT]^2 \vec{C}\vec{C} f_{eq}(\vec{c})(1/V) \tag{4.12}$$

$$< \delta \overset{\circ}{P}{}^{v}_{ij} \delta \overset{\circ}{P}{}^{v}_{ij} >= \int d\vec{c} m^2 C_i^2 C_j^2 f_{eq}(\vec{c})(1/V) \tag{4.13}$$

When Maxwell- Boltzmann distribution function for f_{eq} is placed in (4.12-13) and the corresponding results are inserted in (3.9) and (3.11) , one obtains

$$\begin{aligned} (\tau_1 v / \lambda T^2) &= (2/5)(p^2 T)^{-1} \\ (\tau_2 v / 2\eta T) &= (1/2)(v/pT) \end{aligned} \tag{4.14}$$

The present theory may be also applied to degenerate ideal gases (Jou, Casas-Vázquez and Lebon 1988), but for simplicity it will not be done here.

On the other hand, it is worth emphasizing that (4.14) may also be obtained from kinetic theory(García-Colín and López de Haro 1982). In that case, one assumes that the nonequilibrium distribution function has the form

$$f = f_{eq}[1 + \phi] \tag{4.15}$$

with ϕ a small correction, and one develops (4.10) up to second order in ϕ . This yields in a straightforward manner

$$\rho s = \rho s_{eq} - (k/2) \int f_{eq} \phi^2 d\vec{c} \tag{4.16}$$

and for the entropy flux

$$\vec{J}^s = T^{-1} q - (1/2)k \int f_{eq} \vec{C} \phi^2 d\vec{c} \tag{4.17}$$

To obtain an expression for ϕ one could solve a kinetic equation for f, as the Boltzmann equation, the linearized Boltzmann equation, the relaxation-time approximation to the Boltzmann equation or other equations. Alternatively, one could use maximum entropy arguments to find an expression for ϕ which, when introduced into (4.17), would give the corrections to the local-equilibrium entropy.

For instance, the Grad thirteen-moments method (Grad 1958) yields for ϕ

$$\phi = (1/2pkT) m C_i C_j \overset{\circ}{P}{}^{v}_{ij} + (2m/5pk^2T^2)[(1/2)mC^2 - (5/2)kT] C_i q_i \tag{4.18}$$

which, when introduced into (4.17) , gives

$$\rho s = \rho s_{eq} - (1/4pT)\ \overset{\circ}{\mathbf{P}}{}^{\mathbf{v}}\!:\!\overset{\circ}{\mathbf{P}}{}^{\mathbf{v}}\ - (m/5pkT^2)\vec{q}\cdot\vec{q} \qquad (4.19)$$

and

$$\vec{J^s}_i = T^{-1}q_i - (2/5pT)\ \overset{\circ}{\mathbf{P}}{}^{\mathbf{v}}_{\mathbf{ij}}\ q_j \qquad (4.20)$$

The results in (4.19) are identical to those obtained in (4.14) and (4.15) . For the moment we have not obtained any expression for the coefficients appearing into the entropy flux. This will be done in the next section.

The reader should be aware that comparison with kinetic theory is richer than fluctuation theory, because it provides not only an expression for the generalized entropy but also for the evolution equations of the fluxes. However, the latter ones may not be necessary for the analysis of several features of nonequilibrium steady states.

Dilute real gases

We consider a nonideal gas whose particles interact through a potential $\phi(r_{ij})$ which depends only on the distance between the molecules, r_{ij} being the vector from particle i to particle j. In this case, one cannot longer work with only $f_1(\vec{r},\vec{c})$, the one-particle distribution function, but the two-particle distribution function $f_2(\vec{r_1},\vec{r_2},\vec{c_1},\vec{c_2})$ is also needed. The latter is related to the pair-correlation function $g(r)$ as $f_2(\vec{r_1},\vec{r_2},\vec{c_1},\vec{c_2}) = f_1(\vec{r_1},\vec{c_1})f_1(\vec{r_2},\vec{c_2})g(r_{12})$

The pressure tensor contains a kinetic part plus an interaction part and is given by (Green 1969)

$$P_{ij} = \int mC_iC_jf_1(c)d\vec{c} - (1/2)n^2\int g(r)\phi'(r)r_ir_jr^{-1}d\vec{r} \qquad (4.21)$$

where ϕ' stands for the derivative of ϕ with respect to r and where the three-particle contributions have been neglected.

Each $f_1(c)$ and $g(r)$ have in fact their own evolution equation, with their own relaxation time τ_c and τ_p . Thus, the generalized entropy should be written in this case as

$$\rho s = \rho s_{eq} - (\tau_c/4\eta_c T)\ \overset{\circ}{\mathbf{P}}{}^{\mathbf{v}}_{\mathbf{c}}\!:\!\overset{\circ}{\mathbf{P}}{}^{\mathbf{v}}_{\mathbf{c}}\ - (\tau_p/4\eta_p T)\ \overset{\circ}{\mathbf{P}}{}^{\mathbf{v}}_{\mathbf{p}}\!:\!\overset{\circ}{\mathbf{P}}{}^{\mathbf{v}}_{\mathbf{p}}\ - (\tau_o/2\varsigma T)p^{v2} \qquad (4.22)$$

Here, $\overset{\circ}{\mathbf{P}}{}^{\mathbf{v}}_{\mathbf{c}}$ and $\overset{\circ}{P}{}^{v}_{p}$ stand respectively for the kinetic and potential contribution to $\overset{\circ}{\mathbf{P}}{}^{\mathbf{v}}$ Since the bulk viscous pressure is identically zero for an ideal gas, no subindex has been written in relation to p^v . Furthermore, the heat flux contributions have been neglected for the sake of simplicity.

It follows from (4.22) and (3.5) that, for unit volume,

$$<\delta\ \overset{\circ}{P}{}^{v}_{12c}\ \delta\ \overset{\circ}{P}{}^{v}_{12c}>= kT\eta_c/\tau_c \qquad (4.23)$$

$$< \delta \overset{\circ}{P}{}^{v}_{12p} \, \delta \overset{\circ}{P}{}^{v}_{12p} >= kT\eta_p/\tau_p \tag{4.24}$$

To obtain the microscopic values of the second moments of the fluctuations of $\overset{\circ}{\mathbf{P}}{}^{\mathbf{v}}_{\mathbf{c}}$ and $\overset{\circ}{\mathbf{P}}{}^{\mathbf{v}}_{\mathbf{p}}$ one needs the second moments of f_1 and of g. The microscopic expression for the entropy is now

$$S = -kV \int f(c) \ln f(c) d\vec{c} - (1/2) n^2 k \int g(r) \ln g(r) d\vec{r} \tag{4.25}$$

The second moments of fluctuations according to (3.5) will thus be

$$< \delta f(\vec{c}) \delta f(\vec{c'}) >= (1/V) f_{eq}(\vec{c}) \delta(\vec{c} - \vec{c'}) \tag{4.26}$$

$$< \delta g(\vec{r}) \delta g(\vec{r'}) >= 2n^{-2} g_{eq}(\vec{r}) \delta(\vec{r} - \vec{r'}) \tag{4.27}$$

Thus, from (4.21) , (4.26) and (4.27) , one gets

$$< \delta \overset{\circ}{P}{}^{v}_{12c} \, \delta \overset{\circ}{P}{}^{v}_{12c} >= \int m^2 C_1^2 C_2^2 f_{eq} d\vec{c} \tag{4.28}$$

$$< \delta \overset{\circ}{P}{}^{v}_{12p} \, \delta \overset{\circ}{P}{}^{v}_{12p} >= (n^2/4) \int \phi'^2(r) r_1^2 r_2^2 r^{-2} (2/n^2) g_{eq}(r) d\vec{r} \tag{4.29}$$

Taking into account that up to first order in the density $g_{eq}(r) = \exp\left[-\phi(r)/kT\right]$, we obtain up to second order in n (Jou and Micenmacher 1987)

$$< \delta \overset{\circ}{P}{}^{v}_{12c} \delta \overset{\circ}{P}{}^{v}_{12c} >= kpT \tag{4.30}$$

$$< \delta \overset{\circ}{P}{}^{v}_{12p} \, \delta \overset{\circ}{P}{}^{v}_{12p} >= kpT[(4/15)Y_1 + (1/15)Y_2] \tag{4.31}$$

with the coefficients Y defined as

$$\begin{aligned} Y_1 &= (2\pi n/kT) \int_0^\infty g_{eq} \phi' r^3 dr \\ Y_2 &= (2\pi n/kT) \int_0^\infty g_{eq} \phi'' r^4 dr \end{aligned} \tag{4.32}$$

Expressions (4.30) and (4.31) , together with (4.23) and (4.24) , allow us to evaluate the coefficients appearing in the generalized entropy and relate the corresponding shear viscosities η_c and η_p to the respective relaxation times. Thus, the generalized entropy may be written as

$$\rho s = \rho s_{eq} - (1/4Tp) \, \overset{\circ}{\mathbf{P}}{}^{\mathbf{v}}_{\mathbf{c}} : \overset{\circ}{\mathbf{P}}{}^{\mathbf{v}}_{\mathbf{c}} - (15/4Tp)[4Y_1 + Y_2]^{-1} \, \overset{\circ}{\mathbf{P}}{}^{\mathbf{v}}_{\mathbf{p}} : \overset{\circ}{\mathbf{P}}{}^{\mathbf{v}}_{\mathbf{p}} - (9/Tp)[4Y_1 + Y_2]^{-1} p^{v2} \tag{4.33}$$

In (4.33) we have taken into account the results by Jou (1989) for the coefficient $\tau_o/2\varsigma T$ in the term in p^{v2} . When the interaction potential is negligible, (4.33) reduces to the expression for ideal gases, because both $\overset{\circ}{\mathbf{P}}{}^{\mathbf{v}}{}_{\mathbf{p}}$ and p^v vanish identically. The contributions of the kinetic and potential part of the heat flux to the entropy, and the contribution of the bulk viscous pressure have been studied by Jou (1989). This is a topic too specific to be given here.

Dilute polymer solutions

A simple and useful model for the description of many features of dilute polymer solutions is the so-called Rouse-Zimm model (Bird et al 1977). The configuration of a polymer with N beads is described by $N-1$ vectors, Q_1 , Q_2 , ...Q_{N-1} giving the relative positions of the consecutive beads. Instead of these vectors, it is more convenient to use the normal cordinates Q'_1 , Q'_2 , ...Q'_{N-1} of the elastic system. The viscous pressure tensor of the solution is given by

$$\overset{\circ}{\mathbf{P}}{}^{\mathbf{v}} = \overset{\circ}{\mathbf{P}}{}^{\mathbf{v}}_{\mathbf{s}} + n(N-1)kT\mathbf{U} - \mathbf{nH}\sum_{\mathbf{k}} < \mathbf{Q}'_{\mathbf{k}}\mathbf{Q}'_{\mathbf{k}} > \tag{4.34}$$

with $\overset{\circ}{\mathbf{P}}{}^{\mathbf{v}}_{\mathbf{s}}$ the solvent contribution, n the number of chains per unit volumen, H the elastic constant describing the intramolecular force between consecutive beads and $< ... >$ standing for an average over the configuration space. The viscous pressure tensor of the system is then a sum of several contributions arising from the different degrees of freedom of the chain. The configuration of the chain has been included many times in thermodynamic descriptions of polymeric systems as an internal variable. In extended thermodynamics, it appears through its contribution to the viscous pressure tensor. Note that in the theory of internal variables, no additional variables would have been considered for ideal gases, in contrast with what is done in EIT.

The contribution of the k-th normal mode to the viscous pressure, $\overset{\circ}{\mathbf{P}}{}^{\mathbf{v}}_{\mathbf{k}}$ is given by

$$\overset{\circ}{\mathbf{P}}{}^{\mathbf{v}}_{\mathbf{k}} = -nH < Q'_k Q'_k > + nkT\mathbf{U} \tag{4.35}$$

The generalized entropy of EIT for situations where the effects of heat flux, diffusion and so on may be neglected will have the form (Camacho and Jou 1990, Lebon this volume), up to second order in P^v

$$\rho s(u, v, \overset{\circ}{\mathbf{P}}{}^{\mathbf{v}}) = \rho s_{eq}(u, v) - \sum_k (\tau_k / 4T\eta_k)\, \overset{\circ}{\mathbf{P}}{}^{\mathbf{v}}_{\mathbf{k}} : \overset{\circ}{\mathbf{P}}{}^{\mathbf{v}}_{\mathbf{k}} \tag{4.36}$$

The coefficients of the nonequilibrium part of s may be obtained from the fluctuation relations

$$< \delta \overset{\circ}{P}{}^{v}_{12k} \delta \overset{\circ}{P}{}^{v}_{12k} >= kT\eta_k/\tau_k \tag{4.37}$$

According to the details presented for the previous systems, we may write

$$< \delta\ \overset{\circ}{P}{}^{v}_{12k}\ \delta\ \overset{\circ}{P}{}^{v}_{12k} >= n^2H^2 < Q'^2_{1k} Q'^2_{2k} > \tag{4.38}$$

The average must now be computed over the equilibrium configuration distribution function given by (Bird et al 1977)

$$\Psi(Q'_1, Q'_2, ...Q'_{N-1}) = \Pi_j \Psi_j(Q'_j) \tag{4.39}$$

with

$$\Psi_k(Q') = (H/2\pi kT)^{3/2} \exp\left[-(H/2kT)Q'_k Q'_k\right] \tag{4.40}$$

which follows directly from Boltzmann's expression for the probability by taking into account that the elastic potential energy is $(1/2)HQ' \cdot Q'$.

Combination of (4.37) and (4.40) yields

$$< \delta P^{v}_{12k} \delta P^{v}_{12k} >= nk^2T^2 \tag{4.41}$$

The generalized entropy may thus be written as (Camacho and Jou 1990)

$$\rho s(u, v, \overset{\circ}{\mathbf{P}}{}^{\mathbf{v}}) = \rho s_{eq}(u, v) - \sum_k (1/4nkT^2)\, \overset{\circ}{\mathbf{P}}{}^{\mathbf{v}}_{\mathbf{k}} : \overset{\circ}{\mathbf{P}}{}^{\mathbf{v}}_{\mathbf{k}} \tag{4.42}$$

Note finally that the partial viscous pressure tensors P^{v}_{k} satisfy evolution equations of the form

$$\tau_k\, \overset{\circ}{\mathbf{P}}{}^{\mathbf{v}}_{\mathbf{k}(1)} + \overset{\circ}{\mathbf{P}}{}^{\mathbf{v}}_{\mathbf{k}} = -2nkT\tau_k\, \overset{\circ}{\mathbf{V}} \tag{4.43}$$

with $\overset{\circ}{\mathbf{P}}{}^{\mathbf{v}}_{\mathbf{k}(1)}$ the contravariant derivative of P^{v}_{k} , namely

$$\overset{\circ}{\mathbf{P}}{}^{\mathbf{v}}_{\mathbf{k}(1)} = (\dot{\overset{\circ}{\mathbf{P}}}{}^{\mathbf{v}})_{\mathbf{k}} - [(\nabla\vec{v}) \cdot \overset{\circ}{\mathbf{P}}{}^{\mathbf{v}}_{\mathbf{k}} + \overset{\circ}{\mathbf{P}}{}^{\mathbf{v}}_{\mathbf{k}} \cdot (\nabla\vec{v})] \tag{4.44}$$

with $\tau_k = \varsigma'[8H\sin^2(k\pi/2N)]^{-1}$, with ς' the friction coefficient between the beads and the solvent. It is not difficult to obtain evolution equations as (4.44) from EIT (Camacho and Jou 1990). Thus, the treatment of situations where the viscous pressure tensor has several different contributions may be dealt with quite directly in EIT.

5 The entropy flux

It is also possible to formulate the coefficients β_0 and β_1 appearing in the entropy flux in terms of moments of fluctuations. This is of interest because it reduces the number of independent parameters of the theory. Furthermore, EIT has been able to establish a connection between the coefficients in the entropy flux and those appearing in the evolution equations of the fluxes (2.26)-(2.28) in such a way that the computation performed here is not reduced to a purely thermodynamical problem, but it is also useful from the dynamical point of view. The procedure is not so straightforward and the final result not so direct as for the coefficients appearing in the entropy. In this section we derive the corresponding expression for the mentioned coefficients and compute them for ideal gases.

Here, instead of assuming a development of the nonequilibrium distribution function in the form (4.16), we analize the fluctuations of f around its equilibrium value. In this way, one avoids the need to obtain the nonequilibrium correction ϕ from any kinetic equation, and one may compute the entropy and the entropy flux coefficients from the fluctuation expressions. To do this, we assume that the fluctuating distribution function is described by

$$f_{eq} + \delta f = \exp\left(\alpha + x^2 + \hat{q}_i \delta\lambda_i + \hat{P}^v_{ij} \delta\lambda_{ij}\right) \tag{5.1}$$

with $x^2 = (mC^2/2kT)$, $\hat{\vec{q}}$ and $\overset{\circ}{\mathbf{P}}{}^{\mathbf{v}}$ the microscopic operators associated to the fluctuations of the fluxes and λ_i and λ_{ij} parameters which are to be related to the fluctuations of the fluxes. Thus, one has for δf

$$\delta f = (\partial f/\partial \lambda_i)\delta\lambda_i + (\partial f/\partial \lambda_{ij})\delta\lambda_{ij} \tag{5.2}$$

According to (5.1), δf may be written as

$$\delta f = f_{eq}(\hat{q}_i \delta\lambda_i + \hat{P}^v_{ij} \delta\lambda_{ij}) \tag{5.3}$$

The macroscopic fluctuations in the heat flux and in the viscous pressure tensor are thus related to $\delta\lambda_i$ and to $\delta\lambda_{ij}$ through

$$\begin{aligned} \delta q_i &= \int f_{eq} \delta\lambda_i \hat{q}^2_j d\vec{c} \\ \delta P^v_{ij} &= \int f_{eq} \delta\lambda_{ij} \hat{P}^2_{kl} d\vec{c} \end{aligned} \tag{5.4}$$

where no summation over repeated indices must be done. As a consequence, the parameters $\delta\lambda_i$ and $\delta\lambda_{ij}$ may be written in terms of the fluctuations of q_i and of P^v_{ij} as

$$\delta\lambda_i = \delta q_i \left[\int f_{eq} \hat{q}^2_i d\vec{c}\right]^{-1} \tag{5.5}$$

$$\delta\lambda_{ij} = \delta P^v_{ij} \left[\int f_{eq} \hat{P}^{v2}_{ij} d\vec{c}\right]^{-1} \tag{5.6}$$

From these expressions, one may calculate the nonequilibrium contribution to the entropy and to the entropy flux as obtained from their microscopic definitions

$$\rho s = -k \int f \ln f d\vec{c} \tag{5.7}$$

$$\vec{J}^s = -k \int \vec{C} f \ln f d\vec{c} \tag{5.8}$$

Up to second order in the fluctuations, one obtains from (5.7) and (5.8)

$$\rho s = \rho s_{eq} - (k/2) \int f_{eq} (\delta f/f)^2 d\vec{c} \tag{5.9}$$

$$\vec{J}^s = T^{-1} \vec{q} - (k/2) \int \vec{C} f_{eq} (\delta f/f)^2 d\vec{c} \tag{5.10}$$

This yields

$$\rho s = \rho s_{eq} - (1/2)k \int f_{eq} (\hat{q}_i \delta\lambda_i + \hat{P}^v_{ij} \delta\lambda_{ij})^2 d\vec{c} \tag{5.11}$$

$$\vec{J}^s = T^{-1} \vec{q} - (1/2)k \int \vec{C} f_{eq} (\hat{q}_i \delta\lambda_i + \hat{P}^v_{ij} \delta\lambda_{ij})^2 d\vec{c} \tag{5.12}$$

The entropy associated to the fluctuations minus the equilibrium reference entropy is then given by

$$\Delta s = -(k/2)\int f_{eq}\hat{q}_i(\delta\lambda_i)\hat{q}_j(\delta\lambda_j)d\vec{c} - (k/2)\int f_{eq}\hat{P}^v_{ij}(\delta\lambda_{ij})\hat{P}^v_{kl}(\delta\lambda_{kl})d\vec{c} \tag{5.13}$$

and one has for the entropy flux

$$J^s_i = T^{-1}q_i - k\int f_{eq}\hat{q}_j\hat{P}^v_{ij}C_i(\delta\lambda_j)(\delta\lambda_{ij})d\vec{c} \tag{5.14}$$

By using relations (5.5-6) , these expressions may be written as

$$\Delta s = -(k/2)\delta\vec{q}\cdot\delta\vec{q}[\int f_{eq}\hat{q}_i\hat{q}_i d\vec{c}]^{-1} - (k/2)\delta\,\overset{\circ}{\mathbf{P}}{}^{\mathbf{v}}:\overset{\circ}{\mathbf{P}}{}^{\mathbf{v}}\,[\int f_{eq}\hat{P}^v_{ij}\hat{P}^v_{ij}d\vec{c}]^{-1} \tag{5.15}$$

and

$$\vec{J}^s = T^{-1}q - k\frac{[\int f_{eq}\hat{q}_i\hat{P}^v_{ij}C_j d\vec{c}]}{[\int f_{eq}\hat{q}_i^2 dc][\int f_{eq}\hat{P}^{v2}_{ij}dc]}\delta\vec{q}_i\cdot\delta\,\overset{\circ}{P}{}^v_{ij} \tag{5.16}$$

with no summation in the denominators. When one compares these expressions to the corresponding macroscopic ones

$$\Delta s = -(1/2)(\alpha_1\delta\vec{q}\cdot\delta\vec{q} + \alpha_2\delta\,\overset{\circ}{\mathbf{P}}{}^{\mathbf{v}}:\delta\,\overset{\circ}{\mathbf{P}}{}^{\mathbf{v}}) \tag{5.17}$$

and

$$\Delta\vec{J} = T^{-1}\delta\vec{q} + \beta_1\delta\,\overset{\circ}{\mathbf{P}}{}^{\mathbf{V}}\cdot\delta\vec{q} \tag{5.18}$$

one obtains

$$\alpha_2 = k[2\int f_{eq}\hat{P}^v_{12}\hat{P}^v_{12}d\vec{c}]^{-1} \tag{5.19}$$

$$\alpha_1 = k[\int f_{eq}\hat{q}_1\hat{q}_1 d\vec{c}]^{-1} \tag{5.20}$$

and

$$\beta_1 = -k[\int f_{eq}\hat{q}_1\hat{P}^v_{12}C_2 d\vec{c}][\int f_{eq}\hat{q}_1^2 d\vec{c}]^{-1}[\int f_{eq}\hat{P}^{v2}_{12}d\vec{c}]^{-1} \tag{5.21}$$

Relations (5.19-20) are identical to those previously used whereas (5.21) is new and allows to compute the coefficient β_1 in the entropy flux in terms of fluctuations. Writing the denominators in terms of the dissipative coefficients and the relaxation times according to (3.20) and (3.21) yields the more compact form

$$\beta_1 = -(\tau_1\tau_2/k\lambda\eta T^3)\int mC_1C_2[(1/2)mC^2 - mh]C_1C_2 f_{eq}d\vec{c} \tag{5.22}$$

This expression is a particular case of a formula obtained by Ernst (1970) from kinetic theory by means of projection-operator methods. More general expressions of β in terms of correlation functions of fluxes have been given by several authors (Jou, Casas-Vázquez and Lebon 1988).

Application of (5.21) to ideal gases is straightforward, and yields

$$\beta_1 = -(2/5)(1/pT) \tag{5.23}$$

a well known result in kinetic theory.

It is easy to apply the above formalism to a radiation gas or a mixture of radiation (photons, neutrinos, gravitons) and material fluid (Pavón, Jou, Casas-Vázquez 1982). The latter is assumed to thermalize in a short time, whereas the mean free time of radiation, τ , is supposed to be much longer. Such systems are of great interest in cosmological and astrophysical problems.

In this case we may interpret δq_i and δP_{ij}^v as

$$\delta q_i = c_i \delta e(rad) \tag{5.24}$$

and

$$\delta P_{ij}^V = (c_i c_j / c^2) \delta e(rad) \tag{5.25}$$

where c_i is the i-th component of the radiation velocity (whose speed is c) and $\delta e(rad)$ is the fluctuation of radiation energy. Since for a radiation gas

$$< \delta e(rad) \delta e(rad) >= 4kaT^5/V \tag{5.26}$$

with a the black body constant, we obtain

$$\lambda = (\tau V/kT^2) < \delta q_1 \delta q_1 >= (4/3)c^2 aT^3 \tau \tag{5.27}$$

$$\eta = (\tau V/kT) < \delta \overset{\circ}{P}{}_{12}^v \, \delta \overset{\circ}{P}{}_{12}^v >= (4/15)aT^4 \tau \tag{5.28}$$

$$\beta = -(\tau^2 V/k\lambda\eta T^3) < \delta q_1 \delta \overset{\circ}{P}{}_{12}^v c_2 >= -(4aT^4)^{-1} \tag{5.29}$$

which coincide with the results obtained from relativistic kinetic theory. For the terms related to the viscous pressure, see the Chapter by Pavón in this volume.

6 Nonequilibrium fluctuations of dissipative fluxes

According to the information theory, fluctuations around nonequilibrium steady states may be described in terms of the second derivatives of a generalized entropy taking account of the nonequilibrium constraints characterizing the steady state in the same way as equilibrium fluctuations. This is not true, of course, when local-equilibrium entropy is used. The contribution of the fluxes to the entropy leads to explicit expressions for the second moments of nonequilibrium fluctuations, which we evaluate in this section. We do not deal, however, with the time-correlation of the fluctuations, a topic which has deserved a wide attention in the context of local-equilibrium hydrodynamics (Tremblay 1984).

EIT predicts nonequilibrium contributions to the fluctuations of variables around nonequilibrium steady states (Jou, Llebot, Casas-Vázquez 1982a, b, Jou and Careta 1982). Such contributions are a consequence of the nonvanishing relaxation time of the fluxes. The argument to retain the local-equilibrium form for the non-equilibrium hydrodynamic noise is that the fluctuations constituting it are so fast that they do not have time enough to be influenced by the nonequilibrium condition of the system. Consistently with this argument, if the decay time of the noise is not negligibly small, it may "know" that the system is out of equilibrium, and consequently the noise becomes modified. In the classical treatments of nonequilibrium fluctuations, the hydrodynamic noise due to fluctuations of the dissipative fluxes is assumed to keep its local-equilibrium form, with coefficients depending on the position. From fluctuating Navier-Stokes equations, one obtains nonequilibrium corrections of purely dynamical origin, due to spatial dependence of the noise. Rather than to the dynamical aspects, we address here to the statistical part of the problem, which is of more fundamental importance because it deals with the very foundations of the local-equilibrium formulation. Though at low wavenumber (light scattering in liquids) the present effects are of smaller order of magnitude than the purely dynamical ones, they may become dominant in a regime of short wavelength (neutron scattering).

As an example, we consider the fluctuations of the heat flux in a rigid conductor around a nonequilibrium steady state characterised by a nonvanishing heat flux q_o. In order to obtain all the second-order contributions to the nonequilibrium second moments of the fluctuations we keep fourth-order terms in the entropy. Thus, instead of the second-order entropy used up to now in this paper, we write

$$s = s_{eq} - (\alpha_1/2)\vec{q}\cdot\vec{q} - (\gamma/4)(\vec{q}\cdot\vec{q})^2 \tag{6.1}$$

The coefficient $\alpha_1 = \tau_1 v/\lambda T^2$ has already been evaluated. The coefficient γ of the fourth-order term may be related to the fourth-order moments of the equilibrium fluctuations of the heat flux according to (Callen 1960)

$$< \delta q_1^4 >= -k^3(\partial^3 q_1/\partial F_1^3) + 3k^2(\partial q_1/\partial F_1)^2 \tag{6.2}$$

where F_1, the thermodynamic conjugate to q_1, is defined as

$$F_1 = (\partial S/\partial q_1) = -\alpha_1 q_1 - \gamma(\vec{q}\cdot\vec{q})q_1 \tag{6.3}$$

It thus follows that

$$\gamma = [3 < \delta q_1^2 > - < \delta q_1^4 >](6k^2 < \delta q_1^2 >)^{-1} \tag{6.4}$$

This relation allows one to compute the coefficient γ in terms of fourth and second moments of equilibrium fluctuations. For a monoatomic ideal gas, and taking into account the comments made in Section 3 concerning $\delta\vec{q}$, (6.4) yields

$$\gamma = -(862/625)(m/p^4 kT^2) \tag{6.5}$$

The second differential of (6.1) is

$$\delta^2 s = (\theta^{-1})_u(\delta u)^2 + 2\theta_v^{-1}\delta u\delta v + (\theta^{-1}\Pi)_v(\delta v)^2 - \alpha_1''\delta\vec{q}\cdot\delta\vec{q} - 2\vec{q}_o\alpha_{1v}\delta v\cdot\delta\vec{q} - 2\vec{q}_o\alpha_{1u}\delta u\cdot\delta\vec{q} \quad (6.6)$$

where the subscripts u and v denote partial derivatives with respect to the corresponding variables, and where α_1'' , θ^{-1} and $\theta^{-1}\Pi$ are given by

$$\alpha_1'' = \alpha_1 + 3\gamma q_o^2 \quad (6.7)$$

$$\theta^{-1} = s_u = T^{-1} - (1/2)\alpha_{1u}q_o^2 \quad (6.8)$$

$$\theta^{-1}\Pi = s_v = T^{-1}p - (1/2)\alpha_{1v}q_o^2 \quad (6.9)$$

We now introduce (6.6) into the Einstein relation (3.5) to compute the second moments of the fluctuations in the nonequilibrium steady state. We restrict ourselves to fluctuations δq of the heat flux along the direction of q_o , the mean heat flux. This results in

$$< \delta q\delta q >= k\Delta[\Delta\alpha_1'' + Aq_o^2]^{-1} \quad (6.10)$$

$$< \delta u\delta q >= -kq_o[\alpha_{1v}(\theta^{-1})_v - \alpha_{1u}(\theta^{-1}\Pi)_v][\Delta\alpha_1'' + Aq_o^2]^{-1} \quad (6.11)$$

$$< \delta v\delta q >= -kq_o[\alpha_{1u}(\theta^{-1})_u - \alpha_{1v}(\theta^{-1})_u][\Delta\alpha_1'' + Aq_o^2]^{-1} \quad (6.12)$$

where

$$\Delta = (\theta^{-1})_u(\theta^{-1}\Pi)_v - (\theta^{-1})_v^2 \quad (6.13)$$

and

$$A = \alpha_{1v}^2(\theta^{-1})_u + \alpha_{1u}^2(\theta^{-1}\Pi)_v - 2\alpha_{1u}\alpha_{1v}(\theta^{-1})_v \quad (6.14)$$

Three points are worth of mention in (6.10)-(6.12):

1) The correlations $< \delta u\delta q >$ and $< \delta v\delta q >$, which vanish in equilibrium because of the different time-reversal parity of u and $\vec{q}$ and of v and $\vec{q}$, are different from zero in nonequilibrium: this expresses the breaking of the time-reversal symmetry in nonequilibrium states.

2) Nonequilibrium contributions of order q_o^2 appear in the second moments $< \delta q\delta q >$ and, as a consequence, in the generalized Landau-Lifshitz formulae for the hydrodynamic noise.

3) The fluctuations may diverge when the denominator vanishes; this is so for a critical heat flux satisfying

$$\Delta(\alpha_1 + 3\gamma q_o^2) + Aq_o^2 = 0 \quad (6.15)$$

This is a third-order equation in q_o^2 because Δ contains second-order terms and A first-order terms. A divergence in the fluctuations has been remarked in the information-theory description of a harmonic heat-conducting chain, but one cannot be sure whether this divergence corresponds to a physical reality, indicates the limits of a validity of the theory or is simply a mathematical artifact.

The corrections resulting from the nonequilibrium contributions are computed for small values of q_o. At the lowest order in q_o, expressions (6.10)-(6.12) read as

$$< \delta q \delta q >= (k\lambda T^2/\tau V)\{1 + [(cT^2/\alpha_1)(\partial\alpha_1/\partial u) - 3(\gamma/\alpha_1)]q_o^2\} \quad (6.16)$$

$$< \delta u \delta q >= -(k\lambda T^2/\Delta\tau V)q_o[T_v^{-1}\alpha_{1v} - (T^{-1}p)_v\alpha_{1u}] \quad (6.17)$$

$$< \delta v \delta q >= -(k\lambda T^2/\Delta\tau V)q_o[T_v^{-1}\alpha_{1u} - T_u^{-1}\alpha_{1v}] \quad (6.18)$$

For ideal monoatomic gases one has $(T^{-1}p)_v = -p(Tv)^{-1}$, $T_v^{-1} = 0$, $T_u^{-1} = -(c_vT^2)^{-1}$ with $c_v = (3k/2m)$. Furthermore, according to kinetic theory of gases $\alpha_1 = (2/5)(k^2T^3n^2)^{-1}$ so that $\alpha_{1u=} = -(4/5)m(n^2T^4k^3)^{-1}$ and $\alpha_{1v} = (4/5)m(nk^2T^3)^{-1}$. The final simplified results for the fluctuations are

$$< \delta q \delta q >= (k\lambda T^2/\tau V)\{1 + 11.205\pi l^2(\nabla \ln T_o)^2\} \quad (6.19)$$

$$< \delta u \delta q >= (3kT/M)q_o \quad (6.20)$$

$$< \delta v \delta q >= 2vq_o \quad (6.21)$$

where l is the mean free path defined as $l = v\tau$ with v the mean velocity $v = (8kT/\pi m)^{1/2}$.

These results are of the same order, but they do not coincide exactly, with those obtained from simple kinetic models in the relaxation-time approximation, which lead to (Jou and Careta 1982)

$$< \delta q \delta q >= (k\lambda T^2/\tau V)[1 + 8.925\pi l^2(\nabla \ln T_o{}^2] \quad (6.22)$$

and

$$< \delta u \delta q >= 4.75(kT/M)q_o \quad (6.23)$$

Nettleton (1984) has evaluated the second moments of the heat flux for a liquid under a temperature gradient from the generalized entropy and from a Fokker-Planck microscopic model and has arrived to the same conclusions: the macroscopic and microscopic results are of the same order but not exactly coincident. This is an open problem in EIT.

It is of interest to note that some attempts have been made to define generalized nonequilibrium potentials by starting from the nonequilibrium expressions for the second moments of the fluctuations of the classical variables out of equilibrium (Keizer 1987). The generalized thermodynamic potentials are defined in such a way that they satisfy the Einstein relation (3.5) . In those attempts, the dissipative fluxes are not taken as independent variables whose evolution must be studied, but simply as external parameters. We believe that our approach is more fundamental, because not only static consequences of the fluxes on the fluctuations and the equations of state are considered, but also their dynamical behaviour is paid a central attention. Note that both second-order and fourth-order terms in the generalized entropy may contribute to the second-order terms of the second moments of the nonequilibrium fluctuations. However, only the second-order terms in the entropy appear in the second-order nonequilibrium contribution to the equations of state.

We finally note that the nonequilibrium fluctuations of the viscous pressure under shear (Jou and Careta 1982), of the heat flux in some models of solid heat conductors (Jou et al 1982a) and of the electric flux under and electric field (Jou et al 1982b) have been evaluated in EIT. The essential ideas are the ones exposed in this chapter.

We acknowledge the financial support of the Dirección General de Investigación Científica y Técnica of the Spanish Ministry of Educación y Ciencia under grant PB/86-0287.

References

Bird RB, Hassager O, Armstrong Rc and Curtiss CF, 1977,Dynamics of Polymeric liquids (vol 2) New York: Wiley

Callen HB 1960 ThermodynamicsNew York: Wiley

Camacho J and Jou D 1990, On the thermodynamics of dilute dumbell solutions under shear, J Chem Phys 92: 1399

Casas-Vázquez, J, Jou, D and Lebon G, eds 1984 Recent Developments in Nonequilibrium Thermodynamics, Berlin:Springer

Casas-Vázquez J and Jou D 1989, Nonequilibrium equations of state and thermal waves, Acta Phys Hungar 66: 99

Cattaneo C 1958 Sur une forme de l'équation de la chaleur éliminant le paradoxe d'une propagation instantanée, Compt.Rend.Acad.Sci.Paris.247:431

Coleman BD, Fabrizio M and Owen DR 1982, On the thermodynamics of the second sound in dielectric crystals, Arch Rat Mech Anal 80:135

De Groot SR and Mazur P 1962, Nonequilibrium Thermodynamics, Amsterdam: North Holland

Ernst MH 1970, The linearized Boltzmann equation. Navier-Stokes and Burnett transport coefficients, Am.J.Phys, 38:908

García-Colín L,Rodriguez RF, López de Haro M, Casas-Vázquez J and Jou D, On the foundations of extended irreversible thermodynamics J.Stat.Phys 37,465-484

García-Colín LS and López de Haro M 1982, The Burnett equations in extended irreversible thermodynamics, J.Non-Equilib.Thermodyn 7:95

García-Colín LS 1988, Extended nonequilibrium thermodynamics: scope and limitations, Rev.Mex.Física 34:344

Green HS 1969, The Molecular Theory of Fluids, New York: Dover

Glansdorff P and Prigogine I 1971, Thermodynamic Theory of Structure, Stability and Fluctuations, New York: Wiley

Gyarmati I 1970, Nonequilibrium Thermodynamics, Berlin: Springer

Jou D and Careta T 1982, Nonequilbrium thermodynamic fluctuations and a generalized entropy,J.Phys.A 15: 3195

Jou D, Llebot JE and Casas-Vázquez J 1982 a, Irreversible thermodynamic approach to nonequilibrium heat fluctuations, Phys Rev A 25: 508

Jou D, Llebot JE and Casas-Vázquez J 1982 b, Thermodynamic aspects of nonequilibrium current fluctuations, Phys Rev A 25: 3277

Jou D and Micenmacher V 1987, Extended thermodynamics of viscous phenomena in real gases, J.Phys.A 20:5619

Jou D, Casas-Vázquez J and Lebon G 1988, Extended irreversible thermodynamics, Rep.Prog.Phys. 51: 1104

Jou D 1989, Fluctuation theory and extended irreversible thermodynamics, Physica A.155: 221

Joseph DD, Riccius O and Arney M, 1986, Shear-wave speeds and elastic moduli for different liquids, J.Fluid Mechanics 171: 309

Joseph DD and Preziosi L 1989, Heat Waves Rev.Mod.Phys. 61:41

Keizer J 1983, On the relationship between fluctuating irreversible thermodynamics and extended irreversible thermodynamics J.Stat.Phys. 31,485-498

Keizer J 1987, Statistical Thermodynamics of Nonequilibrium Processes, Berlin: Springer

Landau LD and Lifshitz E 1980 Statistical Mechanics (3rd edition, vol 1), Oxford: Pergamon

Landau LD and Lifshitz E 1980 Fluid Mechanics, Oxford: Pergamon

Lax M 1960, Fluctuations from nonequilibrium steady states, Rev.Mod.Phys. 32 25

Lebon G 1990, Extended irreversible thermodynamics of viscoelastic systems, this volume

Llebot JE, Jou D and Casas-Vázquez J 1983, A thermodynamic approach to heat and electric conduction in solids, Physica A 121: 552

Llebot JE 1990, Extended irreversible thermodynamics of electrical phenomena, this volume

Müller I 1985, Thermodynamics, London: Pitman

Nettleton RE 1984, Coupled fluctuations in temperature and heat flux from a non-equilibrium state, J.Chem.Phys. 81: 2458

Pavón D, Jou D and Casas-Vázquez J 1985, Equilibrium fluctuations in relativistic thermoelectric fluids, Ann.Inst.H.Poncairé 42: 31

Pavón D 1990, Astrophysical and cosmological appplications of extended relativistic thermodynamics, this volume

Pérez-García C and Jou D 1990, Thermodynamic aspects of generalized tranport coefficients, this volume

Tremblay AMS 1984,Theories of fluctuations in nonequilibrium systems, in J. Casas-Vázquez, D. Jou and G. Lebon, eds,Recent Developments in Nonequilibrium Thermodynamics, Berlin: Springer

Thermodynamic Aspects of Memory Functions for Transport Coefficients

Carlos Pérez-García* and David Jou
Departamento de Física, Facultat de Ciències
Universitat Autònoma de Barcelona
08193 Bellaterra, Catalonia, Spain

**Departamento de Física y Matemática Aplicada*
Facultad de Ciencias, Universidad de Navarra
31080 Pamplona, Spain

ABSTRACT

It is shown that EIT provides a suitable macroscopic framework for the interpretation of generalized transport coefficients. As specific applications we analyse the form of the longitudinal and the transverse velocity correlation functions, and ultrafast thermometry. The parallelism between the macroscopic development and Mori's microscopic development for the continued- fraction expansion of transport coefficients is also pointed out.

1. Introduction

Some experimental techniques, such as neutron or light scattering, and molecular dynamics simulations provide an insight on the response of the systems to short wavelength and high-frequency perturbations in liquids (Boon and Yip 1980). In the hydrodynamic limit $(k,\omega \to 0)$, classical irreversible thermodynamics (CIT) is a suitable framework for the analysis of such phenomena in macroscopic systems (de Groot and Mazur 1962). In order to interpret in a correct way the information in the wider range of (k, ω) accesible to experiments, the so-called generalized hydrodynamics (GH) replaces the dissipative transport coefficients by general memory functions that are justified by some microscopic models. Although initially this study was limited to simple liquids it can be easily extended to situations where the relaxation phenomena are important (heat conduction in low temperature solids, polymeric liquids, superconductors or superfluids in nonequilibrium, glass-forming systems, etc.)

The memory functions introduced in GH gather phenomenologically the non-local response, both in space and in time, when a perturbation acts on the system. Nevertheless, GH is compatible with CIT only when the memory functions reduce to delta functions. This is a consequence of the *local equilibrium hypothesis* assumed in CIT. Therefore, the question is: what kind of thermodynamics corresponds to the $k \neq 0$ and $\omega \neq 0$ regime, i.e., when the memory functions do not reduce to delta functions? The answer to this question is far from trivial because it needs a thermodynamical framework more general than CIT. In fact the k and ω dependence obtained in experiments cannot be justified by taking ad hoc dependences of the transport coefficients. The linear relations (or their nonlinear generalizations) between thermodynamical fluxes and forces in CIT must be replaced by some new spatio-temporal evolution equations for these fast variables. Some microscopic formulations (Mc Quarrie 1975) exploit these points. But from the macroscopic point of view this can be done in the framework of the so-called extended irreversible thermodynamics (EIT) (Casas-Vazquez et al 1984).

We use in the following the particular formulation of the EIT by Jou, Casas-Vazquez and Lebon (1988), in which the local equilibrium hypothesis is relaxed by assuming that the dissipative fluxes enter as natural variables in the thermodynamical description. By using a development similar to that in CIT, evolution equations for these fluxes are derived, which have the form of the so-called Maxwell-Cattaneo equations (Maxwell 1965; Cattaneo 1948). As shown by several authors, this scheme is coherent with the Grad's thirteen-moments kinetic method (Grad 1958) of solving Boltzmann equation, and with the fluctuation- dissipation theorem in the formulation of Green and Kubo (Mc Quarrie 1975).

In the present paper we provide a phenomenological theory which includes higher-order dissipative fluxes into the thermodynamical description of a nonequilibrium system. This generalization allows to account for some memory functions in GH, and also for some phenomena related to phonon hydrodynamics and ultrafast thermometry (Jou and Pérez-García 1981; Bampi et al. 1981), ultrasound attenuation in metals (Jou et al. 1982), critical phenomena (Zubarev and Tishenko 1972) and also for polymeric systems (Lebon et al. 1986; 1988; Pérez-García et al. 1989). The present development is original, in that it recovers on purely macroscopic grounds results similar, at least in form, to those obtained by microscopic theories. This brings the limits of the nonequilibrium thermodynamics, up to now too restricted to the local equilibrium hypothesis, a new step towards a wider experimental and theoretical domain presently reserved to statistical mechanics.

In section 2 we present a general scheme of the macroscopic formalism and some aspects of its implications in fluctuations around equilibrium states. In section 3 the evolution equations are specified for a simple fluid system. In section 4 the transverse and longitudinal velocity autocorrelation functions are analysed and compared with the memory functions used in GH. In section 5 we show that, under some restrictions, a continued-fraction expansion for the transport coefficients results in this framework. Finally, section 6 is devoted to discuss some new possible developments and applications of this formalism and to the conclusions.

2. General formalism of the Extended Irreversible Thermodynamics

A general problem in dealing with systems out of equilibrium is that one has more variables than equations, because the dissipative fluxes (heat flux, momentum flux, etc.) remain, in principle, unspecified. Usually, in the framework of CIT, this problem is solved simply by taking linear relations between these fluxes and the associated themodynamical forces (Newton, Fourier, Fick, Ohm laws). But it is well known that these are only approximated laws that a) lead to some inconsistencies (for example, an infinite velocity of propagation for thermal signals) and b) do not cover the wide variety of behaviours encountered in nature (for instance, rheology).

In EIT, these variables are considered as independent variables. The starting point in this theory is the existence of a nonequilibrium thermodynamical potential in which the dissipative fluxes are included. (As the general procedure and discussion is described in detail in the preceding article we give here only a general sketch). Often the chosen thermodynamical potential is a generalized entropy per unit mass s, which is postulated to have the form

$$s = s_{eq} + \frac{1}{2} v \sum \alpha_n a_n \overset{n)}{\cdot} a_n \tag{2.1}$$

where s_{eq} is the specific entropy in the equilibrium reference state, v stands for the specific volume and the variables a_n are dissipative fluxes of tensorial order n. The physical meaning of the coefficients α_n (≥ 0) will be explicited later on. The symbol $\overset{n)}{\cdot}$ denotes the total contraction of a_n with a_n. The choice of this bilinear form (2.1) for the nonequilibrium entropy s is based on the Grad development of the Boltzmann equation in which a hierarchy of moments of increasing tensorial order enter in the description (Grad 1958).

Following the general ideas of CIT we assume a general form for the evolution of s

$$\rho(ds/dt) + \nabla\cdot J_s = \sigma_s \tag{2.2}$$

where J_s and σ_s are the entropy flux and entropy production, respectively. For the entropy flux J_s we assume the form

$$J_s = \sum_n \beta_n a_{n+1} \overset{n)}{\cdot} a_n \tag{2.3}$$

where the coefficients β_n will be interpreted in the following. From (2.1)-(2.3) one obtains the following expression for the entropy production

$$\sigma = \sum_n a_n \overset{n)}{\cdot} [\alpha_n (da_n/dt) + \beta_n \nabla\cdot a_{n+1} + \beta_{n-1} \nabla a_{n-1}] \tag{2.4}$$

The positive character of σ is preserved by the simplest choice

$$\alpha_n (da_n/dt) + \beta_n \nabla \cdot a_{n+1} + \beta_{n-1} \nabla a_{n-1} = \mu_n a_n \tag{2.5}$$

with $\mu_n \geq 0$. Then, one obtains linear evolution equations for the fluxes a_n in the form

$$\alpha_n (da_n/dt) = \mu_n a_n + \beta_n \nabla \cdot a_{n+1} + \beta_{n-1} \nabla a_{n-1} \tag{2.6}$$

This set of equations is precisely of the form predicted by kinetic theory (McQuarrie 1975; Roldughin 1984) in the linear approximation, but we stress here that they have been obtained using only macroscopic arguments.

The evolution equations (2.6) are interpreted in the following way. In the particular case $\beta_n = \beta_{n-1} = 0$, (2.6) reduce to relaxation equations and, therefore, α_n/μ_n can be identified as a relaxation time $\tau_n = \alpha_n/\mu_n$. The variable a_{n+1} can be interpreted as the flux of the variable a_n, and therefore one can take $-\beta_n/\rho\alpha_n = 1$. This relations reduce the independent parameters from 3n to 2n: the relaxation times τ_n and the dissipative coefficients μ_n.

On the other hand, it is possible to generalize the Boltzmann - Einstein formula for the probability of a nonequilibrium fluctuation

$$\Pr \propto \exp\ (\delta^2 S/2k) \tag{2.7}$$

k being the Boltzmann constant and $\delta^2 S$ the second differential of the entropy (2.1) integrated to the whole system. In this manner one can deduce the second moments of the dissipative fluxes (Jou et al. 1980)

$$\langle \delta a_n \delta a_n \rangle = \frac{k\ \mu_n}{\tau_n} \tag{2.8}$$

This constitutes a generalization of fluctuation-dissipation theorem linking the dissipative coefficients to equilibrium fluctuations of the corresponding dissipative fluxes.

3. E.I.T. for a fluid system

In this section we apply the general ideas of section 2 to the particular case of a compressible non-conducting fluid. (For the sake of simplicity we assume that the heat flux is zero). Therefore the generalized nonequilibrium entropy takes the form

$$s = s(u, \rho, p^v, \mathring{\mathcal{P}}^v, \mathbf{r}, \mathfrak{R}) \tag{3.1}$$

where u and ρ are the internal energy and density (conserved variables), respectively. As usual the pressure tensor can be split in a conservative part and a dissipative part $\mathcal{P} = p\,\mathcal{U} + \mathcal{P}^v$, where p is the thermodynamic pressure, $\mathcal{U}$ the unitary tensor and the superscript ^{v} denotes the viscous part. The last one can be separated in two independent terms $\mathcal{P} = p^v\,\mathcal{U} + \mathring{\mathcal{P}}^v$, where p^v is related to the trace of the dissipative part of the pressure, and, therefore, it is responsible for the dissipative effects in global compressions or expansions; $\mathring{\mathcal{P}}^v$ is the traceless part of the pressure tensor, related to shear effects. The vector $\mathbf{r}$ and the third order tensor $\mathfrak{R}$, are the corresponding fluxes of the dissipative fluxes p^v and $\mathring{\mathcal{P}}^v$.

In this case the generalized entropy allows to write

$$T\frac{ds}{dt} = \frac{du}{dt} + p\frac{dv}{dt} + \alpha_1 v\, p^v \frac{dp^v}{dt} + \alpha_2 v\, \mathring{\mathcal{P}}^v : \frac{d\mathring{\mathcal{P}}^v}{dt} + $$

$$+ \alpha_3 v\, \mathbf{r}\cdot \frac{d\mathbf{r}}{dt} + \alpha_4 v\, \mathfrak{R} \,\therefore\, \frac{d\mathfrak{R}}{dt} \tag{3.2}$$

where T is the local equilibrium temperature and α_i (i = 1,2,3) are coefficients which depend only on u and v ($v = \rho^{-1}$). The balance equations for ρ, the velocity $\mathbf{v}$ and u are

$$\frac{d\rho}{dt} = -\rho\, \nabla\cdot\mathbf{v}$$

$$\frac{d\mathbf{v}}{dt} = -\nabla p - \nabla p^v - \nabla\cdot\mathring{\mathcal{P}}^v \tag{3.3}$$

$$\frac{du}{dt} = -(p + p^v)\nabla\cdot\mathbf{v} - \mathring{\mathcal{P}}^v : (\nabla^{\circ}\mathbf{v})$$

Following the procedure described in section 2, we take for the entropy flux $\mathbf{J}_s$ the most general vector that can be constructed with the dissipative variables (Pérez-García et al. 1986)

$$\mathbf{J}_s = \beta_1 p^v \mathbf{r} + \beta_2 \mathring{\mathcal{P}}^v : \mathfrak{R} + \beta_3\, \mathring{\mathcal{P}}^v \cdot \mathbf{r} \tag{3.4}$$

where β_i (i = 1,2,3) are functions of u and v only. We now compute the divergence of $\mathbf{J}_s$. To this end we further simplify the calculations by assuming that the β coefficients are constant. By taking the corresponding equation, together with the general entropy balance equation (2.2) and the balance equations (3.3), one arrives to

$$T\sigma_s = -p^v\nabla\cdot\mathbf{v} - \mathcal{P}^v:(\nabla^{\circ}\mathbf{v}) + \alpha_1 p^v \frac{dp^v}{dt} + \alpha_2 \mathcal{P}^v : \frac{d\mathcal{P}^v}{dt} + \alpha_3 \mathbf{r}\cdot\frac{d\mathbf{r}}{dt} +$$
$$+ \alpha_4 \mathfrak{R} \therefore \frac{d\mathfrak{R}}{dt} + \beta_1 \mathbf{r}\cdot\nabla p^v + \beta_1 p^v \nabla\cdot\mathbf{r} + \beta_2 \mathfrak{R} \therefore \nabla\mathcal{P}^v +$$
$$+ \beta_2 \mathcal{P}^v:\nabla\cdot\mathfrak{R} + \beta_3 \mathbf{r}\cdot\nabla\cdot\mathcal{P}^v + \beta_3 \mathcal{P}^v:(\nabla^{\circ}\mathbf{r}) \tag{3.5}$$

As discussed in the preceding article, the simplest form of the scalar σ_s is

$$\sigma_s = \mu_1 p^v p^v + \mu_2 \mathcal{P}^v:\mathcal{P}^v + \mu_3 \mathbf{r}\cdot\mathbf{r} + \mu_4 \mathfrak{R} \therefore \mathfrak{R} \tag{3.6}$$

Since these two expressions for σ_s, (3.5) and (3.6) must be equal to each other, this condition leads to time evolution equations for the nonconserved variables, which, after some rearrangement, read as follow

$$\begin{aligned}
\frac{dp^v}{dt} &= \frac{T\mu_1}{\alpha_1}\left[p^v + \frac{1}{\mu_1 T}\nabla\cdot\mathbf{v} - \frac{\beta_1}{\mu_1}\nabla\cdot\mathbf{r}\right] \\
\frac{d\mathcal{P}^v}{dt} &= \frac{T\mu_2}{\alpha_2}\left[\mathcal{P}^v + \frac{1}{\mu_2 T}(\nabla^{\circ}\mathbf{v}) - \frac{\beta_2}{\mu_2}(\nabla^{\circ}\mathfrak{R}) - \frac{\beta_3}{\mu_2}(\nabla^{\circ}\mathbf{r})\right] \\
\frac{d\mathbf{r}}{dt} &= \frac{T\mu_3}{\alpha_3}\left[\mathbf{r} - \frac{\beta_1}{\mu_3}\nabla p^v - \frac{\beta_3}{\mu_3}\nabla\cdot\mathcal{P}^v\right] \\
\frac{d\mathfrak{R}}{dt} &= \frac{T\mu_4}{\alpha_4}\left[\mathfrak{R} - \frac{\beta_2}{\mu_4}(\nabla\,\mathcal{P}^v)\right]
\end{aligned} \tag{3.6}$$

Notice that consistency with CIT implies that for the particular case β_i $(i = 1,2,3) = 0$

$$\begin{aligned}
p^v &= -\frac{1}{\mu_1 T}\nabla\cdot\mathbf{v} = -\zeta\,\nabla\cdot\mathbf{v} \\
\mathcal{P}^v &= -\frac{1}{\mu_2 T}(\nabla^{\circ}\mathbf{v}) = -2\eta\,(\nabla^{\circ}\mathbf{v})
\end{aligned} \tag{3.7}$$

where ζ and η are the bulk and shear viscosities, respectively. Thus, μ_1 and μ_3 are nonegative quantities. Introducing the relaxation times

$$\tau_i = -\alpha_i/T\mu_i \qquad (i = 1,2,3,4) \tag{3.8}$$

(3.6) can be written in the form

$$\tau_1 \frac{dp^v}{dt} = - p^v - \zeta \nabla\cdot\mathbf{v} + T\zeta\beta_1 \nabla\cdot\mathbf{r}$$

$$\tau_2 \frac{d\dot{\mathcal{P}}^v}{dt} = -\dot{\mathcal{P}}^v - 2\eta(\overset{\bullet}{\nabla}\mathbf{v}) + 2\eta T\beta_2 (\nabla\overset{\bullet}{\cdot}\mathfrak{R}) + 2\eta T\beta_3 (\overset{\bullet}{\nabla}\mathbf{r}) \tag{3.9}$$

$$\tau_3 \frac{d\mathbf{r}}{dt} = -\mathbf{r} + \frac{\beta_1}{\mu_3}\nabla p^v + \frac{\beta_3}{\mu_3}\nabla\cdot\dot{\mathcal{P}}^v$$

$$\tau_4 \frac{d\mathfrak{R}}{dt} = - \mathfrak{R} + \frac{\beta_2}{\mu_4}(\nabla\,\dot{\mathcal{P}}^v)$$

It is important to stress that (3.9) are a consequence of the structure proposed for the space of state variables of the system and the general postulates of EIT only. In this sense they are unquestionable mathematical objects. However, if a physical interpretation is sought for $\mathbf{r}$ and $\mathfrak{R}$, one has to resort either to experiments or to a microscopic theory. In the second case, a moment analysis of the single particle distribution function for a dilute gas obeying the Boltzmann equation shows that $\mathbf{r}$ and $\mathfrak{R}$ play the rọle of the "currents" associated to the quantities p^v and $\dot{\mathcal{P}}^v$, respectively.

A particular case of physical interest is obtained in a quasistatic assumption, which states that τ_3 and τ_4 are much smaller than τ_1 and τ_2 and, hence, the "fast" variables $\mathbf{r}$ and $\mathfrak{R}$ can be eliminated adiabatically from (3.9.c) and (3.9.d), thus leading to

$$\mathbf{r} = \frac{\beta_1}{\mu_3}\nabla p^v + \frac{\beta_3}{\mu_3}\nabla\cdot\dot{\mathcal{P}}^v$$

$$\mathfrak{R} = \frac{\beta_2}{\mu_4}(\nabla\,\dot{\mathcal{P}}^v) \tag{3.10}$$

When introduced into (3.9), this yields

$$\tau_1 \frac{dp^v}{dt} = - p^v - \zeta\nabla\cdot\mathbf{v} + T\zeta\frac{\beta_1^2}{\mu_3}\nabla^2 p^v + T\zeta\frac{\beta_1\beta_3}{\mu_3}\nabla\cdot(\nabla\cdot\dot{\mathcal{P}}^v)$$

$$\tau_2 \frac{d\dot{\mathcal{P}}^v}{dt} = - \dot{\mathcal{P}}^v - 2\eta(\overset{\bullet}{\nabla}\mathbf{v}) + 2\eta T\frac{\beta_2^2}{\mu_4}\nabla^2\dot{\mathcal{P}}^v + 2\eta T\frac{\beta_1\beta_3}{\mu_3}(\overset{\bullet}{\nabla}\mathbf{r}) +$$

$$+ 2\eta T\frac{\beta_3^2}{\mu_3}\overset{\bullet}{[\nabla(\nabla\cdot\dot{\mathcal{P}}^v)]^s} \tag{3.11}$$

If we simplify further by assuming that $\beta_3 = 0$ these equations can be written in a form like linearized Ginzburg-Landau equations

$$\tau_1 \frac{dp^v}{dt} = - p^v - \zeta\nabla\cdot\mathbf{v} + \ell_1^2 \nabla^2 p^v$$
$$\tau_2 \frac{d\overset{\circ}{\mathcal{P}}{}^v}{dt} = - \overset{\circ}{\mathcal{P}}{}^v - 2\eta(\overset{\circ}{\nabla}\mathbf{v}) + \ell_2^2 \nabla^2 \overset{\circ}{\mathcal{P}}{}^v \qquad (3.12)$$

where $\ell_1^2 = \zeta T\beta_1^2/\mu_3$ and $\ell_2^2 = 2\eta T\beta_2^2/\mu_4$ can be identified with correlation lenghts that account for non-local effects in space of the corresponding variables. Equations of this kind have been used to study ultrafast heat conduction (Pérez-García and Jou 1980; Jou and Pérez-García 1981; Bampi et al. 1981) and ultrasonic absorption in metals (Jou et al. 1982).

4. Velocity autocorrelation functions

We shall now proceed to calculate the time autocorrelation functions of the velocity fluctuations around equilibrium from the general equations (3.3) and the relaxation equations (3.9). The velocity fluctuation around a reference state is defined as

$$\delta\mathbf{v} = \mathbf{v} - \langle\mathbf{v}\rangle \qquad (4.1)$$

where brackets denote an equilibrium average. This fluctuations are split in two components

$$\delta\mathbf{v} = \delta v_{\parallel}\mathbf{a} + \delta v_{\perp}\mathbf{b} \qquad (4.2)$$

where **a** and **b** are vectors parallel and perpendicular to the propagation direction of fluctuations. Corresponding to these components one can define longitudinal and transverse correlation functions in the form

$$C_{\parallel}(\mathbf{r},t;\mathbf{r}',t') = \langle\delta v_{\parallel}(\mathbf{r},t)\,\delta v_{\parallel}(\mathbf{r}',t')\rangle$$
$$C_{\perp}(\mathbf{r},t;\mathbf{r}',t') = \langle\delta v_{\perp}(\mathbf{r},t)\,\delta v_{\perp}(\mathbf{r}',t')\rangle \qquad (4.3)$$

If the fluid considered is also isotropic it will exhibit spatial and temporal invariance properties and these correlations functions write as

$$C_{\parallel,\perp}(\mathbf{r},t;\mathbf{r}',t') = C_{\parallel,\perp}(|\mathbf{r}-\mathbf{r}'|,t-t') \qquad (4.4)$$

Moreover, for the general discussion and actual calculations it is convenient to use the spatial Fourier transform of $C_{\parallel,\perp}(\mathbf{r},t)$, which is taken as

$$J_{\parallel,\perp}(\mathbf{k},t) = \int d\mathbf{r}\ C_{\parallel,\perp}(\mathbf{r},t) \exp(i\mathbf{k}\cdot\mathbf{r}) = \langle \delta v_{\parallel,\perp}(\mathbf{k},t)\ \delta v_{\parallel,\perp}(-\mathbf{k},t)\rangle \tag{4.5}$$

For the sake of clarity we separate the calculations for transverse and longitudinal parts.

A. Transverse velocity autocorrelation function

It is important to point out that although this correlation function cannot be obtained in experiments. its calculation and analysis is worth pursuing since, on one hand, it has been extensively studied by computer simulations (Boon and Yip 1980) and, on the other hand, it plays an essential role in establishing the connection between EIT and GH (Pérez-García 1984; Jou et al. 1985)

The transverse part of (3.3.c) is simply

$$dv_{\perp}/dt = -\nu(\nabla\cdot\dot{\mathcal{P}}^{v})_{\perp} \tag{4.6}$$

where $\nu = \eta/\rho$ is the kinematic viscosity. We notice that this component is decoupled from density or energy variations (which are related only with the longitudinal component, in the linear approximation). We start with the simple particular case that leads to the relaxation equation (3.12.b). Taking the transverse part of this equation and the definition (4.5) we arrive to

$$\tau_2 \frac{\partial^2}{\partial t^2} J_{\perp} + (1 + \ell_2^2 k^2) \frac{\partial}{\partial t} J_{\perp} = -\nu\ k^2\ J_{\perp} \tag{4.7}$$

The initial value $J_{\perp}(k,0)$ is given by the equipartition theorem, whereas $\partial J_{\perp}(k,0)/\partial t$ vanishes identically owing to the general time-reversal invariance properties of any equilibrium autocorrelation function. By taking the Laplace transform in time, eq. (4.7) may be rewritten in the form

$$\frac{\hat{J}_{\perp}(k,s)}{J_{\perp}(k,0)} = \frac{\tau_2 s + (1 + \ell_2^2 k^2)}{\tau_2 s^2 + (1 + \ell_2^2 k^2)s + \nu k^2} \tag{4.8}$$

which in the limit τ_2, $\ell_2 \to 0$ leads to the classical expression

$$\hat{J}_{\perp}(k,\ s) = J_{\perp}(k,\ 0)\ \frac{1}{s + \nu k^2} \tag{4.9}$$

The present analysis may be compared with that based on the microscopic Mori-Zwanzig theory (Berne and Pecora 1976) which gives an exact formal equation for the time evolution of a correlation function in terms of a memory function, namely

$$\frac{\partial}{\partial t}J_\perp(k, t) = -\int_0^t K_\perp(k, t-t')J_\perp(k, t')\, dt' \tag{4.10}$$

It should be remarked that the memory kernel $K_\perp(k,t)$ has a well-defined and unique expression in terms of projection operators. It is usual to write this kernel in the form

$$K_\perp(k, t) = k^2 K'_\perp(k,t) \tag{4.11}$$

In this case $K'_\perp(k,t)$ is called the *shear viscosity function* or the memory function for the transverse velocity autocorrelation. Usually this memory function is modelled mathematically. However, in our case we can compare the Laplace transform of (4.11) with the expression (4.8) to have

$$\hat{K}'(k, s) = \nu\,[\tau_2 s + (1 + \ell_2^2 k^2)]^{-1} \tag{4.12}$$

or antitransforming

$$K'_\perp(k, t) = \nu\tau_2^{-1} \exp\,[-\tau_2^{-1}(1 + \ell_2^2 k^2)\, t] \tag{4.13}$$

This equation shows that the relaxation time for the memory function is given by

$$\tau_\perp(k) = \tau_2(1 + \ell_2^2 k^2)^{-1} \tag{4.14}$$

The memory kernel $K'_\perp(k, t)$ must satisfy some general restrictions, the so-called *sum rules*. The second sum rule implies that (Copley and Lovesey 1975)

$$K'_\perp(k, t=0) = \rho^{-1} G_\infty(k) \tag{4.15}$$

where $G_\infty(k)$ is the high-frequency shear modulus. Also consistency with hydrodynamics requires that

$$\lim_{k\to 0}\lim_{s\to 0} \hat{K}'(k,s) = \nu \tag{4.16}$$

This general condition is obviously satisfied by (4.12). On the other hand, from (4.13) and (4.15) we find

$$G_\infty(k) = \eta\, \tau_2^{-1} \tag{4.17}$$

and since τ_2 is independent of k, in our case the high-frequency shear modulus must be k-independent as well. Furthermore, the ratio $\tau_2^{-1} = G_\infty(0)/\eta$ is identified with the Maxwell relaxation time of the usual viscoelastic theory.

The model (4.12) is precisely the exponential model introduced in GH for the memory function which allows shear-wave propagation. The difference is that here this model arises directly from the postulates of EIT.

In recent computer simulations (Alley and Alder 1983) the generalized shear viscosity coefficient $\eta(k,s)$ of a fluid of hard spheres have been computed. They show that the zero-frequency and small k molecular dynamics data may be fitted with an expression of the form

$$\eta(k,0) = \eta(0,0)[1 + a^2k^2]^{-1} \tag{4.18}$$

where a is an adjustable parameter with the dimensions of length. A physical interpretation for this parameter is obtained by considering a test particle of variable size immersed in the fluid. In this context, a turns out to be a measure of the deviation from the Stokes friction law

$$\mathcal{F}_{drag}/\mathcal{F}_{Stokes} \cong 1 - a\, \sigma_{12}^{-1} \tag{4.19}$$

where $\sigma_{12} = (\sigma_1 + \sigma_2)/2$ is the average diameter of the test particle and a bath particle. In this sense a incorporates the effects of spatial inhomogeneities into the hydrodynamical description. We notice that with (4.12)

$$\nu(k,0) \equiv \rho^{-1}\eta(k,0) = \lim_{s\to 0} \hat{K}'(k,s) = \nu(1 + \ell_2^2k^2)^{-1} \tag{4.20}$$

Therefore, ℓ_2 has the same sense as a, but it has been obtained on purely macroscopic grounds.

B. <u>Longitudinal velocity autocorrelation functions</u>

In this case the memory function is accesible experimentally, because the longitudinal part of the velocity correlation function is easily related with the dynamic structure factor S(k,s), i.e., the density autocorrelation function that is accesible from light and neutron scattering experiments (Forster 1975; Berne and Pecora 1976).

The hydrostatic pressure and the internal energy variations can be written in the form

$$dp = \frac{1}{\rho \varkappa_T} d\rho + \frac{\alpha}{\varkappa_T} dT$$
$$du = -\left(\frac{\alpha T}{\varkappa_T} - p \right) \frac{d\rho}{\rho^2} + c_v dT \tag{4.21}$$

where $\varkappa_T$ is the isothermal compressibility coefficient, α the thermal expansion coefficient, c_v the specific heat at constant volume. Considering the fluctuations around an equilibrium state, $\rho = \rho_o + \rho'$, $T = T_o + T'$, $\mathbf{v} = \mathbf{v}'$, and linearizing around equilibrium (3.3) and (4.21) lead to

$$\frac{\partial \rho'}{\partial t} = - \rho_o \nabla \cdot \mathbf{v}'$$
$$\rho_o \frac{\partial \mathbf{v}'}{\partial t} = - \frac{1}{\rho_o \varkappa_T} \nabla \rho' - \frac{\alpha}{\varkappa_T} \nabla T' - \nabla p^v - \nabla \cdot \overset{\circ}{\mathcal{P}}{}^v \tag{4.22}$$
$$\rho_o c_v \frac{\partial T'}{\partial t} = - \frac{a T_o}{\varkappa_T} \nabla \cdot \mathbf{v}$$

Now for the sake of simplicity we drop the primes and we take the spatial Fourier transform of these quantities (denoted by a subindex $_k$) in (4.22). These expressions are rewritten in the form

$$\frac{\partial \rho_k}{\partial t} = -\rho_o \mathbf{k} \cdot \mathbf{v}_k$$
$$\rho_o \frac{\partial \mathbf{v}_k}{\partial t} = \frac{1}{\rho_o \varkappa_T} \mathbf{k} \rho_k - \frac{\alpha}{\varkappa_T} \mathbf{k} T_k - \mathbf{k} p_k^v - \mathbf{k} \cdot \overset{\circ}{\mathcal{P}}{}_k^v \tag{4.23}$$
$$\rho_o c_v \frac{\partial T_k}{\partial t} = - \frac{a T_o}{\varkappa_T} \mathbf{k} \cdot \mathbf{v}$$

In order to simplify the calculations we take the particular case given in (3.12), that, after linearization and applying the spatial Fourier transform reads as

$$\tau_1 \frac{\partial p_k^v}{\partial t} = - p_k^v - \zeta \mathbf{k} \cdot \mathbf{v}_k + \ell_1^2 k^2 p_k^v$$

$$\tau_2 \frac{\partial \dot{\mathcal{P}}^{v}_k}{\partial t} = - \dot{\mathcal{P}}^{v}_k - 2\eta(\mathbf{k}\cdot\mathbf{v}_k) + \ell_2^2 k^2 \dot{\mathcal{P}}^{v}_k \tag{4.24}$$

The longitudinal part is simply obtained by projection on the direction of $\mathbf{k}$, i.e., $f_{\parallel} = \mathbf{f}_k \cdot \mathbf{k}/|\mathbf{k}|$. The quantity considered here is the longitudinal velocity autocorrelation function $J_{\parallel}(\mathbf{k}, t)$

$$J_{\parallel}(\mathbf{k}, t) = \langle v_{\parallel}(\mathbf{k},t)\, v_{\parallel}(-\mathbf{k}, t)\rangle \tag{4.25}$$

Following an analysis similar to that in the transverse case and after some lengthy algebra that includes the Laplace temporal transform of these equations, one arrives to

$$\frac{\hat{J}_{\parallel}(k,s)}{J_{\parallel}(k,0)} = s + \frac{k^2}{\varkappa_T \rho_o s}\left[1 + \frac{\alpha^2 T_o}{\rho_o \varkappa_T c_v}\right] + \frac{k^2}{\rho_o}\left[\frac{\zeta}{\tau_1' s + 1} + \frac{4\eta}{3(\tau_2' s + 1)}\right] \tag{4.26}$$

where $\tau_i' = \tau_i(0)(1 + \ell_i^2 k^2)^{-1}$ $(i = 1, 2)$. This expression can be further simplified by taking into account the thermodynamical relation

$$\gamma \rho_o c_T^2 = \frac{1}{\varkappa_T \rho_o}\left[1 + \frac{\alpha^2 T_o}{\rho_o \varkappa_T c_v}\right] \tag{4.27}$$

where $\gamma = c_p/c_v$ is the adiabatic coefficient and $c_T = (\rho_o \varkappa_T)^{-2}$ the isothermal sound propagation velocity.

In the GH formalism the memory function associated with the longitudinal velocity autocorrelation function is known to satisfy the equation (Boon and Yip 1980)

$$\hat{J}_{\parallel}(k, s) = J_{\parallel}(k, 0)\,[s + k^2 \hat{K}_{\parallel}'(k, s)] \tag{4.28}$$

Where $\hat{K}_{\parallel}'(k, s)$ is the memory function for the longitudinal velocity autocorrelation function. Comparison of this formal expression and equation (4.26) allows to identify in our case

$$\hat{K}_{\parallel}'(k, s) = c_T^2 \frac{\gamma}{s} + \frac{1}{\rho}\left[\frac{\zeta}{\tau_1' s + 1} + \frac{4\eta}{3(\tau_2' s + 1)}\right] \tag{4.29}$$

from where it follows that

$$K_{\parallel}'(k,\ t) = \gamma c_T^2 + \frac{\zeta}{\rho_0 \tau_1'(k)} \exp\left[- \frac{t}{\tau_1'(k)}\right] + \frac{4\eta}{3\rho_0 \tau_2'(k)} \exp\left[- \frac{t}{\tau_2'(k)}\right] \tag{4.30}$$

This shows explicitely that in the absence of heat flux there are two exponentially decaying modes associated with the bulk and shear viscosities, respectively, each one characterized by a k-dependent relaxation time. This equation is far simpler than the one proposed in GH although the general idea that $K_{\parallel}(k,\ t)$ is governed by two relaxation time model is recovered (Boon and Yip 1980).

5. Continued fraction expansions for transport coefficients

After the discussion of the correlation functions we return to the general scheme introduced in section 2. After taking the Fourier transform, both in space and time, in (2.6), one can write the correlation function of the fluctuations δa_1

$$C a_1(k,\ \omega) = \langle \delta a_1(k, w)\ \delta a_1(-k, -w) \rangle \tag{5.1}$$

in the form of a continued fraction given by (Pérez-García and Jou 1986)

$$C a_1(k,\ \omega) = \cfrac{\langle \delta a_1(0,0)\ \delta a_1(0,0) \rangle}{i\omega\alpha_1 + \mu_1 + \cfrac{\beta_1 k^2}{i\omega\alpha_2 + \mu_2 + \cfrac{\beta_2 k^2}{i\omega\alpha_3 + \mu_3 \ldots}}} \tag{5.2}$$

A. Comparison with the Mori expansion

This form is similar to that obtained by Mori (1965) with a very different method. The projection-operator technique developed by Mori and Zwanzig (Mc Quarrie 1976) shows that, starting from the Liouville equation, the evolution equations for a dynamical variable may be exactly written in the form

$$\frac{da}{dt} - i\Omega a(t) + \int_0^t \mathcal{K}(t-t')a(t')dt' = f(t) \tag{5.3}$$

which is known as the generalized Langevin equation. The random force f(t) introduced in (5.3) is in fact a dynamical variable resulting from the projection of the dynamics of a into a subspace of the fast variables of the system. Though it has not a properly stochastic character, this is assigned to it by analogy with the Langevin equation of Brownian motion. Then the stochastic character of f(t) is not a consequence of an exact equation, but just a choice in (5.3). From this equation it may be seen that

$$\mathcal{K}(t-t') = \langle f(t)f^*(t')\rangle \tag{5.4}$$

This relation is a form of the fluctuation-dissipation theorem (Forster 1975) and may be deduced from the projection-operator technique. Furthermore, if f(t) is a white noise, $\mathcal{K}(t)$ turns out to be a delta function in time, but if this is not the case, (5.3) is a non-Markovian stochastic equation. It is known that such an equation way be written as a hierarchy of Markovian equations if one adds more variables in the description. Mori (1965) has shown that f(t) obeys in turn an exact generalized equation of the form (5.3). Therefore, one arrives to a hierarchy of equations

$$\frac{df_n}{dt} - i\Omega f_n(t) + \int_0^t \mathcal{K}(t-t')f_n(t')dt' = f_{n+1}(t) \tag{5.5}$$

where $f_{n+1}(t)$ is the corresponding noise of the $f_n(t)$ variable and Ω_n and $\mathcal{K}_n$ the corresponding frequency and memory functions. Then, one has an infinite set of equations that give more and more detailed description of the system. This is equivalent to adding more and more variables in the description. It may be truncated in some step by assuming the Markovianicity of the last equation, which closes the hierarchy. If this closure is made at the n-th equation, the description of the system is based on the independent variables (a, $f_1, \ldots, f_n$) with a white noise f_{n+1}. This Markovian assumption on $f_{n+1}(t)$ is equivalent to giving up the possibility of knowing more details on the system: when we cannot keep up the description, we close the hierarchy of evolution equations assuming the Markovian character of the last "random force".

Such an idea is reflected in the continued-fraction due to Mori (1965). From the Fourier transform of (5.3) and (5.5) one can write for the correlation function of a ($C_a(\omega) = \langle a(\omega)a^*(0)\rangle$) the following expansion

$$C_a(\omega) = \cfrac{\langle a(0)\, a^*(0)\rangle}{i\omega - i\Omega_0 + \cfrac{\mathcal{K}_0(0)}{i\omega - i\Omega_1 + \cfrac{\mathcal{K}_1(0)}{i\omega - i\Omega_2 \ldots}}} \tag{5.6}$$

where $\mathcal{K}_i(0)$ are the memory functions at $\omega = 0$, which can be expressed by means of equilibrium correlation functions at equal times. Of course a similar expansion is valid for each memory function $\mathcal{K}_j(\omega)$ (Copley and Lovesey 1975), because, as we have said, $\mathcal{K}_j$ is in turn a time correlation function, that of f_{n+1}. Note that when the property in question has a well defined time-reversal parity, the frequency terms Ω_i vanish in equilibrium. If we assume that f_{n+1} is a white noise, $\mathcal{K}_n(t) = \lambda_n\delta(t)$ and the continued-fraction expansion truncates at the n-th level.

However, although some similarities exist, the two continued-fraction expansions (5.2) and (5.6) differ on fundamental grounds. The first has been obtained from macroscopic arguments and macroscopic variables and no hypothesis has been made about the random nature of the variables, though a particular hydrodynamical form is assumed. The second is fully general, but it is a formidable task to calculate the coefficients from first principles. Moreover, though we can take $i\Omega_0 = -\mu_1/\alpha_1$ to make expressions (5.2) and (5.6) more similar, this identification is inconsistent from the microscopic point of view because the term $i\Omega_0$ has a reversible meaning, while the relaxation terms in (5.2) are clearly irreversible. In fact the Mori expansion is more general because it does not specify the dependence of the $\mathcal{K}_n$ in terms of k, although it seems very difficult to determine this dependence from first principles.

B. Some examples

Now the formalism of the continued fraction expansion is applied to some examples. We start with a rigid conductor where a heat flux $\mathbf{q}$ exists. Following the general analysis of section 2 one arrives to the following couple of equations for the heat flux $\mathbf{q}$ and for the higher order flux Q (Pérez-García and Jou 1986)

$$\tau_1 \frac{d\mathbf{q}}{dt} = -\mathbf{q} - \lambda\,\nabla T + \gamma_1\,\nabla\cdot Q$$
$$\tau_2 \frac{dQ}{dt} = -Q + \gamma_2\,\nabla\,\mathbf{q} \qquad (5.7)$$

where λ is the thermal conductivity and now τ_1 and τ_2 are the relaxation times of $\mathbf{q}$ and Q, respectively, and the coefficients γ, analogous to the β/μ in eqs. (2.6) accounts for the coupling between these variables.

In this particular case the fluctuation-dissipation theorem states that

$$\hat{\lambda}\,(k,\omega) = (kT^2)^{-1}\langle\delta\mathbf{q}(k,\omega)\delta\mathbf{q}(-k,-\omega)\rangle \qquad (5.8)$$

On the other hand, taking the longitudinal part of the double Fourier transform of eqs (5.7) one arrives to

$$i\omega\tau_1\hat{q}_{\parallel} = -\hat{q}_{\parallel} - i\gamma_1 k\,\hat{Q}_{\parallel} + i\lambda_0 k\hat{T}$$
$$i\omega\tau_2\hat{Q}_{\parallel} = -\hat{Q}_{\parallel} - i\gamma_2 k\hat{q}_{\parallel} \qquad (5.9)$$

The expression obtained from these equations can be compared with a generalized Fourier law

$$\hat{q}_{\parallel} = i\, k\, \hat{\lambda}(k,\omega)\, \hat{T} \tag{5.10}$$

to give

$$\hat{\lambda}(k,\omega) = \frac{\lambda_0 \tau_1^{-1}}{i\omega + \tau_1^{-1} + \dfrac{\gamma_1 \gamma_2 \tau_2^{-1} \tau_1^{-1} k^2}{i\omega + \tau_2^{-1}}} \tag{5.11}$$

Of course, this equation has the form of the first terms in the development of a continued-fraction expansion for $\lambda\,(k,\omega)$, that can be easily completed to a form like (5.2) by adding higher moment variables in the description.

An interesting feature in the case of the heat conductivity is that in the frequencies or the relaxation times go to zero the continued fraction expansion for $\lambda\,(k)$ reduces to

$$\lambda(k) = \frac{\lambda_0}{1 + \dfrac{\gamma_1 \gamma_2 k^2}{1 + \dfrac{\gamma_2 \gamma_3 k^2}{1 + \dots}}} \tag{5.12}$$

which in the first approximation ($\gamma_i = 0$ for $i \geq 3$) has the form obtained by Alley and Alder (1983) in molecular dynamics simulations. The form differs considerably from the standard power expansion in the Chapman-Enskog method (Chapman and Cowling 1970)

$$\lambda(k) = \lambda_0 (1 + \Lambda_1 k^2 + \Lambda_2 k^4 + \Lambda_3 k^6 + \dots) \tag{5.13}$$

where the Λ_i are higher-order Burnett coefficients. The convergence of this series for transport coefficients presents some difficulties (Dorfmann 1980), whereas (5.12) is well defined for all values of k.

Starting from a method due to Waldmann (1963), which is a modification of Grad's (1958), a similar expression for the diffusion coefficients has been obtained by Hess (1977). This author shows that in this particular case the continued-fraction expansion has a fast convergence towards the known solutions, whereas a power expansion as (5.13) for the diffusion coefficients does not show this feature.

The practical applications of this formalism can be found in the domain of ultrafast thermometry, in experiments on temperature waves (second sound) and in phonon hydrodynamic theory of solids (Chester 1966; Guyer and Krumhansl 1966; Roger 1971; Beck et al 1974; Fekete 1981). However, the present experimental data are not sufficient to decide what is the most suitable model to describe these phenomena.

The time-dependent form of (5.12), namely

$$\hat{\lambda}(k,\omega) = \cfrac{\lambda_o}{1 + i\omega\tau_1 + \cfrac{\ell_1^2 k^2}{1 + i\omega\tau_2 + \cfrac{\ell_2^2 k^2}{1 + i\omega\tau_3 + \dots}}} \tag{5.14}$$

with $\ell_i^2 \equiv \gamma_i \gamma_{i+1}$ is relevant in the analysis of thermal waves, a subject which as received a wide attention (Jou et al. 1988; Joseph and Preziosi 1989). The energy balance equation may be written in terms of $\lambda(k,\omega)$ as

$$\rho c i\omega \hat{T}(k,\omega) = -k^2 \hat{\lambda}(k,\omega) \hat{T}(k,\omega) \tag{5.15}$$

Introduction of (5.14) into (5.15) leads to the dispersion relation for thermal waves. The equation for the high-frequency limit of the phase speed (i.e., the speed of thermal pulses), depends on the order of approximation. At the first order ($\ell_n = 0$ for $n \geq 1$), second order ($\ell_n = 0$ for $n \geq 2$) and third order ($\ell_n = 0$ for $n \geq 3$), the corresponding expressions are

$$\begin{aligned} v_{p_1}^2(\infty) &= \frac{\lambda}{\rho c \tau_1} \\ v_{p_2}^2(\infty) &= \frac{\lambda}{\rho c \tau_1} + \frac{\ell_1^2}{\tau_1 \tau_2} \\ v_{p_2}^4(\infty) &= \left[\frac{\lambda}{\rho c \tau_1} + \frac{\ell_1^2}{\tau_1 \tau_2} + \frac{\ell_2^2}{\tau_2 \tau_3}\right] v_{p_2}^2(\infty) - \frac{\lambda}{\rho c \tau_1} \frac{\ell_2^2}{\tau_2 \tau_3} \end{aligned} \tag{5.15}$$

These expressions are of interest because they show the conditions on which present analyses of thermal waves are valid. A simple relaxational equation for the heat flux, of the form of the Maxwell-Cattaneo law, is only valid when there is a temporal domain in which $\tau_1 > \omega^{-1} >> \tau_n$ $(n \geq 2)$. This is so in some dielectric crystals at low temperature. However, this is not so in monatomic ideal cases, where all the relaxation times are of the same order as the mean collision time. Thus, for an analysis of thermal waves in monatomic gases one should consider the asymptotic expansions for (5.14) in the limit of an infinite order of approximation. This is not an easy task, yet to be done.

Finally, let us remark that, although in this subsection we have treated the heat conduction case, a similar analysis can be repeated to obtain a continued fraction expansion for the shear (Pérez-García and Jou 1985) and bulk viscosities, or for the electrical conductivity (Jou et al 1988).

6. Conclusions and new perspectives.

Several of the points raised in this paper deserve some additional comments stressing their physical content. The linearized equations of motion of the state variables in section 2 were shown to give rise to some results of GH as it is handled at present, consistent with experimental and computer simulation results. The ability to obtain this specific information on memory functions from thermodynamic arguments only is *per se* an interesting feature. The extreme difficulty of developing explicit results from the statistical mechanical expressions must be circumvented by means of modelling, for which EIT could provide some systematic criteria.

In this paper we have shown the possibility to interpret from purely thermodynamical arguments, based on a natural generalization of the usual EIT (Jou et al. 1988), memory functions and continued-fraction expansions for transport coefficients closer to the present microscopic developments than those achieved by CIT, based on the local equilibrium hypothesis.

The idea that some of the dynamics in macroscopic system can be described for some *fractal processes* opens some new perspectives in the case treated here, because our description is based on a *hierarchical* dynamics, based on including new variables at different scales which are self-similar in form. These ideas are under development to treat glass-forming and polymer systems, and seems a promising topic for future investigations.

Another interesting point for future developments is the possibility to include nonlinearities in the present developments of EIT in order to describe from thermodynamics grounds some rheological models that are in good agreement with experiments, and to analyse their consequences on the dynamics and statistics of nonequilibrium steady states.

Acknowledgments

We thank Professors J. Casas-Vázquez (Universitat Autònoma de Barcelona), L. S. García-Colín (Universidad Autónoma de México), G. Lebon (Université de Liège) and M. López de Haro (Universidad Autónoma de México) for valuable suggestions and comments. This work has been partially supported by the DGICYT (Spanish Government) under grants PB-86-0287 and CE89-0002. One of us (C.P-G) has been also supported by a grant of the European Community (SCI-0035-C(CD)).

References

Alley, W.C.and B.J. Alder. 1983. Phys. Rev. **A27** : 3158

Bampi, F., A. Morro and D. Jou. 1981. Physica **107A** : 393

Beck, H., P.F. Meier and A. Tellung. 1974. Phys. Status Solidi **a24** : 11

Berne, B.J. and R. Pecora. 1976. **Dynamic Light Scattering with Applications to Chemistry, Biology and Physics**. New-York: Wiley.

Boon, J.P. and S. Yip. 1980. **Molecular Hydrodynamics**. New York: McGraw-Hill.

Casas-Vázquez, J., D. Jou and G. Lebon, eds. 1984. **Recent Developments in Nonequilibrium Themodynamics**. Lecture Notes in Physics **199**. Berlin: Springer.

Cattaneo, C. 1948. Atti Sem. Mat. Fis. Univ. Modena **3** : 3

Chapman, S. and T.G. Cowling. 1970. **The Mathematical Theory of Nonuniform Gases**. Cambridge: Cambridge Univ. Press.

Chester, M. 1966. Phys. Rev. **145** : 76

Copley, J.R.D. and S.W. Lovesey. 1975. Rep. Prog. Phys. **38** : 461

De Groot, S.R. and P. Mazur. 1962. **NonEquilibrium Thermodynamics**. Amsterdam : North Holland.

Dorfmann, J.R. 1980. In **Systems far from Equilibrium**, ed., L.M. Garrido. Berlin : Springer.

Fekete, D. 1981. Physica Status Solidi b **105** : 161

Forster, D. 1975. **Hydrodynamic Fluctuations, Broken Symmetries and Correlation Functions.** Reading, MA : Benjamin.

Grad, H. 1958. **Principles of Kinetic Theory of Gases** (Handbuch der Physik **12**), ed,. S. Flugge. Berlin : Springer.

Guyer R.A. and R. Krumhansl. 1966. Phys. Rev. **148** : 766

Hess, S. 1977. Z. Naturf. **A32** : 678

Joseph, D.D. and Preziosi L. 1989. Rev. Mod. Phys. **61** : 41

Jou, D., M. Rubí and J. Casas-Vázquez. 1980. Physica **101A** : 588

Jou, D. and C. Pérez-García. 1980. Physica **104A** : 320

Jou, D., F. Bampi and A. Morro. 1982. J. Non-Equilib. Thermodyn. **7** : 201

Jou, D., C. Pérez-García, L. García-Colín, M. López de Haro and R.F. Rodríguez. 1985. Phys. Rev **A31** : 2502

Jou,D., J. Casas-Vázquez and G. Lebon. 1988. Rep. Prog. Phys. **51** :1105

Lebon, G., C. Pérez-García and J. Casas-Vázquez. 1986. Physica **137A** : 531

Lebon, G., C. Pérez-García and J. Casas-Vázquez. 1988. J. Chem. Phys. **88** : 5068

Mc Quarrie,D. 1975. **Statistical Mechanics**. New-York : Harper and Row.

Maxwell, J.C. 1965. **Collected Papers, vol 2** ed. M.D. Niven. New-York : Dover.

Mori, H. 1965. Progr. Theor. Phys. **33** : 432; **34** : 399

Pérez-García,C. and D. Jou. 1980. Phys. Lett. **75A** : 469

Pérez-García, C. 1984. In **Recent Developments in Nonequilibrium Themodynamics**, eds., J.Casas-Vazquez, D. Jou and G. Lebon. Berlin : Springer

Pérez-García, C. and D. Jou. 1985. Phys. Lett. **107A** : 390

Pérez-García.C, D. Jou, M. López de Haro, R.F. Rodríguez and L. García-Colín. 1986. Physica **135A** : 251

Pérez-García,C. and D. Jou. 1986. J. Phys. A: Math. Gen. **19** : 2881

Pérez-García, C., J. Casas-Vázquez and G. Lebon. 1989. J. Polym. Sci B: Polym. Phys. **27** : 1807

Roger, S.J. 1971. Phys. Rev. B **3** : 1440
Roldughin, V.I. 1984. J. Nonequilib. Thermodyn. **9** : 71
Waldmann, L.1963. Z. Naturf. **18A** :1033
Zubarev, D.N. and S.V. Tishenko. 1972. Physica **59A** : 285

Extended Irreversible Thermodynamics of Rheological Materials

G. Lebon
Liège University
Institute of Physics
B5 Sart Tilman, B-4000 Liège, Belgium

ABSTRACT.
The purpose of the present work is to show that extended irreversible thermodynamics may be viewed as an unifying formalism which is able to generate a wide variety of viscoelastic constitutive relations. The first part of this paper is concerned with linear viscoelasiticity; it is shown that extended irreversible thermodynamics provides a natural thermodynamic framework for describing the classical models of Maxwell, Kelvin-Voigt, Poynting-Thomson and Jeffreys. The relation with the hidden variable theory is also briefly commented. The second part of the work is devoted to an analysis of the linearized Rouse and Zimm models. It is examined to what extent extended irreversible thermodynamics is capable to cope with a relaxational spectrum of viscous modes as introduced by Rouse and Zimm. In the last part, a non linear analysis encompassing second order non-Newtonian fluids is presented. It is in particular shown that a wide class of models, like these of Rivlin-Ericksen, Maxwell generalized, Giesekus may be derived in a simple and coherent way from the postulates of extended irreversible thermodynamics.

TABLE OF CONTENTS.

1. INTRODUCTION

Extended irreversible thermodynamics (EIT) has met a growing interest during the last decade as recalled earlier in this book (Sieniutycz and Salomon 1990, Casas and Jou 1990). The pillars on which the theory is founded can be summarized as follows.

1) Extension of the space of state variables by including non-conserved dissipative fluxes (like the heat flux, the flux of matter, the viscous pressure tensor, ...) in addition to the classical conserved state variables like mass, momentum and energy.

2) Formulation of evolution equations for these new flux-like variables, these equations generalize the usual steady Fourier, Fick and Newton constitutive laws.

3) Restrictions on the form of the evolution equations are placed by the second law of thermodynamics, the convexity property of entropy and the objectivity requirement.

The purpose of the present work is to show that extended irreversible thermodynamics may be viewed as a unifying formalism which is able to generate a wide variety of constitutive equations useful in rheology.

In ordinary incompressible fluids the flow and transport phenomena are fairly well described by Newton's linear constitutive equation

$$P_{ij} = p\delta_{ij} - 2\eta V_{ij} \quad , \tag{1.1}$$

the notation is classical : P_{ij} is the pressure tensor, δ_{ij} the Kronecker symbol, p the hydrodynamic pressure, V_{ij} the symmetric velocity gradient tensor, η the viscosity, which may depend on the temperature and the pressure, but not on V_{ij}. However, it was observed that there exists a wide class of materials, like polymers, soap solutions, asphalts, physiological fluids,... that fail to obey the linear Newton law (1.1) : these materials are generally referred to as viscoelastic materials. These materials behave as fluids with a behaviour reminiscent of solids by exhibiting elastic effects. In ordinary fluids the relaxation time of the pressure tensor is very short; in elastic bodies it is infinite so that no relaxation is observed : viscoelastic materials are characterized by relaxation times between these two limits. Systems with the afore-mentioned property are also called non-Newtonian in the technical literature. The terms "viscoelastic" and "non-Newtonian" are used rather loosely; here, we shall reserve the term "non-Newtonian" for any fluid described by a non-linear constitutive equation, the term "linear viscoelastic" will be used for systems exhibiting both viscous and elastic effects in the linear regime and described by material coefficients which are independent of the velocity gradient.

This work will be divided into three parts. The first one concerns linear viscoelasticity. Historically, the linear viscoelastic models were of pure mechanical nature and constructed by combining linear springs and dashpots. Although very simple, they have proven to be useful for describing a wide class of phenomena and for providing the first step towards more realistic non-linear descriptions. It is our purpose in section 2 to show that linear viscoelasticity is easily interpreted within the framework of EIT. It is true that the description of viscoelastic materials by means of thermodynamic theories is not new. Several papers have been devoted to the subject but the majority of them is inspired either from classical irreversible thermodynamics (e.g. Meixner 1949, Kluitenberg 1962, Bataille and Kestin 1980) or rational thermodynamics (e.g. Rivlin-Ericksen 1955, Koh and Eringen 1963, Huilgol and Phan-Thien 1986). Here we propose a derivation based on extended irreversible thermodynamics. It is seen that the classical models of Maxwell, Kelvin-Voigt, Poynting-Thomson and Jeffreys are recovered as particular cases of the formalism. It may also be asked what are the consequences of introducing a whole spectrum of relaxational modes for the pressure tensor, instead of working with one single mode. This question is dealt with in section 3, wherein the Rouse (1953) and Zimm (1954) models are discussed. These molecular models, based on a whole relaxational spectrum, are very useful for describing dilute polymer solutions. It is examined to what extent Rouse-Zimm models can be incorporated into an EIT description. In the last part (section 4) of this paper, a non-linear analysis is performed to include non-Newtonian fluids. After recalling the properties of the main non-Newtonian fluids, two simple quadratic EIT models are proposed : despite its crudeness, the first model is seen to reproduce the correct signs of the viscometric coefficients while the second model is nothing but Giesekus'model.

The results presented here review and unify past contributions to the subject (Lebon et al 1986, 1988, Lebon and Cloot 1988, Lebon and Casas 1989, Perez et al 1989, Lebon et al 1990).

2. LINEAR VISCOELASTICITY

In this section, the classical theory of linear viscoelasticity is revisited. A method for determining the evolution equations of the pressure tensor components is proposed. It is seen that EIT provides a simple and systematic way to derive the main results of linear viscoelasticity.

2.1. Constitutive and evolution equations

We take for granted the following hypotheses : (a) the deformations are infinitesimally small; (b) the body is at uniform temperature and heat effects are neglected; (c) the material is isotropic.

Generalization to anisotropic systems undergoing large deformations and submitted to temperature gradients, although not a trivial matter, should not raise fundamental difficulties.

The choice of the variables is inspired by earlier results derived for viscous fluids (Lebon 1978, Lebon et al 1980, Lebon and Boukary 1988, Jou et al 1988). Let us recall that in the latter case, the basic parameters are the velocity v_i, the internal energy u and the viscous part P^v_{ij} of the pressure tensor defined through

$$P_{ij} = p\delta_{ij} + P^v_{ij} \quad . \tag{2.1}$$

By analogy with (2.1) we shall in the case of viscoelastic materials, decompose the pressure tensor, assumed symmetric, into an elastic P'_{ij} and an inelastic part P''_{ij} :

$$P_{ij} = P'_{ij} + P''_{ij} \quad , \tag{2.2}$$

P'_{ij} obeys Hooke's law

$$P'_{ij} = -2G\epsilon_{ij} - \lambda\epsilon_{kk}\delta_{ij} \quad , \tag{2.3}$$

with λ and G the Lamé coefficients and ϵ_{ij} the symmetric strain tensor, defined in Cartesian coordinates by

$$\epsilon_{ij} = 1/2(u_{i,j} + u_{j,i}) \quad , \tag{2.4}$$

where u_i is the deformation vector, a comma means derivation with respect to the spatial coordinates.

In parallel to the treatment of a viscous fluid, we choose as basic variables the rate of deformation $\delta_t u_i$, the internal energy u and the inelastic pressure P''_{ij}. The reasons for selecting P''_{ij} as supplementary variable are found in the general axioms underlying EIT. Indeed, the latter demands that the set of conserved classical variables be completed by variables taking the form of dissipative fluxes, vanishing at equilibrium. Clearly, P''_{ij} meets these requirements; this would not be the case for the total pressure tensor P_{ij} as it does not vanish at equilibrium. It is also emphasized that in contrast with earlier works, (Kluitenberg 1962, Palumbo and Valenti 1985), it is not required to decompose the strain tensor into an elastic and an inelastic part.

The behaviour of the classical variable u_i and u is governed by the usual balance laws of momentum and energy; in cartesian coordinates they are

$$\rho \delta_t^2 u_i = -P_{ij,j} + \rho f_i \quad , \tag{2.5}$$

$$\rho \delta_t u = -P_{ij} \delta_t \epsilon_{ij} \quad , \tag{2.6}$$

where ρ is the density and f_i the specific body force. The heat flux does not appear in the energy balance (2.6) as heat effects are omitted.

From now on, any second order tensor a_{ij} will be decomposed into an isotropic part a and a traceless deviatoric part $\overset{0}{a}_{ij}$

$$a_{ij} = 1/3 a \delta_{ij} + \overset{0}{a}_{ij} \quad \left(\overset{0}{a}_{kk} = 0, \quad a = a_{kk}\right) \quad . \tag{2.7}$$

By strict analogy with the classical balance laws (2.5) and (2.6), it is assumed that the supplementary variable P''_{ij} or, equivalently, its isotropic and deviatoric parts P'' and $\overset{0}{P}_{ij}$ satisfy balance equations of the general form

$$\delta_t P'' = -J_{k,k} + S, \tag{2.8}$$

$$\delta_t \overset{0}{P}''_{ij} = -J_{(ij)k,k} + S_{(ij)}, \tag{2.9}$$

wherein J_k and J_{ijk} are flux terms while S and S_{ij} are source terms, round brackets mean a traceless symmetrization. In order that the description be complete, it remains to express these fluxes and as functions of the basic set of variables by means of constitutive equations :

$$
\begin{aligned}
J_k &= J_k\left(u, \delta_t u_i, P'', \overset{0}{P}''_{ij}\right) \quad , \\
J_{ijk} &= J_{ijk}\left(u, \delta_t u_i, P'', \overset{0}{P}''_{ij}\right) \quad , \\
S &= S\left(u, \delta_t u_i, P'', \overset{0}{P}''_{ij}\right) \quad ,
\end{aligned} \tag{2.10}
$$

$$S_{ij} = S_{ij}\left(u, \delta_t u_i, P'', \overset{0}{P}''_{ij}\right) \quad ; \tag{2.11}$$

the most general relations compatible with a linear analysis are

$$J_k = \frac{\zeta}{\tau''_\epsilon}\delta_t u_k \quad , \tag{2.12}$$

$$J_{ijk} = \frac{\eta}{\tau_\epsilon}(\delta_t u_i \delta_{jk} + \delta_t u_j \delta_{ik}) \quad , \tag{2.13}$$

$$S = -\frac{1}{\tau_{\epsilon'}}P'' \quad , \tag{2.14}$$

$$S_{ij} = -\frac{1}{\tau_\epsilon}\overset{0}{P}''_{ij} \quad , \tag{2.15}$$

where τ'_ϵ, τ_ϵ, ζ and η are undetermined coefficients. Substitution of (2.12) and (2.14) in (2.8), (2.13) and (2.15) in (2.9) respectively, results in the following field equations for the extra variables

$$\tau_\epsilon \delta_t \overset{0}{P}''_{ij} = -\overset{0}{P}''_{ij} - 2\eta\delta_t \overset{0}{\epsilon}_{ij} \quad , \tag{2.16}$$

$$\tau_{\epsilon'}\delta_t P'' = -P'' - \zeta\delta_t\epsilon \quad . \tag{2.17}$$

η and ζ are generally called shear and bulk viscosity, respectively, τ_ϵ and $\tau_{\epsilon'}$ are relaxation times. Equations $(2.16)_0$ and (2.17) are the required linear evolution equations of the new variables $\overset{0}{P}_{ij}''$ and P''. It is possible to obtain evolution equations for the total pressure components by elimination of $\overset{0}{P}_{ij}''$ and P'' between (2.16), (2.17) and (2.2). This operations leads to the following results

$$\tau_\epsilon \delta_t \overset{0}{P}_{ij} + \overset{0}{P}_{ij} = -2G\left(\overset{0}{\epsilon}_{ij} + \tau_\sigma \delta_t \overset{0}{\epsilon}_{ij}\right) \quad , \tag{2.18}$$

$$\tau'_\epsilon \delta_t P + P = 3K(\epsilon + \tau'_\sigma \delta_t \epsilon) \quad , \tag{2.19}$$

wherein K, τ_σ, and τ'_σ stand respectively for

$$K = \lambda + \frac{2}{3}G \quad , \tag{2.20}$$

$$\tau_\sigma = \frac{\eta}{G} + \tau_\epsilon \quad , \tag{2.21}$$

$$\tau'_\sigma = \frac{\zeta}{3K} + \tau'_\epsilon \quad . \tag{2.22}$$

It is worth noticing that the relations (2.18) and (2.19) are the constitutive equations for a Poynting-Thomson body and that these relations arise naturally from extended irreversible thermodynamics.

The following particular cases are also of interest. By setting $\tau_\epsilon = 0$ in (2.18), one obtains

$$\overset{0}{P}_{ij} = 2G\overset{0}{\epsilon}_{ij} - 2\eta\delta_t\overset{0}{\epsilon}_{ij}. \tag{2.23}$$

Equation (2.23) is representative of a Kelvin-Voigt body. If one assumes that in equation (2.18) $G = 0$ which means that the total stress has only an inelastic contribution, one recovers the basic equation of Maxwell's model, namely,

$$\tau_\epsilon\delta_t\overset{0}{P}_{ij} + \overset{0}{P}_{ij} = -2\eta\delta_t\overset{0}{\epsilon}_{ij} \quad . \tag{2.24}$$

An interesting generalization is provided by the following model which is a coupling of Newton's viscous fluid and the material described by equation (2.18). Let us write for the stress tensor a relation of the form

$$P_{ij} = P^s_{ij} + P^p_{ij} \quad , \tag{2.25}$$

P^s_{ij} is the viscous pressure tensor whose traceless part is given by

$$\overset{0}{P}{}^s_{ij} = -2\eta^s\delta_t\overset{0}{\epsilon}_{ij} \quad , \qquad \text{(Newton's law)}$$

with η^s the shear viscosity, while P^p_{ij} is supposed to be formed of an elastic and an inelastic part :

$$P^p_{ij} = -2G\epsilon_{ij} + P''_{ij} \quad .$$

For the sake of simplicity bulk effects are omitted here ($P^p_{kk} = P^s_{kk} = 0$).

The decomposition (2.25) is useful for describing dilute solutions of polymers (Bird et al 1987), P^s_{ij} is the contribution of the solvent and P^p_{ij} is the mean pressure tensor of the interacting polymer molecules. Repeating the reasoning leading to (2.16), with P''_{ij} selected as basic variables, we are led to the relaxational equation

$$\tau\delta_t P''_{ij} = -P''_{ij} - 2\eta^p\delta_t\epsilon_{ij} \quad , \tag{2.26}$$

η^p represents the polymer contribution to the viscosity. By taking the time derivative of equation (2.25) and eliminating P''_{ij} by means of (2.26), one obtains

$$\tau\delta_t P_{ij} + P_{ij} = -2G\epsilon_{ij} - 2G\left(\tau + 2\frac{\eta^p + \eta^s}{2G}\right)\delta_t\epsilon_{ij} + 2\tau\eta^s\delta_t^2\epsilon_{ij} \quad . \tag{2.27}$$

Setting $G = 0$ in (2.27) yields

$$\tau\delta_t P_{ij} + P_{ij} = -2(\eta^p + \eta^s)\delta_t\epsilon_{ij} + 2\tau\eta^s\delta_t^2\epsilon_{ij} \quad ,$$

which is nothing but Jeffrey's model while by letting $\eta^s = 0$, one recovers the previous model expressed by equation (2.18). Jeffrey's equation is conveniently cast into the form

$$\tau\delta_t P_{ij} + P_{ij} = -2\eta(V_{ij} + \tau_0\delta_t V_{ij}) \quad , \tag{2.28}$$

with

$$\eta = \eta^p + \eta^s \quad , \quad \tau_0 = \frac{\tau\eta^s}{\eta^p + \eta^s} \quad , \quad V_{ij} = \delta_t\epsilon_{ij} \quad . \tag{2.29}$$

Although in ordinary fluids the relaxation times τ_ϵ and τ'_ϵ are very short (10^{-11} to $10^{-13}s$) so that it is justified to use the classical Newton equation, this is no longer true in viscoelastic materials for which the relaxation times are very much longer; as a matter of fact, for a high density polyethylene at 200°C, τ_ϵ is $10^{-2}s$ while for a polystyrene at 200°C, τ_ϵ is of the order of unit and it may even rise up to $\tau_\epsilon = 10^2 s$ for a polyisobutylene solution in oil.

It should also be emphasized that the material coefficients appearing in the afore-mentioned evolution equations for the pressure tensor components are subjected to some constraints imposed by the second law of thermodynamics, the criterion of objectivity, and the condition that entropy is a convex function at equilibrium. These restrictions are examined in the next subsection.

2.2. Restrictions imposed by the second law of thermodynamics, objectivity, and the convexity requirement.

We postulate that there exists a regular and continuous function, the specific entropy, given by the constitutive relation

$$s = s(u, \epsilon_{ij}, P''_{ij}) \quad , \tag{2.30}$$

and obeying a balance equation of the form

$$\rho\delta_t s = -J^s_{i,i} + \sigma^s(\sigma^s > 0) \quad , \tag{2.31}$$

with a non-negative production term σ^s. Objectivity prevents the entropy to depend on the velocity because the latter is a non-objective quantity. In the absence of heat flux, it is easily shown (e.g. Lebon 1978, Lebon-Boukary 1988, Jou et al 1988) that the entropy flux J^s_i is zero so that

$$\sigma^s = \rho\delta_t s \quad . \tag{2.32}$$

Using the chain differentiation rule to calculate $\delta_t s$, one obtains

$$\sigma^s = \rho\left(\frac{\delta s}{\delta u}\delta_t u + \frac{\delta s}{\delta \overset{0}{\epsilon}_{ij}}\delta_t \overset{0}{\epsilon}_{ij} + \frac{\delta s}{\delta \epsilon}\delta_t \epsilon\right.$$

$$\left. + \frac{\delta s}{\delta \overset{0}{P}{}''_{ij}}\delta_t \overset{0}{P}{}''_{ij} + \frac{\delta s}{\delta P''}\delta_t P''\right) \geq 0 \quad . \tag{2.33}$$

Define as usual the absolute temperature T and the components of the elastic pressure P'_{ij} tensor by the derivative of s with respect to the internal energy and the strain tensor components respectively,

$$\frac{\delta s}{\delta u} = \frac{1}{T} \quad , \tag{2.34}$$

$$\frac{\delta s}{\delta \epsilon} = \frac{1}{3\rho T}P' \quad , \tag{2.35}$$

$$\frac{\delta s}{\delta \overset{0}{\epsilon}_{ij}} = \frac{1}{\rho T}\overset{0}{P}{}'_{ij} \quad . \tag{2.36}$$

Making use of the energy balance (2.6), the entropy inequality (2.33) reads as

$$\sigma^s = -\frac{1}{3}\frac{\rho}{T}P''\delta_t\epsilon - \frac{\rho}{T}\overset{0}{P}{}''_{ij}\delta_t\overset{0}{\epsilon}_{ij} - \frac{\rho}{\tau_\epsilon}\left(\overset{0}{P}{}''_{ij} + 2\eta\delta_t\overset{0}{\epsilon}_{ij}\right)\frac{\delta s}{\delta \overset{0}{P}{}''_{ij}}$$

$$-\frac{\rho}{\tau'_\epsilon}(P'' + \rho\delta_t\epsilon)\frac{\delta s}{\delta P''} \geq 0 \quad . \tag{2.37}$$

For isotropic systems, the most general form for the derivatives of s with respect to P'' and $\overset{0}{P}{}''_{ij}$ is given, in the linear approximation, by

$$\frac{\delta s}{\delta P''} = -LP'' - \gamma\epsilon \quad , \tag{2.38a}$$

$$\frac{\delta s}{\delta \overset{0}{P}{}''_{ij}} = -d\overset{0}{P}{}''_{ij} - \Lambda\overset{0}{\epsilon}_{ij} \quad , \tag{2.38b}$$

By taking the mixed derivatives of (2.38a) and (2.38b) with respect to ϵ and $\overset{0}{\epsilon}_{ij}$, respectively and comparing with the mixed derivatives of (2.35) and (2.31) with respect to P'' and $\overset{0}{P}{}''_{ij}$, it is found that the coefficients γ and Λ are zero. Substitution of equations (2.38a,b) in (2.37) results then in

$$\sigma^s = \rho\frac{d}{\tau_\epsilon}\overset{0}{P}{}''_{ij}\overset{0}{P}{}''_{ij} - \overset{0}{P}{}''_{ij}\delta_t\overset{0}{\epsilon}_{ij}\left(\frac{1}{T} - 2\frac{d\eta}{\tau_\epsilon}\right) + \rho\frac{L}{\tau'_\epsilon}P''P''$$

$$-\frac{1}{3}P''\delta_{l}\epsilon\left(\frac{1}{T}-\frac{3L\zeta}{\tau'_{\epsilon}}\right)\geq 0 \quad . \tag{2.39}$$

Positiveness of expression (2.39) demands that

$$2\eta=\frac{\tau_{\epsilon}}{Td}>0 \quad , \quad \zeta=\frac{\tau'_{\epsilon}}{3LT}>0 \quad . \tag{2.40}$$

We turn now our attention to the consequences stemming from the convexity requirement of entropy. Expanding s around equilibrium, for fixed values of energy and strain, one obtains

$$s\left(u,\epsilon_{ij},P'',\overset{0}{P}_{ij}\right)=s_{eq}(u,\epsilon_{ij})+\left(\frac{\delta s}{\delta \overset{0}{P''}_{ij}}\right)_{eq}\overset{0}{P''}_{ij}$$

$$+\frac{1}{2}\left(\frac{\delta^2 s}{\delta \overset{0}{P''}_{ij}\delta \overset{0}{P''}_{ij}}\right)_{eq}\overset{0}{P''}_{ij}\overset{0}{P''}_{ij}$$

$$+\left(\frac{\delta s}{\delta P''}\right)P''+\frac{1}{2}\left(\frac{\delta^2 s}{\delta P''\delta P''}\right)_{eq}P''P''+O(3) \quad . \tag{2.41}$$

From equations (2.38a,b) and the property that at equilibrium s is maximum, relation (2.41) takes the form

$$s=s_{eq}-\frac{1}{2}d\overset{0}{P''}_{ij}\overset{0}{P''}_{ij}-\frac{1}{2}LP''P''+O(3) \quad . \tag{2.42}$$

with

$$d>0 \quad , \quad L>0 \quad . \tag{2.43}$$

By combining these results with the inequalities (2.40), it is found that

$$\tau_{\epsilon}>0 \quad , \tag{2.44}$$

$$\tau'_{\epsilon}>0 \quad . \tag{2.45}$$

The requirement that the entropy production is positive definite has thus led to the important results that the viscosity coefficients η and ζ are positive while from the convexity property of entropy it is concluded that the relaxation times τ_{ϵ} and τ'_{ϵ} are positive quantities.

2.3. The Gibbs equation for viscoelastic bodies

In classical irreversible thermodynamics the corner-stone is the Gibbs equation which is postulated a priori. In contrast, in the present formalism Gibbs' relation is derived : in view of the results (2.34), (2.36) and (2.40), it can be written in the form

$$Tds = du + \frac{1}{\rho}P'_{ij}d\epsilon_{ij} - \frac{\tau_\epsilon}{2\rho\eta}\overset{0}{P}''_{ij}d\overset{0}{P}''_{ij} - \frac{\tau'_\epsilon}{\rho\zeta}P''dP'' \quad . \tag{2.46}$$

It is interesting to formulate the Gibbs equation in the particular cases of Kelvin-Voigt, Maxwell and Poynting-Thomson bodies. A Kelvin-Voigt material being defined by $\tau_\epsilon = \tau'_\epsilon = 0$, expression (2.46) reduces to

$$Tds = du + \frac{1}{\rho}P'_{ij}d\epsilon_{ij} \quad . \tag{2.47}$$

In order to obtain Gibbs equation for a Maxwell body, it must be recalled that the latter is defined by $G = 0$, or equivalently, by identifying the inelastic stress tensor with the total stress. For the sake of simplicity, bulk effects are ignored. At this point, we introduce an inelastic strain ϵ''_{ij} defined by

$$\delta_t\epsilon''_{ij} = -\frac{1}{2\eta}P_{ij} \quad , \tag{2.48}$$

and an elastic strain ϵ'_{ij} given by

$$\epsilon'_{ij} = \epsilon_{ij} - \epsilon''_{ij} \quad . \tag{2.49}$$

Substituting the evolution equations (2.24) in (2.46) and making use of the above definitions of ϵ'_{ij} and ϵ''_{ij}, the Gibbs equation for a Maxwell body reads

$$Tds = du + \frac{1}{\rho}P_{ij}d\epsilon'_{ij} \quad . \tag{2.50}$$

The results (2.47) and (2.50) are classical and were derived a few years ago by Kluitenberg (1966) and Lambermont and Lebon (1973) on completely different bases. It was also shown by Lambermont and Lebon that for a Poynting-Thomson body Gibbs' equation is given by

$$Tds = du + \frac{1}{\rho}P_{ij}d\epsilon_{ij} - \frac{1}{\rho}P''_{ij}d\epsilon''_{ij} \quad . \tag{2.51}$$

It is an easy task to show that this result can be determined from (2.46) by introducing an inelastic strain tensor by

$$\delta_t\epsilon''_{ij} = -\frac{1}{2\eta}P''_{ij} \quad . \tag{2.52}$$

It may be instructive to compare the merits of extended irreversible thermodynamics with the hidden variable theory (HVT) which fuelled much interest during the three last decades (e.g. Meixner 1954, Coleman and Gurtin 1964, Kestin and Bataille 1980, Maugin and Drouot 1983, Bampi and Morro 1984). In EIT, the physical nature of the extra variables is known from the outset while in HVT the variables are generally identified at the end of the procedure. In contrast with HVT, the variables introduced in EIT take part in the mechanical work and the heat input; in HVT the hidden variables do not appear in the balance equations.

It is also worth mentioning that EIT does not duplicate the results of HVT. In this respect, we refer to a work by Kestin and Bataille (1980) whose objective is, like ours, to provide a thermodynamic description of viscoelasticity. In particular, Kestin and Bataille derive a relation between the relaxation times τ_σ and τ_ϵ of the form

$$\tau_\sigma = \frac{a}{G}\tau_\epsilon \quad ,$$

where a is a constant coefficient. In EIT one obtains

$$\tau_\sigma = \frac{\eta}{G} + \tau_\epsilon \quad ,$$

which differs from the HVT result by the property that τ_σ is non-zero even for a vanishing τ_ϵ.

The aim of this section was to show that linear viscoelasticity is easily and naturally incorporated into EIT. It is worth stressing that EIT encompasses a wider class of materials than Maxwell's bodies; the Kelvin-Voigt, Poynting-Thomson models are also recovered and at the same time the formalism may generate more complex models useful for the description of polymer solutions. The various material coefficients can be determined experimentally by measurements of wave velocity and wave attenuation (Lebon et al, 1988). Generalisation to non-linear models involving non-linear objective time derivatives and non-linear constitutive equations will not raise fundamental difficulties as shown in section 4, and in recent papers by Lebon and Cloot (1988) and Rodriguez et al (1988).

3. THE ROUSE-ZIMM RELAXATIONAL MODEL

Kinetic models are very helpful in gaining information about the type of rheological equation that is needed. It is expected that a kinetic interpretation enlightens and complements a purely continuum approach. Recently, there has been a vast amount of works devoted to this kind of description : impressive results are collected in the masterly books by Doi and Edwards (1986), Bird et al. (1987). It is not our purpose to review the most significative kinetic models but we wish only to stress that most of them exhibit a whole relaxational spectrum for the pressure tensor. This important property is fairly well described by the beads-springs chain models associated with the names of Rouse (1953) and

Zimm (1956). For many years these models have played a key role in the interpretation of rheological phenomena. There have been many extensions of Rouse and Zimm's works but although most of these generalizations give a better fitting of the experimental data by introducing more parameters, they do not alter radically the foundations underlying the Zimm and Rouse theory. Our purpose here is to show that EIT is capable of coping with the relaxational spectrum of the pressure tensor exhibited in the Rouse-Zimm model.

3.1. The Rouse-Zimm model

The current molecular models view the polymer macromolecules as formed by assemblages of beads, springs and rods. In the Rouse model the molecules are modeled by a chain of N point-mass beads connected by N-1 Hookean springs. The forces acting on each bead are

1. The hydrodynamic drag force which is the force experienced by the bead as it moves through the polymeric solution. According to Stokes, it is proportional to the difference between the bead velocity and the center-of-mass velocity of the solution. By hydrodynamic interaction is meant the perturbation of the velocity field of one particular bead by the motion of the other beads. In Rouse model wherein it is admitted that the motions of the N beads are uncorrelated, this kind of interaction is ignored; it is accounted for in the Zimm model.

2. The Brownian force caused by the thermal fluctuations in the solution.

3. The intramolecular force which is the force exerted on one bead through the spring force.

4. The external forces like the gravitational or electrical forces.

The Rouse and Zimm models are appropriate for describing dilute polymer solutions. It is supposed that the roles of the solvent and the polymer can be separated. Accordingly, the pressure tensor is decomposed into two parts

$$P_{ij} = P^{1}_{ij} + P^{p}_{ij} \quad , \tag{3.1}$$

where superscripts 1 and p denote the contribution of the solvent and the polymer respectively; next, P^{p}_{ij} is decomposed into normal modes

$$P^{p}_{ij} = \sum_{\alpha=2}^{N} P^{\alpha}_{ij} \quad , \tag{3.2}$$

and each individual mode α is assumed to satisfy, in the linear regime, a relaxational dynamics of the form

$$\tau^{\alpha}\delta_t P_{ij}^{\alpha} + P_{ij}^{\alpha} = -\eta^{\alpha} V_{ij} \quad , \tag{3.3}$$

τ^{α} is the relaxational time of the α mode, V_{ij} the usual symmetric rate-of-strain tensor.

In the Rouse and Zimm models, the coefficient η^{α} is found to be given by

$$\eta^{\alpha} = nkT\tau^{\alpha} \quad , \tag{3.4}$$

with the relaxation time τ^{α} expressed by

$$\tau^{\alpha} = \frac{\xi}{2Ha^{\alpha}} \qquad (Rouse) \quad , \quad (3.5) \quad , \quad \tau^{\alpha} = \frac{\xi}{2H\bar{a}^{\alpha}} \qquad (Zimm) \quad , \quad (3.6)$$

k is the Boltzmann constant, T the absolute temperature, n the number of molecules per unit volume, ξ the friction coefficient related to the drag force experimented by the bead, H stands for the Hookean spring constant while a^{α} and $\bar{a}^{\alpha}$ are the eigenvalues of the so-called Rouse and modified Rouse matrices, the a^{α} are given by

$$a^{\alpha} = 4\sin^2\left(\frac{\pi\alpha}{2N}\right) \quad ,$$

while the $\bar{a}^{\alpha}$ are found in tabular form or given by approximate expressions (Bird et al., 1987).

3.2. EIT description of the Rouse-Zimm model

We follow the same procedure as in chapter 2, but the set of independent variables is now extended in order to include the contributions arising from the normal modes :

$$u \quad , \quad v_i \quad , \quad \overset{0}{P}{}_{ij}^{\alpha} \quad , \quad (\alpha = 1, 2.N) \quad . \tag{3.7}$$

The internal energy u and the average local velocity v_i of a fluid element are the classical variables, they are supplemented by the components of the viscous pressure tensors associated to the solution ($\alpha = 1$) and the normal modes ($\alpha = 2, 3, \ldots, N$). The viscous pressure is assumed to be symmetric and traceless (no bulk effects) :

$$\overset{0}{P}{}_{kk}^{\alpha} = 0 \quad . \tag{3.8}$$

The average mass density of the solution and the polymer concentration are supposed to remain constant and therefore they have not been introduced into the set of variables. In accordance with the EIT scheme, the extra variables $\overset{0}{P}{}^{\alpha}_{ij}$ are assumed to satisfy evolution equations of the form

$$\delta_t \overset{0}{P}{}^{\alpha}_{ij} = -\overset{0}{J}{}^{\alpha}_{(ij)k,k} + \overset{0}{S}{}^{\alpha}_{(ij)} \quad (\alpha = 1,\ldots,N) \ . \tag{3.9}$$

At the lowest order of approximation, the constitutive equations of the source and flux terms are expressed by

$$\overset{0}{S}{}^{\alpha}_{ij} = -\sum_{\beta=1}^{N} \frac{1}{\tau^{\alpha\beta}} \overset{0}{P}{}^{\beta}_{ij} \ , \tag{3.10}$$

$$\overset{0}{J}{}^{\alpha}_{(ij)k} = 2\frac{\eta^{\alpha}}{\tau^{\alpha\alpha}} v_{(i}\,\delta_{j)k} \ , \tag{3.11}$$

where η^{α} is the shear viscosity associated to the mode α and $\tau^{\alpha\beta}$ are material coefficients. After substitution of the results (3.10) and (3.11) in the evolution equation (3.9), one obtains

$$\delta_t \overset{0}{P}{}^{\alpha}_{ij} = -2\frac{\eta^{\alpha}}{\tau^{\alpha\alpha}} \overset{0}{V}_{ij} - \sum_{\beta=1}^{N} \frac{1}{\tau^{\alpha\beta}} \overset{0}{P}{}^{\beta}_{ij} \ . \tag{3.12}$$

Equation (3.12) contains the coupling between the different modes. The material functions η^{α} and $\tau^{\alpha\beta}$ depend generally on the temperature, at fixed mass density and polymer concentration. The coupling between the modes is removed by assuming that the coefficient $\tau^{\alpha\beta}$ reduce to

$$\tau^{\alpha\beta} = \tau^{\alpha}\delta_{\alpha\beta} \ . \tag{3.13}$$

Under this hypothesis, the evolution equation (3.12) take the familiar Maxwell form

$$\tau^{\alpha}\delta_t \overset{0}{P}{}^{\alpha}_{ij} = -2\eta^{\alpha} \overset{0}{V}_{ij} - \overset{0}{P}{}^{\alpha}_{ij} \quad (\alpha = 1,\ldots,N) \ . \tag{3.14}$$

Assuming with Rouse that the solvent may be considered as a Newtonian fluid which means that $\tau^{1} \ll \tau^{\alpha}$ $(\alpha = 2,\ldots,N)$, one obtains finally the following set :

$$\overset{0}{P}{}^{1}_{ij} = -2\eta^{1} \overset{0}{V}_{ij} \ , \tag{3.15}$$

$$\tau^{\alpha}\delta_t \overset{0}{P}{}^{\alpha}_{ij} + \overset{0}{P}{}^{\alpha}_{ij} = -2\eta^{\alpha} \overset{0}{V}_{ij} \quad (\alpha = 2,\ldots,N) \ , \tag{3.16}$$

where now α runs from 2 to N. Expressions (3.15) and (3.16) are identical with Rouse-Zimm equations which were derived from kinetic arguments at the condition to make the following identification :

$$\eta^{\alpha} = nkT\tau^{\alpha} \quad ,$$

$$\tau^{\alpha} = \frac{\xi}{2Ha^{\alpha}} \quad , \qquad (Rouse \ model)$$

$$\tau^{\alpha} = \frac{\xi}{2H\bar{a}^{\alpha}} \quad . \qquad (Zimm \ model)$$

4.NON-NEWTONIAN FLUIDS

It is shown that EIT provides also a simple and coherent modelling of non-linear systems, like non-Newtonian fluids. The basic hypothesis is still to raise the viscous pressure tensor to the status of independent variable. Applying the technique developed previously, constitutive equations for second-order non-Newtonian fluids are established. For clarity, the EIT description is preceded by a brief summary of the main properties of some current rheological models.

4.1. Rheological models

Polymer solutions and polymer melts are the most familiar examples of fluids which cannot be described by Newton's standard law (1.1). For the sake of simplicity, the materials are supposed to be at constant temperature, incompressible, and without bulk effects ($P^{v}_{kk} = 0$). For further purpose, it is interesting to examine the response of a fluid in a simple shearing flow between two parallel horizontal planes separated by a distance h. If one of the planes, for instance the upper one, is moved in the x direction with a constant velocity v_0 , the velocity field of the fluid is

$$v_x = \dot{\gamma} y \quad , \quad v_y = 0 \quad , \quad v_z = 0 \quad , \quad \dot{\gamma} = \frac{v_0}{h} \quad , \qquad (4.1)$$

with y in the direction normal to the planes and z the neutral direction, $\dot{\gamma}$ is the constant shear rate.

For an incompressible Newtonian fluid, the components of the stress tensor are

$$P_{xz} = P_{yz} = 0 \quad , \qquad (4.2)$$

$$P_{xx} = P_{yy} = P_{zz} = p \quad , \qquad (4.3)$$

$$P_{xy} = -\eta\dot{\gamma} \quad . \qquad (4.4)$$

Denoting the first normal stress by N_1 and the second normal stress by N_2,

$$N_1 = P_{xx} - P_{yy} \quad , \qquad (4.5)$$

$$N_2 = P_{yy} - P_{zz} \quad , \tag{4.6}$$

it is directly seen that for a Newtonian fluid both N_1 and N_2 are zero. The main properties of a Newtonian fluid are a constant viscosity and no normal stresses. In contrast, non-Newtonian fluids are characterized by a shear-rate dependent viscosity and non-vanishing normal stresses. It is common to introduce the "apparent" viscosity μ, the first and second normal stress coefficients ψ_1 and ψ_2 through

$$P_{xy} = -\mu(\dot{\gamma})\dot{\gamma} \quad (\mu > 0) \quad , \tag{4.7}$$

$$N_1 = -\psi_1(\dot{\gamma})\dot{\gamma}^2 \quad (\psi_1 > 0) \quad , \tag{4.8}$$

$$N_2 = -\psi_2(\dot{\gamma})\dot{\gamma}^2 \quad (\psi_2 < 0) \quad , \tag{4.9}$$

μ, ψ_1 and ψ_2 are referred to as the viscometric functions. Experimental data indicate that ψ_1 is positive while ψ_2 is generally found to be negative; ψ_2 is smaller that ψ_1, more or less about 10% of ψ_1. While the dependence of μ with respect to γ gives rise to velocity profiles totally unusual in Newtonian flows, the normal stress coefficients are responsible for the well-known Weissenberg rod-climbing effect (e.g. Lodge 1964, Coleman et al. 1966, Ferry 1980, Tanner 1985).

The main problem in non-Newtonian fluid mechanics is that of formulating a constitutive equation for the pressure tensor P_{ij}, containing the smallest number of parameters and applicable to the widest range of fluid responses. The simplest generalization of Newton's law is the Reiner-Rivlin non-linear model :

$$P_{ij} = p\delta_{ij} - 2\eta V_{ij} - 4\alpha_1 V_{ik} V_{kj} \quad , \tag{4.10}$$

it contains two material functions η and α_1 depending in general on the principal invariants of V_{ij} :

$$I_v = v_{i,i} \quad , \tag{4.11}$$

$$II_v = V_{ik} V_{ik} \quad , \tag{4.12}$$

$$III_v = \det \; V_{ij} \quad . \tag{4.13}$$

Here η and α_1 do not depend on the first invariant I_v as it is zero for an incompressible fluid. It can be shown from continuum mechanics arguments that η and α_1 cannot both be taken constant, unless $\alpha_1 = 0$ in which case Newton's law is recovered (Astarita and Marucci 1974). In the particular case that α_1 is put equal to zero, the most widely used model for η is the Ostwald power law

$$\eta = m\dot{\gamma}^{n-1} \quad , \tag{4.14}$$

wherein $\dot{\gamma}$ stands for

$$\dot{\gamma} = (2V_{ij}V_{ij})^{1/2} \quad .$$

While m and n are parameters characteristic of a given fluid, they may depend on the temperature, concentration and pressure. However, the power law (4.14) is not realistic because it gives $\eta = \infty$ for $\dot{\gamma} = 0$ when $n < 1$ which is the most usual situation. To circumvent this difficulty, Carreau (1979) has proposed the following viscosity equation

$$\frac{\eta - \eta_\infty}{\eta_0 - \eta_\infty} = (1 + (\tau\dot{\gamma})^2)^{\frac{n-1}{2}} \quad .$$

involving four parameters : a zero-shear-rate viscosity η_0 , an infinite shear-rate viscosity η_∞ , a time constant τ and a supplementary constant n. This formula fairly well describes most engineering experiments.

Going back to Reiner-Rivlin model, it is directly shown that for a steady shear flow, the viscometric functions are related to the material coefficients η and α_1 by

$$\mu = \eta \quad , \tag{4.15}$$

$$\psi_1 = 0 \quad , \tag{4.16}$$

$$\psi_2 = \alpha_1 \quad . \tag{4.17}$$

The result (4.16) is at variance with experimental observations and this means that Reiner-Rivlin model does not provide a realistic description of non-Newtonian fluids.

To correct for these undesirable features, one can use the more sophisticated model of Rivlin-Ericksen (1955). To meet the objectivity criterion, Rivlin and Ericksen have introduced the following symmetric tensors, which reflect the above requirement

$$\overset{0}{A}_{ij} = \delta_{ij} \quad , \tag{4.18}$$

$$A_{ij}^{(1)} = 2V_{ij} \quad , \tag{4.19}$$

$$A_{ij}^{(n)} = d_t A_{ij}^{(n-1)} + v_{k,i}A_{kj}^{(n-1)} + A_{ik}^{(n-1)}v_{k,j} \quad (n = 2,3,\dots) \quad . \tag{4.20}$$

The tensors $A_{ij}^{(n)}$ are called the Rivlin-Ericksen tensors. The simplest Rivlin-Eriksen fluid is the second order fluid

$$P_{ij} = p\delta_{ij} - \eta A_{ij}^{(1)} - \alpha_2 A_{ij}^{(2)} - \alpha_1 A_{ik}^{(1)} A_{kj}^{(1)} \quad , \tag{4.21}$$

depending on three material coefficients, η, α_1 and α_2 which are supposed to be constants. By setting in (4.21), α_2 equal to zero, one obtains again Reiner-Rivlin's model. It is instructing to consider what predictions are coming from (4.21) in a steady shearing flow. It is seen that the material constants, η, α_1 and α_2 are related to the usual viscometric functions by

$$\mu = \eta \quad , \tag{4.22}$$

$$\psi_1 = -2\alpha_2 \quad , \tag{4.23}$$

$$\psi_2 = \alpha_1 + 2\alpha_2 \quad . \tag{4.24}$$

Working within the framework of rational thermodynamics, Dunn and Fosdick (1974) have shown that the second principle of thermodynamics and stability of equilibrium require that

$$\alpha_2 > 0 \quad , \quad \alpha_1 + \alpha_2 = 0 \quad . \tag{4.25}$$

However, it follows both from experimental data and theoretical considerations that α_2 should be negative in order to ensure the correct sign for ψ_1 and that the sum of α_1 and α_2 is generally not zero :

$$\alpha_2 < 0 \quad , \quad \alpha_1 + \alpha_2 \neq 0 \quad . \tag{4.26}$$

In view of these contradictory results, we are faced with two alternatives. Either Rivlin-Ericksen model is in contradiction with thermodynamics and cannot pretend to model real fluids for which $\alpha_2 < 0$; or rational thermodynamics contains some deficiencies ruling out a realistic physical model. We shall come back to this problem at the end of this section.

The Rivlin-Ericksen second-order fluid has been extended in several ways. A natural generalization is that of Criminale, Ericksen and Filbey (1958) who used the same equation (4.21) but with shear-rate depending viscometric coefficients. Higher order expansions in terms of the Rivlin-Ericksen tensors $A^{(n)} (n > 2)$ have also been proposed but they lead to a rapid proliferation of the parameters and are not very useful in practical problems.

All the aforementioned models are explicit in the pressure tensor meaning that P^v_{ij} can be substituted in the momentum equation giving a flow problem in the variable v_i. Another possibility is to describe the behaviour of fluids exhibiting elastic effects by means of rate equations relating algebraically the pressure and rate of deformation tensors and some of their time-derivatives. Examples of such rate-equation models are provided by the generalized Maxwell equation

$$\tau D P_{ij} = -2\eta V_{ij} - P_{ij} \quad , \tag{4.27}$$

where D is an objective time derivative : the replacement of δ_t by D is dictated by the objectivity criterion. A frequently used objective derivative is the Jaumann or corotational time derivative :

$$D_J P_{ij} = d_t P_{ij} - W_{ik} P_{kj} + P_{ik} W_{kj} \quad , \tag{4.28}$$

where W_{ij} is the skew-symmetric velocity gradient tensor and d_t the material time-derivative. Clearly, equation (4.27) is not linear due to the occurrence of the non-linear terms in the definition of D.

We shall no longer discuss the properties of the several rheological models found in the literature. We wish only to stress that none of these models is applicable to study all the states of motion of a given particular non-Newtonian fluid. Several constitutive equations may be needed to describe the behaviour of one single fluid, depending on the particular circumstances under which the motion takes place.

4.2. Extended thermodynamics of second-order non-Newtonian fluids

An extended thermodynamic description of viscoelastic materials was given in section 2; in section 3 the Rouse-Zimm bead-springs model was examined within the same framework but only linear situations were investigated. However, EIT is by no means restricted to the linear range. Our purpose is to show that second order non-Newtonian fluids stem out naturally from EIT. Two particular models are derived : a simple 3-parameters model and a Giesekus 4-parameters model. It is seen that the simple 3-parameters description gives the correct signs of the material coefficients of the Rivlin-Ericksen model.

The basic variables are selected to be the same as for an ordinary Newtonian fluid, namely (Lebon and Cloot, 1988).

$$v_i \quad , \quad u \quad , \quad P^v_{ij} \quad , \tag{4.29}$$

P^v_{ij} is the deviatoric symmetric viscous tensor since bulk effects are ignored ($P^v_{kk} = 0$), v_i the velocity and u the internal energy. We know that the behaviour of the classical variables v_i and u is governed by the usual balance laws of momentum and energy; by analogy, we write for P^v_{ij} a balance relation of the form

$$d_t P^v_{ij} = -J_{(ij)k,k} + S_{(ij)} \quad , \tag{4.30}$$

J_{ijk} is the flux of the viscous pressure and S_{ij} the corresponding source term. These quantities are unknowns and will be expressed by means of constitutive relations :

$$J_{ijk} = J_{ijk}(v_i, u, P^v_{ij}) \quad , \tag{4.31}$$

$$S_{ij} = S_{ij}(v_i, u, P^v_{ij}) \quad . \tag{4.32}$$

The evolution equation (4.30) as well as the constitutive equations (4.31) and (4.32) must obey the constraints imposed by the general laws of continuum physics. In particular, they have to comply with the second law of thermodynamics and the criterion of objectivity. The latter demands that the material time derivative be replaced by an objective time derivative, for instance, Jaumann's derivative D_J. The source term in (4.30) must be a traceless symmetric objective quantity, it can be cast in the general form

$$S_{(ij)} = -AP^v_{ij} - B(P^v_{ik}P^v_{kj} - \pi^v\delta_{ij}) \quad , \tag{4.33}$$

where π^v stands for

$$\pi^v = \frac{1}{3}P^v_{kl}P^v_{kl} \quad .$$

The most general form for the flux J_{ijk}, symmetric with respect to the indices i and j, is given by

$$J_{(ij)k} = C(v_i\delta_{jk} + v_j\delta_{ik}) + E(v_iP^v_{jk} + v_jP^v_{ik}) \quad . \tag{4.34}$$

A, B, C, E are arbitrary coefficients. Substitution of (4.33) and (4.34) in (4.30) yields the following evolution equation for P^v_{ik} :

$$DP^v_{ij} = -CV_{ij} - AP^v_{ij} - B(P^v_{ik}P^v_{kj} - \pi^v\delta_{ij}) \quad . \tag{4.35}$$

In order to fulfil the objectivity requirement, the coefficient C has to be taken constant while E must be zero : otherwise $J_{ijk,k}$ and consequently the r.h.s. of (4.35) would contain undesirable non-objective terms in v_i. Equation (4.35) was also derived by Müller and Wilmanski (1986) who based their analysis on another version of EIT (Liu and Müller 1983). For further purpose, it is interesting to put $1/A = \tau$, $C/A = 2\eta$, $B/A = \alpha$ so that (4.35) takes the more familiar form

$$\tau DP^v_{ij} = -2\eta V_{ij} - P^v_{ij} - \alpha(P^v_{ik}P^v_{kj} - \pi^v\delta_{ij}) \quad , \tag{4.36}$$

wherein τ has the dimension of a time and η the dimension of a viscosity. The relation (4.36) is the keystone of the model. For negligible values of τ (in ordinary fluids, τ is of the order of the collision time, i.e. (10^{-12s}) and $\alpha = 0$, one recovers Newton's constitutive relation; in the linear approximation (i.e. $\alpha = 0$), expression (4.36) reduces to the Maxwell model.

We now examine the restrictions placed by the second law of thermodynamics. We still postulate the existence of an entropy function s depending on the set of variables defined by (4.29) and obeying the balance equation

$$\rho \dot{s} + J^s_{k,k} = \sigma^s \quad , \tag{4.37}$$

with

$$\sigma^s \geq 0 \quad . \tag{4.38}$$

The entropy production will be computed by carrying out the operations indicated on the left of equation (4.37). The semi-positive definite property of σ^s is assumed to hold at least at the lowest quadratic order of approximation; when σ^s is developed at higher order in the constitutive variables, say $\sigma^s = \sigma^s_{(2)} + \sigma^s_{(3)} + \ldots$, no general conclusion can be drawn about the cubic term $\sigma^s_{(3)}$ because it may have either sign, a result confirmed by the kinetic theory of gases (de Groot and Mazur 1962, Woods 1981) and the information theory (Vasconcellos et al. 1990). The statement $\sigma^s_{(2)} \geq 0$ is not less general than the restriction imposed earlier. Indeed in the previous sections, σ^s was automatically of second order due to the linearization procedure. In (4.37), s and J^S_i must be given by constitutive relations

$$s = s(u, P^v_{ij}) \quad , \quad J^s_i = J^s_i(u, v_i, P^v_{ij}) \quad , \tag{4.39}$$

s is an objective scalar and therefore does not depend on the non-objective vector v_i, s will depend on P^v_{ij} through its principal invariants. Here $J^s_i = 0$ because of the absence of heat flow. Under these circumstances, the entropy production is simply given by

$$\sigma^s = \rho\left(\frac{\delta s}{\delta u} d_t u + \frac{\delta s}{\delta P^v_{ij}} d_t P^v_{ij}\right) \quad , \tag{4.40}$$

We now use the property that s is a scalar valued function of its arguments, from which follows that

$$\frac{\delta s}{\delta P^v_{ij}} = -\beta P^v_{ij} + O(2) \quad , \tag{4.41}$$

and we set, as usual,

$$\frac{\delta s}{\delta u} = \frac{1}{T} \quad . \tag{4.42}$$

The coefficient β in (4.41) is arbitrary and the minus sign in front of β has been introduced for convenience. Substitution of (4.41) and (4.42) in (4.40) yields, when third order terms are omitted,

$$\sigma^s_{(2)} = -\left(\frac{1}{T} - 2\frac{\rho\eta\beta}{\tau}\right) P^v_{ij} V_{ij} + \frac{\rho\beta}{\tau} P^v_{ij} P^v_{ij} \quad . \tag{4.43}$$

Positiveness of $\sigma^s_{(2)}$ demands that

$$\frac{\beta}{\tau} > 0 \quad , \tag{4.44}$$

$$\beta = \frac{\tau}{2\rho\eta T} \quad . \tag{4.45}$$

It remains to explore the consequence of the convexity of s at equilibrium. Developing s around equilibrium, at fixed equilibrium value of u, one obtains by virtue of (4.41) :

$$s(u_{eq}, P^v_{ij}) - s_{eq}(u_{eq}) = -\frac{1}{2}\beta P^v_{ij} P^v_{ij} \quad . \tag{4.46}$$

Convexity of s requires that

$$\beta > 0 \quad . \tag{4.47}$$

By combining this result with (4.44) and (4.45) it is concluded that

$$\tau > 0 \quad , \quad \eta > 0 \tag{4.48}$$

indicating that P^v_{ij} satisfies an evolution equation with a positive relaxation time and a positive viscosity.

The properties that η and τ are positive coupled with the rate-type equation (4.36) provide the essential features of the model. At this point of the analysis, it is interesting to investigate the transition from rate-type equations like (4.36) to constitutive relations like the Reiner-Rivlin or the Rivlin-Ericksen equations. To this end, let us rewrite equation (4.36) in term of Rivlin-Ericksen's time derivative $P^{(2)}_{ij}$ defined by (4.20) : one obtains

$$\begin{aligned} P^v_{ij} &= -2\eta V_{ij} - \tau P^{(2)}_{ij} - a(P^v_{ik} P^v_{kj} - \pi^v \delta_{ij}) \\ &\quad + A\tau(P^v_{ik} V_{kj} + V_{ik} P^v_{kj}) \quad , \end{aligned} \tag{4.49}$$

wherein use has been made of the result

$$DP^v_{ij} = P^{(2)}_{ij} - F(P^v_{ik} V_{kj} + V_{ik} P^v_{kj}) \quad . \tag{4.50}$$

For $F = 0, 1$ and 2 , one recovers respectively the lower convected $D_{\downarrow}$, the Jaumann D_J and the upper convected derivative $D_{\uparrow}$ (Astarita and Marucci 1974, Lodge 1964).

At the first order approximation and in the limit $\tau = 0$, (4.49) reduces to

$$ {}_{(1)}P^{v}_{ij} = -2\eta V_{ij} \quad , \tag{4.51} $$

which is the Newton relation, subscript (1) refers to first order approximation.

For a non-vanishing relaxation time, (4.49) becomes

$$ {}_{(1)}P^{v}_{ij} = -2\eta V_{ij} - \tau\delta_{t(1)}P^{v}_{ij} \quad , \tag{4.52} $$

and this is the basic equation of Maxwell's model.

The second order approximation with $\tau = 0$ is obtained by substituting in the right hand side of (4.49), P^{v}_{ij} by its value (4.51), this operation leads to the Reiner-Rivlin relation

$$ {}_{(2)}P^{v}_{ij} = -2\eta V_{ij} - 4\eta^2 \alpha V_{ij} V_{kj} \quad . \tag{4.53} $$

If the relaxation time τ is not zero, one obtains from (4.49)

$$ {}_{(2)}P^{v}_{ij} = -2\eta V_{ij} + 2\tau\eta V^{(2)}_{ij} - 4\eta(\eta\alpha + F\tau) V_{ik} V_{kj} \quad . \tag{4.54} $$

Comparison between (4.54) and the Rivlin-Ericksen equation (4.21) allows to express the coefficients α_1 and α_2 in terms of the parameters η, τ and α, namely

$$ \alpha_1 = F\eta\tau + \eta^2\alpha \quad , \tag{4.55} $$

$$ \alpha_2 = -\eta\tau \quad . \tag{4.56} $$

Recalling that τ and η are positive, it is clear from (4.56) that α_2 is a negative quantity in accordance with experiments. Moreover, addition of (4.55) and (4.56) yields

$$ \alpha_1 + \alpha_2 = \mu\tau(F - 1) + \eta^2\alpha \quad , \tag{4.57} $$

showing that the sum of the coefficients α_1 and α_2 is generally non-zero. In particular, when Jaumann's derivative is used, the sum α_1 and α_2 is simply equal to $\eta^2\alpha$.

Let us recall that Dunn and Fosdick have found $\alpha_2 > 0$ and $\alpha_1 + \alpha_2 = 0$. It can thus be concluded that by considering the Rivlin-Ericksen equation as an approximation of the more general rate-type model (4.37), one avoids the contradictions raised by Dunn and Fosdick's work. The essential conclusion is that $\alpha_2 < 0$ is not in contradiction with thermodynamics. The above model contains only three parameters which may not be sufficient to fit experimental data.

A more complicated model involving more parameters may be easily generated if it is admitted that the total viscous pressure tensor P^{v}_{ij} is the sum of contributions P^{s}_{ij} from the solvent and P^{p}_{ij} from the polymer chain :

$$P^{v}_{ij} = P^{s}_{ij} + P^{p}_{ij} \quad . \tag{4.58}$$

It is assumed that the solvent is an incompressible Newtonian fluid so that

$$P^{s}_{ij} = -2\eta^{s} V_{ij} \tag{4.59}$$

and that the space of the independent variables is formed by v_i, u and P^{p}_{ij}. Repeating the procedure of section 4.2 one obtains the next evolution equation for P^{p}_{ij} :

$$\tau^{p} d_t P^{p}_{ij} = -2\eta^{p} V_{ij} - P^{p}_{ij} - a^{p}(P^{p}_{ik} P^{p}_{kj} - \pi^{p}\delta_{ij}) \quad , \quad \left(\pi^{p} = \frac{1}{3} P^{p}_{ij} P^{p}_{ij}\right) \quad , \tag{4.60}$$

which contains three parameters τ^{p}, η^{p} and a^{p}. After elimination of P^{p}_{ij} between (4.58) and (4.60) and use of (4.59), one obtains the following rheological equation for the total pressure tensor :

$$\tau^{p} d_t P^{v}_{ij} + P^{v}_{ij} - \frac{a\tau^{p}}{\eta}(P^{v}_{ik} P^{v}_{kj} - \pi^{v}\delta_{ij}) - a\lambda(V_{ik} P^{v}_{kj} + P^{v}_{ik} V_{kj} - \bar{\pi}\delta_{ij})$$

$$= -2\eta\left[V_{ij} + \lambda d_t V_{ij} - \frac{2a\lambda^{2}}{\tau^{p}}\left(V_{ik} V_{kj} - \bar{\bar{\pi}}\,\delta_{ij}\right)\right] \tag{4.61}$$

Where the non-identified coefficients η, a, λ, $\bar{\pi}$ and $\bar{\bar{\pi}}$ stand for

$$\eta = \eta^{s} + \eta^{p} \quad , \quad a = \frac{-\eta a^{p}}{\tau^{p}} \quad , \quad \lambda = \frac{\eta^{s}\tau^{p}}{\eta}$$

$$\bar{\pi} = \frac{2}{3} P^{v}_{ij} V_{ij} \quad , \quad \bar{\bar{\pi}} = \frac{1}{3} V_{ij} V_{ij} \quad . \tag{4.62}$$

Expression (4.61) is the Giesekus constitutive equation (Giesekus 1982) except for the terms in π^{v}, $\bar{\pi}$ and $\bar{\bar{\pi}}$ which appear as a consequence of the no bulk viscous pressure assumtion. Equation (4.61) contains four independent parameters that can be indifferently chosen as τ^{p}, a^{p}, η^{s} and η^{p} or τ^{p}, a, η and λ. The result (4.61) is particularly promising as it allows to derive the Giesekus model from very simple macroscopic considerations. Here we have followed a continuum method while Giesekus original derivation was based on a microscopic approach.

It is worth stressing that the above results were obtained from very simple considerations : it was only required that the pressure tensor be selected as independent variable and that it obeys a second-order non-linear evolution equation of the relaxation type. Of course, it is still possible to improve the quality of the model by introducing supplementary variables, for instance, the conformation tensor. This would certainly result in a more realistic description of the geometrical configuration of the polymer chains and to a reduction of the number of undetermined parameters. But this is a further step in the formulation of the theory that will be developed in the future.

REFERENCES

Astarita, G., and G. Marucci. 1974. Principles of Non-Newtonian Fluid Mechanics. London : Mac Graw Hill.

Bampi, F., and A. Morro. 1984. Non-equilibrium Thermodynamics : a Hidden Variable approach. In Lecture Notes in Physics, Vol 199, eds. Casas-Vazquez, J., D. Jou, and G. Lebon, 211-232. Berlin : Springer.

Bataille, J., and J. Kestin. 1975. L'interprétation physique de la mécanique rationnelle. J. de Mécanique 14 : 365-384.

Bird, R.B., C.F. Curtiss, R.C. Armstrong, and O. Hassager. 1987. Dynamics of Polymeric Liquids. Vol. I and II, 2nd ed., New York : Wiley.

Carreau, P.J. 1972. Rheological equations from molecular network theories. Trans. Soc. Rheol. 16 : 99-127.

Casas-Vazquez, J., and D. Jou. 1990. Extended thermodynamics and fluctuation theory. In Non-equilibrium Thermodynamics, eds Sieniutycz. S., and P. Salomon, New York : Taylor and Francis.

Coleman, B.D., and M. Gurtin. 1964. Thermodynamics with internal variables. J.Chem. Phys. 47 : 597-613.

Coleman, B.D., H. Markowitz, and W. Noll. 1966. Viscometric flows of non-Newtonian Fluids. New York : Springer.

Criminale, W.O., J.L. Ericksen, and G.K. Filbey. 1958. Steady shear flow of non-Newtonian fluids. Arch. Rat. Mech. Anal. 1 : 410-417.

de Groot, S.R., and P. Mazur. 1962. Non-equilibrium Thermodynamics. Amsterdam : North-Holland Publ.

Doi, M., and S.F. Edwards. 1986. The theory of polymer Dynamics. Oxford : Clarendon Press.

Dunn, J.E., and R.L. Fosdick. 1974. Thermodynamics, stability, and boundedness of fluids of complexity 2 and fluids of second grade, Arch. Rat. Mech. Anal. 56 : 191-252.

Eu, B. 1985. Theory of viscoelasticity of fluids. J. Chem. Phys. 82 : 4683-4689.

Ferry, J.D. 1980. Viscoelastic Properties of Polymers, 3rd ed., New York : Wiley.

Giesekus, H. 1982. A simple constitutive equation for polymer fluids based on the concept of deformation dependent mobility. J. Non Newt. Fluid Mech. 11 : 69-109.

Huilgol, R.R., and N. Phan-Thien. 1986. Recent advances in the continuum mechanics of viscoelastic liquids, Int. J. Engng. Sci 24 : 161-261.

Jou, D., J. Casas-Vazquez, and G. Lebon. 1988. Extended irreversible thermodynamics. Rep. Prog. Phys. 51 : 1105-1179.

Kestin, J., and J. Bataille. 1980. Thermodynamics of solids. In Continuum Models of Discrete Systems, 99-147. Univ. of Waterloo Press.

Kluitenberg, G. 1966. Application of the theory of irreversible processus to continuum mechanics. In Non-equilibrium Thermodynamics, Variational Techniques, and Stability. eds. R. Donnelly, R. Herman, I. Prigonine, 91-100. Chicago : University Chicago Press.

Kluitenberg, G. 1984. Plasticity and non-Equilibrium Thermodynamics, In CISM Course Vol. 281, Wien : Springer.

Koh, S., and C. Eringen. 1963. On the fondations of non-linear thermo-viscoelasticity. Int. J. Engng. Sci. 1 : 199-229.

Lambermont, J., and G. Lebon. 1974. On the derivation of the Gibbs equation for a class of rheological bodies. Int. J. Non-linear Mech. 9 : 55-74.

Lebon, G. 1978. Derivation of generalized Fourier and Navier-Stokes equations, based on the thermodynamics of irreversible processes. Bull. Acad. Roy. Sc. Belgique 64 : 456-472.

Lebon, G., D. Jou, and J. Casas-Vazquez. 1980. An extension of the local equilibrium hypothesis. J. of Phys. A13 : 275-290.

Lebon, G., C. Perez-Garcia, and J. Casas-Vazquez. 1986. On a new thermodynamic description of viscoelastic materials. Physica 137A : 531-545.

Lebon, G., C. Perez-Garcia, and J. Casas-Vazquez. 1988. On the thermodynamic foundations of viscoelasticity. J. Chem. Phys. 88 : 5068-5075.

Lebon, G., and A. Cloot. 1988. An extended thermodynamic approach of non-Newtonian fluids and related results in Marangoni instability problem. J. Non Newt. Fluid Mech. 28 : 61-76.

Lebon, G., and M.S. Boukary. 1988. Objectivity, kinetic theory and extended irreversible thermodynamics. Int. J. Engng. Sc. 26 : 471-483.

Lebon, G., P.C. Dauby, A. Palumbo, and G. Valenti. 1990. Rheological properties of dilute polymer solutions : an extended thermodynamic approach. Rheol. Acta (to appear).

Liu-I-Shih, and I. Müller. 1983. Extended thermodynamics of classical and degenerate ideal gases. Arch. Ration. Mech. Anal. 83 : 285-332.

Lodge, A.S. 1964. Elastic Liquids. New York : Academic Press.

Maugin, G., and R. Drouot. 1983. Internal variables and the thermodynamics of macromolecule solutions. Int. J. Engng. Sci. 21 : 705-724.

Maxwell, J.C. 1867. On the dynamical of theory of gases. Philos. Trans. Roy. Soc. London 157A : 49-88.

Meixner, J. 1949. Thermodynamik und Relaxation serscheinungen. Z. Naturforschung 4a : 594-600.

Meixner, J. 1954. Thermodynamische Theorie der elastischen Relaxation. Z. Naturforschung 9a : 654-663.

Müller, I., and K. Wilmanski. 1986. Extended thermodynamics of a non-Newtonian fluid. Rheol. Acta 25 : 335-350.

Palumbo, A., and G. Valenti. 1985. Extended irreversible thermodynamics of Maxwell bodies. J. Non-Equil. Thermo. 10 : 209-219.

Perez-Garcia, C., J. Casas-Vazquez, and G. Lebon. 1989. A non equilibrium thermodynamic description of dilute polymer solutions. J. Polym. Sci., B-Polym. Phys. 27 : 1807-1821.

Rivlin, R.S., and J.L. Ericksen. 1955. Stress deformations relations for isotropic materials. J. Rat. Mech. Anal. 4 : 323-425.

Rodriguez, R.F., M. Lopez de Haro, and O. Manero. 1988. A thermodynamic approcach to non-linear viscoelasticity. Rheol. Acta 27 : 217-223.

Rouse, P.E. 1953. A theory of the linear viscoelastic properties of dilute solutions of coiling polymers. J. Chem. Phys. 21 : 1272-1280.

Sieniutycz, S., and P. Salomon. 1990. Diversity of non-equilibrium thermodynamics theories. In Non-equilibrium Thermodynamics, eds. Sieniutycz, S., and P. Salomon. New York : Taylor and Francis.

Tanner, K. 1985. Engineering Rheology. Oxford : Clarendon Press.

Vasconcellos, R., R. Luzzi, and L.S. Garcia-Colin. 1990. On the statistical foundations of extended irreversible thermodynamics. Phys. Rev. A : to appear.

Woods, L.C. 1981. The bogus axioms of continuum mechamics; 1981. Inst. Math. and Appl. 17 : 98-102.

Zimm, B.H. 1956. Dynamics of polymer molecules in dilute solution : viscoelasticity, flow birefringence and dielectric loss. J. Chem. Phys. 24 : 269-278.

Extended Irreversible Thermodynamics of Electrical Phenomena

Josep Enric Llebot
Departament de Física
Universitat Autònoma de Barcelona
08193 Bellaterra, Catalonia, Spain

Abstract

In this paper, it is reviewed extended irreversible thermodynamics applied to electric phenomena. When the electric field is zero, the classical results are obtained. For non zero values of the electric field a phenomenological result for the current fluctuations is derived. These results are compared with microscopic calculations . The stability of a steady state is also studied for a completely ionized plasma under and external field, introducing fourth order terms in the electric current for the generalized entropy. A good agreement is obtained with a kinetic evaluation of the critical value for the electric field.

Introduction

One of the main problems related to the description of non equilibrium systems is the choice of fundamental variables. While in equilibrium this problem is well known, and there are diverse equivalent formulations, in non-equilibrium this is not the case since does not exist a general criterion for this choice. In classical irreversible thermodynamics fundamental variables are the same as in equilibrium, i.e, the conserved ones. Many possible different choices arise in non-equilibrium thermodynamics. One may take as fundamental variables, for instance, the temperature and the temperature gradient, as is usually done in rational thermodynamics, or the temperature and its time derivative. This choice has a direct physical sense and it has a definite experimental meaning. In the so-called extended irreversible thermodynamics, both the usual thermodynamic variables and the dissipative fluxes are taken as independent variables. Therefore a generalized entropy is postulated (Jou et al. 1988) In this volume, some other works are included that treat the same starting point and we remit there for further discussions about the convenience or not of such an approach. In this work, we use extended irreversible thermodynamics to study a rigid heat and electric conductor, and we explore some of the consequences which are derived from the use of this formalism.

Extended Irreversible Thermodynamics for a Electric and Heat Conductor

The system we will study is a solid conductor, rigid and isotropic. In classical thermodynamics the entropy per unit mass s is a function of u, the internal energy per unit mass, and of c_e, the electron density per unit mass. In Extended Irreversible Thermodynamics, it is postulated a non-equilibrium entropy which depends on u, c_e and on the dissipative fluxes, $\vec{q}$ (heat flux) and $\vec{J}_e$ (electric current). When studying a system in an equilibrium state, the non-equilibrium entropy reduces to the entropy given by the classical theory, since both $\vec{q}$ and $\vec{J}_e$ vanish.

We neglect, for the sake of simplicity viscous effects. In this case, the balance equations of the classical variables read as

$$\rho \, \dot{u} = \vec{J}_e . \vec{E} - \nabla . \vec{q} \tag{1}$$

$$\rho\,\dot{c}_e = -\nabla.\vec{J}_e \qquad (2)$$

with $\vec{E}$ the electric field and $\vec{J}_e.\vec{E}$, the Joule heating. To find the evolution equation for $\vec{q}$ and $\vec{J}_e$ we proceed following the standard procedure in extended irreversible thermodynamics (Jou et al 1989).

We assume that if s is differentiable enough the following Gibbs equation can be written

$$ds = \theta^{-1}du - \theta^{-1}\eta\, dc_e + T^{-1}v(\alpha_1\vec{q} + \alpha_2\vec{J}_e).d\vec{q} + T^{-1}v(\gamma_1\vec{q} + \gamma_2\vec{J}_e).d\vec{J}_e \qquad (3)$$

T is the equilibrium absolute temperature, v the specific volume, α_i and γ_i coefficients which will be identified in the coarse of our development and

$$\theta^{-1}(u,c_e,\vec{q},\vec{J}_e) = (\partial s/\partial u)_{c_e,\vec{q},\vec{J}_e} \qquad (4)$$

$$-\theta^{-1}\eta(u,c_e,\vec{q},\vec{J}_e) = (\partial s/\partial c_e)_{u,\vec{q},\vec{J}_e} \qquad (5)$$

In analogy with classical theory θ is a generalized non-equilibrium temperature and η is a generalized non-equilibrium electrochemical potential which, when developed in series up to second order in the fluxes, have the following form

$$\theta^{-1}(u,c_e,\vec{q},\vec{J}_e) = T^{-1}(u,c_e) + (1/2)(\partial^2\theta^{-1}/\partial\vec{q}^2)\vec{q}^2 + (1/2)(\partial^2\theta^{-1}/\partial\vec{J}_e^2)\vec{J}_e^2 + (\partial^2\theta^{-1}/\partial\vec{q}\partial\vec{J}_e)\vec{q}.\vec{J}_e \qquad (6)$$

$$\theta^{-1}\eta(u,c_e,\vec{q},\vec{J}_e) = T^{-1}\mu_e(u,c_e) + (1/2)(\partial^2\theta^{-1}\eta/\partial\vec{q}^2)\vec{q}^2 + (1/2)(\partial^2\theta^{-1}\eta//\partial\vec{J}_e^2)\vec{J}_e^2 + (\partial^2\theta^{-1}\eta/\partial\vec{q}\partial\vec{J}_e)\vec{q}.\vec{J}_e \tag{7}$$

with μ_e the equilibrium chemical potential of electrons. Some of the terms which appear in eqs. (4) and (5) have been evaluated for a metallic rigid heat conductor (Casas-Vazquez and Jou 1981) and for an electric conductor (Llebot 1981) ; the values of the correction terms to the equilibrium temperature and electrochemical potential are very small and therefore we will omit them here, considering only the classical local equilibrium values for the temperature and the chemical potential of the electrons. These terms, however, may be important in the study of fluctuations of the dissipative fluxes in a non equilibrium state.

From the Gibbs equation (3) can be derived two state equations more, that up second order write

$$(\partial s/\partial\vec{q})_{u,c_e,\vec{J}_e} = T^{-1}v(\alpha_1\vec{q}+\alpha_2\vec{J}_e) \tag{8}$$

$$(\partial s/\partial\vec{J}_e)_{u,c_e,\vec{J}_e} = T^{-1}v(\gamma_1\vec{q}+\gamma_2\vec{J}_e) \tag{9}$$

The coefficients α_i and γ_i depend on u and c_e. Taking into account the equality between the crossed derivatives:

$$(\partial^2 s/\partial\vec{q}\partial\vec{J}_e) = (\partial^2 s/\partial\vec{J}_e\partial\vec{q}) \tag{10}$$

from equation (8) and (9) one gets the equality between the coefficients α_2 and γ_1.

Taking the time derivative in (3), one easily gets

$$\rho\dot{s}=\rho T^{-1}\dot{u}-\rho T^{-1}\mu_e\dot{c}_e+T^{-1}(\alpha_1\vec{q}+\alpha_2\vec{J}_e).\dot{\vec{q}}+T^{-1}(\gamma_1\vec{q}+\gamma_2\vec{J}_e).\dot{\vec{J}}_e \tag{11}$$

Introducing the balance equations (1), (2) in (11), after some easy transformations, one gets

$$\rho\dot{s} + \nabla.(T^{-1}\vec{q} - \mu_e T^{-1}\vec{J}_e) = \vec{q}.(\nabla T^{-1} + T^{-1}\alpha_1\dot{\vec{q}} + T^{-1}\gamma_1\dot{\vec{J}}_e) + \vec{J}_e.[T^{-1}\vec{E} - \nabla(\mu_e T^{-1}) + T^{-1}\alpha_2\dot{\vec{q}} + T^{-1}\gamma_2\dot{\vec{J}}_e] \tag{12}$$

This equation can be considered as a balance equation for the entropy,

$$\rho\dot{s} + \nabla.\vec{J}_s = \sigma \tag{13}$$

In order to identify the entropy production from (12), we need an expression for the entropy flux. Here we assume that it depends on all the dissipative fluxes, so the most general vector depending on all the fluxes is, up to the second order terms

$$\vec{J}_s = T^{-1}\vec{q} - \mu_e T^{-1}\vec{J}_e \tag{14}$$

that has the same form that the classical expression for the entropy flux. Moreover, σ is identified as follows

$$\sigma=\vec{q}.(\nabla T^{-1}+T^{-1}\alpha_1\dot{\vec{q}}+T^{-1}\gamma_1\dot{\vec{J}}_e)+\vec{J}_e.[T^{-1}\vec{E}-\nabla(\mu_e T^{-1})+T^{-1}\alpha_2\dot{\vec{q}}+T^{-1}\gamma_2\dot{\vec{J}}_e] \tag{15}$$

The entropy production has structure of a bilinear form

$$\sigma = \vec{X}_1.\vec{q} + \vec{X}_2.\vec{J}_e \tag{16}$$

where $\vec{X}_1$ and $\vec{X}_2$, depend on u, c_e, $\vec{q}$ and $\vec{J}_e$. Comparing equations

(15)- (16):

$$\vec{X}_1 = \nabla T^{-1} + T\alpha_1 \dot{\vec{q}} + T^{-1}\gamma_1 \dot{\vec{J}}_e \tag{17}$$

$$\vec{X}_2 = T^{-1}\vec{E} - \nabla(\mu_e T^{-1}) + T^{-1}\alpha_2 \dot{\vec{q}} + T^{-1}\gamma_2 \dot{\vec{J}}_e \tag{18}$$

In defining the proper form of the forces $\vec{X}_1$ and $\vec{X}_2$ it must be noted that there exists a class of transformations of the time derivatives which leave invariant the entropy production. This point is studied in other works in this volume and, for example, in Jou et al 1988, so here no more attention will be devoted .

Up to the second order, the most general form for $\vec{X}_1$ and $\vec{X}_2$ reads

$$\vec{X}_1 = \mu_{11}\vec{q} + \mu_{12}\vec{J}_e \tag{19}$$

$$\vec{X}_2 = \mu_{21}\vec{q} + \mu_{22}\vec{J}_e \tag{20}$$

where μ_{ij} are phenomenological coefficients. From the positive character of σ, some restrictions are imposed upon the possible values of these magnitudes, which can be written as

$$\mu_{11} \geq 0; \ \mu_{22} \geq 0; \ \mu_{11}\mu_{22} \geq \ (1/4)(\mu_{12}+\mu_{21})^2 \tag{21}$$

From (15), (19) and (20) one easily arrives to a system of evolution equations for the dissipative fluxes which can be written in compact form as

$$\begin{bmatrix} \dot{\vec{q}} \\ \dot{\vec{J}}_e \end{bmatrix} = \mathbb{M} \begin{bmatrix} \vec{q} \\ \vec{J}_e \end{bmatrix} + \mathbb{N} \tag{22}$$

with $\mathbb{M}$ given by

$$\mathbb{M} = T(\alpha_1\gamma_2-\alpha_2\gamma_1)^{-1} \begin{pmatrix} \gamma_2\mu_{11}-\gamma_1\mu_{21} & \gamma_2\mu_{12}-\gamma_1\mu_{22} \\ \alpha_1\mu_{21}-\alpha_2\mu_{11} & \alpha_1\mu_{22}-\alpha_2\mu_{12} \end{pmatrix} \tag{23}$$

and $\mathbb{N}$, a columm vector given by

$$\mathbb{N} = T(\alpha_1\gamma_2-\alpha_2\gamma_1)^{-1} \begin{bmatrix} T^{-1}\gamma_1\vec{E}-\gamma_2\nabla T^{-1}-\gamma_1\nabla(\mu_e T^{-1}) \\ \alpha_2\nabla T^{-1}-\alpha_1 T^{-1}\vec{E}+\alpha_1\nabla(\mu_e T^{-1}) \end{bmatrix} \tag{24}$$

In order for eqs (22)-(24) to be completely determined we should know the expressions of the coefficients μ_{ij}, α_i, γ_i.

Coefficients μ_{ij} are determined by comparing expressions (22)-(24) in the stationary state with the well known phenomenological equations (De Groot and Mazur 1962)

$$\nabla T = -(1/\lambda)\vec{q} + (1/\lambda)\ (\pi+\mu_e)\vec{J}_e \tag{25}$$

$$\vec{E}-\nabla\mu_e = (\varepsilon/\lambda)\vec{q}-[(1/\lambda)\varepsilon(\pi+\mu_e)-R]\vec{J}_e \tag{26}$$

with λ the heat conductivity at zero current, ε the thermo-electric power, π the Peltier coefficient and R the isothermal resistivity. Equation (22) in the stationary state, rearranged, gives the two following expressions

$$\nabla T = -\mu_{11}T^2\vec{q}-\mu_{12}T^2\vec{J}_e \tag{27}$$

$$E-\nabla\mu_e = T(\mu_{11}\mu_e+\mu_{21})\vec{q} + T(\mu_{12}\mu_e+\mu_{22})\vec{J}_e \tag{28}$$

which together with (25), (26) drive to the following identifications

$$\begin{aligned}&\mu_{11}= (1/\lambda T^2); \quad \mu_{22}=(R/T)+(1/\lambda T^2)[(\mu_e-\varepsilon T)(\pi+\mu_e)];\\ &\mu_{12}=(-1/\lambda T^2)(\pi+\mu_e); \quad \mu_{21}=(1/\lambda T^2)(\varepsilon T-\mu_e)\end{aligned} \tag{29}$$

In the nonstationary case, one can derive the following equations

$$(T^{-1}\alpha_1\dot{\vec{q}}-\mu_{11}\vec{q})+(T^{-1}\gamma_1\dot{\vec{J}}_e-\mu_{12}\vec{J}_e)=-\nabla T^{-1} \tag{30}$$

$$(T^{-1}\alpha_2\dot{\vec{q}}-\mu_{21}\vec{q})+(T^{-1}\gamma_2\dot{\vec{J}}_e-\mu_{22}\vec{J}_e)=-T^{-1}E+\nabla(\mu_e T^{-1}) \tag{31}$$

On the other side, in view of eqs. (22)-(24), and of the identification taken for α_1 and γ_2 in others works (Jou and Llebot 1980; Casas-Vázquez and Jou 1981), we consider

$$\begin{aligned}&\alpha_1= -\tau_1\mu_{11}T; \quad \alpha_2=-\tau_2\mu_{21}T; \quad \gamma_1=-\tau_3\mu_{12}T\\ &\gamma_2=-\tau_4\mu_{22}T\end{aligned} \tag{32}$$

τ_i are the relaxation times, which for i=1,4 refer to the heat and electric conduction, whereas for i= 2,3 they are linked to the cross effects between electric and heat conduction.

The relaxation times, in general, are difficult to measure. However, when studying the propagation of thermal waves with high-frequency thermometry in dielectric solids, a relation is found between τ_1 and quantities such the density, thermal conductivity and velocity of propagation of thermal waves, magnitudes which can be measured experimentally (Chester 1966). Determinations of τ_4 can be attained from measurements of the electrical conductivity (Wannier 1966) Provided with the last identifications the Gibbs equation (3) can be written as follows

$$ds = T^{-1}du-\mu_e T^{-1}dc_e-(\tau_1 v/\lambda T^2)\vec{q}.d\vec{q}- (\tau_2 v/\lambda T^2)(\varepsilon T-\mu_e)\vec{J}_e.d\vec{q} + \\ + (\tau_3 v/\lambda T^2)(\pi+\mu_e)\vec{q}.d\vec{J}_e- (\tau_4 v/T)[R+(1/\lambda T)(\mu_e- \varepsilon T)(\pi+\mu_e)]\vec{J}_e.d\vec{J}_e \qquad (33)$$

We have obtained a set of equations which provide a complete description of the system upon study, using a development very close to the procedure used in linear irreversible thermodynamics. Whereas constitutive equations of hyperbolic character, i.e., with finite velocities for the propagation of thermal and electric signals, have been used before, no generalization of the Gibbs equation was used. Here we have seen a thermodynamic framework where both points have been put together.

ELECTRIC CONDUCTION

In the simple case with no heat conduction, the evolution equation for the electric flux can be written as:

$$\tau_4\dot{\vec{J}}_e= -\vec{J}_e- (\tau_4/\gamma_2)[\vec{E} - T\nabla(\mu_e T^{-1})] \qquad (34)$$

which can be re-written

$$\tau_4 \dot{J}_e = -(J_e - \sigma_e E') \tag{35}$$

with $E' = T\nabla(\mu_e T^{-1}) - E$. σ_e the electrical conductivity, is related to the relaxation time τ_4 and to the coefficient γ_2 through the following expression

$$\sigma_e = \tau_4 / \gamma_2 \tag{36}$$

Equation (35) generalizes the usual Ohm law and is often used in plasma physics and in the analysis of high-frequency currents (Krall and Trivelpiece, 1973). The theory here but apports a thermodynamic justification that often is not treated in this kind of studies. From the kinetic theory one can obtain the same equations in the relaxation time approximation.

Assume now, that is imposed on the system an alternating electric field:

$$E' = E_o \cos wt \tag{37}$$

Taking into account eqs (35) and (37), a expression for the electric current is obtained. The solution for the electric current has two parts: a transient one, which usually is neglected in front of the sinusoidal signal

$$J_e = J_1 \cos wt + J_2 \sin wt \tag{38}$$

with

$$J_1 = (\sigma_e / 1 + \omega^2 \tau^2) E_0 \quad ; \quad J_2 = (\sigma_e \omega\tau / 1 + \omega^2 \tau^2) E_0 \tag{39}$$

since the characteristic time is very small.

In the classical theory, the entropy production is given by

$$\sigma = (1/T)\vec{J}_e.\vec{E} = (1/T)[\sigma_e E_0^2/(1 + \omega^2\tau^2)](\cos^2 \omega t + \omega\tau \cos \omega t \sin \omega t) \tag{40}$$

The first term in (40) is always positive, whereas the second one is alternating in sign. At low frequency, the first term is dominating and σ is positive. For $w\tau > 2$, the second term dominates and σ becomes negative during a part of each cyde. In contrast, the average of σ over a cycle, $\bar{\sigma}$, is

$$\bar{\sigma} = (1/2T)[\sigma_e/(1+\omega^2\tau^2)]\ E_o^2 \tag{41}$$

that is always positive. Note that $\bar{\sigma}$ (w) becomes zero in the high frequency limit, in contrast with the classical theory, with τ=0, where it is different of zero.

In extended irreversible thermodynamics, the entropy production is instead of (40)

$$\sigma = (1/T\sigma_e)J_e^2 \tag{42}$$

which is always positive. Given the classical form for σ, one could ask wheter the entropy production should be positive only as an average over a cicle, or wheter it should be instantaneously positive. However, extended irreversible thermodynamics, the entropy production is always positive.

Onsager reciprocal relations

One of the most important ingredients of classical non-equilibrium thermodynamics are the Onsager reciprocal relations, that cannot be obtained from purely macrocopic arguments, but only with the use of other additional hypothesis like time reversal symmetry of equilibrium correllation functions and regression of fluctuations.

In the classical derivation of the Onsager reciprocal relations, the case we are dealing with, neither the flux nor the electric current are time derivative of state variables. In the framework of the classical theory (De Groot and Mazur 1962), when studying heat conduction in anisotropic rigid solids, this difficulty is solved in the following way:

The entropy production written as

$$dS/dt = \int_V \vec{q}.\nabla T^{-1} dV \tag{43}$$

can be integrated by parts, neglecting thermal dilatation, taking into account that the energy remains constant for an adiabatically isolated system of constant volume and shape. Up to the second order in T

$$dS/dt= -(\rho/T_0^2)\int_v [T(\vec{r})-T_0](\partial u/\partial t)dV \tag{44}$$

where $T(\vec{r})$ is the temperature in r and T_0 the temperature in r_0, a reference point. Expression (44) is a bilinear form, where the thermodynamic force can be assumed to be $-\rho/T_0^2\ [T(\vec{r}) - T_0]$ and the corresponding flux $\partial u/\partial t$. If one states

$$\partial u/\partial t = -(\rho/T_0^2)\int K(\vec{r},\vec{r}')[T(\vec{r}) - T_0]\, d\vec{r}' \tag{45}$$

then, the Onsager relations take the form $K(\vec{r},\vec{r}') = K(\vec{r}',\vec{r})$. This equality, in fact, drives to the symmetry of the thermal conductivity tensor. There are several objections to this procedure, based mainly in the legitimacy of going from (43) to (44) (Coleman and Truesdell, 1960).

However, when one uses extended irreversible thermodynamics to derive Onsager relations, no difficulty is found in applying the classical Onsager's treatment. In an equilibrium state, where there is not an electric field nor temperature gradient, nor a gradient in the electrochemical potential, equations (8) and (9) can be written as follows:

$$\vec{F} = \mathbb{G}.\vec{\alpha} \tag{46}$$

where $\vec{F}$, $\mathbb{G}$, and $\vec{\alpha}$ are respectively

$$\vec{F} = \begin{bmatrix} \partial s/\partial \vec{q} \\ \partial s/\partial \vec{J}_e \end{bmatrix} \quad ; \quad \mathbb{G} = T^{-1}v \begin{bmatrix} \alpha_1 & \alpha_2 \\ \gamma_1 & \gamma_2 \end{bmatrix} \quad ; \quad \vec{\alpha} = \begin{bmatrix} \vec{q} \\ \vec{J}_e \end{bmatrix} \tag{47}$$

The classical theory states that if a set of variables $\vec{\alpha}$ verifies equation (43), and if the one-time correlation of these variables satisfies

$$< \vec{\alpha} \quad \vec{\alpha} > = k.\mathbb{G}^{-1} \tag{48}$$

where k is the Boltzmann constant, then the principle of microscopic reversibility implies that the matrix defined as

$$\mathbb{L} = \mathbb{M} \cdot \mathbb{G}^{-1} \tag{49}$$

($\mathbb{M}$ is defined in eq. (23)) is symmetrical. In our case, we obtain

$$(\alpha_1\gamma_2 - \gamma_1^2)\mu_{12} = (\alpha_1\gamma_2 - \gamma_1^2)\mu_{21} \tag{50}$$

let us say, $\mu_{12} = \mu_{21}$ or its equivalency

$$-(\pi + \mu_e) = (\varepsilon T - \mu_e) \tag{51}$$

so that the Thomson relation $\varepsilon T = -\pi$ is recovered. If we consider in (33) the equality between the cross derivatives of s

$$-\tau_2(\varepsilon T - \mu_e) = \tau_3(\pi + \mu_3) \tag{52}$$

In order for eqs (50) and (51) to be compatible we get $\tau_2 = \tau_3$. This equality has been obtained from the standard original development of the classical Onsager theory.Also some authors have assumed that the equality of the crossed derivatives of non-equilibrium entropy leads directly to Onsager relations (Eu, 1980). As can be seen above this is not exactly true. The equality leads to eq (51) but the microscopic argument of time reversal symmetry can not be avoided even in a more general theory than the classical one.

Non equilibrium fluctuations of the electric current

The central point of our development has been the obtention of a generalized Gibbs equation. Now we will show that the generalized entropy plays an important role in the description of fluctuations around equilibrium states of the conserved variables and also of the dissipative fluxes. Moreover, our results will greatly support the convenience of such a generalized entropy.

We assume the relation between entropy and probability in equilibrium can be extended to a non equilibrium situation using a generalized entropy. We will show the good agreement between our results with previous ones obtained from microscopic theories.

The probability of fluctuations of thermodynamic variables with respect to a equilibrium reference state for an isolate system is related to the change of entropy associated with the fluctuations through the Boltzmann Planck formula

$$P \sim \exp\left(\Delta S/k\right) \tag{53}$$

where P is the probability of a macrostate, ΔS the change of entropy associated to the separation of the equilibrium reference state and k the Boltzmann constant. The entropy of the new macrostate can be developed, if it is near the equilibrium reference state, through

$$S = S_{eq} + (\delta S)_{eq} + \frac{1}{2}(\delta^2 S)_{eq} \tag{54}$$

where δS and $\delta^2 S$ are the first and the second differentials of the entropy. Given the maximum character of the entropy for an isolated system, it follows that $(\delta S)_{eq} = 0$. Therefore

$$P \sim \exp(\delta^2 S/2k) \tag{55}$$

This expression is known as the Einstein formula for the

probability of fluctuations. The validity of this expression is restricted to up to second order in the moments (Jou et al. 1988).

In the following developments for the sake of maximum simplicity, we will restrict ourselves to the study of a rigid electric conductor. The non equilibrium entropy stated in the framework of extended irreversible thermodynamics can be written as

$$\rho s = \rho s_{eq} - (\tau/2\sigma_e T)\vec{J}_e.\vec{J}_e \tag{56}$$

where τ is the relaxation time for the electric flux and σ_e is the electrical conductivity.

The second differential of the generalized entropy, at constant electron density is given by

$$\begin{aligned}\delta^2 s = &- [(cT^2)^{-1} + (1/2)(\partial^2\alpha/\partial u^2)\vec{J}_0^2](\delta u)^2 \\ &- \alpha\, \delta\vec{J}_e.\delta\vec{J}_e - 2(\partial\alpha/\partial u)\vec{J}_0.\delta\vec{J}_e\delta u\end{aligned} \tag{57}$$

where c is the specific heat per unit mass, $\vec{J}_0$ is the mean non equilibrium value for $\vec{J}_e$ given by $\vec{J}_o = \sigma_e.\vec{E}$, and α is given by

$$\alpha = (\tau v/\sigma_e T) \tag{58}$$

The introduction of (57) into the Einstein formula (55) allows the calculation of the second moments of the fluctuations of the internal energy and the electric current:

$$\langle\delta u\ \delta u\rangle = kcT^2(1 + \Lambda\vec{J}_0^2)^{-1} \tag{59}$$

$$\langle \delta u \; \delta J_{ei} \rangle = (kcT^2/\alpha)(\partial\alpha/\partial u)J_{0i}(1 + \Lambda\vec{J}_0^2)^{-1} \tag{60}$$

$$\langle \delta J_{ei} \; \delta J_{ej} \rangle = (kcT/\tau v)(1 + \Lambda\vec{J}_0^2)^{-1}[\delta_{ij} + (cT^2/2)(\partial^2\alpha/\partial u^2)J_{0i}J_{0j}] \tag{61}$$

with $\Lambda = (cT^2/2)(\partial^2\alpha/\partial u^2) - (cT^2/\alpha)(\partial\alpha/\delta u)^2$. If the applied electric field is low, the term $(1 + \Lambda\vec{J}_0^2)^{-1}$ can be developed in serie. The corresponding expressions for (59)-(61) are:

$$\langle \delta u \; \delta u \rangle = kcT^2(1 - \Lambda\vec{J}_0^2)^{-1} \tag{62}$$

$$\langle \delta u \; \delta J_{ei} \rangle = (kcT^2/\alpha)(\partial\alpha/\partial u)J_{0i}(1 - \Lambda\vec{J}_0^2)^{-1} \tag{63}$$

$$\langle \delta J_{ei} \; \delta J_{ej} \rangle = (kcT/\tau v)(1 - \Lambda\vec{J}_0)^{-1}[\delta_{ij} + (cT^2/2)(\partial^2\alpha/\partial u^2)J_{0i}J_{0j}] \tag{64}$$

Up to the second order, the current-current fluctuation have the following correlations

$$\langle \delta J_{ei} \; \delta J_{ei} \rangle = (k\sigma_e T/\tau v)[1 + (cT^2/\alpha)(\partial\alpha/\partial u)^2 J_0^2] \tag{65}$$

Introducing in (65) the well-known relation

$$\sigma_e = (c_e e^2/m)\tau \tag{66}$$

with c_e the electron density, e the electron charge and m the electron mass, we get

$$\langle \delta J_{ei} \; \delta J_{ei} \rangle = (k\sigma_e T/\tau v)[1 + (c_e e^2 T/mc)v(E\tau/T)^2] \tag{67}$$

In equilibrium $\vec{E} = 0$ and (67) reduces to

$$\langle \delta J_{ei}\ \delta J_{ei} \rangle = (k\sigma_e T/\tau v) \qquad (68)$$

which is the expression of the well-known Nyquist formula for current fluctuations (Jou and Llebot, 1980). This expression also corresponds to the Green-Kubo expression for the conductivity for an exponential decay of the current fluctuations.

The relaxation time τ is related to the electron mean free path l by

$$\tau = l v_F^{-1} = l m (k_F h)^{-1} \qquad (69)$$

where k_F is the Fermi wave vector of the system. Taking into account the specific heat per unit mass for the electron gas (Wannier, 1966)

$$c = (\pi^2/2)(k^2 T/\varepsilon_F m) \qquad (70)$$

with the Fermi energy given by $\varepsilon_F = (hk_F)^2/2m$, we obtain from (67)

$$\langle \delta J_{ei}\ \delta J_{ei} \rangle = (k\sigma_e T/\tau v)[1 + \pi^{-2}(eEl/kT)^2] \qquad (71)$$

It is common to express this result in terms of the total intensity I through the conductor. In a resistor of cross section A and length L

$$I = A.J_e \qquad (72)$$

and $R = L/\sigma_e A$. If also it is assumed that the evolution of the current fluctuations is described by an equation like

$$\dot{\vec{J}}_e = = \tau^{-1}(\vec{J}_e - \sigma_e\vec{E}) \tag{73}$$

one gets for the spectral distribution of the time correlation function

$$<\delta I\ \delta I>_\omega = (2kT/R)[1 + \pi^{-2}(eEl/kT)^2](1+\omega^2\tau^2)^{-1} \tag{74}$$

If $w\tau<<1$, (74) reads

$$<\delta I\delta I>_w = (2kT/R)[1 + \pi^{-2}(eEl/kT)^2] \tag{75}$$

expression that gives the second moments of the electric current fluctuations near $\vec{J}_o=\sigma_e\vec{E}$. We may compare our result with the autocorrelation function obtained by other authors (Tremblay et al. in 1979; Tremblay and Vidal 1982) by means of a diagrammatic perturbation theory. They obtain, for a system of metalic electrons that interact only with dilute, static and isotropically scattering impurities, after removing some divergences arising from the Joule heating

$$<\delta I\delta I>_w = (2kT/R)[1 + 0.156(eEl/kT)^2] \tag{76}$$

As it may be seen, expression (75) is of the same order than expression (76) but our result has been obtained from purely macroscopic methods. From the phenomenogical entropy we also obtain some results of nonequilibrium fluctuations which are clearly beyond the scope of classical irreversible thermodynamics.

Electric current in a classical plasma

Finally, we study an instability of the electric current in a classical plasma under an electric field, as predicted by extended irreversible thermodynamics. Thus, we will show that our theory does not uniquely correct the standard thermodynamic theory, but may be useful for the description of quite relevant phenomena. (Jou et al 1989) The stability analysis is a crucial point in the understanding of the generalized equations of state, so that it deserves a careful analysis either from numerical and experimental perpectives.

As before, the generalized entropy per unit mass depends of the classical variables, in this case u,v and c_e plus the electric current density $\vec{J}_e$. Up to the fourth order in the electric current

$$s\ (u,\ v,\ c_e,\ \vec{J}_e) = s_{eq}\ (u,\ v,\ c_e) - \alpha\vec{J}_e.\vec{J}_e - \beta(\vec{J}_e.\vec{J}_e)^2 \qquad (77)$$

where the coefficients α and β may be calculated from fluctuation theory.

The second differential of the entropy, if, as before, an electric field $\vec{E}$ is present, reads

$$(\partial^2 s/\partial \vec{J}_e^{\,2}) = -2(\alpha + 6\beta\vec{J}_o^{\,2}) \qquad (78)$$

the thermodynamic stability theory requires (Casas-Vazquez 1982) this derivative to be negative. From (78) it can be easily seen that the derivative becomes positive when $\vec{E}$ becomes higher than the critical value

$$E_{crit} = (1/\sigma_e)(-\alpha/6\beta)^{1/2} \qquad (79)$$

In view of (79) a explicit calculation for α and β it is neededfrom equilibrium statistical mechanics. For the moments of the fluctuations of the fluxes we have

$$<(\delta J_{ei})^2> = \int d\vec{c}\int d\vec{c}' e^2 c_i c_i' <\delta f(\vec{c})\delta f(\vec{c}')> \tag{80}$$

and

$$<(\delta J_{ei})^4> = \\ = \int d\vec{c}\int d\vec{c}'\int d\vec{c}''\int d\vec{c}''' e^4 c_i c_i' c_i'' c_i''' <\delta f(\vec{c})\delta f(\vec{c}')\delta f(\vec{c}'')\delta f(\vec{c}''')> \tag{81}$$

where $\vec{c}$ is the relative particle velocity with respect to the barycentric motion of the system and $f(\vec{c})$ is the velocity distribution function. We have also taken into account that

$$\vec{J}_e = \int e\vec{c}\ f(\vec{c})\ d\vec{c} \tag{82}$$

From Callen's theory (Callen 1961), the moments of the fluctuations in the velocity distribution function can be calculated by starting from the Boltzmann definition of the entropy

$$s = -kv\int f(\vec{c})\ \ln f(\vec{c})\ d\vec{c} \tag{83}$$

Applying Callen's formalism, we define the thermodynamic conjugate of $f(\vec{c})$ as

$$F(\vec{c},\vec{c}) = -kv[1 + \ln f(\vec{c})]\delta(\vec{c}-\vec{c}') \tag{84}$$

where $\delta(\vec{c}-\vec{c}')$ is the Dirac's delta. For the second and fourth moments of fluctuatuations it is obtained

$$<\delta f(\vec{c})\ \delta f(\vec{c}\,')> = (1/v) f(\vec{c})\delta(\vec{c}-\vec{c}\,') \qquad (85)$$

and

$$\begin{aligned} <\delta f(\vec{c})\delta f(\vec{c}\,')\delta f(\vec{c}\,'')\delta f(\vec{c}\,''')> &= \\ &= (1/v^3) f(\vec{c})\delta(\vec{c}-\vec{c}\,')\delta(\vec{c}\,'-\vec{c}\,'')\delta(\vec{c}\,''-\vec{c}\,''') \\ &+ (1/v^2) f(\vec{c}) f(\vec{c}\,')\delta(\vec{c}-\vec{c}\,''')\delta(\vec{c}\,'-\vec{c}\,'') \\ &+ (1/v^2) f(\vec{c}) f(\vec{c}\,'')\delta(\vec{c}\,''-\vec{c}\,''')\delta(\vec{c}-\vec{c}\,') \\ &+ (1/v^2) f(\vec{c}) f(\vec{c}\,''')\delta(\vec{c}-\vec{c}\,'')\delta(\vec{c}\,'-\vec{c}\,''') \end{aligned} \qquad (86)$$

Equations (85) and (86) are introduced into ther macroscopic relations obtained acording the Callen's theory of fluctuations

$$<(\delta J_{ei})^2> = k/2\alpha \qquad (87)$$

$$<(\delta J_{ei})^4> = -(3\beta k^3/2\alpha^4) + (3k^2/4\alpha^2) \qquad (88)$$

taking for $f(\vec{c})$ the Maxwell-Boltzmann equilibrium distribution function. Thus, one obtains for the parameters α and β the follo values

$$\alpha = k(2e^2c_e^2kT)^{-1}\ ;\quad \beta = -km(8e^4c_e^4k^2T^2)^{-1} \qquad (89)$$

From this result, the critical value of the electric field results from (79), turns out to be

$$E_{crit} = (2/3)^{1/2}(c_e e/\sigma_e)(kT/m)^{1/2} \qquad (90)$$

To compare this result with a different independent calculation we turn our attention to plasma physics. When it is studied a fully ionised classical plasma under an external electric field (Krall and Trivelpiece, 1973), a starting point is the Fokker-Planck equation. If the electric field is small enough, it may be assumed that the acceleration of the particles impulsed by the electric field $\vec{E}$ is balanced by the collisional drag. If, in the contrary, the electric field is too strong, no steady state can be reached and the electrons acquire increasing speed and run away. It can be calculated, from the Fokker-Planck equation, the critical value of the electric field beyond which there is no steady state balance between collisions and the driving force, which results to be

$$E_{crit} = 0.99(2/3)^{1/2}(c_e e/\sigma_e)(kT/m)^{1/2} \tag{91}$$

The comparison between eqs (90) and (91) shows a good agreement between the macroscopic and the microscopic calculations.

Extended irreversible thermodynamics does not merely lead to minor corrections to the standard thermodynamic theory, but may be useful for the description of relevant phenomena, as the stability of the steady state of a completely ionized plasma. To accomplish this objective, the use of fourth order terms in the non equilibrium entropy has been essential. Furthermore, fluctuation theory reveals itself as a useful tool for the calculation of the coefficients appearing in the entropy, which are also essential for the numerical evaluation of our results.

Acknowledgements

This work has been partially supported by the Direccion General de Investigacion Cientifica y Tecnica of the spanish ministry of education under grant PB86-0287

References

Callen, H.B. 1961, Thermodynamics, New York: Wiley

Casas-Vazquez,J 1982, Stability of Thermodynamic Systems,(Lecture Notes in Physisc, 164) Berlin:Springer

Casas-Vazquez, J and D.Jou, 1981, On a phenomenological non equilibrium entropy for a class of rigid heat conductors, J.Phys.A, 14: 1225

Chester, M 1966, High-Frequency thermometry , Phys.Rev. 145: 76

Coleman, B. and C.Truesdell 1960, On the reciprocal relations of Onsager ,J.Chem.Phys. 33: 28

De Groot and P. Mazur 1962, Non equilibrium Thermodynamics, Amsterdam:North Holland

Eu, B.C. 1980, A modified method in irreversible thermodynamics, J.Chem. Phys. 73: 225

Jou, D, J.Casas-Vázquez and G. Lebon 1988, Extended irreversible thermodynamics, Rep.Prog.Phys. 51:1105

Jou, D, M. Ferrer and J.E.Llebot 1988, Extended irreversible thermodynamics and runaway electrons in plasma J.Phys.A:Math Gen. 21: L1039

Jou, D and J.E.Llebot 1980, Electric current fluctuations in extended irreversible thermodynamics, J.Phys A:Math Gen. 13: L47

Jou, D, J.E.Llebot and J.Casas-Vázquez 1982, Irreversible - thermodynamic approach to nonequilibrium heat fluctuations, Phys.Rev.A 25: 508

Jou, D, J.E.Llebot and J.Casas-Vázquez 1982, Thermodynamic aspects of nonequilibrium current fluctuations, Phys.Rev.A 25: 3277

Krall, A.N. and A.W.Trivelpiece 1973, Principles of plasma physics, New York: McGraw-Hill

Llebot, J.E. 1981, Termodinámica Extendida de sólidos rígidos conductores, PhD thesis, Bellaterra: U.A.B.

Llebot,J.E., D.Jou and J.Casas-Vázquez 1983, A thermodynamic approach to heat and electric conduction in solids, Physica 121A: 552

Tremblay A.-M.S., B.Patton, P.C.Martin and P.F.Maldague 1979, Microscopic calculation of the non linear fluctuations of a metallic resistor: the problem of heating in perturbation theory, Phys. Rev.A, 19: 1721

Tremblay A.-M.S. and F.Vidal 1982, Fluctuations in dissipative steady states of thin metallic films, Phys. Rev.B 23: 7562

Wannier, G.H. 1966, Statistical Physics, New York: Wiley

Chemically Reacting Systems in Extended Irreversible Thermodynamics

L. S. García-Colín
Departamento de Física
Universidad Autónoma Metropolitana-Iztapalapa
Apdo. Postal 55-534, 09340 México, D.F.

and

El Colegio Nacional, Luis González Obregón 23
06240, México, D.F.

ABSTRACT

Chemically reacting systems are long known to obey constitutive equations which are highly non linear in the independent variables chosen to describe the states of the system. The kinetic mass action law (KMAL) formulated over one hundred years ago by Guldberg and Waage expressing the relatioship between the chemical flux in terms of the concentrations of the intervening chemical species in single step homogeneous reactions is a typical example. It also gives rise to a non-linear relationship between the flux and the thermodynamic force, namely, the chemical affinity. Therefore, its content falls beyond the scope of linear non-equilibrium thermodynamics (LIT). The question then arises concerning the nature of a thermodynamical framework into which this and possibly other non-linear phenomena may be incorporated. In this paper I shall review some efforts that have been done to accomplish such a goal through one version of what is now referred to as Extended Irreversible Thermodynamics (EIT). The subject of fluctuations around a suitably defined stationary state in a chemically reactive fluid is also discussed in the light shed by EIT. Lastly, recent empirical work aimed at establishing a unique relationship between the KMAL and non-equilibrium thermodynamics is also critically discussed. The main outcome of all these ideas leads us to think that although the KMAL may be understood as a general result steming from the prionciples of EIT, after a short transient its well known form as proposed by Guldberg and Waage governs the time evolution of single homogeneous chemical reactions. This result is consistent with the local equilibrium assumption but not with the other postulates of LIT, clearly emphasizing the non-linear nature of chemical reactions.

INTRODUCTION

The formulation of a thermodynamic theory capable of describing processes ocurring in systems far from equilibrium is in the stage of development. Beyond the domain of the well established theory

now referred to as Linear Irreversible Thermodynamics (LIT), much work has been done to incorporate several features of processes not adequately described by this theory, but so far we still lack of an adequate formalism deeply rooted in the laws of microscopic physics. Among the processes which depart from the framework of LIT are chemical reactions. The macroscopic laws governing them have been a subject of close study for many years but still defy physical chemists in developing the thermodynamic frame to which they naturally belong. In this paper I wish to review the work which has been done in the past few years related to the description of some of the features characteristic of chemical reactions using Extended Irreversible Thermodynamics (EIT). Since the underlying basis of this theory has been accounted for by many other authors in this volume I will restrain myself here in explaining only those aspects of EIT which are relevant for the specific purposes of the paper.
Chemical reactions are rather complex phenomena, too complex to attempt to deal with them in the most general sense. Here we will think on single step reactions*, the intervening reactants not necessarily assumed to be ideal systems and occurring in fluid phases. Therefore, the reactions are governed by external.

*For simplicity we may write for species A_i,

$$\nu_1 A_1 + \ldots + \nu_r A_r \underset{k_b}{\overset{k_f}{\leftarrow}} \nu_{r+1} A_{r+1} + \ldots \nu_{r+p} A_{r+p}$$

constraints imposed by the observer which act on thermodynamic variables such as pressure (p), volume (V), temperature (T), etc. The variations in the concentrations (C_i for species i) of reactives and products is undertaken by internal constraints which, following Kestin, (Kestin 1978), may be conceived through the existence of anticatalysts or catalysts whose function are to arrest the reaction completely or to induce it to its full completion, respectively. In the particular case of homogeneous reactions, these internal constraints manifest themselves through Dalton's law of definite proportions and they appear in the thermodynamic equations through the variable known as the "degree of advancement", which is denoted by ξ. Following well known thermodynamic arguments (Kestin 1978), we remind the reader that the Gibbs equation may be then cast in the usual way,

$$dG=-SdT+Vdp-Ad\xi \tag{1.1}$$

where G is Gibb's free energy, S the entropy and A is the variable known as the affinity (de Donder 1923), which is defined by

$$A\equiv\sum_i \nu_i \mu_i \tag{1.2}$$

where the νi's are the stoichiometric coefficients taken to be positive numbers for the product and negative for the reactives and μ_i is the chemical potential of species *i*. In spite of the fact that eq. (1.1), may be justified using equilibrium thermodynamics arguments (Kestin 1978), the reader is encouraged to meditate about the distinct origin of the independent variables appearing therein. The temperature T and the pressure p correspond to constraints imposed by an extermal observer, where as ξ arises from the inevitable action of an internal constraint. Moreover, in order to integrate eq. (1.1), along a reversible path one has to restore to artificial methods because for any equilibrium state $A^{eq}=0$ and $\xi^{eq}=\xi^{eq}(p,T)$ is a well defined function of T and p. Therefore, in spite of the fact that the term $Ad\xi$ allows for formal equilibrium thermodynamics manipulations, its physical origin is inescapable and leads to rather confusing statements about its nature. Let us not forget that it was introduced by de Donder (de Donder 1928; García-Colín 1984) as the term that accounts for the Clausius uncompensated heat term in the case of chemical reactions conceived as irreversible processes. (Prigogine and Defay 1954).

Notwithstanding the subtleties mentioned above, the crucial question about the macroscopic description of chemical reactions arises when they are conceived as what they really are, namely, irreversible processes. Let us first examine this question by restoring to LIT. The local equilibrium assumption implies that, after taking a suitable Legendre transformation of eq. (1.1), and taking all extensive quantities per unit of total mass, the time rate of change of the local entropy density s $(\vec{r},t)$, is given by:

$$T\frac{ds}{dt}=\frac{de}{dt} - \frac{p}{\rho^2}\frac{d\rho}{dt} - A\frac{d\xi}{dt} \tag{1.3}$$

where ρ, is the local mass density ($\rho = v^{-1}$), and e is the local energy density. Here ξ now refers to the degree of advancement when the chemical potentials are taken per unit of total mass but we won't bother to use a different symbol for it. Also $\frac{d}{dt}=\frac{\partial}{\partial t} + u\cdot$ grad, u being the hydrodynamic velocity. In order to compute the hydrodynamic equations for a reactive mixture one then has to appeal to the well known conservation equations for e and ρ which must be supplemented with constitutive equations for the fluxes, q the heat flux and ℓi the diffusive fluxes. However, in order to compute $\frac{d\xi}{dt}$ since ξ is not a locally conserved density but a local variable arising from the internal constraints imposed by the reaction one has to restore to different arguments. In this sense, and only in this sense, it has a different status as its two other partners. How to evaluate $\frac{d\xi}{dt}$ then becomes an entirely different problem, which has been extensively dwelt with in the literature (de Groot-Mazur 1974, Kestin 1978, Pri gogine 1971, Eigen 1968, de la Selva and García-Colín, 1980 García-Colín 1984

Sieniutycz 1987). In particular I want to refer myself here to one of the possibilities to compute this term namely, that steming out from the rate laws. If the macroscopic rate laws governing the chemical reactions are assumed to be known then this term, in the linear approximation, simply reduces to a typical relaxation equation with a relaxation time which may be computed from the rate constants and the equilibrium concentrations. Once this is done the whole hydrodynamic picture of a reactive mixture follows in a straightforward manner. But the macroscopic rate laws are, as it is well known, non linear relations between the chemical flux and the concentrations of the species involved in the reaction. When such a relationship is written in terms of the thermodynamic force $^{A}/_{T}$ it reads as,

$$J=\ell(fe^{A/RT}-1) \tag{1.4}$$

where $J\equiv\frac{d\xi}{dt}$ is the chemical flux and ℓ and f are two factors whose form depends on the nature of the intervening species. For ideal systems, f=1 and $\ell=k_b\prod_{i=1}^{\nu}C_i^{\nu_i}$, k_b being the backward rate constant and the product is taken over all the products, C_i being the concentration of species i [Priogogine 1967, de Groot Mazur 1974]. After this brief excursion we may pose our objective in a rather precise language. Ordinary constitutive equations such as Fourier's law relating the heat flux to the temperature gradient, Fick's law relating the mass flux to the gradient of the chemical potential and so on, are known to be rather simple first approximations of more complicated equations which may be derived using the assumptions of EIT. This work has been discussed at length in the literature [García-Colín and López de Haro 1982, García-Colín 1988(a), López de Haro and del Rio 1990]. One may therefore pose the rather legitimate question of wondering if eq. (1.4), may not also find its place within some extended framework of irreversible thermodynamics. It is precisely the main objective of this paper, to review the work that has been done in providing a precise answer to this question. This does not mean that chemical reactions need for an extension of irreversible thermodynamics for their description. Once eq. (1.4) is accepted the local equilibrium assumption as stated in eq. (1.3) holds true although, the flux-force relation is not linear and therefore Onsager's reciprocity relations will only be valid in the linear regime. Thus LIT is not fully adequate to deal with those systems. We shall come back to this question later on. Moreover, once we accept the fact that the macroscopic rate laws belong to the formalism of EIT other interesting results may be derived from them such as some general statements about hydrodynamic fluctuations in chemically reactive fluids. This work will also be discussed in a later section. Finally a rather interesting consequence steming directly from the very intrinsic nature of eq. (1.4), indicating the precise physical nature of a somewhat novel empirical approach summarizing the behavior of chemical reactions ocurring in closed, isothermal vessels is also included here.

2. The Mass Action Law and EIT.

The underlying idea behind EIT [Jou et al. 1988] [García-Colín 1988a], is to enlarge the space of state variables used to describe a non-equilibrium state of a system by adding to the usual locally conserved densities others whose nature and number will depend on the characteristics of the non-equilibrium state. The rather difficult question that arises here is what variables are to be selected to broaden the state space. According to Grad [Grad 58], the fluxes appearing in the balance equations satisfied by the locally conserved densities are good candidates. This step was formally undertaken by Müller [Müller 67], who raised the heat, momentum and other fluxes to the status of independent space variables. The function which results and now substitutes the ordinary local entropy in LIT will no longer be such as was clearly manifested in Grad's work although it reduces to such a function when all the additional variables, the fluxes, become irrelevant in the description of the states of the system. Naturally, from the strict operational point of view it is difficult to recognize fluxes as thermodynamic variables. They are the response of the system to the action of a force (gradient) and not the measure of inherent internal constraints characterizing the system itself. Just as the degree of advancement in a chemical reaction is a meassure of an internal contraint reflecting the properties of the substances participating in it, one would like to imagine that other equally important internal constraints may be characterized by similar variables which could be named "internal affinities", (Kestin 1985). But so far this task has remained unaccomplished and the only practical alternative we have available is the Grad-Müller proposal.

Thus for a chemically reacting system, the first postulate of EIT demands the existence of a scalar, continous and at least twice differentiable function η of the ordinary conserved variables, $v(r,t)$, $e(r,t)$, the mass fractions $C_i(r,t)$ of species i, plus a subset of non conserved variables which are arbitrarily chosen to be the heat flux q, the diffusive fluxes ℓ_i and the chemical flux. J. This set of variables spans a space which we call the G space and specifies the state of a fluid particle. Thus $\eta=\eta(e,\nu,C_i,q,\ell_i,J)$. Since it is differentable, we may write that

$$d\eta=\left(\frac{\partial\eta}{\partial e}\right)de+\left(\frac{\partial\eta}{\partial\nu}\right)d\nu+\sum_i\frac{\partial\eta}{\partial C_i}dC_i+\frac{\partial\eta}{\partial J}dJ+\frac{\partial\eta}{\partial q}dq+\sum_i\frac{\partial\eta}{\partial\ell_i}d\ell_i \qquad (2.1)$$

Each term in eq. (2.1) must comply with the tensorial requirements pertinent to its property of belonging to G space in order to produce a scalar $d\eta$. Thus the sclars $\frac{\partial\eta}{\partial e}$, etc. are functions of all scalar invariants which may be contructed by the variables defining the space. Also, $\frac{\partial\eta}{\partial q}$ and $\frac{\partial\eta}{\partial \ell_i}$ are vectors in such a space and so on. In all the differential coefficients appearing in eq. (2.1.) it is understood that all other variables in the space are kept fixed. Moreover, since we want this theory to reduce to ordinary LIT when the subset of non conserved variables is irrelevant in describing the states of the system, we further assume that each of these quantities may be expanded around the local equilibrium state characterized by e, v and C_i.

If in this expansion we keep only those terms which are of first order in the non-conserved quantities $\frac{\partial\eta}{\partial e}$ reduces to the inverse local equilibrium temperature, T^{-1}, $\frac{\partial\eta}{\partial v}$ to the local equilibrium presure times T^{-1} and $\frac{\partial\eta}{\partial C_i}$ to the local chemical potential times $T^{-1}M_i^{-1}$ where M_i is the molar mass of species *i*. Furthermore,

$$\frac{\partial\eta}{\partial q} \equiv T^{-1}v\left(\alpha_o q+\sum_{\ell=1}^{n} \alpha_{oi}\ell_i\right)$$

$$\frac{\partial\eta}{\partial \ell i} \equiv T^{-1}v\ i=T^{-1}v\left(\gamma o_i q+\sum_{k=1}^{n} \gamma_{ik}\ell_k\right) \tag{2.2}$$

where the coefficients $\omega_o, \alpha_o, \alpha_{oi}$, γ_o and γ_{ij} are all evaluated at the local equilibrium state.

Under these conditions the expression for $\frac{d\eta}{dt}$ which will now play the role of the Gibbs equation (1.3), now reads as

$$T\frac{d\eta}{dt}=\frac{de}{dt} - \frac{p}{\rho^2}\frac{d\rho}{dt} - \sum_{i=1}^{n}\frac{\mu_i dC_i}{M_i\, dt} + \nu\alpha_o\frac{dq}{dt} + \nu\sum_{i=1}^{n}\gamma_{io}\frac{d\ell_i}{dt} + v\omega_o J\frac{dJ}{dt} \tag{2.3}$$

Emphasis should be made on the fact that in eq. (2.3) the time rate of change of the locally conserved variables is given through the well known balance equations whereas $\frac{dq}{dt}$, $\frac{d\ell_i}{dt}$ and $\frac{dJ}{dt}$ are unknown and must be provided by the formalism itself.

To answer the question raised in this last statement, one introduces the second postulate of EIT namely, the requirement that η satisfies a balance type equation of the form.

$$\rho\frac{d\eta}{dt} + \text{div } J_\eta = \sigma_\eta \tag{2.4}$$

where $J\eta$ and σ_η are a vector and a scalar defined in G space. Notice should be made that eq. (2.4), is in fact playing the role of a closure property requiring that all operations involved give rise to terms which belong to G space. Its nature has been profusely discussed and illustrated in a recent paper (Rodríguez and López de Haro 1989), although it is not relevant in our case. Since J_η is a vector in G space its must be of the form:

$$J_\eta = \beta q + \sum_{i=1}^{n} \beta_i \ell_i \tag{2.5}$$

where clearly the coefficients β and β_1 are functions of all scalar invariants in G space. By performing the same operations with these quantities as we outlined before for the coefficients of $d\eta$, keeping in mind that J_η has to reduce to the ordinary entropy flux vector of LIT when the non conserved variables are irrelevant for the description of the system, we may write eq. (2.5) as follows:

$$J_\eta = \frac{-}{T} q + \beta_o(\text{scal. inv})q - \sum_{i=1}^{n} \frac{i}{TM_i} \ell_i + \sum_{i=1} \beta_{io}(\text{scal. inv.})\ell_i \tag{2.6}$$

If we now calculate the divergence of J_η we add it to $\rho\frac{d\eta}{dt}$ obtained through eq. (3.3) and use,

$$\rho\frac{dC_i}{dt} = - \text{div } \ell_i \pm \nu_i M_i J \tag{2.7}$$

$$\rho\frac{de}{dt} = - \frac{1}{\rho} \text{div } q - p\frac{dv}{dt} + \frac{1}{\rho} \sum_i \ell_i \circ F_i \tag{2.8}$$

$$\frac{d\rho}{dt} = -\rho \text{div } u \tag{2.9}$$

the well known balance equations for the conserved quantities, F_i denoting the external force on species i, we obtain after a bit of algebra and convenient rearrangement of the resulting terms, that

$$\sigma_\eta = q \cdot \left[\text{grad } T^{-1} + \text{grad}\beta_o + \frac{\alpha_o}{T} \frac{dq}{dt} + \sum_{i=1}^{n} \frac{\gamma_{oi}}{T} \frac{d\ell_i}{dt} \right]$$

$$+\sum_{i=1}^{n} \ell_i \cdot \left[\text{grad} \left(-\frac{\mu_i}{TM_i} \right) + \text{grad}\, \beta_{io} + \frac{\alpha_{oi}}{T} \frac{dq}{dt} + \sum_{i=1}^{n} \frac{\gamma_{ij}}{T} \frac{d\ell_j}{dt} + \frac{1}{T} F_i \right]$$

$$+\beta_o \text{div}\, q + \sum_{i=1}^{n} \beta_{io} \text{div}.\, \ell_i + \frac{AJ}{T} + \omega_o \frac{J}{T} \frac{dJ}{dt} \tag{2.10}$$

where A is the chemical affinity as defined in eq. (1.2). Eq. (2.10) is the expression for σ_η as it comes from the operations indicated in the left hand side of eq. (2.4).

On the other hand, σ_η is a scalar function of all the variables defining the G space. Thus we note that its most general form is

$$\sigma_\eta = P'(\text{scal. inv.}) + q \cdot Xq + \sum_{i=1} Xi \cdot \ell i \tag{2.11}$$

where P' is a function of all scalar invariants and Xq, X_i for the same reasons as those argued for σ_η have a structure identical to eq. (2.5) namely,

$$X_q = \xi_i q + \sum_{i=1}^{n} \xi_i \ell_i$$
$$\tag{2.12}$$
$$X_i = \lambda_i q + \sum_{j=1}^{n} \lambda_{ij} \ell_j$$

In order to compare the two expression for σ_η which must be identical, we recall that the quantities $\beta o, \beta o_i$ and P' are functions of <u>all</u> scalar invariants in G space, J included. Thus we may as well write that

$$\beta_o = \beta_o \ J$$
$$\tag{2.13}$$
$$\beta_{io} = \beta_{io}^{1} \ J$$

$$P'(\text{scal. inv.}) = JP(J, \text{scal. inv.})$$

If we now substitute (2.13) into eq. (2.10) and equate the corresponding coefficients of $q, \sum_i \ell_i$ and J we obtain the desired time evolution equations for the fluxes, namely:

$$\frac{\alpha_o}{T}\frac{dq}{dt} + \sum_{i=1}^{n} \frac{\gamma_{oi}}{T}\frac{d\ell_i}{dt} = - \text{grad}\ T^{-1} - \text{grad}\ (\beta_o^1 J) + \xi_o q + \sum \xi_i \ell_i \tag{2.14}$$

$$\frac{\alpha_{oi}}{T}\frac{dq}{dt} + \sum_{i=1}^{n} \frac{\gamma_{ij}}{T}\frac{d\ell_j}{dt} = \text{grad}\left(\frac{U_i}{M_i T}\right) - \text{grad}\ (\beta_{io}^1 J) + \lambda_i q + \sum_j \lambda_{ij}\ell_j - \frac{1}{T} F_i \tag{2.15}$$

$$\frac{\omega_o}{T}\frac{dJ}{dt} = - \frac{A}{T} - \beta_o^1\ \text{div}\ q - \sum_{i=1}^{n} \beta_{oi}^1\ \text{div}\ \ell_i + P(J\ldots) \tag{2.16}$$

These equations constitute the main result of EIT. They represent the time evolution equations for the nonconserved variables $q, \ell_1, \ldots, \ell_n$ and J which together with eqs. (2.7)-(2.9) form a complete set of equations which may in principle be solved if all the coefficients appearing in them are known from experiment or from microscopic theories. The system because of the restrictions imposed in keeping only terms which are linear in the non-conserved variables and for a homogeneous system in which $\frac{d}{dt}=\frac{\partial}{\partial t}$, is a linear set of coupled equations. It is true that as it stands it is a rather complicated one to analyze in full detail. But as we shall see below, it already contains the result we are searching for, namely, a rather general statement about the kinetic mass action law (García-Colín et al. 1985, 1986).

Let us restrict ourselves to the case of a homogeneous reactive fluid in which there is no heat conduction nor diffusion. Then eq. (2.16) reduces to the expression

$$\frac{\omega_o}{T}\frac{dJ}{dt} = - \frac{A}{T} + P(J,\ldots) \tag{2.17}$$

Eq. (2.17) implies that according to EIT the time rate of change of the chemical flux can only be a function of the local state variables T, p, etc. through ω_o, A and of J itself. This is in fact the rather general content of the mass action law. If is also important to emphasize that a general macroscopic theory such as EIT cannot go beyond a statement of this nature, a fact that has been often misunderstood in the literature (Lebon et al 1987, García-Colín 1988b, Lebon et al 1989, Ross and García-Colín 1989). The specific dependence of ω_o and P(J) with J itself and the local variables can only be obtained through experiment or a microscopic model (Ross and Mazur 1961), in much the same way that the equation of state is a general consequence of the zeroth law of thermostatics, not its analytical form. Still as it was shown in the original paper (García-Colín et al 1985-1986), an additional ansatz does provide same enlightment about this question. For the simple situation under examination, σ_η simply becomes equal to JP(J). We also know that in the specific case of LIT, its counterpart, the entropy production is a positive definite quantity. In such case $\omega_o T^{-1}\frac{dJ}{dt}=0$ and $P(J)=\frac{A}{T}$ so that $\sigma=AJT^{-1}$, the well known expression for the entropy production (de Groot Mazur 1974, Prigogine 1967). With this in mind let us now assume that for a single step reaction P(J) is of the form

$$P(J)=R\ell n(1+{}^{J}/_{\ell}) \tag{2.18}$$

where R is the gas constant and ℓ a positive undetermined parameter. Thus, JP(P) is a positive quantity and furthermore, eq. (2.17) now reads as

$$\omega_o T^{-1}\frac{dT}{dt}= - AT^{-1}+R\ell n(1+{}^{J}/_{\ell}) \tag{2.19}$$

Eq. (3.14), is an equation of the Maxwell-Vernotte-Cattaneo type of relaxation equations with ${}^{\omega o}/_{T}$ playing the role of the relaxation time. Those equations have played a crucial role in

generalizing thermodynamics beyond its linear regime and are extensively discussed in the literature (Gyarmati 1977, Jou et al. 1988, García-Colín 1988a). When ${}^{\omega o}/_{T}$ is very small so that ${}^{\omega o}/_{T}\,{}^{dJ}/_{dt}$ 0, that is the chemical flux, after a very small transient looses its status of independent variable, eq. (2.19) reduces to

$$J=\ell(e^{A/RT}-1) \tag{2.20}$$

which is the well known expression for the kinetic mass action law KMAL. Notice however that in the derivation of eq. (2.20) no assumption has been made whatsoever on the nature of the reacting species wether liquids or gases, ideal or not. When such species are ideal then ℓ reduces to its ordinary expression as pointed out in section 1. Moreover, once the regime in which eq. (2.20) holds is established, chemical reactions are at grips with the local equilibrium assumption (Ross and García-Colín 1989), but not with the whole framework of LIT.

Therefore we may conclude that the KMAL and indeed the whole set of equations describing the behavior of one-step chemical reactions ocurring in fluid systems may be looked upon as consequence of the postulates of EIT. In the case in which the heat and diffussive fluxes are present the resulting equations are more complicated leading to many other interesting features and we refer the reader to the original sources for all the details (García-Colín et al. 1986, Sieniutycz 1987).

3. Chemical Fluctuations in Extended Thermodynamics.

The theory of fluctuations around equilibrium states is by now a rather well developed subject. The main ideas behind it date back to Einstein and Smoluchowski (Einstein 1956), and they constitute the starting point of the Onsagerian approach to LIT (Onsager 1931, Onsager and Machlup 1953, 1954). For the case of chemically inert fluids they have been systematically analyzed by Fox and Uhlenbeck (Fox 1978), and extended to binary mixtures by Foch (Foch 1971). For fluctuations around equilibrium one assume that local equilibrium holds and with this assumption one can write the change of entropy of the system caused by the fluctuations in terms of the thermostatic properties. Second, the entropy change is assumed to obey the Boltzmann-Einstein-Smoluchowski formula for the probability of a state for which fluctuations in the variables around their respective averages occur. Third, the fluctuations are even in time and their regression equations are assumed to be of the same linear form as the one describing the evolution of macroscopic quantities towards equilibrium. These three hypothesis lead to a fluctuation-dissipation expression from which the non local in space and time correlation functions are obtained. More generally, it has been shown (Fox 1978), that independently of the parity in time of the variables, if the fluctuations obey a linear regression law and are regarded as a Gaussian and Markovian process, then a fluctuation-dissipation relation is satisfied from which the correlations are obtained. This approach was extended to a chemically reactive fluid by de la Selva (de la Selva 1986), and has also been widely used for reactive and other type of systems by Keizer (Keizer 1987).

Here we wish to discuss how these ideas have been used within the context of EIT and the requirements that have to be imposed on the underlying equations. For this purpose we shall restrict ourselves to a two component fluid in which a single chemical reaction takes place disregarding viscous processes and heat conduction so that eq. (2.3), reduces to

$$T_o d\eta = de + p_o dv - \mu_A dC_A - \mu_B dC_B + \omega_o vJdJ + \gamma_{01} vL\cdot dL \qquad (3.1)$$

where the subscripts naught label the equilibrium variables of the corresponding quantities and μ is the chemical potential per unit mass. The single diffusive flux present in the system is $\ell_1 \equiv L$ Correspondingly, eqs. (2.14) and (2.16) reduce to

$$\frac{\gamma_{o1}}{T_o}\frac{dL}{dt} = \text{grad}\,\frac{\mu}{T_o} - \text{grad}\,\beta_{01}J + \lambda_1 L \qquad (3.2)$$

$$\frac{\omega_o}{T}\frac{dJ}{dt} = -\frac{A}{T_o} - \beta_{01}\ \mathrm{div}\ L + P\ (J,\dots) \tag{3.3}$$

where $\mu=\mu_A-\mu_B$ and the coefficients γ_{01}, ω_o and β_{o1} are scalar functions depending on local equilibrium variables. In order to proceed with the theory, we introduce the ansatz stated in eq. (2.18), for P(J) but we shall assume that J/ℓ is a small number so that $\ell n\ (1+x) \simeq x$ and that grad J is negligible. Therefore,

$$\frac{dJ}{dt} = \frac{T_o}{\omega_o}\left(\frac{RJ}{\ell} - \frac{A}{T_o}\right) \tag{3.4}$$

$$\frac{dL}{dt} = \frac{T_o}{\gamma_{o1}}\left(T_o^{-1}\mathrm{grad}\ \mu+\lambda_1 L\right) \tag{3.5}$$

are now two linear uncoupled equations for the two "fast" variables J and $\vec{L}$. Denoting by J_o and $\vec{L}_o$ the steady state solutions of eqs. (3.4 and (3.5), which characterize the stationary state around which $\vec{\tilde{L}}$ and $\tilde{J}$ will represent their fluctuations, we find that:

$$\tilde{J}(t)=\tilde{J}_o \exp\left\{\frac{RT_o}{\omega_o \ell}\right\} t \tag{3.6}$$

$$\tilde{L}(t)=L_o \exp\left\{\frac{T_o\lambda_1}{\gamma_{01}}\right\} t \tag{3.7}$$

where

$$\tau_{ch} \equiv -\frac{\omega_o \ell}{RT_o} \qquad \tau_d = -\frac{\gamma_{01}}{T_o\lambda_1} \tag{3.8}$$

are the relaxation times for the fluxes. Recalling that from LIT, $J_o = \frac{\ell rr}{\rho_o T_o}$ and $\vec{L}_0$=-D grad μ where D is the diffusion coefficient, we get that $\ell_{rr}=\ell R^{-1}$ and $\lambda_1=(T_o D)^{-1}$ so that in eqs. (3.8) the only unknown parameters are ω_o and γ_{01}. If we now use eqs. (3.6) and (3.7) to construct the corresponding correlation functions averaging over and ensemble characterizing the stationary state, we get that

$$\left\langle \tilde{J}(r,t)\tilde{J}(r,t)\right\rangle = \left\langle \tilde{J}_o(r,o)\tilde{J}_o(r,o)\right\rangle e^{-t/\tau_{ch}} \tag{3.9}$$

$$\left\langle \tilde{L}(r,t)\tilde{L}(r,t)\right\rangle = \left\langle \tilde{L}(r,o)\tilde{L}(r,o)\right\rangle e^{-t/\tau_d} \tag{3.10}$$

so that in principle, ω_o and γ_{01}, may be determined from the mesoscopic information provided by the correlation functions appearing in eqs. (3.9) and (3.10). Using these results we are now in a position to discuss the possibility of computing correlations in the realm of EIT.

Going back to eq. (3.1), we notice that the first four terms in the right hand side represent the change ΔS in the entropy of the whole isolated system which is also equal to $\Delta\eta$ minus the two terms containing the contribution in the fast variables. Since these terms are quadratic in the fluxes, we assume that the Einstein-Smoluchowski formula for the probability of fluctuations arround a global equilibrium state (Callen 1985), holds in the form,

$$W(\tilde{\vec{L}}, \tilde{J})=\Omega_o \exp \frac{1}{k}\left[\Delta\eta - \frac{\omega_o}{T_o} J_o\Delta J - \frac{\gamma_{01}}{T_o} \tilde{L}_o \cdot \Delta L \right] \tag{3.11}$$

where Ω_o is a normalization constant and Jo, $\vec{L}o$ are the stationary state values of the fluxes occuring between a fluid particle and rest of the system which thus plays the role of surroundings. Also, $\Delta X \equiv \tilde{X}$. Switching to the language of availability and omitting the work of expansion term (3.11) is readily rewritten as,

$$W(\tilde{L}, \tilde{J})=\Omega_o^1 \exp\left\{ - \frac{1}{kT_o} \left[\Delta e - T_o\Delta\eta + \omega_o J_o \Delta J + \gamma_{01}\vec{L}_o \cdot \Delta L - \mu_o \Delta C \right]\right\} \tag{3.12}$$

Following now the standard procedure, (Callen 1985, Landau & Lifshitz 1958) expanding e in terms of L, J, η and C($\equiv C_A$) and keeping terms to second order around the stationary state, eq. (3.12) is cast into the form

$$W=\Omega_0^1 \exp \left\{ - \frac{1}{2kT_o} \left[\Delta T\Delta\eta - \Delta(\omega_o J)\Delta J + \Delta\mu\Delta C - \Delta(\gamma_{01} L_\alpha)\ \Delta L_\alpha \right] \right\} \tag{3.13}$$

where $\propto \equiv (x,y,z)$ Eq. (3.13) may be regarded as the extension of the canonical form for the probability that a fluctuation occurs in a small subsystem of a closed thermodynamical system, to the domain of chemically reacting systems within the framework of EIT. From here on the computation of the ensemble averages of the fluctuating quantities, here depending on space and time coordinates follows in the same way as in the case of fluctuations around thermodynamic equilibrium. Therefore, if we regard $\Delta\eta$, $\Delta(\omega_o J)$ ΔC and $\Delta(\gamma_{01} L_\propto)$ as functions of ΔT, ΔJ, $\Delta L_\propto$ and $\Delta\mu$ and follow the standard methods (Landau and Lifshitz 1958), then after some laborious but straightforward algebra we get that,

$$\left\langle \tilde{J}(r,t)\ \tilde{J}(r,t) \right\rangle = -\frac{kT_o}{\omega_o} \tag{3.14a}$$

$$\left\langle L^{\sim}_{\alpha}(r,t)\ L^{\sim}_{\beta}(r,t) \right\rangle = -\frac{kT}{\gamma_{01}}\ \delta_{\alpha\beta} \tag{3.14b}$$

$$\left\langle \tilde{L}_{\alpha}(r,t)\tilde{J}(r,t) \right\rangle = 0 \tag{3.14c}$$

$$\left\langle \tilde{T}(r,t)\ \tilde{T}\ (r,t) \right\rangle \tag{3.14d}$$

$$\left\langle \tilde{e}\ (r,t)\ \tilde{e}\ (r,t) \right\rangle = kT_o^2\ C_{JL\mu} + kT_o\left(\frac{\partial e}{\partial \mu}\right)\left[\left(\frac{\partial e}{\partial T}\right) - T_o\left(\frac{\partial \mu}{\partial T}\right) + \mu\right]$$
$$- kT_o\left[\frac{1}{\omega_o}\left(\frac{\partial e}{\partial J}\right)^2_{T\mu J} + \frac{1}{\gamma_{01}}\left(\frac{\partial e}{\partial L_\alpha}\right)_{T\mu J}\right] \tag{3.14e}$$

$$\left\langle \tilde{C}\ (r,t)\ \tilde{T}\ (r,t) = \left\langle \tilde{J}\ (r,t)\ \tilde{T}\ (r,t) \right\rangle = \left\langle \tilde{J}\ (r,t)\ \tilde{\mu}\ (r,t) \right\rangle\right.$$
$$= \left\langle \tilde{L}_\alpha(r,t)\ \tilde{\mu}\ (r,t) \right\rangle = \left\langle \tilde{L}_\alpha(r,t)\ \tilde{T}\ (r,t) \right\rangle = 0 \tag{3.14f}$$

Some remarks regarding the nature of the set of eqs. (3.14) are pertinent. In the first place mention should be made of the fact that the coefficients ω_o and γ_{01}, have been taken strictly as constants. Secondly, the auto correlation function for the local energy density differs from its conventional expression given by the first term of eq. (3.14e), if the presence of fast variables J and L are taken into account and the fluctuating subsystem is taken as an open one. Thirdly, the results expressed in equation (3.14f), are new. For small fluctuations around the stationary state defined through eqs. (3.4) the cross correlations of the variables appearing therein are zero. This result was first proved by de la Selva en 1986, (de la Selva 1986).

Let us now concentrate in equations (3.14 a,b). Making use of eqs. (3.6), (3.7) and (3.8), they may be readily rewritten as follows,

$$\left\langle \tilde{J}(r,t)\tilde{J}(r^1,t^1) \right\rangle = k\ \ell_{rr}\tau^{-1}_{ch}\delta(r-r^1)e^{-(t-t^1)\tau_{ch}} \tag{3.15}$$

and

$$\left\langle \tilde{L}_\alpha(r,t)\tilde{L}_\beta(r^1_,\ t^1) \right\rangle = k\ T_o D\tau_d^{-1}\delta_{\alpha\beta}\delta(r-r^1)e^{-(t-t^1)/\tau_d} \tag{3.16}$$

where the delta correlation in configuration space has been tacitly assumed. Eqs. (3.15), and (3.16), are the main results of this discussion. They imply that in a linear regime taken around the steady state resulting from eqs. (3.4) and (3.5), the "noise" associated with the chemical and diffusive fluxes in a chemically reacting fluid is non-gaussian. This fact stems directly from the fact that the time evolution equations for these quantities here raised to the status of independent variables are linear and in homogeneous in the fluxes themselves and therefore foreign to the linear regression assumption of Onsagerian thermodynamics. Clearly, when τ_d and τ_{ch} tend to zero we recover the well known results of ordinary LIT (Fox 1978).

Summarizing, we have shown here that fluctuations in chemically reactive systems can be taken into account in the Maxwell-Cattaneo-Vernotte regime of EIT. These fluctuations contain the important ingredient that their time decay obeys an exponential regression instead of the linear regression of LIT. Furthermore the still undetermined parameters ω_o and γ_{01} appearing in the relaxation times of such fluctuations may be in principle determined from knowledge, experimental or theoretical, of the self correlation functions for the chemical and the diffusive fluxes.

4. <u>Recent Results</u>

Although EIT provides for some enlightment to understand the general macroscopic framework in which chemical kinetics may be placed, once we accept the validity of the kinetic mass action law (KMAL), as given by eq. (2.20), or in its equivalent form applicable to ideal substances as it was formulated by Guldberg & Waage (Glasstone 1947), the whole time evolution of chemical reactions is consistent with the local equilibrium assumption (Ross 1989). Yet, the KMAL is mechanistic in nature and not thermodynamic so that one might still ask if there is a general thermodynamic criteria which is suitable for predicting how a chemical reaction evolves in time. This question has been tackled in great detail in the past few years by (Garfinkle 1989), and poses some interesting questions.

Consider the class of homogeneous chemical reactions ocurring in closed vessels and at constant temperature. If X_i denotes de mass fraction of the $i^{\underline{th}}$ species, the affinity A may be written as

$$A=A^{\circ}+RT\sum_{i=1}^{n} \ell n\, X_i^{\nu_i} \tag{4.1}$$

where A° is the affinity referred to a standard state and ν_i the stoichiometric coefficient of species i. Recalling that in equilibrium $A^{eq}=0$ so that $A^{\circ}= -RT\ell nK^*$ where K^* is the equilibrium constant, eq. (4.1), may be trivially rewritten as

$$A = -RT\,\ell n\xi_Q \qquad (4.2)$$

where $\xi_Q = K^* / \prod_i (X_i)^{\nu_i}$ and clearly $0 \le \xi_Q \le 1$, ξ_Q is an adimensional paramenter measuring the degree of advancement of the reaction from the initial arbitrary state until equilibrium is reached. If we now invoke the local equilibrium assumption thus asserting that eq. (4.2), is valid at any time t, we may write that for any homogeneous isothermal reaction occurring in a closed vessel,

$$\dot{A}_{T,V} < 0 \qquad (4.3)$$

where $\mathring{A} \equiv \frac{\partial A}{\partial t}$. This result is the ultimate consequence that one may draw from macroscopic thermodynamics. Since A the affinity is a state variable, eq. (4.3), is valid independently of the path actually traversed by the reacting system. One should also remember at this stage that for a given reaction mechanism the rate laws provide also for an expression for the affinity from which $\mathring{A}$ can be computed and the result compared with the thermodynamic inequality (4.3). We shall come back to this question in a moment.

What eq. (4.3), suggests is that from the thermodynamic point of view every homogeneous reaction, taking place in a closed vessel and at constant temperature must follow a unique path in the space whose coordinates are the time and the affinity such that $\mathring{A}$ can never increase. This behavior must occur in the time interval starting with the time at which the reaction is initiated and ending after the finite time at which the reaction is completed at least from the statistical point of view. Yet thermodynamics is unable to provide the "dynamic equation of state", $\mathring{A}=f(t)$ satistfying these requeriments. This fact was clearly recognized by Garfinkle (Garfinkle 1982-1983a), who by close examination of over one hundred reactions for which the kinetic data are given in terms of concentration of partial pressures to comply with the use of eq. (3), reached the conclusion that the empirical data for such systems can be summarized in a single equation namely,

$$\dot{A}_{T,V} = A_r\left(\frac{1}{t} - \frac{1}{t_k}\right) \qquad (4.4)$$

where $A_r < 0$ and t_k are two parameters characteristic of each reaction. Moreover to check the consistency of eq. (4.4), with the KMAL, noticing that for ideal systems $X_i \sim C_i$ and using the standard definition for the chemical flux $J \equiv \frac{1}{\nu_i}\frac{dC_i}{dt}$ one gets by time differentiation of eq. (4.1), that

$$-J=\frac{\dot{A}_{T,V}/RT}{\sum_i \nu_i^{2}/C_i} \tag{4.5}$$

Eq. (4.5), is the most general relationship which may be obtained between the chemical flux and $\mathring{A}_{T,V}$ from thermodynamic considerations. Together with eq. (4.4) it allows to establish a consistency criterion between the KMAL and thermodynamics. Indeed $\mathring{A}_{T,V}$ may be determined from eq. (4.4), without apparently restoring to kinetic considerations, provided of course A_r and t_k are known for a given reaction, a determination that implicitly depends on the existence of reliable kinetic data. On the other hand for such reaction J is known, for each possible reaction mechanism through the KMAL. Then, for each set of concentrations determined from kinetics and J, $\mathring{A}_{T,V}$ itself may be determined using eq. (4.5), and the results compared with those provided by the rate laws. The details of this lengthy analysis is given in the literature so that we urge the interested reader to consult the original papers (Garfinkle 1983, 1984). What one finds is that indeed the data obtained from eq. (4.4), agrees very well with the points for $\mathring{A}_{T,V}$ computed using eq. (4.5), for one of the many mechanisms proposed for each reaction. One is therefore lead to think that indeed eq. (4.4) seems to provide for the sought thermodynamic criteria explicitly exhibiting the nature of inequality (4.3). And furthermore since the linear relationship between J and A is clearly seen not to hold true (Garfinkle 1983, 1989), showing once more that chemical reactions lie outside the realm of LIT one immediately wonders if eq. (4.4), may be well fitted into a generalized version of thermodynamics, in particular into EIT.

Surprisingly enough, this is not the case. The details of the ensuing discussion will soon be published (García-Colín et al 1990), so I shall only underline the main issues. Going back to eq. (4.5), one might conceive the outcome of computing $\mathring{A}$ from the general expression for the chemical flux as provided by EIT namely from Eq. (2.20). If we call $\mathring{A}_k$ this affinity, we get that

$$\dot{A}_K = - RT \sum_i \frac{\nu_i^{2}}{C_i}\, \ell \left[e^{A/RT}-1\right] \tag{4.6}$$

Recalling that for ideal systems $\ell = k_b \pi C_j^{\nu}{}_j$, j ranging over all the products, that $K_{eq} = \frac{k_b}{k_f}$ the ratio between the backward and forward rate constants and that A is given by eq. (4.2), one finds after a few algebraic steps that

$$\mathring{A}_K = \mathring{A}_{T,V} \qquad (4.7)$$

which simply implies that $\mathring{A}_{T,V}$ is uniquely given through the KMAL. Furthermore the explicit analytical dependece of $\mathring{A}_{T,V}$ with time will of course be different for different reaction mechanisms. What is then the significance of eq. (4.4)? As we see from eq. (4.7), it cannot be placed in EIT. The answer came after some thought and turned out to be rather simple.

Starting from Dalton's law of definite propertions we may write that

$$\frac{1}{\nu_i}\left[C_i(t)-C_i(0) \right] = \frac{1}{\nu_j}\left[C_j(t)-C_j(0) \right] \qquad (4.8)$$

for arbitrary i and j and any time t.

Let us now assume that

$$C_r(0) \not\equiv 0; \text{ r (reactives)}$$

$$Cp(0) = 0; \text{ p (products)} \qquad (4.9)$$

and $C_i(t)$ is an analytic function of time. Therefore $C_i(t)$ may be expanded in a Taylor series around t=0 to read

$$\frac{1}{\nu_i}\left[C_i(t)-C_i(0)(\text{only reactives}) \right] = \sum_{\ell=1}^{\infty} F_\ell t^\ell \qquad (4.10)$$

where for simplicity we assume that F_ℓ=0 for all ℓ's. If we now compute $\mathring{A}$ from eq. (4.1) with $X_i \sim C_i$, substitute eq. (4.10), and rearrange the resulting expression in a convenient way, we get that

$$\dot{A} = -RT\left\{ \sum_{i=1}^{r} \nu_i \left(\frac{1}{t} \right) + \frac{F_2}{F_1} \sum_{j=r+1}^{r+p} \nu_j + F_1 \sum_{i=1}^{r} \frac{\nu_i^2}{C_i(0)} \right\} + 0(t) \qquad (4.11)$$

where (t) represents all those terms which go to zero when t 0. The comparison of eq. (4.11) with eq. (4.4) simply shows that

$$Ar = - RT \sum_{i=1}^{r} \nu_i \qquad (4.11)$$

$$\frac{A_r}{t_k} = RT \left\{ \frac{F_2}{F_1} \sum_{i=r+1}^{r+p} \nu_i + F_1 \sum_{i=1}^{r} \frac{\nu_i^2}{C_i(0)} \right\} \tag{4.12}$$

if we ignore all terms O(t). This is course is not strictly correct. As it has been shown in specific examples where the coefficients F_ℓ can be explicitly computed from the rate laws, the values of A_r obtained from eq. (4.12), agree rather well with those obtained by Garfinkle in according to eq. (4.4), but the error found between the terms A_r/t_k is not necessarily negligible. This is clear if we recall that eqs. (4.12), ignore the presence of all those terms which appear as powers of t^ℓ $\ell \geq 1$. The conclusion is that in the numerical analysis that led him to eq. (4.4), Garfinkle picked up the dominant or leading term $^1/_t$ and gave the best fit of the data by using only two parameters A_r and t_K. Consistently with the analysis here presented in all the reactions he examined the initial concentrations of the products are stricly zero. As the reader may check for himself the whole argument breaks down if eq. (4.9), is not satisfied, eq. (4.11) having a much more complicated structure.

We are thus led to think that for the whole class of chemical reactions governed by rate laws of the general form given in eq. (2.20), their time evolution towards equilibrium is consistent with the local equilibrium assumption and the quest of finding a general thermodynamic criterium as suggested by eq. (4.4), from which one could single out a priori a the reaction mechanism consistent with thermodynamics, remains an open one.

REFERENCES

1. Callen, H.B. 1985. Thermodynamics and an Introduction to Thermostatistics. 2nd ed. New York: John Wiley and Sons.

2. de Donder, Th. 1928, L'Affinité. Paris: Gauthier-Villars.

3. de Groot, S.R.; Mazur, P. 1974. Non-equilibrium Thermodynamics. New York: Dover Publications, Inc.

4. del Rio, J.A. López de Haro, M. 1990. On the criteria for deriving approximations of different orders in Extended Irreversible Thermodynamics. J. Non-equilib. Thermodyn. 15, (in press).

5. de la Selva S.M.T., García-Colín L.S. 1980. The Bulk Viscosity and the chemical relaxation measurements within the context of thermodynamics. KINAM (México) 2: 435-453.

6. de la Selva S.M.T., García-Colín, L.S. 1986. Hydrodynamic Fluctuations in a Chemically reacting fluid. J. Chem. Phys. 85: 2140-2146.

7. Eigen, M. de Maeyer L. 1968. "Relaxation Methods in Techniques in Organic Chemistry. A. Weissberger, S.L. Friess and E.S. Lewis, eds. Vol. III Chap 18, New York: John Wiley & Sons.

8. Einstein, A. 1956. Investigations on the Theory of the Brownian Movement. New York: Dover Publications, Inc.

9. Foch, J. 1971. Stochastic Equations for Fluid Mixtures. Phys. Fluids 14: 893-897.

10. Fox, R.F. 1978. Gaussian Stochastic Processes in Physics. Phys. Rep. C. 48: 179-283.

11. García-Colín L.S., López de Haro M. 1982. The Burnett Equations in Extended Irreversible Thermodynamics. J. of Non-Equilib. Thermodyn. 7: 95-104.

12. García-Colín L.S. 1987. Extended Irreversible Thermodynamics and Chemical Kinetics. in Lecture Notes in Physics 199 Berlin: Springer-Verlag. 144-165.

13. García-Colín L.S., de la Selva S.M.T., Piña E. 1985. The Mass Action Law Extended Irreversible Thermodynamis. Phys. Letters. 110A: 363-365.

14. García-Colín L.S., de la Selva S.M.T., Piña E. 1986 Consistency of the Kinetic Mass Action Law with Thermodynamics. J. Phys. Chem. 90: 953-956.

15. García-Colín L.S., 1988a. Extended Irreversible Thermodynamics: Scope and Limitations. Rev. Mex. Fís 34: 344-366.

16. García-Colín L.S.1988b. Comments on The Kinetic Mass Action Law Revisted, by Thermodynamics J. Phys. Chem. 92: 3017-3018.

17. García-Colín L.S., de la Selva S.M.T., Piña E., 1990. On the Thermodynamic basis of the affinity decay rate. J. Chem. Phys. (in press).

18. Garfinkle, M. 1982., Non-equilibrium Thermodynamics of Closed-System Reactions. Materials Chem. and Phys. 7:, 359-394.

19. Garfinkle, M. 1983. Reaction Velocities and the Affinity decay rate. J. Chem. Phys. 79: 2779-2785.

20. Garfinkle, M. 1984. Temperature dependency of the affinity decay rate. J. Chem. Phys. 80: 3640-3643.

21. Garfinkle, M. 1989. Natural Path in Chemical Thermodynamics. J. Phys. Chem. 93: 2158-2164.

22. Glasstone, S. 1947. Textbook of Physical Chemistry. New York: Van Nostrand.

23. Grad, H. 1958. Principles of the Kinetic Theory of Gases in Hand buch der Physik Band XII, S. Flügge, ed. Berlin: Springer Verlag. 205-294.

24. Gyarmati, I. 1977: On the Wave Approach of Thermodynamics and some problems of non-linear theories. J. Non-equilib. Thermodyn 2: 233-

25. Jou, D., Casas-Vázquez, J. Lebón, G. 1988. Extended Irreversible Thermodynamics Rep. Prog. Phys. 51: 1105-1179.

26. Keizer, J. 1987. Statistical Thermodynamics of Non equilibrium Processes. Berlin-New York: Springer-Verlag.

27. Kestin, J. 1978. A Course in Thermodynamics Vol. II. Chap. 21 New York: Hemisphere.

28. Kestin, J. 1985. Irreversible Thermodynamics in Jahrbuch 84/85 Wissenschaftskolleg zu Berlin, P. Wapnewski, ed. Berlin: Siedler, Verlag.

29. Landau, L, Lifshitz, E.M. 1958. Statistical Physics. London: Pergamon Press Ltd.

30. Lebón, G., Torrisi. M., Valenti, A. 1987. The Kinetic Mass Action Law Revisted by Thermodynamics J. Phys. Chem 91,: 5103-5106.

31. Lebón G., Torrisi, M. and Valenti A.,1989. Reply to Comments on the Kinetic Mass Action Law Revisted by Thermodynamics J. Phys. Chem. 93:, 4382.

32. Müller, I. 1967- Zum Paradoxen der Wärmeleitungs theorie. Zeits-für Physik 198: 329-336.

33. Onsager, L. 1931. Reciprocal Relations in irreversible processes. I and II. Phys. Rev. 37,:. 405-426; ibid 38: 2265-2279.

34. Onsager, L., Machlup. S. 1953. Fluctuations and Irreversible Processes I, II. Phys. Rev. 91: 1505-1512. (ibid) 91, 1512-1515.

35. Prigogine I., Defay R. 1954. Chemical Thermodynamics. London: Longmass-Green.

36. Prigogine I. 1967. Introduction to Thermodynamics of Irreversible Processes. New York: John Wiley and Sons.

37. Rodríguez R.F., López de Haro, M. 1989. On the closure assumption in Extended Irreversible Thermodynamics. J. of Non-equilib. Thermodyn 14: 37-44.

38. Ross, J. Mazur, P. 1961. Some deductions from a formal Statistical Mechanical Theory of Chemical Kinetics. J. Chem. Phys. 35: 19-28.

39. Ross, J., García-Colín L.S., 1989. Thermodynamics of Chemical Systems far from Equilibrium. J. Phys. Chem. 93: 2091-2092.

40. Sieniutycz S. 1987. From a least action principle to massaction law and extended affinity. Chem. Eng. Sciencie 42: 2697-2710.

Astrophysical and Cosmological Applications of Extended Relativistic Thermodynamics

Diego Pavón
Department of Physics, Faculty of Sciences
Autonomous University of Barcelona
08193 Bellaterra, Spain

Abstract

We review some consequences of the extended relativistic thermodynamics theory of irreversible processes (ERT) on several fields of application in astrophysics and cosmology. This theory is more general than its conventional counterpart and comply with causality, which is essential in the relativistic context.

1. INTRODUCTION

Extended relativistic thermodynamics (ERT) such as is formulated in a series of papers by Israel (1976, 1979) and Pavón et al. (1982) was initially conceived to overcome the causality problem inherent in the conventional theories of Eckart (1940) and Landau and Lifshitz (1980), which rest on the local equilibrium hypothesis. This problem is nothing but the allowance of infinite speed of propagation for the dissipative signals, an unpleasant feature in the Newtonian context and an inadmissible one in the relativistic framework.

However, ERT not only succeeded in bringing the predicted values of the velocity of the dissipative effects below the speed of light in vacuum but also obtained, as a by-product, some additional advantages. These as listed by Hiscock and Lindblom (1983) are the following: i) For stable fluid configurations, the dissipative signals propagate causally. ii) Unlike Eckart's theory, there is not generic short-wavelength secular instability inherent in ERT. iii) Even for rotating fluids, the perturbations have a well-posed initial value problem.

Therefore, Hiscock and Lindblom conclude that the new theory should replace the conventional ones as the standard theory of relativistic irreversible thermodynamics.

In actual fact, ERT has been applied to the analysis of some astrophysical and cosmological problems and it should also be applied when dealing with situations in which rapid rotations or steep gradients or both are involved.

In particular, for a viscous heat conducting fluid, ERT (Pavón et al. 1982) hypothesizes, up to second order in the dissipative quantities (namely heat flux q^{μ}, bulk viscous pressure $\sigma^{\mu\nu}$), the generalized Gibbs equation

$$T\dot{s} = \dot{\varepsilon} + p\dot{\upsilon} + \upsilon\, \alpha_1\, q^{\mu}\, \dot{q}_{\mu} + \upsilon\, \alpha_2\, \sigma\dot{\sigma} + \upsilon\, \alpha_3\, \sigma^{\mu\nu}\, \dot{\sigma}_{\mu\nu}\,, \tag{1.1}$$

where the upper dot means differentiation along the fluid flux. Furthermore, for the spacelike part of the generalized entropy flux vector, the expression

$$I^{\mu} = \beta_1\, q^{\mu} + \beta_2\, \sigma q^{\mu} + \beta_3\, \sigma^{\mu}{}_{\nu}\, q^{\nu} \tag{1.2}$$

is assumed. In the above equations s, ε and υ denote the specific values of entropy, energy, and volume, and p stands for the hydrostatic scalar pressure and T for the absolute temperature. The α_i parameters are related to the corresponding first-order transport coefficients of heat conductivity k, bulk viscosity ζ, and shear viscosity ξ, through $\alpha_1 = -\tau_1/kT$, $\alpha_2 = -\tau_2/\zeta$, and $\alpha_3 = -\tau_3/2\xi$, the τ_i being the proper relaxation time of each individual dissipative phenomenon. The parameter β_1 is easily identified as 1/cT, but the determination of β_2 and β_3 is harder, and it seems beyond the scope of any phenomenological theory. In fact both parameters depend on the specific fluid under consideration.

If the system neither experiences rapid rotations nor is submitted to strong gravitational tidal forces, linear transport equations take the form:

$$q_{\mu} = kcT\Delta_{\mu}{}^{\nu}\left(-\frac{1}{c\,T}\,(T_{,\nu} + T\dot{u}_{\nu}) - \frac{\tau_1}{k\,T}\,\dot{q}_{\nu} + \gamma_1\,\sigma_{,\nu} + \gamma_2\,\sigma^{\lambda}{}_{\nu;\lambda}\right), \tag{1.3}$$

$$\sigma = -c\zeta\,\left(u^{\mu}{}_{;\mu} + \frac{\tau_2}{\zeta}\,\dot{\sigma} + \gamma_3\, q^{\mu}{}_{;\mu}\right) \tag{1.4}$$

$$\sigma_{\mu\nu} = -2c\xi\Big([[\Delta^{\lambda}{}_{(\mu}\,\Delta_{\nu)}{}^{\rho}\ u_{(\lambda;\rho)}]] + \frac{\tau_3}{2\,\xi}\,\dot{\sigma}_{\mu\nu} + \gamma_4 [[\Delta^{\lambda}{}_{(\mu}\,\Delta_{\nu)}{}^{\rho}\, q_{(\lambda;\rho)}]]\Big) \tag{1.5}$$

where u^{μ} denotes the four-velocity of an arbitrary volume element of the fluid normalized so that $u^{\mu}\,u_{\mu} = -1$, and $\Delta^{\mu\nu}$ is the spatial projector $g^{\mu\nu} + u^{\mu}\,u^{\nu}$. The double bracket [[...]] indicates that we must take only the traceless part of the enclosed quantity. A comma means partial derivation and the semicolon covariant differentiation. Those readers interested in the relativistic transformation of the thermodynamic quantities (the Planck-Ott controversy) are referred to Israel(1981).

2. ASTROPHYSICAL APPLICATIONS OF ERT

In order to apply the ERT theory outlined above it is needed to determine the coefficients appearing in the transport equations (1.2)-(1.6). For some fluids of interest this can be done by means a mesoscopic approach (Pavón et al.1983, Jou and Pavón 1985) based in the analysis of the fluctuations of the dissipative quantities around equilibrium.

2.1 Linear transport coefficients of first order

As is well known the probability of spontaneous thermodynamic fluctuation around the equilibrium state of a given physical system is governed by the Einstein-Boltzmann formula (Landau and Lifshitz 1980)

$$\Pr \propto \exp[(M/2k_B)\,\delta^2 s] \tag{2.1}$$

where k_B denotes the Boltzmann constant and M the mass of the system. When this expression is combined with the generalized Gibbs equation (1.1) one follows for the fluctuations of the dissipative quantities

$$\Pr(\delta q\ ,\ \delta\sigma,\ \delta\sigma^{\lambda\rho}) \propto \exp\Big((V/2k_BT)\,[\alpha_1\ \delta q^{\mu}\delta q_{\mu} + \alpha_2(\delta\sigma)^2 + \alpha_3\ \delta\sigma_{\mu\nu}\delta\sigma^{\mu\nu}]\Big). \tag{2.2}$$

From last expression it is immediate to obtain the second moments in an arbitrary event point,

$$<\delta q^{\mu}\, \delta q_{\mu}> = (k\, k_B T^2/V\tau_1)\Delta^{\mu}_{\nu} \quad , \qquad (2.3)$$

$$<(\delta\sigma)^2> = \zeta\, k_B T/V\tau_2 \quad , \qquad (2.4)$$

$$<\delta\sigma^{\mu\nu}\, \delta\sigma_{\lambda\rho}> = (2\xi k_B T/V\tau_3)\, [\Delta^{\mu}_{(\lambda}\, \Delta^{\nu}_{\rho)} - \frac{1}{3}\, \Delta^{\mu\nu}\, \Delta_{\lambda\rho}] \quad , \qquad (2.5)$$

where the angle brackets mean equilibrium statistical average. These relationships are nothing but expressions of the fluctuation-dissipation theorem relating fluctuations around the equilibrium with the corresponding coefficients of transport. Next, we use them to derive the transport coefficients of heat conductivity, bulk and shear viscosity of a radiative fluid (Pavon et al. 1983).

To this end, let us consider a mixture of material fluid -with very short mean free times- and radiation (photons, neutrinos, gravitons), with a finite mean free time τ. The radiation is mainly responsible for the dissipation, as its mean free time is much longer than the material fluid one. Hence, any component of fluctuation of heat flux, say δq_1, reduces to $(c/\sqrt{3})\delta\rho_r$, with ρ_r the energy density of the radiation. Introducing the well known relationship (see Landau and Lifshitz 1980)

$$<(\delta\rho_r)^2> = (k_B C_V/V^2)T^2, \qquad (2.6)$$

in the expression for the second moments of δq_1 and then solving for the coefficient of heat conductivity we obtain

$$k = (4/3)c^2 a T^3 \tau_1 \, , \qquad (2.7)$$

where C_V (= $4aVT^3$) denotes the heat capacity at constant volume of the radiation (photons for the sake of conciseness), and a the black-body constant.

To determine the coefficient of shear viscosity, ξ, we recall that any component of the viscous pressure tensor represents a flux of momemt, i.e. $\sigma_{12} = \rho_r\, c_1 c_2/c^2$. Using the same procedure as before we get

$$\xi = (4/15)a\, T^4 \tau_3 . \qquad (2.8)$$

The calculation of ζ is more involved and requires a more subtle argument. As the radiative fluid is slightly removed from thermodynamic equilibrium a bulk viscous pressure σ, related to the equilibrium one trough $\sigma = P_{tot} - p$, arises. P_{tot} is the new total scalar pressure, and we may assume that this quantity obeys the relation $P_{tot} = n\ aT^4$, n being a numerical coefficient not depending on υ or T. After these considerations we can write

$$\delta\sigma = \Omega\ \delta\rho_r, \qquad \Omega \equiv (\partial P_{tot}/\partial\rho)_{\bar{\upsilon}} - (1/3)\ . \tag{2.9}$$

Using (2.9) together with (2.4) and (2.6) one follows

$$\zeta = 4\Omega^2 aT^4\tau_2. \tag{2.10}$$

Inspection of the expressions (2.9) shows that if the particles making up the material fluid are very relativistic ζ vanishes.

Our approach has the advantage of simplicity, as compared with the rather heavy calculations involved in prior procedures (Weinberg 1971, Udey and Israel 1982). Furthermore, in the expression for k, ζ and ξ by previous authors there appears a single relaxation proper time τ, whereas in ours, there appear three different times, each corresponding to a different dissipative process, which is more satisfactory. This is a direct consequence of the generalized Gibbs equation (1.1).

2.2 Linear transport coefficients of second order

The task of calculating the six remaining transport coefficients occurring in equations (1.2)-(1.5) is greatly simplified by the Newtonian version of the extended thermodynamic theory. This has provided us with the relationships (Jou and Casas-Vazquez 1983)

$$\gamma_1 = -\gamma_3 = T\ \beta_2\ , \tag{2.11}$$

$$\gamma_2 = -\gamma_4 = T\ \beta_3\ , \tag{2.12}$$

hence we only need to calculate γ_1 and γ_2 . This can be done by means of the method of microscopic projection operators due to Ernst (1970). Effectivaly, according to Jou and Casas-V zquez (1983) if the dissipative quantities decay exponentially the method mentioned allows us to relate γ_1 (and hence γ_4) to the fluctuations around equilibrium of the heat flux and the shear viscous stress tensor trough the expression

$$\gamma_2 = -\langle c_1\ \delta\sigma^{12}\delta q_2\rangle \frac{\tau_1\tau_3\ V}{k\ k_B\ \xi\ T^2}\ . \tag{2.13}$$

Reasoning along parallel lines, an analogous expression for γ_1 (and γ_3), namely

$$\gamma_1 = -<c_1 \delta q_1 \delta\sigma> \frac{\tau_1 \tau_2 V}{k\, k_B\, \zeta\, T^2}, \tag{2.14}$$

can be derived.

Inserting the above expressions for the fluctuations of the dissipative quantities in terms of $\delta\rho_r$ in the equilibrium averages occurring in (2.13) and (2.14) and using the set of equations (2.6)-(2.8) as well as (2.10)-(2.14), it follows (Jou and PavÂn 1985)

$$\gamma_1 = \gamma_3 = -(4a\Omega T^4)^{-1} \tag{2.15}$$

$$\gamma_2 = \gamma_4 = -(4aT^4)^{-1} \tag{2.16}$$

Our expressions for γ_2 and γ_4 coincide with those of previous literature (Schweizer 1982, Udey and Israel 1982). However, they differ from γ_1 (and γ_3) by the multiplicative factor $\chi \equiv [\partial\rho_m/\partial\rho]_\upsilon$, which does not appear in our expression (2.15).

Here ρ_m denotes the energy density of the material medium and ρ the total energy density of the radiative fluid. This factor accounts for the influence of the temperature used in the description of the fluid (the initial equilibrium temperature of the material medium, or the final equilibrium temperature of the material plus radiation, Schweizer (1982)). On physical grounds, the latter temperature is the more suitable one, and the factor χ appears in the transformation of the first temperature to the latter one. However, in our formalism both temperatures appear confused. This is due to the fact that we are considering small fluctuations around the equilibrium state of material plus radiation. Such a transformation of temperatures does not affect either heat flux or the traceless viscous pressure tensor, but it appears in the definition of bulk viscosity. This fact explains the exact coincidence of our results for γ_2 (and γ_4), and the discrepancy of our expressions for γ_1 (and γ_3), with the mentioned prior results. At any rate, the discrepancy will be small when the thermal energy and pressure of the fluid are dominated by the material component.

From equation (2.16) it is evident that the importance of

coefficients γ_2 and γ_4 diminishes as T increases. But this is not necessarily so for the coefficients γ_1 and γ_3 (equation (2.15)) since for sufficiently high values of T the particles composing the material fluid become more relativistic, and consequently Ω approaches zero. In the limit of an ultrarelativistic material medium, the bulk viscosity coefficient ζ (which varies as Ω^2T^4) vanishes identically (Weinberg 1971). But to our knowledge no general conclusions exist about the behavior of $\Omega\ T^4$ at very high temperatures. That behavior must depend on the precise state equation of the material fluid component of the radiating fluid. In some cases, for instance for neutrinos in the central regions of collapsing stellar cores, Ω behaves as T^{-3} for very high temperatures (Kazanas 1978) so that Ω^2T^4 vanishes but ΩT^4 diverges as T tends to infinite. However other kinds of behavior are, in principle, conceivable.

2.3 Nonequilibrium fluctuations

Expression (2.1) although originally intended for equilibrium situations (Landau and Lifshitz 1980) has also been applied to situations slightly away from equilibrium (Jou et al 1982). The basis for this nonequilibrium application may be found in information theoretical arguments, which use a nonequilibrium distribution function which maximizes the entropy subject to the requirements imposed on the system: energy, volume and dissipative fluxes (Casas-Vázquez and Jou 1990). In this subsection we use (2.1) to obtain explicit though approximated values for the nonequilibrium corrections of the bulk viscous pressure of some particular relativistic fluids in which the only dissipative process taking place is the one due to the bulk viscosity (PavÂn et al. 1983). In this case the generalized Gibbs equation (1.1) reduces to

$$T\dot{s} = \dot{\epsilon} + p\dot{\upsilon} + \upsilon\,\alpha_2\,\sigma\dot{\sigma} \quad . \tag{2.17}$$

Assuming that the system experiences a slowly expansion, its mean viscous pressure σ_0 (<<1) is given by $\sigma_0 = -c\zeta\, u^{\mu}{}_{;\mu}$ and the second moment, up to second order in σ_0, can be expressed as

$$<(\delta\sigma)^2> \approx |k_BT/V\alpha_2|(1 - \sigma_0^2\ A/\Delta) \tag{2.18}$$

where A and Δ are rather complicated functions of α_2 and the equilibrium variables, so we will not explicit them here. From these and (2.18) explicit expressions for $<(\delta\sigma)^2>$ are readily obtainable for different kind of fluids.

In the particular case of the fluid filling the early universe during the leptonic period - a mixture of neutrinos and electrons with temperature varying between 1.5×10^{12} K and 6×10^{9} K- the energy density may be approximated by $\rho \approx a'T^4$, where a' differs from a by a factor of the order of unity (see e.g. Weinberg 1972, p 536), although for practical purposes we shall not make a distinction amongst them, and $p = (\gamma - 1)\rho$, where the adiabatic index, γ, of the fluid lies in the range $1\le\gamma\le2$. As for ζ we will consider two possibilities. In the first place we take the expression (2.10). Then we have

$$<(\delta\sigma)^2> \approx \left|\frac{k_B T}{V\,\alpha_2}\right| \left[1 - \frac{2241}{(3\gamma-4)^3} \left(\frac{\sigma_0}{4\rho}\right)^2 \right] . \qquad (2.19)$$

In the second place, we consider ζ given by (De Groot et al. 1980, p 231)

$$\zeta = \frac{1}{4\pi} \left(\frac{k_B T}{c\omega(T)}\right)^4 \frac{\kappa_\nu \kappa_e}{432(1+C)+126C^2} . \qquad (2.20)$$

This expression can be cast into a more convenient form by substituting the cross section $\omega(T)$, in the function of the mean free time and particle density. Then

$$\zeta = (m^4c^5/7020\pi\hbar^3)\tau_2 . \qquad (2.21)$$

Here we have considered the particle fractions κ_ν and κ_e being 1/3 and 2/3 respectively, and the parameter $C \approx 1$. In this way we have

$$<(\delta\rho)^2> \approx \left|\frac{k_B T}{V\,\alpha_2}\right| \left(1 - \frac{90 - 3(\gamma-1)}{4(3\gamma-4)} \frac{\sigma_0^2}{B\varepsilon} \right) , \qquad (2.22)$$

where the quantity B stands for the coefficient of τ_2 on the right hand side of (2.21).

Finally, as a third example, we consider a fluid of massive neutrinos. Its bulk viscosity coefficient was determined by

Calkoen and De Groot (1981). Proceeding as before we obtain, for $mc^2 << k_B T$

$$<(\delta\sigma)^2> \approx \left|\frac{k_B T}{V\, \alpha_2}\right| \left(1 - \frac{4\phi(y)\sigma_0^2}{(3\gamma-4)B\varepsilon}\right), \quad y = \ln(k_B T/mc^2) \qquad (2.23)$$

where $B = m^4c^8/(2^6 3^2 \pi\hbar^3 c^3 \ln 10)$ and $\phi(y) = -(5y+1)^2 + \frac{3}{16}(\gamma-1)[(y+1)^2 + \frac{1}{2}(y+1)(5y+1)]$.

A quick inspection shows that (2.18) and as a consequence (2.19) as well as (2.22) and (2.23) reduce to their equilibrium expression (2.4) when σ_0 tends to zero.

Apparently, in the limit $\gamma \longrightarrow \frac{4}{3}$ the right hand sides of (2.19), (2.22) and (2.23) diverge, but this is not so since $\gamma = \frac{4}{3}$ means a fluid of massless particles and hence both ζ and σ_0 vanishes.

2.4 Survival of protogalaxies

Israel's version of ERT has been utilized by Schweizer (1984) to study the compressional modes in radiative plasmas. The main results have been applied to the analysis of survival of protogalaxies in an expanding universe during the acoustic phase.

In the regime where the photons dominate the thermal energy and pressure of the two component fluid (i.e., $\rho = \rho_0 + (1/3)\, aT^4$) three non-propagating modes and one propagating mode (with characteristic velocity $\pm\sqrt{3/5}$) arise from the plane wave dispersion relation associated with shear modes.

For spherically symmetric modes shear stresses vanish identically and the resulting dispersion relation is cubic allowing a fully analytical treatment. Furthermore, a zero-sound band of finite area and triangular shape appears. This contrasts with the issues of conventional theory (Weinberg 1971) as the corresponding cubic dispersion relation leads to a zero-sound band of infinite area. This reveals the fact that zero-sound bands depend sensitively on the chosen thermodynamic theory.

Concerning the survival of protogalaxies during the acoustic phase the ERT does not introduce substantial changes with respect to standard theory. This is so since protogalactic fluctuations drop into a domain where the long wavelength approximation holds. The long wavelength approximation follows from both, the conventional and the ERT theory. It is, however, reasonable to expect that the inclusion of self-gravitating effects should make the modifications more pronounced. This view is supported by the fact that zero-sound bands play an important role in standard thermodynamic approaches to this topic, and as we have pointed out before the detailed appearance of zero-sound bands depends sensitively on the selected thermodynamic theory.

3. COSMOLOGICAL APPLICATIONS OF ERT

It has been recognized since long that nonequilibrium effects might play a non-negligible role, even a significant one, in the cosmic evolution of the primeval universe. However, in studying these effects many cosmologists have resorted to Ecktart's theory of dissipative fluids, thereby overlooking the problems mentioned in Section 1 (see e.g., Weinberg 1971, Caderni and Fabri 1979, Barrow 1988). Nevertheless, it turns out that very often the predictions of ERT, when applied to cosmic evolution, differ in a non-negligible respect from those based on Eckart's.

3.1 Bianchi type-I cosmological model and ERT

Ten years ago Belinskii et al (1979) analyzed the influence of viscosity on cosmological evolution using Israel's causal theory. They focused their attention on the homogeneous Bianchi type-I model of the universe. In that model Einstein field equations can be expressed according to

$$3\dot{H} + 3\ \Sigma H_k^{\,2} = -\tfrac{1}{2}\{\rho + 3\ (p+\sigma)\}\ , \qquad (k = 1,2,3) \qquad (3.1)$$

$$3H_k\ H + \dot{H}_k = \tfrac{1}{2}\{2\sigma^k + \rho - (\rho+\sigma)\}\ , \qquad (3.2)$$

where $H_k \equiv \dot{R}_k/R$, $H \equiv (1/3)\Sigma H_k$, and σ^k denote the diagonal components of the three-dimensional tensor $\sigma^k{}_l$. Units have been chosen so that c = G = 1.

Since this cosmological model does not admit heat flux the

above equations reduce to

$$\sigma + \tau_2 \dot{\sigma} = -3\zeta H \quad , \tag{3.3}$$

$$\sigma^k + \tau_3 \dot{\sigma}^k = -2\zeta(H_k - H) \quad . \tag{3.4}$$

For low and high values of the energy density the equation of state is taken to be $p = (\gamma-1)\rho$. For intermediate values of ρ the only assumption on p is that its dependence on ρ is smooth and has no zeros.

The solutions differ greatly from those obtained via conventional thermodynamics. In particular the isotropic solution is stable under certain conditions. However, the effect of matter creation near the singularity is preserved.

Belinskii et al (1979) also investigated the evolution of weak perturbations on Friedmann metric. It turns out that in the vicinity of the initial singularity Friedmann solutions are stable only if superluminal velocities are allowed.

4.2 Production of entropy in the leptonic period

As is well known one of the most interesting unsolved problems is the high value of the ratio number of photons, N_γ, over number of baryons N_B. That ratio, about 10^9 was noticed after the discovery of the 3 K cosmic background radiation.

Weinberg (1971) made an attempt to explain that ratio by assuming that a very copious number of photons were produced via bulk viscosity in the very early universe. Adopting the Friedmann-Robertson-Walker (FRW for short) model for the primeval expansion and the conventional thermodynamic theory, Weinberg obtained for the entropy production the expression

$$\Pi_{st} = (9\zeta/T)H^2 \quad , \quad H \equiv \dot{R}/R \quad . \tag{3.5}$$

After exploring the contribution to Π_{st} from different kind of fluids -photons, neutrinos, gravitons, ...- Weinberg concluded that the bulk viscosity is unable to account for the observed ratio N_γ/N_B.

Some years ago Fustero and Pavón (1982) used ERT to study the entropy production via bulk viscosity in the leptonic era of the universe. In that period -temperature ranging between 6×10^9 and 1.5×10^{12} K- the main constituents of the fluid filling the universe were photons, electrons and neutrinos, and the dissipation was chiefly caused by the neutrinos because of their long mean free paths.

It turned out that in some instant of that period the relaxation term $\tau_2\dot{\sigma}$ occurring in the causal transport equation $\sigma + \tau_2\dot{\sigma} = -c\zeta\, u^{\mu}{}_{;\mu}$ becomes of the same order of magnitude than the leading one σ. In particular, for $T = 10^{11}$ K one has $|\sigma/\tau_2\dot{\sigma}| \approx 3$. This outcome reveals that the omission of the relaxation term in the calculation of the entropy implies an underestimation -which can be significative- of this quantity. Anyway, the entropy production calculated via ERT differs from Π_{st} by a factor of the order of unity at most. Hence, the main conclusions of Weinberg remain valid.

Currently the common feeling is that the ratio N_γ/N_B could be explained by grand unification theories.

3.3 The inflationary universe

At the beginning of last decade Guth (1981) proposed a cosmological model aimed to overcome some shortcomings of the standard hot big-bang scenario, apparently linked to its initial conditions. These are the horizon and the flatness problem. The former refers to the high degree of homogeneity and regularity of the many different regions in which we can mentally divide the observable universe. The problem arises when one realizes that according to the standard scenario these regions never were in causal contact. The latter refers to the initial value of the energy density of the universe which, in order the hot big-bang model be viable, should coincide with its initial critical value to an incredible degree of fine tuning.

Both problems can be solved by a single stroke if one allows that the primeval universe undergoes a period of exponential inflation (i.e., $H = \text{const}$, whence $R \propto \exp(Ht)$). It would result in an extraordinary increase of the scale factor. After that the universe would start its conventional radiation dominated expansion. This process of inflationary expansion can be implemented by allowing that the effective pressure be negative. Of course, ordinary matter under normal conditions always exhibits positive pressure . However, several mechanisms, in connection with the inflationary model, able to generate negative pressures have been advanced in the literature (for a review see the article by Brandenberger (1985)). Our purpose in this subsection is to review the inflationary model of Lima et al (1988), proposed in the context of ERT.

These authors assume a FRW universe with flat spatial sections and filled with a viscous relativistic fluid whose equation of state reads

$$p = [\gamma(T)-1]\rho \quad , \tag{3.6}$$

where the parameter γ is allowed to vary slowly with temperature. Moreover, the universe is supposed to be asymmetric in the sense

of possessing a net bosonic charge $q(>10^6 T^3)$ so that the cosmic fluid be dense enough. Under these circumstances the negative pressure associated to the bulk viscosity (shear stresses vanish in homogeneous universes) might dominate over the hydrostatic pressure and drive exponential inflation. The condition for this will be explicited below. Units have been chosen so that $c = G = \hbar = k_B = 1$.

The basic dynamic equations of Lima et al are the Einstein field equation

$$\rho = \frac{3}{8\pi} H^2 \quad , \tag{3.7}$$

and the conservation of energy

$$\dot{\rho} + 3(\rho + p + \sigma) H = 0 \quad . \tag{3.8}$$

These together with the conservation of bosonic charge

$$\dot{q} + 3qH = 0 \quad , \tag{3.9}$$

and the thermodynamic equations, namely the equation of state (3.6), the production of entropy

$$\Pi = -\frac{\sigma}{T}\left(3H + \frac{\tau}{\zeta}\dot{\sigma}\right) \quad , \tag{3.10}$$

the relationship for viscous transport

$$\sigma + \tau\dot{\sigma} = -3\zeta H \quad , \tag{3.11}$$

and the Gibbs-Duhem equation ,

$$dp = q\ \eta\ dT + q\ d\mu \quad , \tag{3.12}$$

where

$$\mu = \frac{\rho + p}{q} - T\eta \tag{3.13}$$

is the relativistic chemical bosonic potential and η the entropy associated to the unit of bosonic charge, permit to show that the condition for exponential inflation is

$$\gamma\rho\left(\frac{\dot{\gamma}}{\gamma} - \frac{\dot{T}}{T}\right) = \mu\ q\left(\frac{\dot{\mu}}{\mu} - \frac{\dot{T}}{T}\right) \quad . \tag{3.14}$$

It can be proved, within this model, that during the phase of exponential inflation the bulk viscous pressure and the energy density are related by

$$\sigma = -\ \gamma\rho \quad . \tag{3.15}$$

This expression when introduced into (3.11) gives

$$\zeta = \zeta^{*}\left(1 + \frac{\tau\ \dot{\gamma}}{\gamma}\right) \quad , \tag{3.16}$$

where ζ^{*} is the bulk viscosity coefficient of the Eckart theory, namely $\gamma\rho/3H$.

Likewise, the entropy density, s*, during the inflation (which begins at t_0), reads

$$s^{*} = T^{-1}\Big(\gamma\rho\ (1 - 3\tau H/2) - \mu q_0\ \exp[-3H(t-t_0)]\Big) \quad , \tag{3.17}$$

while the duration of the inflation is

$$\Delta t \cong \frac{1}{3\ H}\ \ln\left(\frac{S_f}{s^{*}\ R_0^{3}}\right) \quad , \tag{3.18}$$

where S_f is the entropy at the end of the inflation. Resorting to the fact that the entropy density is nearly constant, numerical estimates lead to the conclusion that the scale factor, R, augments by a factor of 10^{28}, which is enough to solve the problems posed by the standard scenario.

It is worth mentioning that in the model by Lima et al (1988) inflation and generation of entropy go together. This is at variance with most conventional approaches in which the increase of entropy comes after the inflationary expansion through a first order thermodynamic phase transition.

3.4 The generalized second law and ERT

Historically, one of the earliest solutions of the Einstein field equations is that due to de Sitter. It corresponds to an isotropic, homogeneous and empty universe dominated by the energy density, ρ_v, and pressure, P_v, of the vacuum (Hawking and Ellis 1973). The scale factor obeys $R(t) \propto \exp(Ht)$ where the parameter of Hubble satisfies H = constant. This universe has the distinguishing feature of possessing a cosmic horizon (which can be understood as a mental border beyond which no information can reach the observer) with area

$$A = 4\pi/H^2 \quad . \tag{3.19}$$

Due to quantum effects this horizon posses an entropy and a temperature (Gibbons and Hawking 1977) given by

$$S_h = 2\pi A \quad , \tag{3.20}$$

$$T_h = H/2\pi \quad , \tag{3.21}$$

respectively. We are using units so that $8\pi G = c = \hbar = k_B = 1$.

So long as the space is exactly de Sitter (H = const.) the temperature and the entropy do not experience variation. However, if a small quantity of material dissipative fluid is allowed within this universe H ceases to be exactly constant, $\dot{H} \neq 0$, whence the entropy of the cosmic horizon can diminish or increase. The crucial point here is to ascertain whether the so called generalized second law (GSL), which states that the entropy of the horizon plus the entropy of the environment cannot decrease with time, i.e.,

$$\dot{S}_h + \dot{S}_f \geq 0 \tag{3.22}$$

holds in this context. Davies (1987) has been able to show that the GSL is satisfied in the case we are considering, a quasi-de Sitter space. He obtained the following expression

$$\dot{S}_h + \dot{S}_f = \frac{8\pi^2}{H^3}\gamma\rho_f \quad , \tag{3.23}$$

which cannot be negative since H is always positive for de Sitter universes, and γ (the adiabatic index of the fluid) and ρ_f the energy density of the fluid are also positive. Nevertheless, his calculations were carried out with the help of the conventional Eckart's theory of irreversible processes. Therefore his result can be viewed as valid only for stationary periods of expansion, but not for the transient ones. Furthermore, the very concept of event horizon breaks down when one permits the propagation of any physical signal with infinite speed.

Recently, in order to overcome these drawbacks Pavón (1990) has analyzed the same problem but in the context of the ERT. In the sequel we sketch his calculations.

Starting from the Einstein field equation

$$\rho = 3H^2 \tag{3.24}$$

and the energy conservation equation

$$\dot{\rho} + 3(\rho + P)\, H = 0 \quad , \tag{3.25}$$

where

$$\rho = \rho_U + \rho_f \quad , \qquad P = P_U + P_f \tag{3.26}$$

and the state equation for the vacuum

$$P_U = -\rho_U \quad , \tag{3.27}$$

and for the material fluid

$$P_f = (\gamma-1)\rho_f + \sigma \ , \tag{3.28}$$

where the bulk viscous pressure obeys the relaxed transport equation

$$\sigma + \tau\, \dot{\sigma} = -3\zeta H \quad , \tag{3.29}$$

with ζ related to ρ_f by

$$\zeta = \alpha\ \rho_f \quad , \qquad (\alpha \geq 0) \tag{3.30}$$

-see Davies (1987) and Murphy (1973)- it can be derived the following equation

$$\dot{\rho} = -3\left((\gamma-3\alpha H)\rho_f - \tau\dot{\sigma} \right) \ . \tag{3.31}$$

Last relationship when combined with $\dot{S}_h$ and $\dot{S}_f(= R^3\sigma^2/T_f\zeta)$ yields

$$\dot{S}_h + \dot{S}_f = \frac{8\pi^2}{H^3}\left(\gamma\rho_f + \frac{\tau^2\sigma^2}{3\alpha\rho_f H} + \tau\dot{\vartheta}\right) . \qquad (3.32)$$

This expression reduces to that by Davies (3.23) in the stationary regime ($\dot{\vartheta} = 0$) only. (Note that, on physical grounds, τ is always positive). However (3.32) differs from (3.23) in several respects. Firstly, in the appearance of the quantity $\tau\dot{\vartheta}$ in the right hand side of (3.32), which permits to extend the calculations of Davies to transient regimes. Secondly, and more importantly, from (3.32) it follows that the GSL can be violated if $\tau\dot{\vartheta}$ becomes extraordinarily negative. Although ordinary fluids are not expected to present very large values for $|\tau\dot{\vartheta}|$ it cannot be ruled out in principle, specially at the very early stages of cosmic expansion where exotic quantum fields might play a key role. Thirdly, if the dominant energy condition (Hawking and Ellis 1973)

$$\rho + P \geq 0 \qquad (3.33)$$

which in the context of the ERT takes the form

$$\gamma\rho_f \geq \tau\dot{\vartheta} + 3\alpha\rho_f H \qquad (3.34)$$

holds, it can be shown by combining last expression with (3.32) that $\dot{S}_h+\dot{S}_f \geq 0$. This is at variance with the result by Davies (1987) since expression (3.23) is positive-semidefinite irrespective of the DEC (which in that context reads $\gamma \geq 3\alpha H$) is fulfilled or not.

3.5 FRW cosmology and ERT

The recent years have witnessed a surge of interest about the role played by the fundamental strings (Witten 1985) in shaping our universe, both before and after the compactification of the spatial extradimensions. In particular, Turok (1988) has suggested that right after the Planck's time ($\approx 10^{-44}$ sec) the very fast rhythm of string creation by the strong quantum fields leads, in a natural way, to an exponential expansion of the non-compactified extradimensions. This is so since for some period of time the production of strings compensated for its dilution, hence the energy density remained nearly constant.

In a recent paper Barrow (1988) has argued that this approach can be phenologically understood in terms of the dynamics of an imperfect fluid possessing a coefficient of bulk viscosity. Starting from this point, he analyzes the dynamics of a FRW universe whose source of the gravitational field is a bulk-viscous fluid whose coefficient of bulk viscosity obeys

$$\zeta = \alpha \rho^{m} , \qquad (\alpha \geq 0) \qquad (3.35)$$

with α and m constants. The resulting dynamic equation can be expressed as

$$\dot{h} + \frac{3\gamma}{2}\left(1 - h^{2m-1}\right) h^{2} = 0 , \qquad (3.36)$$

where $h \equiv H/H_0$ denotes the reduced Hubble parameter, H_0 being a constant. The analysis of this equation for different values of γ and m leads to a variety of results. However, one of the main ingredients of Barrow work is the conventional Eckart's theory of nonequilibrium processes. This automatically reduces the applicability of last equation to those stages of cosmic evolution compatible with $\dot{\sigma} = 0$.

In a recent work Pavon et al (1990) have taken up the same problem but using ERT instead. The corresponding dynamic equation

$$\gamma h^{2(m-1)} \ddot{h} + 3\left(1 + \Gamma \dot{h}^{2m-1}\right)\dot{h} + \frac{9\gamma}{2}\left(1 - h^{2m-1}\right)h^{2} = 0, \qquad (3.37)$$

with $\Gamma \equiv \gamma^{2}$, has two stationary and trivial solutions, namely $h = 1$ and $h = 0$. Nevertheless, in general it can be solved only through numerical integration. Its solutions strongly depend on the initial values of $h(0)$ and $\dot{h}(0)$. Of course, last equation reduces to (3.36) under the unphysical assumption of vanishing relaxation proper time. Among the main conclusions of Pavón et al are the following:

A) Analytical, though approximated, integration of (3.37) in the neighborhood of the stationary solutions reveals that:

i) In the vicinity of $h = 1$ and for $m < -(1+\Gamma^{2})/4\Gamma$, one has an oscillatory damped behavior leading to an exact exponential inflation for $t \longrightarrow \infty$. Curiously enough the DEC is violated and fulfilled in an alternated way.

ii) In the range $-(1+\Gamma^{2})/4\Gamma < m < 1/2$ there are two solutions. Both of them tend asymptotically to $h = 1$ and, as before, one shows exponential inflation. A solution happens to obey the DEC while the other not.

iii) For $m>1/2$ the stationary solution is, in principle, unstable. However, for $t>0$ there is again exponential inflation as $t \longrightarrow \infty$ if a certain initial condition holds. Likewise, for $t<0$ exponential inflation occurs as $t \longrightarrow -\infty$ when a particular different initial condition is met instead.

iv) In the vicinity of $h = 0$ we have inflation from the start, for $m<1/2$, and Friedmann-like behavior for $m>1/2$.

B) Numerical integration of equation (3.37) shows that:

i) For short times there are quantitative and qualitative differences between the solutions of equation (3.36) and those of (3.37). For long times these differences persist when both solutions diverge. Otherwise they tend to the same asymptotic value.

ii) If $m>1/2$ and the initial value of $|\dot{h}|$ is not very high, the solutions tend to the constant value $h = 1$ whether or not the initial value of h is greater or lower than unity.

iii) If $m>0$ and the initial value of $|\dot{h}|$ is large enough the solution of (3.37) diverges regardless the initial value of his greater or lower than unity.

iv) For $m>1/2$ and $h(0)>1$ the solutions diverge if $|\dot{h}|$ is large enough, otherwise they tend to $h = 0$.

v) Some solutions involve a strong violation of the GSL at very early times, and some others at late times. However, most of the solutions respect the GSL.

4. CONCLUDING REMARKS

Before closing this paper it seems worthwhile to briefly mention some additional astrophysical and cosmological problems to which it could be interesting to apply ERT.

On th e astrophysical side to supernova explosions. This process is so fast and separated from thermodynamic equilibrium that the emission of energy and particles could be analyzed with the help of ERT.

Likewise, the release of energy in the last stages of the

gravitational collapse of massive astrophysical objects leading to neutron stars or to black holes.

On the cosmological side, the separation of quark-gluon

phases at the transition temperature $T_t \approx 10^{10}$ K (Witten 1984). This separation of phases results in a rapid diffusion of neutrons which could be studied via the generalized relaxed Fick's law. Furthermore, the interaction between the phases could give rise to bulk and shear stresses whose impact on the expanding cosmic fluid could be analyzed via ERT. This process might have importance with regard to the cosmic abundance of light elements (Applegate et al 1988).

Acknowledgments

This work has been partially supported by the Spanish Ministry of Education under grant PB/86-0287.

REFERENCES

Applegate, J.H., C.H. Hogan and R.J. Scherrer. 1988. Cosmological quantum chromodynamics, neutron diffusion and the production of primordial heavy elements. *Astrophys. J. 329:* 572-579.

Barrow, J.D. 1988. String driven-inflationary and deflationary cosmological models. *Nucl. Phys. B310:* 743-763.

Belinskii, V.A., E.S. Nikomarov and I.M. Khalatnikov. 1979. Investigation of the cosmological evolution of viscoelastic matter with causal thermodynamics. *Sov. Phys. JETP 50:* 213-221.

Brandenberger, R.H. 1985. Quantum field theory methods and inflationary universe models. *Rev. Mod. Phys. 57:* 1-60.

Caderni, N. and R. Fabri. 1979. Neutrino viscosity and isotropization of anisotropic-curvature cosmological models. *Phys. Rev. D 20: 1251-1259.*

Calkoen, Ch.J. and S.R. De Groot. 1981. The volume viscosity of a massive neutrino gas. *Phys. Lett. 83 A:* 319-320.

Casas-Vázquez, J. and D. Jou 1990. Extended irreversible thermodynamics and fluctuations. *This volume.*
Davies, P.C.W. 1987. Cosmological horizons and the generalized second law of thermodynamics. *Class. Quantum Grav. 4:* L225-L228.

De Groot, S.R., W.A. van Leeuwen and Ch.G.van Weert. 1980. *Relativistic kinetic theory.* Amsterdam: North Holland.

Eckart, C. 1940. The thermodynamics of irreversible processes. III. Relativistic theory of the simple fluid. *Phys. Rev.58:* 919-924.

Ernst, M.H. 1970. The linearized Boltzmann equation. Navier-Stokes and Burnett transport coefficients. *Am. J. Phys. 38:* 908-914.

Fustero X. and D. Pavón. 1982. Relaxation terms and entropy production in a cosmological model. *Lettere al Nuovo Cimento 35:* 427-429.

Gibbons, G. and S.W. Hawking. 1977. Cosmological event horizons, thermodynamics and particle creation. *Phys. Rev. D 15:* 2738-2751.

Guth, A.H. 1981. Inflationary universe: A possible solution to the horizon and flatness problem. *Phys. Rev. D 23:* 347-356.

Hawking, S.W. and G.F.R. Ellis. 1973. *The large scale structure of space time.* Cambridge: Cambridge University Press.

Hiscock, W.A. and L.A. Lindblom. 1983. Stability and causality in dissipative relativistic fluids. *Ann. Phys. (N.Y.) 151:* 466-496.

Israel, W. 1976. Nonstationary irreversible thermodynamics: a causal theory. *Ann. Phys. (N.Y.) 100:* 310-313.

Israel, W. and J.M. Stewart. 1980. Progress in relativistic thermodynamics and electrodynamics of continuous media. In *General Relativity and Gravitation.* Vol. 2, ed. A. Held, 491-525. New York: Plenum Press.

Israel, W. 1981. Thermodynamics of relativistic systems. *Physica A 106: 204-214.*

Jou D. and J. Casas-Vázquez.1983. Equilibrium third moments and non-equilibrium second moments of fluctuations of hydrodynamic dissipative fluxes. *J. Nonequilib. Thermodyn. 8:* 127-142.

Jou, D. and D. Pavón. 1985. Second order coefficients for radiating fluids. *Astrophys. J. 291:* 447-449.

Kazanas, D. 1978. On neutrino viscosity in collapsing stellar cores. *Astrophys. J. (Letters) 222:* L109-L111.

Landau, L. and E.M. Lifshitz. 1980. *Statistical Physics.* 3rd ed. New York: Pergamon Press.

Lima, J.A.S., R. Portugal and I. Waga. 1988. Bulk viscosity-driven asymmetric inflationary universe. *Phys. Rev. D 37:* 2755-2760.

Murphy, G.L. 1973. Big-bang model without singularities. *Phys. Rev. D 8:* 549-552.

Pavón, D., D. Jou and J. Casas-Vázquez. 1982. On a covariant formulation of dissipative phenomena. *Ann. Inst. H. Poincar 36:* 79-88.

Pavón, D., D. Jou and J. Casas-Vázquez. 1983. Equilibrium and nonequilibrium fluctuations in relativistic fluids. *J. Phys. A 16:* 775-782.

Pavón, D. 1990. The generalized second law and extended thermodynamics. *Class. Quantum Grav. 7:* 487-491.

Pavón, D. J. Bafaluy and D. Jou. 1990. FRW cosmology and extended irreversible thermodynamics. (Paper in preparation).

Schweizer, M.A. 1982. Transient transport coefficients for radiative fluids. *Astrophys. J. 258:* 798-811.

Schweizer, M.A. 1984. On compressional modes in radiative plasmas. *Mon. Notic. R. astr. Soc. 210:* 303-322.

Turok, N. 1988. String driven inflation. *Phys. Rev. Lett. 60:* 549-552.

Udey, N. and W. Israel. 1982. General relativistic radiation transfer: the 14-moment approximation. *Mon. Notic. R. astr. Soc. 199:* 1137-1147.

Weinberg, S.W. 1971. Entropy generation and the survival of protogalaxies in an expanding universe. *Astrophys. J. 168:* 175-194.

Weinberg, S.W. 1972. *Gravitation and cosmology.* New York: John Wiley.

Witten, E. 1984. Cosmic separation of phases. *Pys. Rev. D 30:* 272-285.

Witten, E. 1985. Cosmic superstrings. *Phys. Lett. 153 B:* 243-246.

Thermal Momentum, Heat Inertia and a Macroscopic Extension of de Broglie Microthermodynamics II. The Conservation Laws for Continuity Equations with Sources

S. Sieniutycz
Department of Chemical Engineering, Warsaw Technical University
1 Warynskiego Street, 00-645 Warsaw, Poland

ABSTRACT
We continue an approach to an extended thermodynamics (Sieniutycz 1990b) regarding this theory as a macroscopic consequence of the recent "micro-thermodynamics" of de Broglie (de Broglie 1964). The author's treatment is in the context of the Hamilton's action and the conservation laws, for multicomponent fluids with sources in the continuity equations. The de Broglie theory is microscopic, quantum and relativistic; however, its nonrelativistic macroscopic counterpart is pursued, which is sufficient to preserve the effect of thermal inertia, inherent in that theory. In this extended thermodynamics, the sources of the entropy and matter result as the partial derivatives of a generalized energy, E, with respect to the action-type variables ("phases") under the condition of constancy of momenta and other natural variables of E. These sources can also be obtained by differentiation of any generalized thermodynamic potential (the suitable Legendre transform of E) with respect to the phases, provided that it deals with its own natural variables. Internal symmetries of the theory are shown to yield the global mass conservation law and the chemical stoichiometry. Vector external fields and charged systems can be considered in a generalized variational scheme. General equations of motion are obtained. Close to equilibrium, they correspond to the phenomenological equations of the irreversible thermodynamics under condition of a suitable kinetic potential or the generalized energy function. Onsagerian dissipation potentials, incorporated in the action integral, play an essential role in a "irreversible" form of the theory. Nonlocality of the irreversible transport and rate processes is a natural feature of the theory. It is shown that the conventional assumption, allowing the energy-momentum tensor to be independent of the sources of the entropy and matter, may not be strictly met for the popular quadratic (Joule-type) dissipation functions, and hence, either the conservation laws of highly irreversible processes must be modified by the presence of dissipation or the quadratic dissipation function is insufficient approximation when used in the context of the action approach.

1. INTRODUCTION

The stationary action principle is as a rule associated with the perfect or nondisssipastive media (Yourgrau and Mandelstamm 1968). It is a frequent although unproved opinion (Landau and Lifshitz 1974b) that the so called dissipative terms, such like the heat flux, mass diffusion fluxes and viscous stress tensor should be only formally added to the macroscopic conservation laws of a perfect medium (obtained from the least action approach) in order to preserve an agreement of the theory with experiments and with the Boltzmann equation. This kinetic equation yields the conservation laws by exploiting conservation of the collision invariants and the perfect medium equations are obtained only in the equilibrium (zero-order) approximation of the distribution function otherwise the "dissipative" terms appear (Huang 1967). The natural appearance of the dissipative terms (or, more precisely, the disequilibrium terms since they may also be present also when the entropy source vanish, Sieniutycz 1990b) in the conservation laws challenged the trials toward incorporating the disequilibrium effects in the least action formalism. It was shown that, at least for the sourceless systems (where the four-divergences of the entropy and matter vanish), the disequilibrium fluxes (of heat, mass and momentum) appear indeed in the conservation laws resulting from the least action as the result of the matter and entropy diffusion (Sieniutycz and Berry 1989, Sieniutycz 1990b, the latter reference called here as part I). Since these systems are "reversible" in a weaker sense than the ordinary perfect fluids (although the entropy itself is conserved, the particle one is not; the particle can exchange the heat with its environment) one can expect that the inclusion of the sources will not destroy the effectiveness of the generalized least action approach. In this work we show that it is indeed so, although the effectiveness is preserved in the framework of the general theory rather than for the examples. Any example must deal with certain particular kinetic potential L, and our results show that it is doubtful whether the popular quadratic (Joule-law type) expressions for the entropy dissipation function are sufficiently exact for the formalism. Actually, we show on the examples that these classical expressions change the analytical form of the conservation laws of the energy and momentum in comparison with those of a related sourceless medium (or a adiabatic perfect fluid). This rises the question of the more exact (nonquadratic) dissipation functions of the nonlinear processes, consistent with the so-called internal kinetic potential L^{int}, as an important problem for the future investigations. In view of these findings, due to the lack of the knowledge of the exact dissipation functions, the general approach constitutes here an exact self-consistent theory whereas the testing of examples involves necessarily approximations. It is shown that the general theory has a canonical Hamiltonian structure in the space of the phases (Clebsch potentials) and densities, for both reversible and irreversible processes and that the basic formal difference between them is that in the reversible case the energy E contains the phases only in terms of their gradients whereas in the irreversible case E contains the phases explicitly with the only restriction on them that the global conservation of mass must hold. The reasons for the appearance of the disequilibrium terms in the formalism are those of Part I: the freedom in the variation of the entropy flux, (i.e. the distinction between the thermal and material paths) and an apparent inertia of the entropy and heat for finite $\theta_a = m_a/k$, predictable also in the framework of Grad's solution of the Boltzmann equation with the same order of the magnitude (Sieniutycz and Berry 1989). Although designed in the framework of the de Broglie concept of the microthermodynamics, and the related heat inertia (de Broglie 1962, 1964, 1967), see Appendix of Part I, the theory can live an independent life when the thermal inertia effect is ignored (i.e. $\theta_a=0$ is put in any of the equations). However, as shown in the part I, the frame of the vanishing momentum corresponds then with the well known barycentric frame, and the so-simplified ("quasiclassical") theory cannot describe the effects of the viscosity and the related selfdiffusion in a single-component fluid, although other irreversible effects (as e.g. heat flux) can still be predicted. In Appendix 2 of this paper it is shown that the macroscopic counterpart of the de Broglie thermodynamics is close to the assumption that a part of the measurable mass of a macroscopic system is of purely thermal origin and that such an assumption removes an unclarity of the original macroscopic approach given in Part I pertaining to the difference between the thermal potentials T_a^- at equilibrium. Namely, these potentials involve the effect of the chemical potentials of the species; whenever this effect is removed, the equality of the partial temperatures T_a at equilibrium is met.

An alternative approach could involve the maximum entropy formalism. Actually, the difficulties do not disappear in any of two approaches; they are always caused by the knowledge of the dissipation potentials, but in the action approach the power of the Nöthers theorem can be exploited to display the problem of the irreversibility in the context of the symmetry principles and the related conservation laws.

2. GENERALIZED VARIATIONAL SCHEME, EQUATIONS OF MOTION AND CONSERVATION LAWS.

In Part I (Sieniutycz 1990b) a suitability of the use in the action formalism of the absolute velocities and absolute fluxes as opposed to the relative (diffusion related) quantities was outlined. This requires the subsequent transformation to the convective and diffusive fluxes if one wants to single out the "dissipative" terms, but the detailed procedure is already known from the part I. We will work here in terms of absolute velocities rather than absolute fluxes, the former being the natural variables of any kinetic potential L and Lagrangian Λ. A generalized action functional A based on the kinetic potential L containing explicitly velocities, rates, densities and phases as the field variables but not their partial derivatives has the form

$$A=\int_{t_1}^{t_2} \Lambda dVdt = \int_{t_1}^{t_2} \{L(\rho_a,\rho_{sa},\mathbf{u}_a,\mathbf{u}_{sa},\eta_a,\phi_a,r_b,\mathbf{x},t) - \sum_a (\rho_a\frac{\partial\phi_a}{\partial t} + \rho_a\mathbf{u}_a.\nabla\phi_a + \rho_{sa}\frac{\partial\eta_a}{\partial t} + \rho_{sa}\mathbf{u}_{sa}.\nabla\eta_a) + \sum_a[\lambda_a\rho_a(\frac{\partial\alpha_a}{\partial t} + \mathbf{u}_a.\nabla\alpha_a) + \lambda_{sa}\rho_{sa}(\frac{\partial\alpha_{sa}}{\partial t} + \mathbf{u}_{sa}.\nabla\alpha_{sa})]\}dVdt \tag{1}$$

where the constancy of the initial Lagrangian coordinates, the labels for thermal and material elements of the fluid, $d\alpha_a/dt$ and $d\alpha_{sa}/dt$, has been incorporated. L may contain the rates r_b which may pertain to the chemical reactions or any relaxing internal variables. The simplified notation in L and other quantities pertains actually to the general case when all the properties of the possible components may influence L. The structure of L is a general one. It involves the sources of the masses and entropies, σ_a and σ_{sa}, and it is related to the "sourceless kinetic potential", L^1, of Part I

$$L^1 = \sum_a [\rho_a \frac{\mathbf{u}_a^2}{2} + \theta \rho_{sa} \frac{\mathbf{u}_{sa}^2}{2} - (\rho_a + \theta \rho_{sa}) \psi(\mathbf{x},t) - \rho_e(\rho_a, \rho_{sa})] \quad (2)$$

by

$$L = L^1(\rho_a, \rho_{sa}, \mathbf{u}_a, \mathbf{u}_{sa})$$
$$- \sum_{a=1}^{n} [\phi_a \, \sigma_a(\rho_a, \rho_{sa}, \mathbf{u}_a, \mathbf{u}_{sa}, \mathbf{r}_b) + \eta_a \, \sigma_{sa}(\rho_a, \rho_{sa}, \mathbf{u}_a, \mathbf{u}_{sa}, \mathbf{r}_b)] \quad (3)$$

Most commonly $\sigma_a = \Sigma \, \nu_{ab} M_a r_b$, where r_b is the rate of b-th chemical reaction, i.e. the mass sources are independent of the velocities. One may note that L with respect to L^1 has the structure of a (negative) hamiltonian. Actually, it is a suitable structure although the physical energy E corresponds to a Legendre transform of L, not to L itself. The thermodynamic kinetic potential L is not directly related to any kinetic and remaining parts of the energy or free energy, the issue already pointed out (Nettleton 1986) in a different context.

We will also modify eq. (3) by a single entropy description with an averaged phase η. The total entropy source σ will be then expressed in the Onsagerian way, as the sum of the two dissipation potentials, the rate dependent Φ^* and the state dependent Ψ^*, such that $\Phi^* = \Psi^*$.

The stationarity conditions of the action A with respect to the multipliers λ_a and λ_{sa} describe the preservation of the identity of the material and thermal trajectories, $d\alpha_a/dt=0$ and $d\alpha_{sa}/dt=0$. Those with respect to the labels describe the "conservation laws for the multipliers λ_a and λ_{sa}" in the case when the elastic effects are absent in the fluid (otherwise L should also contain the gradients and time derivatives of the labels). The stationarity conditions of A with respect to phases are the balance laws for masses and entropies of the species

$$\frac{\partial \rho_a}{\partial t} + \nabla.(\rho_a \mathbf{u}_a) = - \frac{\partial L}{\partial \phi_a} = \sigma_a(\rho_a, \rho_{sa}, \mathbf{u}_a, \mathbf{u}_{sa}, \mathbf{r}_b) \quad (4)$$

$$\frac{\partial \rho_{sa}}{\partial t} + \nabla.(\rho_{sa} \mathbf{u}_{sa}) = - \frac{\partial L}{\partial \eta_a} = \sigma_{sa}(\rho_a, \rho_{sa}, \mathbf{u}_a, \mathbf{u}_{sa}, \mathbf{r}_b) \quad (5)$$

They indicate that an explicit presence of the *derivatives* of the phases in L, eq. (1), is inadmissible since it would result in violation of the basic four-divergence structures on the left hand sides of these equations. In examples, the cases of the identical temperature of each of the species will only be considered, which will allow to neglect the thermal relaxation and to use the single entropy equation, treated formally as an equation describing a zero-th component, of the density $\rho_s = \rho_0$ and the velocity $\mathbf{u}_s = \mathbf{u}_0$. Its source depends then on all the ρ_a, $\mathbf{u}_a$, and r_b as well as $\rho_s = \rho_0$ and $\mathbf{u}_s = \mathbf{u}_0$

$$\frac{\partial \rho_s}{\partial t} + \nabla.(\rho_s \mathbf{u}_s) = - \frac{\partial L}{\partial \eta} = \sigma_s(\rho_a, \rho_s, \mathbf{u}_a, \mathbf{u}_s, \mathbf{r}_b) \quad (5a)$$

Actually, the structures of the source functions are not arbitrary. In a nonrelativistic theory, the mass source functions in eqs. (3) through (5a) are practically dependent on the rates r_b only; we put on other variables only for the sake of the generality. Moreover, one may expect that a certain combinations of the squared differences between the velocities can play a role in the entropy source rather than the absolute velocities themselves, making the entropy source invariant with respect to the Galilei transformation and the change in the sign of the relative velocity. However, our general formulae are not restricted explicitly by these requirements. The latter play a role in the applications.

As shown in the Section 7 the chemical rates r_b result as the stationarity conditions of A or L with respect to r_b in a form generalizing the kinetic law of mass action for the unsteady state processes if a suitable formula for chemical resistance is applied in the rate-dependent entropy dissipation function.

The stationarity conditions of A with respect to the densities express the canonical intensities - the partial derivatives of negative of L with respect to the corresponding densities - in terms of the gradients and time derivatives of corresponding phases and labels:

$$\widetilde{T}_a = - \partial L(\rho_a,\rho_{sa},\mathbf{u}_a,\mathbf{u}_{sa},\eta_a,\phi_a,r_b, \mathbf{x},t)/\partial\rho_{sa} = -\left(\frac{\partial\eta_a}{\partial t} + \mathbf{u}_{sa}.\nabla\eta_a\right) \tag{6}$$

$$\widetilde{\mu}_a = - \partial L(\rho_a,\rho_{sa},\mathbf{u}_a,\mathbf{u}_{sa},\eta_a,\phi_a,r_b, \mathbf{x},t)/\partial\rho_a = -\left(\frac{\partial\phi_a}{\partial t} + \mathbf{u}_a.\nabla\phi_a\right) \tag{7}$$

This occurs since all α_a and α_{sa} are constant along the path. The canonical intensities are then simply the total time derivatives of the corresponding phases. The nonequilibrium temperatures and chemical potentials defined above in terms of L are the phase dependent quantities having a nonlocal nature. The above definitions include the external potentials (hence the explicit presence of x and t in the equations in question) in the way similar to that encountered in the definition of the electrochemical potential. However, the external fields can be excluded from these definitions if one wants to match requirements of gauge invariance of the resulting energy-momentum tensor and to work with the kinetic potential L without external fields. These will then appear separately in the so-called interaction Lagrangian.

The stationarity conditions of A with respect to the absolute velocities express the generalized (phase dependent) momentum densities - the derivatives of L with respect to the transfer velocities - in terms of the gradients of the phases and labels:

$$\mathbf{K}_{sa} = \partial L(\rho_a,\rho_{sa},\mathbf{u}_a,\mathbf{u}_{sa},\eta_a,\phi_a,r_b, \mathbf{x}, t)/\partial\mathbf{u}_{sa} = \rho_{sa}(\nabla\eta_a - \lambda_{sa}\nabla\alpha_{sa}) \tag{8}$$

$$\mathbf{K}_a = \partial L(\rho_a,\rho_{sa},\mathbf{u}_a,\mathbf{u}_{sa},\eta_a,\phi_a,\mathbf{r}_b, \mathbf{x}, t)/\partial \mathbf{u}_a = \rho_a(\nabla\phi_a - \lambda_a\nabla\alpha_a) \tag{9}$$

These momentum densities are no longer so simply related to the velocities as in the case of reversible processes.

The stationary Lagrangian L results from eqs. (1), (6) and (7) as the following Legendre transform of the kinetic potential L

$$\Lambda = L - \sum_a \left(\rho_a \frac{\partial L}{\partial \rho_a} + \rho_{sa}\frac{\partial L}{\partial \rho_{sa}}\right) \tag{10}$$

and it equals to a nonequilibrium pressure P which depends on the transfer velocities, intensities and phases as its natural variables.

The perfect differentials of the thermodynamic potentials (E, P, H, G etc.) obey the formulae already developed for the reversible (sourceless) processes. This means that eqs. (117) and (118) of Part I hold, describing the non-equilibrium phase-dependent thermodynamics with the entropy and matter sources as the thermodynamic variables. This also means that the generalized momentum densities, eqs. (8) and (9) can be obtained not only from the potential L but also from other thermodynamic potentials containing transfer velocities amongst their independent variables, which have to be their natural variables. It is convenient to refer these momenta to the unit of entropies or masses as to the basic entities of the stationary action formalism:

$$\kappa_{sa} = \rho_{sa}^{-1}\mathbf{K}_{sa} = \frac{\partial L}{\rho_{sa}\partial \mathbf{u}_{sa}} = \frac{\partial P}{\rho_{sa}\partial \mathbf{u}_{sa}} = \frac{\partial \Phi}{\rho_{sa}\partial \mathbf{u}_{sa}} .... = \nabla\eta_a - \lambda_{sa}\nabla\alpha_{sa} \tag{11}$$

$$\kappa_a = \rho_a^{-1}\mathbf{K}_a = \frac{\partial L}{\rho_a\partial \mathbf{u}_a} = \frac{\partial P}{\rho_a\partial \mathbf{u}_a} = \frac{\partial \Phi}{\rho_a\partial \mathbf{u}_a} .... = \nabla\phi_a - \lambda_a\nabla\alpha_{sa} \tag{12}$$

Routhian type thermodynamic potentials can also be defined which contain the part of the transfer variables as the velocities and the remaining part as the momenta. This may be useful in the case of the so called cyclic coordinates.

Similarly, any of the canonical intensities can be expressed as the derivative of the arbitrary thermodynamic potential involving corresponding density expressed in terms of its natural variables:

$$\tilde{\tilde{T}}_a = -\frac{\partial L}{\partial \rho_{sa}} = \frac{\partial E}{\partial \rho_{sa}} = \frac{\partial H}{\partial \rho_{sa}} = ... \tag{13}$$

$$\tilde{\mu}_a = -\frac{\partial L}{\partial \rho_a} = \frac{\partial E}{\partial \rho_a} = \frac{\partial F}{\partial \rho_a} = \frac{\partial G}{\partial \rho_a} = \frac{\partial \Phi}{\partial \rho_a} \ldots \tag{14}$$

(See Part I for the definition of the phase-dependent thermodynamic potentials.) As pointed out earlier, these "cooler" intensities are the most fundamental ingredients of the canonical formalism. However, they are the "differential" measures of the partial (negative) Lagrangians rather than the energies of the flowing elements. The corresponding "differential energies" result from eqs. (6), (7), (11) and (12) as the derivatives

$$\tilde{T}_a^{+} = -\left(\frac{\partial \eta_a}{\partial t} - \lambda_{sa}\frac{\partial \alpha_{sa}}{\partial t}\right) = \frac{\partial L}{\rho_{sa}\partial \mathbf{u}_{sa}}.\mathbf{u}_{sa} - \frac{\partial L}{\partial \rho_{sa}} = \kappa_{sa}.\mathbf{u}_{sa} + \tilde{T}_a^{-} \tag{15}$$

$$\tilde{\mu}_a^{+} = -\left(\frac{\partial \phi_a}{\partial t} - \lambda_a\frac{\partial \alpha_a}{\partial t}\right) = \frac{\partial L}{\rho_a\partial \mathbf{u}_a}.\mathbf{u}_a - \frac{\partial L}{\partial \rho_a} = \kappa_a.\mathbf{u}_a + \tilde{\mu}_a \tag{16}$$

Kupershmidt (1990) pointed out that the dynamics of the extended nonequilibrium fluids should admit (like that of the classical fluids) the canonical Hamiltonian structures. Indeed, after constructing the energy function E, eq. (27), for our problem, in the space of the Clebsch potentials and densities, the canonical Hamiltonian structure can be generated if one substitutes into E the velocities and reaction rates obtained as the results of the stationarity conditions of Λ. Substituting into the general action functional (1) the energy definition, eq. (27), momentum representations, eqs. (11) and (12) and the stationarity condition of Λ with respect to the rates r_b, eq. (75), the following important expression for the action is obtained

$$A = \int_{t_1}^{t_2} \{-E(\rho_a, \rho_{sa}, \rho_a(\nabla\phi_a - \lambda_a\nabla\alpha_a), \rho_{sa}(\nabla\eta_a - \lambda_{sa}\nabla\alpha_{sa}), \eta_a, \phi_a, \mathbf{x}, t) - \sum_a [\rho_{sa}(\frac{\partial \eta_a}{\partial t} - \lambda_{sa}\frac{\partial \alpha_{sa}}{\partial t}) + \rho_a(\frac{\partial \phi_a}{\partial t} - \lambda_a\frac{\partial \alpha_a}{\partial t})]\} dV dt \tag{17}$$

Here, the thermodynamic potential of the energy E rather than L is used and the unit mass and entropy momenta rather than the momentum densities are incorporated into action in terms of their gradient representation. In this form a strong analogy with a variational principle of quantum mechanics is achieved.

The variations of phases in eq. (17) lead to the balance laws for the entropies and masses, eqs. (4) and (5) or (5a). Those of densities yield the equations of the Hamilton-Jacobi type for all phases which are in fact eqs. (15) and (16), in which all the quantities are expressed in terms of the densities, phases and their gradients.

The above formulae are reminiscent of the well known relations linking the energy, momenta and kinetic potential in the mechanics of material points, provided that the partial derivatives of the density L with respect to densities are interpreted as the "partial lagrangians" in the sense described above. These formulae contain the partial derivatives of L with respect to the densities and velocities which appear in the generalized equations of motions of the fluid. They constitute the set of equations describing transfer of species and entropy (entropies). Hence, they are also related to the flow of heat, diffusion of the matter and the frictional behavior, especially in the forms where the total currents are split into convective and diffusive contributions (see also Part I). This description of the phenomenology is a nonstandard one, as it uses the information about resistance data contained in the rate dissipation function of the entropy source to yield the additional (frictional type) forces **X** in the equations of motion.

Although the complete information about the motion is contained in all of the stationarity conditions obtained, we will reserve the term "equations of motion" to a particular combined form of these conditions, resembling the Euler equations of a perfect fluid with the additional forces **X**. These equations obtained by the partial eliminating of the time derivatives and gradients of the phases and labels from eqs. (6) through (9). (It is essential that the procedure eliminates the phases completely only in the case of a perfect fluid as the equations in question become phase independent just in that limiting case). The procedure involves adding the gradient of eq. (6) to the partial derivative of eq. (8) and the gradient of eq. (7) to the partial derivative of eq. (9), expressing the remaining time derivatives in terms of gradients, and then applying the well-known formula of the vector calculus to express the gradient of the scalar product of two vectors in terms of the gradients of the individual vectors and their rotations. Since it preserves all basic ingredients of the similar procedure known in the theory of perfect fluid, its details are relegated to the Appendix 1; see also perfect fluid counterpart in Yourgrau and Mandelstamm (1969).

For the arbitrary kinetic potential L, eq. (3), containing the phases but not their derivatives, the procedure yields the following general equations of motion:

$$\frac{\partial}{\partial t}\left(\frac{\partial L}{\rho_a \partial \mathbf{u}_a}\right) + \nabla.\left(\frac{\partial L}{\rho_a \partial \mathbf{u}_a}.\mathbf{u}_a - \frac{\partial L}{\partial \rho_a}\right) - \mathbf{u}_a \times \left(\nabla \times \frac{\partial L}{\rho_a \partial \mathbf{u}_a}\right) + \frac{\sigma_a}{\rho_a}\left(\frac{\partial L}{\rho_a \partial \mathbf{u}_a} - \nabla\phi_a\right) = 0 \qquad (18)$$

$$\frac{\partial}{\partial t}\left(\frac{\partial L}{\rho_{sa} \partial \mathbf{u}_{sa}}\right) + \nabla.\left(\frac{\partial L}{\rho_{sa} \partial \mathbf{u}_{sa}}.\mathbf{u}_{sa} - \frac{\partial L}{\partial \rho_{sa}}\right) - \mathbf{u}_{sa} \times \left(\nabla \times \frac{\partial L}{\rho_{sa} \partial \mathbf{u}_{sa}}\right) + \frac{\sigma_{sa}}{\rho_{sa}}\left(\frac{\partial L}{\partial \mathbf{u}_{sa}} - \nabla\eta_a\right) = 0 \qquad (19)$$

with the sources of mass and entropy,

$$\sigma_a = -\frac{\partial L}{\partial \phi_a} \qquad \sigma_{sa} = -\frac{\partial L}{\partial \eta_a} \qquad (20)$$

The presence of the generalized momenta and energy type ("hot") intensities (15) and (16) in equations of motion is easily seen. When the well known formula for the gradient of the scalar product of any two vectors is used, these equations can be written in the alternative forms operating with the total time derivatives of the momenta. They involve corresponding velocities; hence superscript a in eq. (17). For instance eq. (18) takes the form

$$\frac{d^a}{dt}\left(\frac{\partial L}{\rho_a \partial \mathbf{u}_a}\right) + \left(\frac{\partial L}{\rho_a \partial \mathbf{u}_a}.\nabla\right)\mathbf{u}_a - \nabla\frac{\partial L}{\partial \rho_a} + \frac{\partial L}{\rho_a \partial \mathbf{u}_a} \times (\nabla \times \mathbf{u}_a) + \frac{\sigma_a}{\rho_a}\left(\frac{\partial L}{\rho_a \partial \mathbf{u}_a} - \nabla\phi_a\right) = 0 \qquad (21)$$

In the sourceless case of the perfect fluid where $\theta_a=0$ and $\mathbf{u}_a=\mathbf{u}_{sa}=\mathbf{u}$ the sum of the products of these equations and corresponding densities yield the Euler equation of motion on the basis of the Gibbs-Duhem equation of classical thermodynamics. The frictional forces, related to the source term, appear only in the case of the phase dependent L.

The perfect differential of the Lagrangian Λ, eq. (1), is

$$\begin{aligned} d\Lambda = \sum_a [\, & \rho_{sa}(\kappa_{sa} - \nabla\eta_a + \lambda_{sa}\nabla\alpha_{sa}).d\mathbf{u}_{sa} + \rho_a(\kappa_a - \nabla\phi_a + \lambda_a\nabla\alpha_a).d\mathbf{u}_a \\ & - (\tilde{T}_a + \frac{d\eta_a}{dt} - \frac{\lambda_{sa}d\alpha_{sa}}{dt})d\rho_{sa} - (\tilde{\mu}_a + \frac{d\phi_a}{dt} - \frac{\lambda_a d\alpha_a}{dt})d\rho_a \\ & - \sigma_{sa}d\eta_a - \rho_{sa}d\dot{\eta}_a - \rho_{sa}\mathbf{u}_{sa}.\mathbf{d}\nabla\eta_a - \sigma_a d\phi_a - \rho_{sa}d\dot{\phi}_a - \rho_a\mathbf{u}_a.\mathbf{d}\nabla\phi_a \\ & + \lambda_a\rho_a d\dot{\alpha}_a + \lambda_a\rho_a\mathbf{u}_a.\mathbf{d}\nabla\alpha_a + \lambda_{sa}\rho_{sa}d\dot{\alpha}_{sa} + \lambda_{sa}\rho_{sa}\mathbf{u}_{sa}.\mathbf{d}\nabla\alpha_{sa} \\ & - \sum_b (\phi_a\frac{\delta\sigma_a}{\delta r_b} - \eta_a\frac{\delta\sigma_a}{\delta r_b})dr_b \,] \end{aligned} \qquad (22)$$

the last line describing the relaxation phenomena. The intensities and momenta are defined here in terms of the derivatives of the kinetic potential L with respect to the densities and velocities. To express them in terms of other variables a suitable thermodynamic potential should be used, eqs. (11) through (14). The vanishing variational derivatives (or in particular cases the partial derivatives) of Λ with respect to independent variables constitute the stationarity conditions of the action A and summarize the theory. The variations of transfer velocities yield the gradient representations of the generalized momenta, those of densities yield the representations of the canonical intensities and those of phases the balance laws for the entropies and masses. The variations of the reaction rates yield the kinetics of the rate (relaxation) phenomena. The associated equations of motions (18) and (19) describe (for the phase dependent L) the irreversible evolutions under the conservation laws resulting from the invariancy of the action A with respect to the space and time translations (conservation laws for the energy and momentum) or with respect to the translations of phases or their combinations (internal symmetries).

Conservation of energy and momentum. Let us find the general form of the conservation or balance laws for the energy and linear momentum, eq. (35) in Part I. Using the general formula for the energy- momentum tensor G^{jk}, eq. (34) of Part I, and the stationarity conditions, eqs. (6) through (8), the stress tensor is (j, k =1, 2, 3)

$$G^{jk} = -\sum_a [(\frac{\partial\eta_a}{\partial x^j} - \frac{\lambda_{sa}\partial\alpha_{sa}}{\partial x^j})\rho_{sa}u^k_{sa} + (\frac{\partial\phi_a}{\partial x^j} - \frac{\lambda_a\partial\alpha_a}{\partial x^j})\rho_a u^k_a] - P\delta^{jk} =$$

$$-\sum_a [\frac{\partial L}{\partial u^j_{sa}}u^k_{sa} + \frac{\partial L}{\partial u^j_a}u^k_a] - P\delta^{jk} \tag{23}$$

The momentum density is:

$$G^{j4} = \Gamma^j = -\sum_a [(\frac{\partial\eta_a}{\partial x^j} - \frac{\lambda_{sa}\partial\alpha_{sa}}{\partial x^j})\rho_{sa} + (\frac{\partial\phi_a}{\partial x^j} - \frac{\lambda_a\partial\alpha_a}{\partial x^j})\rho_a] = -\sum_a [\frac{\partial L}{\partial u^j_{sa}} + \frac{\partial L}{\partial u^j_a}] \tag{24}$$

It is instructive to apply this formula in the case of the kinetic potential L^1 expressed in terms of the absolute velocities and then verify that the expression resulting from eq. (24) is equivalent to eqs. (43) through (44) of Part I when its condition (5) describing the vanishing of the sum of the diffusional momenta is exploited. This example clearly shows that the nonequilibrium stresses are created due to the diffusion of mass and entropy in the fluid frame.

The density of the energy flux is on the basis of eqs. (15) and (16)

$$G^{4k} = Q^{4k} = -\sum_a [(\frac{\partial\eta_a}{\partial t} - \frac{\lambda_{sa}\partial\alpha_{sa}}{\partial t})\rho_{sa}u^k_{sa} + (\frac{\partial\phi_a}{\partial t} - \frac{\lambda_a\partial\alpha_a}{\partial t})\rho_a u^k_a] =$$

$$= \sum_a [(\frac{\partial L}{\rho_{sa}\partial \mathbf{u}_{sa}}.\mathbf{u}_{sa} - \frac{\partial L}{\partial\rho_{sa}})\rho_{sa}\mathbf{u}^k_{sa} + (\frac{\partial L}{\rho_a\partial \mathbf{u}_a}.\mathbf{u}_a - \frac{\partial L}{\partial\rho_a})\rho_a\mathbf{u}^k_a] \tag{25}$$

and the equivalent formula is

$$G^{4k} = Q^{4k} = \sum_a [\tilde{T}^+_a \rho_{sa}\mathbf{u}^k_{sa} + \tilde{\mu}^+_a \rho_a \mathbf{u}^k_a] \tag{26}$$

The simplicity of the above equation is due to inclusion of the external fields into the intensities and to the operating with the absolute velocities. Again, it may be instructive to verify that in the case of the kinetic potential L^1 expressed in the form containing the absolute velocities, the equation (26) yields precisely eq. (47) of Part I and its equivalents if the transition is made to the description in terms of the mean hydrodynamic velocity and the sum of all the diffusion fluxes vanish.

The energy density is:

$$G^{44} = E = -\sum_a [(\frac{\partial \eta_a}{\partial t} - \frac{\lambda_{sa}\partial \alpha_{sa}}{\partial t})\rho_{sa} + (\frac{\partial \phi_a}{\partial t} - \frac{\lambda_a \partial \alpha_a}{\partial t})\rho_a] - P =$$

$$\sum_a [- \frac{\partial L}{\partial \rho_{sa}}\rho_{sa} - \frac{\partial L}{\partial \rho_a}\rho_a + \frac{\partial L}{\partial \mathbf{u}_{sa}}.\mathbf{u}_{sa} + \frac{\partial L}{\partial \mathbf{u}_a}.\mathbf{u}_a] - P \tag{27}$$

$$= \sum_a [\frac{\partial L}{\partial \mathbf{u}_{sa}}.\mathbf{u}_{sa} + \frac{\partial L}{\partial \mathbf{u}_a}.\mathbf{u}_a] - L$$

where eqs. (15) and (16) as well as eq. (10) with Λ=P were used.

The theory, eqs. (1) through (27), can be used in quite involved cases of various external vector and scalar fields and for complex dissipative systems. Although formulae such as eqs. (24) and (27) resemble the definitions of the momentum and energy well known from the mechanics of material points, the kinetic potential L can be phase dependent and the definitions of E and Γ can be more complicated than classical ones, corresponding with the *generalized* momenta and energy.

3. INTRINSIC SYMMETRIES AND IRREVERSIBILITY

Intrinsic (or internal) symmetries are the result of the invariancy of the action A with respect to the changes of phases when changing them by some functions or parameters. From the mathematical viewpoint they arise as the consequence of the second Nöther theorem which states that between the right hand sides of the Euler equations, associated with the variational problems invariant with respect to the transformation group dependent on the p arbitrary functions, there exist p identities, which means that p of these equations result from the remaining ones. The most common example of such invariance are the extremum length problems in parametric form where the transformations which leave lengths of the curves unchanged are all possible changes of the parameter. In this case one of the (parametric) Euler's equations results from the remaining ones. However, the symmetries we are interested in here are those resulting from the continuous changes of phases and thus related to the balance laws of matter and entropy.

As opposed to the conservation laws of the momentum and energy resulting from very general properties of the nonobservability of the absolute time and of absolute space, the most of the potentially possible internal symmetries are of an approximate nature or are simply broken in the real word. The masses of the individual species are not conserved in the chemical reactions and this obvious fact must be reflected by the kinetic potential L which requires an explicit presence of the phase of any nonconserved component in the formula describing L, see eqs. (4) - (5a) and (20). The entropies of the individual species and the total entropy are not conserved either, hence the thermal phases have to be explicitly present in L, allowing the sources.

Let us compute the variation of the action A, eq. (1), with respect to the translation of the phases written in the form

$$\bar{\eta}'_a = \eta'_a + \delta \qquad \text{and} \qquad \bar{\phi}_a = \phi_a + \delta \tag{28}$$

where $\eta'_a = \eta_a/\theta_a$. Each of the phases has the same dimension of the action per unit of mass, and the gradient of each term has the meaning of a unit mass momentum. This pertains to the thermodynamics associated with the de Broglie theory. In the classical theory, where the phases η' cannot be introduced, only the invariance of A with respect the shift of all the modified phases ϕ_a (related to the effective densities) is required, but the mode of the approach and the final result are the same (see below).

The variation of the action A is

$$\delta A = \int \sum_a \left[\frac{\partial \Lambda}{\partial \eta_{sa}} + \frac{\partial \Lambda}{\partial \phi_a}\right] \delta dVdt = \int \sum_a \left[\frac{\partial L}{\partial \eta_{sa}} + \frac{\partial L}{\partial \phi_a}\right] \delta dVdt = 0 \qquad (29)$$

It is sufficient to investigate the invariance of the kinetic potential L only, or the invariance of any thermodynamic potential expressed in its natural variables. In order to make L invariant with respect to the variations of the phases, the sum of the partial derivatives of L with respect to all ϕ_a and η_a (appearing in Λ, L or any thermodynamic potential) should vanish. But these derivatives represent the negative sources of the bare or apparent masses, on the basis of eqs. (4) and (5). Hence the total mass source should vanish for any physical form of L. One may note that the *entropy may not be conserved in this scheme and it can exhibit a source.* According to the second law, whatever the sources of the partial entropies are assumed in L, eq. (3), their total source has to be positive. The sum of the products $\partial L/\partial \eta'_a = \theta_a \sigma_{sa}$ in a more exact description of a multicomponent case, or the product $\theta \sigma_s$ in a single entropy theory, i.e. the entropy effect expressed in the mass units, compensate with the sum of the resulting sources of the bare masses, which is a nonvanishing quantity. But the total measurable mass is conserved.

In terms of the classical theory, only the invariancy of A with respect to the shift of the modified phases ϕ_a (associated with the effective not bare densities) is preserved but the symmetry with respect to the independent shift of η_a is broken, i.e. L must depend explicitly on η in order to have a nonvanishing source of the total entropy. If one stays on the ground of the invariance of A with respect to all the phase shifts (including the shift of η) then, classically, such invariance will allow only processes conserving the total entropy. For example, the invariance requirement for A of the classical theory in the case of the one-component fluid leads to the conservation of the mass and and entropy separately, whereas the same requirement in the extended theory considered here leads to the conservation of the total mass only, with the positive source for the entropy σ_s and the negative for the bare mass, the latter equal to $- \sigma_s \theta$, so that the total source of mass $\sigma_s + \theta \sigma_m$ vanish. When $\theta = 0$ only conservations of the entropy and mass may result when the invariance of A is required.

As a second example consider a closed, classically decribed, system with a chemical reaction. For a single one-step chemical reaction involving various species, the changes in the actions related to the conversions of the (effective not bare) masses and entropies can be written in terms of the corresponding phases in the form representing the *action of the chemical reaction* per one moleequivalent or the so-called *actinity*

$$\Delta\Phi = \sum_{\substack{a \\ \text{products}}} \nu_a^- \widetilde{M}_a \phi_a - \sum_{\substack{a \\ \text{substracts}}} \nu_a^+ \widetilde{M}_a \phi_a = \sum_a \nu_a \widetilde{M}_a \phi_a \tag{30}$$

Here the superscripts + and - pertain to the forward and backward reactions and the resulting stoichiometric coefficient is $\nu_a = \nu_a^- - \nu_a^+$ (entering the thermodynamic relations). The conservation law for mass results from eq. (30) by taking $\delta\,\Delta\Phi = 0$ for the same phase shift $\delta\phi_a = \delta$. The production of the entropy in this (classical) reaction system is indirectly related to the invariance of eq. (30).

One can formulate an explicit functional for this reaction neglecting flow terms and operating with the single thermal phase η and the entropy density ρ_s. The observed experimental rate of reaction $r(C_a, P, T)$ in moleequivalents per unit volume and time is given standardly in terms of the molar concentrations $C_a = \rho_a/\widetilde{M}_a$, where $\widetilde{M}_a$ are molar masses. This function can be considered both as unknown i.e. yet undetermined from the vanishing variation of A, (Section 7) or in a given form $r'(\rho_a, \rho_s, \phi_a, \eta)$ where the natural variables of Lagrangian appear. A truncated action (1) for a closed, well-mixed, system (the flow terms neglected) is:

$$A = \int [(-\sum_a \sum_b r_b \nu'_{ab} \phi_a - \eta\sigma_s)V - \sum_a M_a \frac{d\phi_a}{dt} - S\frac{d\eta}{dt}]dt \tag{31}$$

where S and V, M_a are the total entropy, volume and masses of a-th species of the system. The invariance condition of A with respect to the simultaneous shift of all the phases $\delta\phi_a = \delta$ yields the condition of the conservation of the total mass and the conservations of masses in each of the chemical reactions

$$\delta A = \int V \sum_a \sum_b r_b \nu'_{ab} \delta = 0 \quad \text{or} \quad \sum_a \nu_{ab} \widetilde{M}_a = \sum_a \nu'_{ab} = 0 \tag{32}$$

and hence the conservation of the total mass in the reaction system. In the classical theory the entropy behavior has no influence on the mass conservation law. The second law of thermodynamics in this insulated system results from the vanishing variation of A with respect to the thermal phase η

$$\frac{dS}{dt} = \sigma_s V \geq 0 \qquad (\mathbf{q} = \mathbf{j}_a \equiv 0; \mathbf{u} = \mathbf{u}_s) \tag{33}$$

Thus, we have arrived at the important conclusion: *the invariance conditions of* A, Λ, L *or any thermodynamic potential with respect to the simultaneous shift of phases yields the conservation law of mass. In the classical theory, the second law inequality is the result of the explicit dependence of* L *on the thermal phase* η *or the broken symmetry of* A *with respect to the shift of* η. Whatever thermodynamic potential is used it must have the form preserving these basic conditions otherwise it would be unphysical.

For those models where A is invariant with respect to the shift of all the (unprimed) thermal phases η_a, the condition of the vanishing total entropy source is obtained, associated with the conservation of the entropy. This can be an approximate result only, since the entropy is produced according to the second law. An explicit dependence of L on the total thermal phase η, for the total entropy, or the change of A with the same shift of all of the thermal phases η_a reflect the irreversibility.

4. IRREVERSIBLE HEAT TRANSPORT AND SELF-DIFFUSION

We will now investigate simultaneous transfer of the entropy and the bare mass (self-diffusion) in a flowing nonreactive fluid under the assumption of a single temperature in the system (rapid thermal relaxation between the species). This allows us to use single thermal phases η or $\eta_0 = \eta/\theta$. Designating $\rho_0 = \theta\rho_s$, the kinetic potential L was tested in a "quasiclassical" form:

$$L = \frac{1}{2}\rho_0 u_0^2 + \sum_{a=1}^{n} \frac{1}{2}\rho_a u_a^2 - \rho_e(\rho_a, \rho_0/\theta) - \sum_{a=0}^{n} \rho_a \psi$$
$$- \eta\{\frac{\rho_0}{4\theta} \sum_{a,b=0}^{n} \rho_b\rho_a(u_b - u_a)^2/\rho^2 \mathbf{D}_{ba} + \Psi^*(\rho_a, \rho_b, \rho_0)\} \tag{34}$$

where the observable, not bare, densities are used and the mass sources are not present. Here $\rho = \rho_0 + \Sigma\rho_a$ is the total density, $\mathbf{D}_{ba} = \mathbf{D}_{ab}$ is a diffusivity matrix, θ is an average value of θ_a, and Ψ^* is the state-dependent dissipation potential. Ψ^* is related to the rate dependent dissipation function Φ^*, i.e. $\mathbf{D}_{ba}$ term of eq. (34), by the important Onsagerian equality

$$\Phi^*(\rho_a, \rho_b, \rho_0, \mathbf{u}_a, \mathbf{u}_b, \mathbf{u}_0) = \Psi^*(\rho_a, \rho_b, \rho_0) \tag{35}$$

Working with the bare densities, an alternative (nonequivalent) form of L was tested

$$L = \sum_{a=0}^{n} \frac{1}{2}\rho_a u_a^2 - \rho_e(\rho_a, \rho_0/\theta) - \sum_{a=0}^{n} \rho_a \psi$$
$$- \sum_{a,b=0}^{n} [\rho_a(\eta_0 - \phi_a)\rho_b\rho_a(u_b - u_a)^2/(4\rho^2 D_{ba}) + \theta\Psi^*] \tag{36}$$

where $D_{ba}/\rho_a = D_{ab}/\rho_b$. We report here only a part of these tests, mainly for a single-component system where a selfdiffusion may occur due to the relative motion of the bare mass and entropy with respect to the convective motion. When $\rho'_{sa} = \rho_{sa}\theta_a$ and $\eta'_a = \eta_a\theta_a^{-1}$, the following relations based on these definitions are useful

$$\phi_{0a} = \eta_a s_a = \eta'_a\theta_a s_a = \eta'_a\theta_a\rho_{sa}/\rho = \eta_a\rho'_{sa}/\rho = \eta'_a\rho'_{sa}/\rho \tag{37a}$$

for a general multitemperature case. In the considered single temperature situation it is yet convenient to make the following identifications and definitions:

$$\eta_a = \eta, \ \eta'_a = \eta' = \eta/\theta = \eta_0, \ \Sigma s_a = s, \qquad \Sigma\rho_{sa} = \rho_s, \ \rho_s\theta = \rho_s' = \rho_0 \ , \ \Sigma\phi_{0a} = \phi_0 \tag{37b}$$

and

$$\phi_0 = s\eta = s\theta\eta_0 = \eta_0\theta\rho_s/\rho = \eta\rho_s/\rho = \eta_0\,\rho_0/\rho \tag{37c}$$

as well as

$$\rho\phi_0 = \eta\rho_s = \eta'\theta^{-1}\rho_0 = \eta_0\theta\rho_s = \eta'\rho_s' = \eta_0\,\rho_0 \tag{37d}$$

The reason for introducing the new symbol η_0 for the phase η' and ρ_0 for ρ'_s is that in the single temperature case the entropy density ρ'_s appears in the kinetic potential L, eqs. (34) and (36), as it would be a zero-th component. If the zero subscript symbols are used in the phase term of L, there is no need for splitting this phase term in the two long separate parts corresponding with diffusion of the entropy and mass.

Here $\phi_0 = s\eta = \theta s\eta'$ is the modified thermal phase expressed as an action per unit mass of the mixture, and the product $\rho\phi_0$ is the thermal phase per unit volume. ρ_s is the total entropy density or the sum of the entropy densities of all the species, ρ is the total mass density. The assumed form of the phase term in L corresponds to the entropy source expressed in terms of the squares of the relative velocities as indicated by experience as well as the Joule and Maxwell-Stefan laws. We will show how the classical expression for the entropy source, related to the "binary" transport of the heat and selfdiffusing mass, can be obtained as a special case from eq. (36).

Consider a nonequilibrium system called customarily the one component with a selfdiffusion and heat transfer. In the present formalism it behaves as a binary system, of the total density ρ, composed of the bare mass (subscript m for bare mass) and the entropy (subscript 0). The internal (phase) part of its kinetic potential, based on a simplified form of eq. (36) for $\mathbf{u}_0 = \mathbf{u}_s$ is

$$L^{int} = -(\eta_0 - \phi_m)\{\tfrac{1}{2}\rho_0\rho_m^2(u_0 - u_m)^2(\rho_0 + \rho_m)^{-2}D_0^{-1} + \theta\Psi^*\} \tag{38}$$

and, since $s\rho_m = \rho_s = \rho_0/\theta$, the obvious transformations yield

$$L^{int} = -(\eta_0 - \phi_m)\{\tfrac{1}{2}\theta^{-1}\rho_0^2\rho_m^2(u_0 - u_m)^2\rho^{-2}(\rho_m D_0 s)^{-1} + \theta\Psi^*\} \tag{39}$$

Taking into account that $\eta_0 = \eta/\theta$ and $\rho_0 = \rho_s\theta$

$$L^{int} = -(\eta - \phi_m\theta)\{\tfrac{1}{2}\rho_s^2\rho_m^2(u_0 - u_m)^2\rho^{-2}(\rho_m D_0 s)^{-1} + \Psi^*\} =$$

$$= -(\eta - \phi_m\theta)(\mathbf{j}_s^2/2\lambda + \Psi^*) = -(\eta - \phi_m\theta)(\mathbf{q}^{,2}/(2\lambda T^2 + \Psi^*) \tag{40}$$

where a thermal conductivity coefficient was defined as

$$\lambda = (\rho/\rho_m)^2\rho_m D_0 s = (\rho/\rho_m)^2\rho_0 D_0\theta^{-1} \tag{41}$$

This definition contains the entropy per unit mass rather than the specific heat as it occurs in the conventional definition, but in the latter case one has to operate with the usual "thermal diffusivity", D_h, rather than with that of the entropy, D_0, to arrive at the same value of λ. Comparison of the two expressions for λ, classical $\rho C_p D_h$ and that of eq. (41), links D_h with D_0.

As $\Phi^*=\Psi^*$, the entropy source found from eqs. (5a) and (40), is

$$\sigma_s = -\theta^{-1}\frac{\partial L}{\partial \eta_0} = -\frac{\partial L}{\partial \eta} = -\frac{\partial L^{int}}{\partial \eta} = \tfrac{1}{2}\mathbf{q}^{,2}/(\lambda T^2) + \Psi^* = \mathbf{q}^{,2}/(\lambda T^2) \geq 0 \tag{42}$$

in agreement with the well-known expression for pure heat transfer. The temperature T in eq. (42) corresponds with its equilibrium approximation.

The role of the phase ϕ_m in eq. (40) can be seen now. The related source of the bare mass is

$$\sigma_m = -\frac{\partial L}{\partial \phi_m} = -\frac{\partial L^{int}}{\partial \phi_m} = -\theta\,\mathbf{q}^{,2}/(\lambda T^2) \leq 0 \tag{43}$$

Thus, due to an inertia of the entropy, the entropy increase corresponds with the decrease of the bare mass in the manner dependent of the inertial coefficient θ. However, the total mass is conserved; from eq. (29) $\sigma_m + \theta\sigma_s = 0$.

According to de Broglie (1964; his formula (22)) the fluctuation of the instantaneous proper mass M_0 of a particle results in the instantaneous entropy of the hidden thermostat, S, where

$$S = S_0 - \frac{M_0}{\theta} \tag{44}$$

Here S_0 is a nonfluctuating ("equilibrium") entropy of the termostat, perhaps more precisely the particle-thermostat system. In terms of densities, eq. (44) reads, $\rho_s = \rho_{s0} - \theta^{-1}\rho_m$, or $\theta\rho_{s0} = \theta\rho_s + \rho_m$. The product $\theta\rho_{s0}$ is then equal to the total (effective) density which is the conserved property; c.f. eq (4) of Part I and eq. (29). Nonfluctuating entropy S_0, like the total mass, is then conserved in the theory studied, whereas it is the instantaneous entropy S which obeys the second law.

5. VECTOR EXTERNAL FIELDS INTERACTING WITH CHARGED PARTICLES.

Let us return to eq. (21)

$$\frac{d^a}{dt}\left(\frac{\partial L}{\rho_a \partial \mathbf{u}_a}\right) + \left(\frac{\partial L}{\rho_a \partial \mathbf{u}_a}.\nabla\right)\mathbf{u}_a - \nabla\frac{\partial L}{\partial \rho_a} + \frac{\partial L}{\rho_a \partial \mathbf{u}_a} \times (\nabla \times \mathbf{u}\,) + \frac{\sigma_a}{\rho_a}\left(\frac{\partial L}{\rho_a \partial \mathbf{u}_a} - \nabla\phi_a\right) = 0 \qquad (21)$$

This equation, and its formally identical entropy counterpart (velocity $\mathbf{u}_{sa}$ instead $\mathbf{u}_a$), constitute a set of equations of motion true for arbitrary kinetic potential L containing the velocities, densities and phases as the state variables but *not* derivatives of these quantities. L may also contain explicitly independent variables ($\mathbf{x}$, t) which means that some external fields are included in its definition. E.g., in eq. (2) it is a scalar gravitational field which is included in L. The power of the lagrangian approach lies, between others, in that the inclusion of the new fields is easy due to the additivity property of L. Thus one can add to the original kinetic potential L^1, (which can be the "reversible" kinetic potential of eq. (2) for instance), a new additive part, say L^2, describing a new interaction or a new subsystem included. The new potential $L = L^1 + L^2$ will still obey eqs. (18), (19) and (21) as well as equations for the components of the energy-momentum tensor of Section 2 in this new physical situation.

To visualize the effect consider the inclusion of external vector fields (electromagnetic fields) in L, eq. (3), in the reversible "sourceless" case. The nonrelativistic kinetic potential L contains then the electromagnetic potentials, the vector potential $\mathbf{A}$ and the scalar potential Φ, is of the form

$$L = \sum_a \Big[\rho_a\frac{\mathbf{u}_a^2}{2} + \theta\rho_{sa}\frac{\mathbf{u}_{sa}^2}{2} - (\rho_a + \theta\rho_{sa})\Big(\psi + \frac{\tilde{e}_a}{\tilde{m}_a}\Phi\Big) + \frac{\tilde{e}_a}{\tilde{m}_a}(\rho_a\mathbf{u}_a + \theta_a\rho_{sa}\mathbf{u}_{sa}).\mathbf{A}\Big] - \rho_e(\rho_a,\rho_{sa}) \qquad (45)$$

The coupling constants $\tilde{e}_a$ are the products of the elementary electric charge (electron charge) and the valencies of the species. The ratios of the charge and mass take into account the fact that the electromagnetic potentials are expressed in a conventional way as the energies per unit of charge rather than per unit of mass. We assume that the scalar potential ψ has the same value for each species which corresponds with the external gravitational field.

Using the well known formula of vector calculus

$$\nabla\frac{\mathbf{u}^2}{2} - (\mathbf{u}.\nabla)\mathbf{u} - \mathbf{u}\times(\nabla\times\mathbf{u}) = 0 \qquad (46)$$

it is easy to show that in the case of the kinetic potential, eq. (45), eqs. (18) and (19) yield the equations of motion

$$\frac{d\mathbf{u}_a}{dt} + \nabla(\mu_a + \psi_a) - \frac{\tilde{e}_a}{\tilde{m}_a}(\mathbf{E} + \mathbf{u}_a \times \mathbf{B}) = 0$$

and

$$\theta\frac{d\mathbf{u}_{sa}}{dt} + \nabla(T_a + \theta\psi_a) - \theta\frac{\tilde{e}_a}{\tilde{m}_a}(\mathbf{E} + \mathbf{u}_{sa} \times \mathbf{B}) = 0 \tag{47}$$

containing the electric and magnetic fields **E** and **B** defined in the standard manner as

$$\mathbf{E} = -\left(\frac{\partial \mathbf{A}}{\partial t} + \nabla\Phi\right) \quad \text{and} \quad \mathbf{B} = \nabla\times\mathbf{A} \tag{48}$$

Equations (47) describe a motion of the thermal and material degrees of freedom separately; they are nondissipative (and hence still restrictive in applications) in the sense that no phase dependent term was assumed in the kinetic potential (45) and hence no related sources of entropy and mass are possible. However, they are capable of yielding a useful nontrivial equation of motion of the *total* fluid, under an assumption that any possible internal forces cancel out. This equation constitutes a generalization of the momentum balance, eq. (43) of Part I, involving the effect of the electromagnetic field. It is obtained by multiplying either of eqs. (47) by the related density, combining the results obtained with the conservation equations for mass and entropy, eqs. (2) and (3), adding all the resulting equations, and using the identity

$$\sum_a (\theta\rho_{sa}\mathbf{u}_{sa}\mathbf{u}_{sa} + \rho_a\mathbf{u}_a\mathbf{u}_a) = \rho\mathbf{u}\mathbf{u} + \sum_a (\theta\rho_{sa}\mathbf{v}_{sa}\mathbf{v}_{sa} + \rho_a\mathbf{v}_a\mathbf{v}_a) = \rho\mathbf{u}\mathbf{u} + \Pi \tag{49}$$

The above formula, [obtained from the definition of the hydrodynamic velocity, or eq. (5) of Part I], yields the momentum balance, eq. (43) of Part I, where only the source term containing the total density of the external forces ($\rho\mathbf{F}$ term) has been changed to the form

$$\rho\mathbf{F} = \sum_a \left\{(\rho_a + \rho_{sa}\theta)\left(\frac{\tilde{e}_a}{\tilde{m}_a}\mathbf{E} - \nabla\psi\right) + (\rho_a\mathbf{u}_a + \rho_{sa}\theta\mathbf{u}_{sa}) \times \frac{\tilde{e}_a}{\tilde{m}_a}\mathbf{B}\right\} \tag{50}$$

This takes into account an external electromagnetic field. The density of the Lorentz force acting on the total fluid can be obtained by expressing the absolute velocities in eq. (50) in terms of the hydrodynamic velocity **u** and the velocities of diffusion $\mathbf{v}_a$ and $\mathbf{v}_{sa}$, part I. The Ohm's currents and heat currents (associated with diffusion of the mass and entropy respectively) can then be singled out in eq. (50). The validity of the momentum source density, eq. (50), in a momentum balance formula, can be confirmed by computing the Nöther's expressions [describing the densities of momentum and momentum flux, eqs. (24) and (25)], by the use of the balance equation (35) of Part I, with the external terms selected.

The effect of the electromagnetic field on the balance equation for the energy can also be determined in two ways, from the equations of motion and from the Nöther's theorem. The equations of motion (47) are multiplied scalarly by the products of the corresponding velocities and densities, transformed with the help of the balance equations for mass an entropy, the equilibrium Gibbs equation, and the results added to arrive at the final result. However, the use of the equations of motion is more cumbersome than the straightforward way based on the Nöther's expression for the energy- momentum tensor and the energy balance formula, eqs. (34) and (35) of Part I. The second method is therefore recommended. The resulting energy source corresponding with the kinetic potential (45) can be written in the form

$$\rho\dot{\varepsilon}=\sum_{a} \{(\rho_a \mathbf{u}_a + \rho_{sa}\theta \mathbf{u}_{sa})(\frac{\tilde{e}_a}{\tilde{m}_a}\mathbf{E} - \nabla\psi)\} \tag{51}$$

Its simplicity results from the independence of the gravitational potential of the unit mass of the species involved and the redundancy of the work of the Lorenz forces. The energy balance containing the source term (51) is

$$\frac{\partial}{\partial t}\{\rho e + \sum_{a} [\rho_a \frac{\mathbf{u}_a^2}{2} + \theta\rho_{sa}\frac{\mathbf{u}_{sa}^2}{2}]\} + \nabla . \sum_{a} [\rho_a \mathbf{u}_a(\frac{\mathbf{u}_a^2}{2} + h_a) + \theta\rho_{sa}\mathbf{u}_{sa}\frac{\mathbf{u}_{sa}^2}{2} + T_a\rho_{sa}(\mathbf{u}_{sa} - \mathbf{u}_a)] = \rho\dot{\varepsilon} \tag{52}$$

the last term on the left hand side being the sensible heat term of eqs. (52) and (53) in Part I. Here the role of the all external fields has been singled out from the expression of any component of the energy-momentum tensor to preserve the gauge invariancy of the theory. It is important to note that in this representation, where only the *absolute* velocities appear, the work of the nonequilibrium (viscous) stresses $\Pi.\mathbf{u}$ is not explicit. This work is manifested only when a passage is made to the description involving diffusive fluxes (see part I). Then, it can be shown that the term under the divergence of eq. (52) is equivalent to any of the energy density fluxes given by eqs. (47) to (55) of Part I if the external scalar field ψ [excluded from the four-divergence of the energy flux of eq. (52)] is excluded from the four-divergences of these equations. To show how the stress work $\Pi.\mathbf{u}$ appears one can also transform the flow term of the kinetic energy density $\mathbf{Q}$ [resulting from eqs. (25) or (26)] according to the identity formula

$$\sum_{a} (\theta\rho_{sa}\mathbf{u}_{sa}\frac{\mathbf{u}_{sa}^2}{2} + \rho_a\mathbf{u}_a\frac{\mathbf{u}_a^2}{2}) = \rho\mathbf{u}\frac{\mathbf{u}^2}{2} + \sum_{a} [\theta\rho_{sa}\mathbf{u}_{sa}\frac{\mathbf{v}_{sa}^2}{2} + \rho_a\mathbf{u}_a\frac{\mathbf{v}_a^2}{2} + (\rho_a\mathbf{v}_a\mathbf{u}_a + \theta\rho_{sa}\mathbf{v}_{sa}\mathbf{u}_{sa}).\mathbf{u}] = \rho\mathbf{u}\frac{\mathbf{u}^2}{2} + \sum_{a} [\theta\rho_{sa}\mathbf{u}_{sa}\frac{\mathbf{v}_{sa}^2}{2} + \rho_a\mathbf{u}_a\frac{\mathbf{v}_a^2}{2}] + \Pi.\mathbf{u} \tag{53}$$

This proves equivalency of the energy four-divergences in eq. (52) and eq. (53) of Part I and shows that the viscosity effects selfdiffusion and the viscosity-related work appear as the result of the mutual diffusion of the material and thermal degrees of freedom. This simple interpretation disappears in the classical irreversible thermodynamics of the one-component fluid where the hydrodynamic velocity coincides with the mass velocity. There, due to the lack of the distinction between the momentum of the unit mass and the velocity, the nonequilibrium stress tensor Π and its work become equal to zero, and the formula (53) becomes the trivial identity describing the equality of the two identical kinetic energies. This fallacy of the classical formalism proves a superiority of the thermodynamics involving the idea of the inertial entropy transferred independently of mass.

Using the following definitions of the unit mass charges, z_a and z, and the conduction current density **i**

$$z_a = \frac{\tilde{e}_a}{\tilde{m}_a} \qquad z = \rho^{-1}\sum_a (\rho_a + \theta\rho_{sa})\frac{\tilde{e}_a}{\tilde{m}_a} \qquad \mathbf{i} = \sum_a \mathbf{i}_a = \sum_a z_a(\mathbf{j}_a + \theta\mathbf{j}_{sa}) \tag{54}$$

which correspond with those known in the classical nonequilibrium theory (de Groot and Mazur 1984), the equation of the total momentum balance, [eq. (43) of Part I with the source term, eq. (50)], can be written in the form of an equation of motion

$$\rho\frac{d\mathbf{u}}{dt} = -\nabla.(1P + \Pi) - \rho\nabla\psi + \rho z\,(\mathbf{E} + \mathbf{u}\times\mathbf{B}) + \mathbf{i}\times\mathbf{B} \tag{55}$$

The corresponding energy equation, in terms of the total specific energy e^0 (containing the specific internal, kinetic, potential and diffusional energies; see the middle sigma term in eq. (57) of Part I) has the form

$$\rho\frac{de^0}{dt} = -\nabla.(P\mathbf{u} + \Pi.\mathbf{u} + \mathbf{q}) - \rho\mathbf{u}.\nabla\psi + (\rho z\mathbf{u} + \mathbf{i}).\mathbf{E} \tag{56}$$

where the heat flux **q** contains the contribution of the kinetic energy of diffusion. In eqs. (55) and (56) there appear all the well-known terms. The hydrodynamic derivative operator is used. The energy and momentum of the electromagnetic field are not included; the equations describe, therefore, an open system. In particular, the terms called commonly irreversible appear, related with the viscosity (Π term), total heat (including the role of the mass diffusion; **q** term), and the electric charge conduction (**i** term). It is not unconceivable that the appearance of the total entropy source in the classical formalism can be related to an effect of the transition from the more detailed description of motion, the set of vector eqs. (47), to the single vector relationship, eq. (55), and the related loss of the information. This, however, will be pursued elsewhere.

An important fact should be noted. In the case considered, the form of the energy and momentum expressions of the system itself as well as the components of the (gauge invariant) energy-momentum tensor (without fields) remain unchanged with the change of the interaction lagrangian of eq. (45). Only the sources of the energy and momentum, i.e. the densities of the external forces and the inputs of the external energy, change when changing the interaction Lagrangian. A consequence of this notion may be a viewpoint of the conventional theory of the irreversible processes where it is *assumed* that the expressions for the total energy density and momentum density remain unchanged when the source terms appear in the mass and entropy balances. Such an important assumption can be tested to determine the class of the possible structures of the internal interaction terms, describing dissipation, such that their effect would not change the components of the energy momentum tensor and the standard conservation laws (obtained from the sourceless analysis of Part I). An analysis of this assumption is done in the next section.

6. CONVENTIONAL POSTULATE CONSTRAINING IRREVERSIBLE PROCESSES

We return for simplicity of the formulas to the electrically neutral systems, as in the Sections 1-5, where the dissipative effects and the corresponding interaction lagrangians will be essential. However, we preserve the gravitational potential ψ to allow the testing of an effect of the external fields on the internal interaction. Such interaction can be described by the phase part of the kinetic potential, L^{int}, which may take into account the exchange phenomena. In the case of the sourceless motions [free of the sources of mass and entropy; the kinetic potential (2)], the equations of motion of the sourceless theory of Part I have the form

$$\frac{d\mathbf{u}_a}{dt} + \nabla(\mu_a + \psi_a) = 0 \qquad (\sigma_a = \sigma_{sa} \equiv 0) \tag{57}$$

$$\theta_a \frac{d\mathbf{u}_{sa}}{dt} + \nabla(T_a + \theta\psi_a) = 0 \qquad (\sigma_a = \sigma_{sa} \equiv 0) \tag{58}$$

These can also be obtained from eqs. (18), (19), (21) or (47) by neglecting the sources σ as well as **E** and **B** terms.

We have already proved that the phase free conservation laws such as eqs. (43) and (58a) of Part I and eqs. (52), (55), (56) of this work, correspond with the standard (phase free) form of the conservation laws and the energy-momentum tensor. The inclusion of any internal exchange may or may not affect the global form of the conservation laws of the energy and momentum. In this regard, the implications of the nonequilibrium statistical mechanics are not obvious, since they are always associated with the change in the accuracy of the description when one passes from, say, fife to thirteen moment approximation. An internal exchange should change effectively the form of the stationarity conditions of the action, and the resulting equations of motion, which will become the dissipative equations.

On the basis of the additivity property of the Lagrangian, one can easily conclude that in the case of any generalized kinetic potential L defined as the sum of the original sourceless L^1 (eq. 2) and with any *additive* interaction part L^{int} (such as in eq. (3)) describing an irreversible exchange of mass, momentum and energy) dependent explicitly on phases, the "sourceless" equations of motion will gain some additional additive parts related to L^{int}. For $L = L^1 + L^{int}$ the equations of motion will have the following structure [resulting from eqs. (18), (19) and (21)]

$$\frac{d\mathbf{u}_a}{dt} + \nabla(\mu_a + \psi_a) = \mathbf{X}_a \tag{59}$$

$$\theta_a \frac{d\mathbf{u}_{sa}}{dt} + \nabla(T_a + \theta\psi_a) = \mathbf{X}_{sa} \tag{60}$$

where $\mathbf{X}_a$ and $\mathbf{X}_{sa}$ are the forces related to L^{int} and the sources $\sigma = -(\partial L/\partial\phi, \partial L/\partial\eta)$ which were not taken into account in the kinetic potential $L = L^1$, eq. (2). These forces result from the comparison of eq. (21) written for both L^1 and $L = L^1 + L^{int}$ as well as eqs. (57) through (60)

$$\mathbf{X}_a = -\sigma_a \frac{\partial L^1}{\rho_a^2 \partial \mathbf{u}_a} - \Big\{ \frac{d^a}{dt}\Big(\frac{\partial L^{int}}{\rho_a \partial \mathbf{u}_a}\Big) + \Big(\frac{\partial L^{int}}{\rho_a \partial \mathbf{u}_a}.\nabla\Big)\mathbf{u}_a - \nabla \frac{\partial L^{int}}{\partial \rho_a} + \frac{\partial L^{int}}{\rho_a \partial \mathbf{u}_a} \times (\nabla \times \mathbf{u}_a) + \frac{\sigma_a}{\rho_a}\Big(\frac{\partial L^{int}}{\rho_a \partial \mathbf{u}_a} - \nabla\phi_a\Big) \Big\} \tag{61}$$

$$\mathbf{X}_{sa} = -\sigma_{sa} \frac{\partial L^1}{\rho_{sa}^2 \partial \mathbf{u}_{sa}} - \Big\{ \frac{d^{sa}}{dt}\Big(\frac{\partial L^{int}}{\rho_{sa} \partial \mathbf{u}_{sa}}\Big) + \Big(\frac{\partial L^{int}}{\rho_{sa} \partial \mathbf{u}_{sa}}.\nabla\Big)\mathbf{u}_{sa} - \nabla \frac{\partial L^{int}}{\partial \rho_{sa}} + \frac{\partial L^{int}}{\rho_{sa} \partial \mathbf{u}_{sa}} \times (\nabla \times \mathbf{u}_{sa}) + \frac{\sigma_{sa}}{\rho_{sa}}\Big(\frac{\partial L^{int}}{\rho_{sa} \partial \mathbf{u}_{sa}} - \nabla\eta_a\Big) \Big\} \tag{62}$$

Eqs. (18) and (19) yield an alternative form of these forces

$$\mathbf{X}_a = -\sigma_a \frac{\partial L^1}{\rho_a^2 \partial \mathbf{u}_a} - \Big\{ \frac{\partial}{\partial t}\Big(\frac{\partial L^{int}}{\rho_a \partial \mathbf{u}_a}\Big) + \nabla.\Big(\frac{\partial L^{int}}{\rho_a \partial \mathbf{u}_a}.\mathbf{u}_a - \frac{\partial L^{int}}{\partial \rho_a}\Big) - \mathbf{u}_a \times \Big(\nabla \times \frac{\partial L^{int}}{\rho_a \partial \mathbf{u}_a}\Big) + \frac{\sigma_a}{\rho_a}\Big(\frac{\partial L^{int}}{\rho_a \partial \mathbf{u}_a} - \nabla\phi_a\Big) \Big\} \tag{63}$$

$$\mathbf{X}_{sa} = -\sigma_{sa} \frac{\partial L^1}{\rho_{sa}^2 \partial \mathbf{u}_{sa}} - \Big\{ \frac{\partial}{\partial t}\Big(\frac{\partial L^{int}}{\rho_{sa} \partial \mathbf{u}_{sa}}\Big) + \nabla.\Big(\frac{\partial L^{int}}{\rho_{sa} \partial \mathbf{u}_{sa}}.\mathbf{u}_{sa} - \frac{\partial L^{int}}{\partial \rho_{sa}}\Big) - \mathbf{u}_{sa} \times \Big(\nabla \times \frac{\partial L^{int}}{\rho_{sa} \partial \mathbf{u}_{sa}}\Big) + \frac{\sigma_{sa}}{\rho_{sa}}\Big(\frac{\partial L^{int}}{\rho_{sa} \partial \mathbf{u}_{sa}} - \nabla\eta_a\Big) \Big\} \tag{64}$$

Multiplying each of the eqs. (59) and (60) by the corresponding densities, adding the results obtained, using the mass and entropy balance equations with sources, eqs. (4) and (5), applying the Gibbs-Duhem equation for individual species, one arrives at a structure generalizing the momentum balance eq. (43) of Part I [or eq. (55) with the electromagnetic terms neglected]. This generalized form of the total momentum balance is eq. (66) below. It contains an additional term $\rho\mathbf{F}^{int}$, pertaining to the internal forces, $\mathbf{X}_a$ and $\mathbf{X}_{sa}$, as the source of the momentum related to the sources of mass and entropy

$$\rho\mathbf{F}^{int} = \sum_a [\rho_a \mathbf{X}_a + \rho_{sa}\mathbf{X}_{sa} - \mathbf{u}_a \frac{\partial L^{int}}{\partial \phi_a} - \theta_a \mathbf{u}_{sa} \frac{\partial L^{int}}{\partial \eta_a}] \tag{65}$$

But the negative derivatives of L^{int} with respect to the phases are the sources of mass and entropy. Hence, in terms of the source functions σ_a and σ_{sa}, the total momentum balance is

$$\frac{\partial \rho \mathbf{u}}{\partial t} + \nabla.(\rho \mathbf{uu}+1P+\Pi)+\rho\nabla\psi = \sum_a [\rho_a \mathbf{X}_a + \rho_{sa}\mathbf{X}_{sa} + \mathbf{u}_a\sigma_a + \theta_a \mathbf{u}_{sa}\sigma_{sa}] \quad (66)$$

At this point, one may *postulate* the right hand sides of eqs. (65) or eq. (66) may vanish, arguing that an internal exchange and the resulting irreversibility should not affect the "classical" form of the balance of total momentum derived from the sourceless kinetic potential, eq. (2),

$$\sum_a [\rho_a \mathbf{X}_a + \rho_{sa}\mathbf{X}_{sa} + \mathbf{u}_a\sigma_a + \theta_a \mathbf{u}_{sa}\sigma_{sa}] = 0 \quad (67)$$

For $\theta_a = 0$, this is usually related to a separate vanishing of the momentum of each chemical reaction and the sum of $\rho_a \mathbf{X}'_a$ where $\mathbf{X}'_a = \mathbf{X}_a + s_a \mathbf{X}_{sa}$ (Baranowski, 1974). A classical counterpart of eq. (67) for θ_a=0, is also known in the literature for the case of absence of the mass sources in a diffusion system (Fitts 1962). However, it pertains to the left hand sides of eqs. (59) and (60) rather than to the right hand sides of these equations, i.e. not to the expressions (61) through (64). Such form do not constitute any new information; it represents only a transformed version of the *classical* momentum balance under an implicit assumption that such balance holds as an exact relationship.

For the theory, eq. (67) is of interest since it determines certain extra conditions imposed on the exchange of matter, heat, energy and momentum in the system with irreversible phenomena in order to preserve the classical form of the momentum balance. However, it should still be completed by another postulated constraint, requiring the balance of the energy in the same form as that resulting from the interaction-free kinetic potential $L = L^1$, eq. (2). Multiplying eqs. (59) and (60) scalarly by the corresponding total fluxes, adding the results obtained, using balance equations for mass and entropy with sources, and then using the Gibbs equation for the internal energy density ρ_e, one arrives at the generalized structure of the energy balance [compare eq. (58a) of Part 1 and eq. (68)]

$$\begin{aligned} &\frac{\partial \rho e^0}{\partial t} + \nabla.(\rho \mathbf{u} e^0 + P\mathbf{u} + \Pi.\mathbf{u} + \mathbf{q}) + \rho \mathbf{u}.\nabla\psi = \\ &\sum_a [\rho_a \mathbf{u}_a.\mathbf{X}_a + \rho_{sa}\mathbf{u}_{sa}.\mathbf{X}_{sa} + (\mu_a + \frac{\mathbf{u}_a^2}{2})\sigma_a + (T_a + \frac{\theta \mathbf{u}_{sa}^2}{2})\sigma_{sa}] \end{aligned} \quad (68)$$

where e^0 involves the specific kinetic and internal energy as well as the kinetic energy of diffusion.

Again, requiring the same analytical form of the energy balance as in the classical case, one can *postulate* a condition

$$\sum_a [\rho_a \mathbf{u}_a.\mathbf{X}_a + \rho_{sa}\mathbf{u}_{sa}.\mathbf{X}_{sa} + (\mu_a + \frac{\mathbf{u}_a^2}{2})\sigma_a + (T_a + \frac{\theta \mathbf{u}_{sa}^2}{2})\sigma_{sa}] = 0 \quad (69)$$

Thus, according to the *conventional postulate*, whatever the form of the sources, eqs. (67) and (69) should be met. It is known from the nonequilibrium statistical mechanics that such condition can actually be assumed, at least not too far from equilibrium where its outcomes are unambiguous. However, eq. (69) with the forces **X** defined as the left hand sides of eqs. (59) and (60) rather than (61) and (62) represents only an identity, resulting from the *classical* energy conservation law transformed to the form allowing to determine the entropy source. This is seen particularly well in the single temperature case where the classical expression for the entropy source is recovered instantly from eq. (69).

When the conventional postulate is not accepted, eqs. (66) and (68) with the forces defined by eqs. (61) through (64) constitute important relationships describing the conservation of the total momentum and energy for an arbitrary kinetic potential L, generalizing eq. (2). It contains the phases since, inherently, there are sources of mass and entropy in the system.

Are there any essential physical reasons to expect that the right hand sides of eq. (66) and (68) will always vanish, as the postulate, eqs (67) and (69), says? One can speculate that the postulate can perhaps be met in the situation when the expression for L^{int} has a special structure, corresponding with the vanishing of the four-divergence of the energy-momentum tensor $G_{ik}{}^{int}$ (related to L^{int}). This would result in the vanishing corrections to the momentum and energy densities of the sourceless case. We do not feel there is a sufficient basis to classify the postulate as an exact law. The postulate may be only a reasonable assumption which frequently yields acceptable approximate results. What is physically necessary is the consistency of the (nontruncated) eq. (66) with the requirement of the invariance of the *total* action. Although the phase part of L, or L^{int} is the only essential part of the action A when the phase shift invariance is considered, c. f. eq. (29), it is the *total* L which has to be taken into account when considering the time and space translations. We return to this question in Section 9.

It may be added that, with the *internal* electric and magnetic fields in the system as the only interactions, the postulate, when applied formally to the interaction part of eq. (45), leads to the conclusion that all electric and magnetic fields should mutually compensate. I.e., the resulting fields should vanish (the associated electromagnetic potentials $\mathbf{A}^{int}$ and Φ^{int} may not vanish but they should preserve vanishing $\mathbf{E}^{int}$ and $\mathbf{B}^{int}$). However, this is a static condition. In a dynamic situation, with macroscopic motions taking place, the resulting fields $\mathbf{E}^{int}$ and $\mathbf{B}^{int}$ may, in principle, persist. Thus, the postulate seems to be a reminiscent of a (quasi)static relationship. One can rest the theory on a given L^{int} and not insist that the resulting equations should meet eqs. (67) and (69).

7. NONLINEAR NONLOCAL CHEMICAL KINETICS

Chemical kinetics of a classical, single temperature and entropy system, corresponding with the Guldberg and Waage kinetic law in the limiting steady-state situation, will be considered. An example will show the role played by the internal kinetic potential L^{int} in the general action functional, eq. (3). When θ_a=0, the following Λ can be considered

$$\Lambda = L^1(\rho, \rho_s, \mathbf{u}_a, \mathbf{u}_s) - \sum_a \rho_a \dot{\phi}_a - \rho_s \dot{\eta}$$
$$- \sum_a [\, \phi_a \sum_b (\overset{-}{\nu}_{ab} - \overset{+}{\nu}_{ab}) \widetilde{M}_a r_b] - \eta [\sum_b \frac{1}{2} R_b r_b^2 + \Psi^*(\rho, \rho_s)] \tag{70}$$

(dots pertain to the substantial derivative approximation). For a finite θ_a, the masses and densities become the bare quantities, but the form of the above equation is unchanged in the single T approximation. The derivatives of the reaction rates dr_b/dt may appear in Ψ^* for the more involved relaxation models. The transport terms are neglected in the entropy source. R_b is the nonlinear resistance function of b-th reaction, eq. (72), involving complex dependence of R_b on the concentrations $C_a = \rho_a / \widetilde{M}_a$ or the corresponding activities (Sieniutycz 1987). Velocity-dependent extensions of the phase term of eq. (70) serve to treat the coupled diffusion and chemical reactions in a unified way. However, for simplicity, we neglect any transfer phenomena restricting η term to the chemical dissipation as the only source of the irreversibility.

The Onsagerian state-dependent potential Ψ^* (Lavenda 1985, Keizer 1977) can be constructed as a quadratic function of the deviations of the densities ρ_a (the components of the vector ρ) from their equilibrium values ρ_a^{eq}, where the latter were expressed as the functions of the entropy density, ρ_s. In terms of the Onsagerian matrix, $\mathcal{L}$, the mass stoichiometric matrix, ν', the entropy hessian, s', and the reaction advancement vector ξ, one has

$$\begin{aligned}\Psi^*(\rho, \rho_s) &= \frac{1}{2}(\nu' \mathbf{s}' \mathcal{L} \mathbf{s}' \nu'):(\xi - \xi^{eq})(\xi - \xi^{eq}) \\ &= (\mathbf{s}' \mathcal{L} \mathbf{s}'):[\rho - \rho^{eq}(\rho_s)][\rho - \rho^{eq}(\rho_s)]\end{aligned} \tag{71}$$

The matrices of eq. (71) can be taken as the state-dependent one, allowing the simplest treatment of the nonlinearities.

The nonlinear resistance in eq. (70), pertaining to the entropy representation rather than that of the free energy, and to the Planck potentials driving force, is given by the formula

$$R_b = kN_{Av} \frac{\ln[\overset{+}{k}_b \prod_a (\rho_a/\widetilde{M}_a)^{\overset{+}{\nu}_{ab}}] - \ln[\overset{-}{k}_b \prod_a (\rho_a/\widetilde{M}_a)^{\overset{-}{\nu}_{ab}}]}{\overset{+}{k}_b \prod_a (\rho_a/\widetilde{M}_a)^{\overset{+}{\nu}_{ab}} - \overset{-}{k}_b \prod_a (\rho_a/\widetilde{M}_a)^{\overset{-}{\nu}_{ab}}} \tag{72}$$

where the Avogadro number N_{Av} and the forward and backward rate constants of b-th reaction, k_b^+ and k_b^-, appear. The latter are the Arrhenius functions of temperature T which should be expressed in terms of ρ and ρ_s.

In the case of variations of the densities of entropy and masses, the stationarity conditions of the action based on eq. (70) yield the relations between the total time derivatives of phases and the nonequilibrium, rate-dependent, temperature and chemical potentials

$$
\begin{aligned}
-\dot{\eta} = \widetilde{T} &\equiv -\frac{\partial L}{\partial \rho_s} = T + \eta\{\sum_b \frac{r_b^2 \partial R_b}{2\,\partial \rho_s} + \frac{\partial \psi^*}{\partial \rho_s}\} \\
-\dot{\phi}_a = \widetilde{\mu} &\equiv -\frac{\partial L}{\partial \rho_a} = \mu_a + \eta\{\sum_b \frac{r_b^2 \partial R_b}{2\,\partial \rho_a} + \frac{\partial \psi^*}{\partial \rho_a}\}
\end{aligned} \tag{73}
$$

The variation of the transfer velocities leads to the gradient representation of the momenta. These are the classical momenta since there are no velocities in the present approximation of the entropy source, eq. (70). With the already known method, these equations can be transformed into a form of equations of motion [a sort of the equations (59) through (64) affected by the chemical terms contained in the forces $\mathbf{X}_a$ and $\mathbf{X}_s$]. This is omitted here, the chemical kinetics being of a primary importance. The stationarity condition of action A corresponding with the phase variations yield the balances of the entropy and species

$$
\begin{aligned}
\frac{\partial \rho_s}{\partial t} + \nabla.\rho_s \mathbf{u}_s &= \sum_b \frac{1}{2} R_b r_b^2 + \Psi^*(\rho, \rho_s) \\
\frac{\partial \rho_a}{\partial t} + \nabla.\rho_a \mathbf{u}_a &= \sum_b \widetilde{M}_a \nu_{ab} r_b
\end{aligned} \tag{74}
$$

The stationarity conditions of action A with respect to variations of reaction rates r_b yield equations of chemical kinetics in terms of the phases ϕ_a and η

$$
r_b = -R_b^{-1} \sum_a \frac{\phi_a \widetilde{M}_a}{\eta} (\nu_{ab}^- - \nu_{ab}^+) \tag{75}
$$

The sum with respect to all the components contains the generalized chemical driving force of b-th reaction in entropy rather than energy representation, the *actinity*, $\Delta\Phi_b$, eq. (30). Eq. (75) has a form of a Ohm law. It says that the chemical reaction proceeds in direction of the decreasing generalized force $\Delta\Phi_b$. However, it involves the ratios of the material phases to the thermal phase rather than the usual ratios of the chemical potentials to the temperature. Since these phases are related to the time integrals of the chemical potentials and temperature, the kinetics obtained is nonlocal in time. The question arises: how does it correspond with the Guldberg and Waage kinetics?

Taking the total hydrodynamic derivative of the product ηr_b in eq. (75), neglecting all diffusion terms, and using eqs. (73) an equation of the nonstationary chemical kinetics is obtained

$$
\eta \frac{d(R_b r_b)}{dt} - \widetilde{T} R_b r_b = \sum_a \widetilde{\mu}_a \widetilde{M}_a (\nu_{ab}^- - \nu_{ab}^+) \equiv \widetilde{A}_b \tag{76}
$$

It contains the η term as a time lag term and the nonequilibrium chemical potentials and temperature, eqs. (73). The product of the temperature and the entropy representation resistance is the free energy representation resistance, (Sieniutycz, 1987), which can alternatively be used in eqs. (70)

through (76). A generalized, rate dependent, affinity A_b appears in eq. (76), (Sieniutycz 1987), which, for small rates, simplifies to the classical chemical affinity, A_b. Since the chemical equilibrium constant K equals to k^+/k^-, the classical affinity can be expressed in the form containing the rate constants

$$A_b = - kN_{Av}T\ln[k_b^+\prod_a (\rho_a/\tilde{M}_a)^{\overset{+}{\nu}_{ab}}] - \ln[k_b^-\prod_a (\rho_a/\tilde{M}_a)^{\overset{-}{\nu}_{ab}}] \tag{77}$$

When the time derivative term in eq. (76) can be neglected and the disequilibrium intensities are replaced by their equilibrium values, the resistance formula, eq. (72), and eqs. (76) and (77) yield the Guldberg and Waage kinetic law

$$r_b = k_b^+\prod_a (C_a)^{\overset{+}{\nu}_{ab}} - k_b^-\prod_a (C_a)^{\overset{-}{\nu}_{ab}} \tag{78}$$

Consequently, the Gulberg and Waage kinetic law is a limiting law neglecting the chemical transients and corresponding to the steady-state situation. It is not difficult to extend this treatment to a nonideal case when the concentrations are replaced by activities (Sieniutycz 1987). It should be recognized that the approach is possible due to the use of the two dissipation functions, Φ^* and Ψ^*, and not only the doubled value of Φ^*, in the action functional. However, the Onsagerian equality $\Phi^* = \Psi^*$ is a consequence of a tacit assumption that the form of the energy conservation law is unaffected by the sources, and, if that assumption does not apply, the approach will not necessarily correspond to the classical expressions of the energy and momentum balances.

8. KINETICS OF ENTROPY AND MASS TRANSFER

For a one-component system the part of internal kinetic potential corresponding with the relative diffusion of the entropy with respect to mass is given by eq. (40) which can be written in the form

$$L^{int} = - (\eta - \theta\phi_m)[\tfrac{1}{2}a(\rho_m,\rho_s)(u_s - u_m)^2 + \Psi^*(\rho_m,\rho_s)] \tag{79}$$

Since $\Phi^* = \Psi^*$, the forces (61) and (62) are

$$\begin{aligned} \mathbf{X}_m = - \Big\{ & \frac{d}{dt}[(\eta - \theta\phi_m)a\rho_m^{-1}(\mathbf{u}_s - \mathbf{u}_m)] + [(\eta - \theta\phi_m)a\rho_m^{-1}(\mathbf{u}_s - \mathbf{u}_m).\nabla]\mathbf{u}_m \\ & + \nabla[\tfrac{1}{2}(\eta - \theta\phi_m)\partial a/\partial\rho_m(\mathbf{u}_s - \mathbf{u}_m)^2 + \partial\Psi^*/\partial\rho_m] \\ & - \mathbf{u}_m\times(\nabla\times[(\eta - \theta\phi_m)a\rho_m^{-1}(\mathbf{u}_s - \mathbf{u}_m)])\Big\} \\ & + \tfrac{1}{2}a\rho_m^{-1}\theta(\mathbf{u}_s - \mathbf{u}_m)^2[\mathbf{u}_m + \rho_m^{-1}(\eta - \theta\phi_m)a(\mathbf{u}_s - \mathbf{u}_m) + \nabla\phi_m] \end{aligned} \tag{80}$$

$$\mathbf{X}_s = \Big\{\frac{d}{dt}[(\eta - \theta\phi_m)a\rho_s^{-1}(\mathbf{u}_s - \mathbf{u}_m)] + [(\eta - \theta\phi_m)a\rho_s^{-1}(\mathbf{u}_s - \mathbf{u}_m).\nabla]\mathbf{u}_s$$
$$+ \nabla[\frac{1}{2}(\eta - \theta\phi_m)\partial a/\partial\rho_s(\mathbf{u}_s - \mathbf{u}_m)^2 + \partial\Psi^*/\partial\rho_s] \tag{81}$$
$$- \mathbf{u}_s \times (\nabla\times[(\eta - \theta\phi_m)a\rho_s^{-1}(\mathbf{u}_s - \mathbf{u}_m)])\Big\}$$
$$- \frac{1}{2}a\rho_s^{-1}(\mathbf{u}_s - \mathbf{u}_m)^2[\mathbf{u}_m + \rho_s^{-1}(\eta - \theta\phi_m)a(\mathbf{u}_s - \mathbf{u}_m) + \nabla\eta]$$

Thus, even in this simple case the forces have a complex form. However, the total time derivative of η can be approximated by -T and that of ϕ_m by $-\mu_m$. Hence, the approximate forces corresponding with low gradients (when only the time derivative terms in the above formulae are essential) are

$$\mathbf{X}_m \cong \Big\{ a\rho_m^{-1}(T - \theta\mu_m)(\mathbf{u}_s - \mathbf{u}_m) - (\eta - \theta\phi_m)\frac{d}{dt}a\rho_m^{-1}(\mathbf{u}_s - \mathbf{u}_m)\Big\} \tag{80a}$$

$$\mathbf{X}_s \cong \Big\{- a\rho_s^{-1}(T - \theta\mu_m)(\mathbf{u}_s - \mathbf{u}_m) + (\eta - \theta\phi_m)\frac{d}{dt}a\rho_s^{-1}(\mathbf{u}_s - \mathbf{u}_m)\Big\} \tag{81a}$$

Therefore, the classical frictional forces appear, proportional to the differences of velocities, as well as some integral type forces describing time lags. In the sourceless case of the constant densities, these classical forces obey the sourceless equality (67), as postulated by the classical theory. However, only a very approximate correspondence of the present theory with the classical one can be shown. This can be explained in a few ways.

First, the "Joule-law" form of the internal kinetic potential, L^{int}, used above, can be only a rough, insufficient, approximation. One can argue that perhaps another, more suitable, form of L^{int} exists, leading to the postulate, eqs. (67) and (69), exactly and making frictional forces linear in the limit of the low slip velocities. The reason why this form must be quite special is that (if it exists at all) it would have to yield vanishing contribution to the classical energy-momentum tensor (Part I), or, at least, vanishing contribution to the four-divergences of this tensor. Such a possibility cannot be excluded from the mathematical point of view. However, the physical reasons for vanishing eqs. (67) and (69), given in the literature, are obscure. Tacitly the same conservation laws for reversible and irreversible processes are used; the reasons are never clearly specified. As pointed out in Section 6, the only well established requirement is that the time and space translations should make invariant the *total* action A; hence, eqs. (66) and (68) should hold in their entirety.

Hence, another explanation evolves, assuming that there is not such L^{int} which provides an exact compensation of the sources of the entropy and mass balance equations and the nonclassical terms on the right hand sides of the momentum and energy balances, eqs. (66) and (68). This statement might serve to claim that the variational description of the processes with sources cannot be organized. However, in our opinion, this could rather be an evidence that the expressions (67) and (69) cannot separately vanish in an irreversible process. In other words we suspect that the difficulties in matching identical balance laws for the reversible and irreversible processes within a Lagrangian formalism constitute an evidence for the invalidity of the conventional postulate requiring the same analytical form of conservation laws for the energy and momentum in both reversible and irreversible cases. While not excluding the possibility of negligible differences close to equilibrium, one can postulate a physical L^{int} contributing to the generalized conservation laws for the momentum and energy, eqs. (66) and (68), but not obeying the separate conditions (67) and (69). Consequently, one may allow the irreversibility to change analytical form of the conservation laws.

9. CONSERVATION LAWS MODIFIED BY DISSIPATION

We will again neglect chemical reactions, i.e., only heat and mass transport will be considered, described by eq. (79). Application of the general formulae for the stationary Lagrangian Λ, eq. (10), and the components of the energy-momentum tensor, eqs. (23) through (27), to the kinetic potential $L = L^1 + L^{int}$, eqs. (2), (3) and (79), yields the total stationary Lagrangian $\Lambda = P$ and the tensor $G^{\alpha\beta}$ as a sum of the classical components [eqs. (31), (38), (40), (47), and (57) of Part I, pertaining to L^1], and the "irreversible" or phase dependent corrections ΔG^{jk}, pertaining to L^{int}. These corrections are specified below, eqs. (82) through (89). Their first lines pertain to the general formulae, whereas the second ones to the heat flow example considered. The superscript T pertains to the thermodynamic Legendre transforms, as in eq. (82).

Nonequilibrium pressure, related to the bulk viscosity:

$$\Delta P = \Delta\Lambda = L^{int^T} = L^{int} - \sum_a (\rho_{sa}\frac{\partial L^{int}}{\partial \rho_{sa}} + \rho_a\frac{\partial L^{int}}{\partial \rho_a}) = \tag{82}$$
$$- (\eta - \theta\phi_m)[\frac{1}{2}(a - \rho_s\partial a/\partial\rho_s - \rho_m\partial a/\partial\rho_m)(u_s - u_m)^2 + \Psi^* - \rho_s\frac{\partial\Psi^*}{\partial\rho_s} - \rho_m\frac{\partial\Psi^*}{\partial\rho_m}]$$

Stress tensor correction:

$$\Delta G^{jk} = - \sum_a [\frac{\partial L^{int}}{\partial u^j_{sa}}u^k_{sa} + \frac{\partial L^{int}}{\partial u^j_a}u^k_a] - \delta^{jk}\Delta P = \tag{83}$$
$$- (\eta - \theta\phi_m)[a(u_s - u_m)^j(u_s - u_m)^k - \delta^{jk}[\frac{1}{2}a^T(u_s - u_m)^2 + \Psi^{*T}]$$

Momentum density correction:

$$\Delta G^{j4} = - \Gamma^j = - \sum_a [\frac{\partial L^{int}}{\partial u^j_{sa}} + \frac{\partial L^{int}}{\partial u^j_a}] = \tag{84}$$
$$- (\eta - \theta\phi_m)[a(u_s - u_m)^j - a(u_s - u_m)^j] = 0$$

Energy density flux correction:

$$\Delta G^{4k} = \Delta Q^{4k} = - \sum_a [(\frac{\partial L^{int}}{\rho_{sa}\partial \mathbf{u}_{sa}}.\mathbf{u}_{sa} - \frac{\partial L^{int}}{\partial \rho_{sa}})\rho_{sa}u_{sa}^k + (\frac{\partial L^{int}}{\rho_a \partial \mathbf{u}_a}.\mathbf{u}_a - \frac{\partial L^{int}}{\partial \rho_a})\rho_a u_a^k] =$$

$$- (\eta - \theta\phi_m)\{[\rho_s^{-1} a(\mathbf{u}_s - \mathbf{u}_m).\mathbf{u}_s - \frac{1}{2}\frac{\partial a}{\partial \rho_s}(u_s - u_a)^2 - \frac{\partial \Psi^*}{\partial \rho_s}]\rho_s u_s^k \tag{85}$$

$$- [\rho_m^{-1} a(\mathbf{u}_s - \mathbf{u}_m).\mathbf{u}_m + \frac{1}{2}\frac{\partial a}{\partial \rho_m}(u_s - u_a)^2 - \frac{\partial \Psi^*}{\partial \rho_m}]\rho_m u_m^k \}$$

or in the abbreviated form corresponding with eq. (26)

$$\Delta G^{4k} = \Delta Q^{4k} = \sum_a [\rho_s u_s^k \Delta T^+ + \rho_m u_m^k \Delta \mu_m^+] \tag{85a}$$

The correction to the total energy density is

$$\Delta E = \sum_a (\frac{\partial L^{int}}{\partial \mathbf{u}_{sa}}.\mathbf{u}_{sa} + \frac{\partial L^{int}}{\partial \mathbf{u}_a}.\mathbf{u}_a) - L^{int} =$$

$$- (\eta - \theta\phi_m)[\frac{1}{2}a(u_s - u_m)^2 - \Psi^*(\rho_s, \rho_m)] = 0 \tag{86}$$

One can note that due to the Onsagerian equality, $\Phi^* = \Psi^*$, and the Joule-law structure of L^{int}, the energy and momentum densities are described by the same formulae as in the dissipationless case. However, other quantities (the fluxes **Q**, P, Π, etc.) change. Therefore, the changes appear also in the balance equations for the momentum and energy. For the definition of the hydrodynamic velocity and momentum unchanged, taking into account the classical components and the phase dependent corrections yields the momentum balance

$$\frac{\partial \rho \mathbf{u}}{\partial t} + \nabla.[\rho \mathbf{u}\mathbf{u} + 1(P+\Delta P) + \Pi + \Delta \Pi] - \rho \mathbf{F} = 0$$

$$\text{with} \tag{87}$$

$$\Delta \Gamma \equiv 0; \quad \Delta \Pi = (\eta - \theta\phi_m)a(\mathbf{u}_s - \mathbf{u}_m)(\mathbf{u}_s - \mathbf{u}_m)$$

The phase contribution to the viscous tensor, $\Delta\Pi$, results from comparison of eqs. (82) and (83). The related energy balance is

$$\frac{\partial(\rho e^0 + \Delta E)}{\partial t} + \nabla.[\rho \mathbf{u}(e^0 + \Delta E) + (P+\Delta P)\mathbf{u} + (\Pi + \Delta\Pi).\mathbf{u} + \mathbf{q} + \Delta \mathbf{q}] - \rho \mathbf{u}.\mathbf{F} = 0$$

$$\text{with} \tag{88}$$

$$\Delta \mathbf{q} \equiv \Delta \mathbf{Q} - \rho \mathbf{u} \Delta E - \mathbf{u}\Delta P - \mathbf{u}.\Delta\Pi$$

The balance laws, eq. (87) and (88), can also be obtained from eqs. (66) and (68) with the forces **X** and sources σ resulting from eqs. (4), (5) and (79) through (81).

The structure of the energy balance, eq. (88), is a general one, corresponding to an arbitrary L^{int} if the definition of the hydrodynamic velocity is unchanged. However, the structure of the momentum balance, eq. (87), and the definition of $\Delta\Pi$ correspond to the case of L^{int}, eq. (79), where $\Delta\Gamma=0$. A general form of the momentum balance equation is

$$\frac{\partial(\rho\mathbf{u} + \Delta\Gamma)}{\partial t} + \nabla.[\,\rho\mathbf{uu} + \mathbf{u}\Delta\Gamma + 1(P+\Delta P) + \Pi + \Delta\Pi\,] - \rho\mathbf{F} = 0$$
$$\text{with} \qquad (89)$$
$$\Delta\Pi = \Delta G - \mathbf{u}\Delta\Gamma - 1\Delta P$$

Here $\Delta\mathbf{G}$ is a general correction to the stress tensor defined in terms of L^{int} by the upper line of eq. (83).

The reader can verify that for the more general L^{int} ("classical case"; $\theta = 0$) corresponding to the Maxwell-Stefan diffusion

$$L^{int} = -\eta\,[\tfrac{1}{2}a_{01}(\mathbf{u}_0 - \mathbf{u}_1)^2 + \tfrac{1}{2}a_{02}(\mathbf{u}_0 - \mathbf{u}_2)^2 + \ldots$$
$$\ldots + \tfrac{1}{2}a_{n-1,n}(\mathbf{u}_{n-1} - \mathbf{u}_n)^2 + \Psi^*(\rho_s, \rho)] \qquad (90)$$

the corrections to the energy and momentum density, ΔE and $\Delta\Gamma$, are, again, the only vanishing corrections when $\Phi^* = \Psi^*$.

10. FINAL REMARKS

The extended thermodynamics which corresponds to the microscopic theory of de Broglie (finite θ_a) and the conventional irreversible theory work in a very similar manner when imbedded in the least action scheme; the differences are mainly caused by the form of L. Therefore, whatever the final assessment of the de Broglie theory will be, the variational and thermodynamic schemes proposed here should survive. Eqs. (88) and (89) constitute a general structure of the balance laws for the energy and momentum in arbitrary fluids not obeying the local equilibrium assumption, immersed in a scalar gravitational field ψ (with neglected electric and elastic effects). The phase corrections Δ pertaining to an arbitrary internal kinetic potential involving phases are described by appropriate expressions in eqs. (82) through (86), containing L^{int}.

An importance of the linear structure of the internal kinetic potential, L^{int}, with respect to all the phases should be pointed out. Thanks to this linearity, the invariance of the action A^{int} and the kinetic potential L^{int} with respect to the simultaneous shift of all the phases and the related conservation laws of mass constitute both a necessary and sufficient condition for the invariance of any correction Δ in the energy-momentum tensor. This occurs because the linear structure is an exclusive one which remains unchanged under operations performed in eqs. (82) through (86). Thus, the total mass conservation exclusively preserves the invariance of the phase corrections to the energy-momentum tensor with respect to the change in the reference frame of the phases.

Moreover, a direct relation of the phase variables ϕ_a to the quantum phases of the species should be underlined. The latter (which are dimensionless) equal simply $m_a\phi_a/h$, where h is the Planck constant. They could appear in our variational formalism if the mass densities ρ_a and fluxes $\rho_a \mathbf{u}_a$ were expressed in the units of action, i. e., the action densities $h\rho_a/m_a$ and action fluxes $h\rho_a \mathbf{u}_a/m_a$ were used. Then, the related four-divergences and the mass sources $h\sigma_a/m_a$ would have the units of the action sources, expressed as the actions created or annihilated per unit time and volume. The passage to the dimensionless thermal phases would require the use of the entropy densities and fluxes in the units of action, $h\theta_a \rho_{sa}/m_a = h\rho_{sa}/k$ and $h\theta_a \rho_{sa} \mathbf{u}_{sa}/m_a = h\rho_{sa} \mathbf{u}_{sa}/k$, as well as the related entropy sources $h\theta_a \sigma_{sa}/m_a = h\sigma_{sa}/k$. While the dimensionless phase of the matter plays the well known fundamental role in the hydrodynamic representation of the Schrödinger equation (Nonnenmacher 1984), the role of the thermal phase has not been recognized so far in the quantum context, and it is one of the purposes of this work to show that, macroscopically, this role is not less fundamental then the role of the matter phase. Perhaps the nonstandard quantum theories (de Brooglie 1976), postulating a coexistence of waves and particles, can contribute to a microscopic interpretation of η rather than the standard quantum theory. In a future research we will proceed to establish a relativistic version of the present theory.

The giving up of the "conventional postulate", according to which the irreversibility (and in particular the chemical reactions and the entropy generation) has no influence on the analytical form of the conservation laws of the energy and momentum, is equivalent to the questioning of the general applicability of the standard bilinear expression for the entropy production (De Groot and Mazur 1984), (Glansdorff and Prigogine 1976), (Fitts 1974), (Baranowski 1974) and others. The standard bilinear expression is, in fact, always derived under a tacit assumption that the sources of the mass and the entropy have no effect on the analytical form of the balances of the total energy and momentum. In this work, we have shown that an "irreversible" form of these balances may only correspond to but not coincide with those derived for the sourceless ("reversible") constraints. In real irreversible processes, especially far from equilibrium, the balances of the energy and momentum may contain the (invariant) phase corrections, resulting from an "internal"part of the kinetic potential, which defines the sources. Although, for a Joule-law type dissipation, and the equality of the Onsagerian potentials, $\Phi^* = \Psi^*$, the corrections to the total energy density, E, and the momentum density, Γ, vanish, those to the total energy and momentum fluxes, **Q** and **G**, do not, and the conventional form of the balance laws is affected by the irreversibility.

More investigation is needed in this area, but in the light of the results of this work, the frequent questioning of the applicability of an least action principle to the irreversible phenomena does not seem to be justified. Namely, it has been shown that the stationarity conditions of a *distributed* action A with respect to the densities of the mass and entropy have a form *generalizing* the classical Hamilton-Jacobi equations in the following sense; the actions of the individual fluid elements, or the phases ϕ_a, and η_a, and not only their derivatives, can explicitly appear in these conditions. Since such explicit presence of the individual actions always makes the related system of the ordinary differential equations (characteristics) a non-Hamiltonian, [the Hamiltonian system corresponding to the case when the Hamilton-Jacobi equation doesn't contain the particle action explicitly (Courant and Hilbert

1949)], it is clear that the trajectories of the individual particles of the fluid obtained from such distributed action can exhibit dissipative properties.

APPENDIX 1: DERIVATION OF THE EQUATIONS OF MOTION

Consider the stationarity conditions of the functional:

$$A = \int \{L(\mathbf{u}, \rho, \Phi, \mathbf{r}, \mathbf{x}, t) - \sum_{k=1}^{2n} [\rho_k(\frac{\partial \Phi_k}{\partial t} + \mathbf{u}_k.\nabla\Phi_k) - (\lambda\rho)_k(\frac{\partial \alpha_k}{\partial t} + \mathbf{u}_k.\nabla\alpha_k)]\}dVdt \qquad \text{(A1-1)}$$

where the subscript k refers to the k-th component of the following 2n dimensional rows

$$\mathbf{u} \equiv (\mathbf{u}_{s1}, \mathbf{u}_1, ...\mathbf{u}_{sa}, \mathbf{u}_a...\mathbf{u}_{sn}, \mathbf{u}_n)$$
$$\rho \equiv (\rho_{s1}, \rho_1, ...\rho_{sa}, \rho_a...\rho_{sn}, \rho_n) \qquad \text{(A1-2)}$$
$$\Phi \equiv (\eta_{s1}, \phi_1, ...\eta_{sa}, \phi_a...\eta_{sn}, \phi_n)$$

and

$$\alpha \equiv (\alpha_{s1}, \alpha_1, ...\alpha_{sa}, \alpha_a...\alpha_{sn}, \alpha_n)$$
$$\lambda \equiv (\lambda_{s1}, \lambda_1, ...\lambda_{sa}, \lambda_a...\lambda_{sn}, \lambda_n) \qquad \text{(A1-3)}$$
$$(\lambda\rho) \equiv (\lambda_{s1}\rho_{s1}, \lambda_1\rho_1, ...\lambda_{sa}\rho_{sa}, \lambda_a\rho_a...\lambda_{sn}\rho_{sn}, \lambda_n\rho_n)$$

Due to an uncoupled structure of the derivatives in (A1-1), the stationarity conditions of (A1-1) for any component or any partial entropy have the same form, independent of k. Therefore the subscript k is neglected in all the formulae below. The stationarity conditions for any k are

$$\delta\alpha: \qquad \frac{\partial(\lambda\rho)}{\partial t} + \nabla.[(\lambda\rho)\mathbf{u}] = 0 \qquad \text{(A1-4)}$$

$$\delta\mathbf{u}: \qquad \frac{\partial L}{\rho\partial \mathbf{u}} + \lambda\nabla\alpha - \nabla\Phi = 0 \qquad \text{(A1-5)}$$

$$\delta\rho: \qquad \frac{\partial L}{\partial \rho} + \lambda(\frac{\partial \alpha}{\partial t} + \mathbf{u}.\nabla\alpha) - \frac{\partial \Phi}{\partial t} - \mathbf{u}.\nabla\Phi = 0 \qquad \text{(A1-6)}$$

(We specify only these Euler equations which are necessary in derivations). Subtracting the gradient of (A1-6) from the partial time derivative of (A5) one obtains

$$\frac{\partial}{\partial t}(\frac{\partial L}{\rho\partial \mathbf{u}}) - \nabla.\frac{\partial L}{\partial \rho} + \frac{\partial \lambda}{\partial t}\nabla\alpha - \frac{\partial \alpha}{\partial t}\nabla\lambda - \nabla(\lambda\mathbf{u}.\nabla\alpha - \mathbf{u}.\nabla\Phi) = 0 \qquad \text{(A1-7)}$$

Using the mass and entropy balances (the stationarity conditions of A with respect to Φ) in eq. (A1-4) yields

$$\lambda\,\sigma + \rho\left(\frac{\partial\lambda}{\partial t} + \mathbf{u}.\nabla\lambda\right) = 0 \tag{A1-8}$$

where σ is the negative derivative of L with respect to Φ. Eliminating the partial time derivative of λ from (A1-7) by using (A1-8) and that of α with the help of the identity condition, $d\alpha/dt=0$, and also using the scalar product of (A1-5) and $\mathbf{u}$ in the last term on the left hand side of (A1-7), yields

$$\frac{\partial}{\partial t}\left(\frac{\partial L}{\rho\partial\mathbf{u}}\right) - \nabla.\frac{\partial L}{\partial\rho} - (\mathbf{u}.\nabla\lambda)\nabla\alpha - (\lambda\sigma/\rho)\nabla\alpha + (\mathbf{u}.\nabla\alpha)\nabla\lambda + \nabla\left(\mathbf{u}.\frac{\partial L}{\rho\partial\mathbf{u}}\right) = 0 \tag{A1-9}$$

From the vector calculus, for the arbitrary scalars α and λ and the vector $\mathbf{u}$ the following holds

$$(\mathbf{u}.\nabla\alpha)\nabla\lambda - (\mathbf{u}.\nabla\lambda)\nabla\alpha = \mathbf{u}\times(\nabla\lambda\times\nabla\alpha) \tag{A1-10}$$

To compute the curl vector of the gradients of λ and α in the above equation, the curl of eq. (A1-5) is taken which yields

$$\nabla\times\left(\frac{\partial L}{\rho\partial\mathbf{u}}\right) = -\nabla\times(\lambda\nabla\alpha) = -\,\nabla\lambda\times\nabla\alpha \tag{A1-11}$$

The term containing source σ in eq. (A1-9) can be transformed with the help of eq. (A1-5) which gives

$$-\,\lambda\sigma\nabla\alpha/\rho = \frac{\sigma}{\rho}\left(\frac{\partial L}{\rho\partial\mathbf{u}} - \nabla\Phi\right) \tag{A1-12}$$

Using (A1-10) through (A1-12) in (A1-9) yields an equation of motion

$$\frac{\partial}{\partial t}\left(\frac{\partial L}{\rho\partial\mathbf{u}}\right) + \nabla\left(\mathbf{u}.\frac{\partial L}{\rho\partial\mathbf{u}} - \frac{\partial L}{\partial\rho}\right) - \mathbf{u}\times\nabla\left(\times\frac{\partial L}{\rho\partial\mathbf{u}}\right) + \frac{\sigma}{\rho}\left(\frac{\partial L}{\rho\partial\mathbf{u}} - \nabla\Phi\right) = 0 \tag{A1-13}$$

where

$$\sigma = -\,\frac{\partial L}{\partial\Phi} \tag{A1-14}$$

is the vector of sources. Equation of motion of an irreversible process contains explicitly the phases Φ. Due to the uncoupled structure of the derivative terms in (A1-1), which is the consequence of the choice of mass and entropies as the most natural variables of the problem, the expression (A1-13) has the same form for every component and every partial entropy. The set of 2n equations (A1-13) taken for $\mathbf{u}_{s1}, \rho_{s1}, \eta_{s1}, ...\mathbf{u}_n, \rho_n, \phi_n$ is coupled due to the presence of the common kinetic potential L in (A13). The single thermal phase approximation involves only one thermal equation, for the total entropy.

APPENDIX 2: RELATION TO HYPOTHESIS OF A THERMAL MASS

The canonical temperatures $T_a{}^-$ of the theory with the heat inertia differ slightly in their properties and definition from the conventional temperatures T_a^{class} even at equilibrium attained in a resting system without external forces. In fact, they should be rather called, more appropriately, thermal potentials. As pointed out in the part I, in the presence of the external fields the canonical temperatures are constant throughout the system only if they incorporate the effect of that field. However, they may also be different for the different species due to the effect of the chemical potential μ contained in them, as shown below. The reason for this difference is that the related Gibbs equation at equilibrium deals with the "bare" densities of mass ρ_{ma} and not for the measurable effective densities, $\rho_a = \rho_{ma} + \theta\rho_{sa}$. (Note that in the text and in the Part I the subscript m was for brevity neglected but the use of bare ρ_{ma} rather than ρ_a is obvious in that context). But the conventional temperatures of the classical Gibbs equation T_a^{class} pertain to the observed densities ρ_a rather than to the bare densities ρ_{ma}. It is instructive to show how the canonical and classical intensities are linked and how one can obtain the equality of the classical temperatures T_a^{class} at equilibrium.

Consider, for instance, the differential of the *relativistic* internal energy density of a one-component system (subscript a neglected) in terms of the canonical intensities and the density of bare mass ρ_m used in the text.

$$d(\rho_e)_{rel} = \frac{\partial(\rho_e)_{rel}}{\partial\rho_s} d\,\rho_s + \frac{\partial(\rho_e)_{rel}}{\partial\rho} d\rho_m = T^-_{rel}\, d\,\rho_s + \mu^-_{rel}\, d\rho_m \tag{A2-1}$$

Transformation to the classical set of variables using the observed mass density ρ and the entropy density ρ_s yields

$$d(\rho_e)_{rel} = T^-_{rel}\, d\,\rho_s + \mu^-_{rel}\, d(\rho - \theta\rho_s) = (T^-_{rel} - \theta\mu^-_{rel})d\,\rho_s + \mu^-_{rel}\, d\rho \tag{A2-2}$$

This links the canonical intensities with those of the conventional description. Comparison of (A2-2) with the classical Gibbs equation (operating with ρ_s and ρ)

$$d(\rho_e)_{rel} = T^{class}_{rel}\, d\,\rho_s + \mu^{class}_{rel}\, d\rho = T^{class} d\,\rho_s + (\mu^{class} + c^2)d\rho \tag{A2-3}$$

yields

$$\mu^-_{rel} = \mu^{class} + c^2 \tag{A2-4}$$

$$T^-_{rel} = T^{class} + \theta\mu^-_{rel} = T^{class} + \theta\mu^{class} + \theta c^2 \tag{A2-5}$$

In Part I the *nonrelativistic* quantities T^- and μ^- are used. In terms of their relativistic counterparts they are defined as

$$\mu^- \equiv \mu^-_{rel} - c^2 \tag{A2-6}$$

$$T^- \equiv T^-_{rel} - \theta c^2 \quad \text{(Part I)} \tag{A2-7}$$

Comparison of (A2-4) and (A2-6) shows that

$$\mu^- = \mu^{class} \tag{A2-8}$$

i.e the nonrelativistic canonical chemical potential coincides with the classical one whereas

$$T^- = T^{class} + \theta\mu^{class} \tag{A2-9}$$

i.e. the nonrelativistic canonical temperature differs from the classical one due to the contribution of the nonrelativistic chemical potential. Of course, when $\theta=0$ both temperatures are identical. However, in order to achieve not only a correspondence but a complete consistency with the classical thermostatics it is better to redefine T^-, i.e. to replace the definition (A2-7) of Part I by the following *modified* definition

$$T^- \equiv T^-_{rel} - \theta c^2 - \theta\mu^{class} \tag{A2-10}$$

Then, instead of eq. (A2-9), the (redefined) nonrelativistic canonical T^- obeys the equation

$$T^- = T^{class} \tag{A2-11}$$

Thus, we have shown the consistency of the canonical temperatures T^- and the canonical chemical potentials μ^- with the classical thermodynamics (applied with the hypothesis that part of the observed mass or density of a macroscopic medium is of thermal origin) provided that eqs. (A2-6) and (A2-10) hold.

The consistency conditions, eqs. (A2-8) and (A2-11), can be substituted into eqs. (A2-6) and (A2-10) which yields the useful relations between the relativistic canonical temperatures and chemical potentials and the classical *statistical* intensities

$$\mu^-_{rel} \equiv \mu^{class} + c^2 \tag{A2-12}$$

$$T^-_{rel} \equiv T^{class} + \theta\mu^{class} + \theta c^2 \tag{A2-13}$$

For the external scalar field ψ their generalization is

$$\tilde{\mu}_{rel} \equiv \mu^{class} + \psi + c^2 \tag{A2-12a}$$

$$\tilde{T}^-_{rel} \equiv T^{class} + \theta\mu^{class} + \theta\psi + \theta c^2 \tag{A2-13a}$$

As shown in Part I, the *relativistic* canonical intensities obey the Einstein formula for the relativistic temperature transformation. Thus, the motion of the matter and entropy with the velocities $\mathbf{u}_m$ and $\mathbf{u}_s$, respectively, obeys the transformations

$$\widetilde{\mu}_{rel}(u_m) \equiv (\mu^{class} + \psi + c^2)\sqrt{1 - u_m^2/c^2} \cong (\mu^{class} + \psi + c^2)(1 - u_m^2/2c^2) \cong \mu^{class} - u_m^2/2 + \psi + c^2 \tag{A2-14}$$

$$\begin{aligned}\widetilde{T}_{rel}(u_s) &\equiv (T^{class} + \theta\mu^{class} + \theta\psi + \theta c^2)\sqrt{1 - u_s^2/c^2} \cong \\ &(T^{class} + \theta\mu^{class} + \theta\psi + \theta c^2)(1 - u_s^2/2c^2) \\ &\cong T^{class} + \theta\mu^{class} - \theta u_s^2/2 + \theta\psi + \theta c^2\end{aligned} \tag{A2-15}$$

The kinetic potential L^1, eq. (1) of Part I and eq. (2) of the text, can be verified. Since the negative of L^1 is the rest energy, i.e., $- L^1 = E^0(\rho_s{}^0, \rho_m{}^0)$ expressed in terms of the laboratory densities, ρ_s and ρ_m, and the velocities, $\mathbf{u}_s$ and $\mathbf{u}_m$, one has

$$\begin{aligned}L^1 &= - E^0 = -\{[\rho_m\sqrt{(1-u_m^2/c^2)} + \theta\rho_s\sqrt{(1-u_s^2/c^2)}]c^2 + \rho\psi + \rho_e^{class}\} \\ &\cong - \{[\rho_m(1-u_m^2/2c^2) + \theta\rho_s(1-u_s^2/2c^2)]c^2 + \rho\psi + \rho_e^{class}\} \\ &\cong \rho_m u_m^2/2 + \theta\rho_s u_s^2/2) - \rho_e^{class} - \rho(\psi + c^2)\end{aligned} \tag{A2-16}$$

where $\rho_e{}^{class}$ is the classical statistical internal energy. Moreover, as eq. (95) of Part I indicates, the same result is obtained by substituting eqs. (A2-12a) and (A2-13a) into the Legendre transform of the static pressure

$$\begin{aligned}- L^1(\rho_s, \rho_m, \mathbf{u}_s, \mathbf{u}_m) &= \frac{\partial P}{\partial \widetilde{T}_{rel}}\widetilde{T}_{rel} + \frac{\partial P}{\partial \widetilde{\mu}_{rel}}\widetilde{\mu}_{rel} - P(\widetilde{T}_{rel}, \widetilde{\mu}_{rel}, \mathbf{u}_s, \mathbf{u}_m) = \\ \rho_s\widetilde{T}_{rel}(\rho_s, \rho_m, \mathbf{u}_s, \mathbf{u}_m) &+ \rho_m\widetilde{\mu}_{rel}(\rho_s, \rho_m, \mathbf{u}_s, \mathbf{u}_m) - P(\rho_s, \rho_m, \mathbf{u}_s, \mathbf{u}_m) \\ &= T^{class}\rho_s + (\mu^{class} + \psi + c^2)(\rho_m + \theta\rho_s) - P - \rho_m u_m^2/2 - \theta\rho_s u_s^2/2 \\ &= \rho_e^{class} + \rho(\psi + c^2) - \rho_m u_m^2/2 - \theta\rho_s u_s^2/2\end{aligned} \tag{A2-17}$$

However, the power of the theory is not restricted to the above sourceless case; when the phases ϕ and η are taken into account, the analogous transformations hold for the generalized kinetic potential L. For the general L the canonical intensities are more comlicated because the role of sources. Since the sources play a role only out of equilibrium, the theory is always consistent with statics (where $\mathbf{u}_m = \mathbf{u}_s$), the role of the heat inertia manifesting only at the kinetic (nonequilibrium) level.

Substituting in place of the (nonrelativistic) temperatures T_a of Part I the sum classical quantities $T_a{}^{class}$ and $\theta_a\mu_a{}^{class}$ (c.f. eqs. A2-7 and A2-13), replacing the symbols ρ_a by more appropriate ρ_{ma} (bare quantities) and then returning to the classical (observable) quantities, exact formulae of the classical irreversible theory can be recovered which deal with the fluxes of the observable and not bare densities. Consider, for example, the energy flux formula, eq. (54) of Part I in order to show that the thermal part of it can be broken down to the well-known expression ($\mu_a = \mu_a^{class}$ remains unchanged, and Π term is the viscous work)

$$\mathbf{Q} = \rho\mathbf{u}\frac{u^2}{2} + \sum_a \left[\frac{\rho_a \mathbf{u}_a^2}{2} + \frac{\theta_a \rho_{sa} \mathbf{u}_{sa}^2}{2}\right] + \Pi.\mathbf{u} + \mathbf{B} \qquad \text{(A2-18)}$$

where $\mathbf{B}$ is the part requiring the transformation to the classical form

$$\begin{aligned}\mathbf{B} &= \sum_a [\rho_{ma}\mathbf{u}_{ma}(\mu_a + \psi_a) + \rho_{sa}\mathbf{u}_{sa}(T_a^{class} + \theta_a\mu_a + \theta_a\psi_a)] \\ &= \sum_a [(\rho_{ma}\mathbf{u}_{ma} + \theta_a\rho_{sa}\mathbf{u}_{sa})(\mu_a + \psi_a) + \rho_{sa}\mathbf{u}_{sa}T_a^{class}] \\ &\sum_a [(\rho_a\mathbf{u}_a)^{class}(\mu_a + \psi_a) + \rho_{sa}\mathbf{u}_{sa}T_a^{class}] = \\ &\sum_a [\mathbf{j}_a^{class}(\mu_a + \psi_a) + \mathbf{j}_{sa}T_a^{class}] + \rho\mathbf{u}(h^{class} + \psi) \\ &= \mathbf{q}^{class} + \rho\mathbf{u}(h^{class} + \psi)\end{aligned} \qquad \text{(A2-19)}$$

where the classically defined diffusion fluxes of the mass and the entropy ($\mathbf{j}_s$) and heat ($\mathbf{q}$) have been introduced. Substituting $\mathbf{B}$, eq. (A2-19), into eq. (A2-18), the well known formula for the energy flux is obtained, corrected by the term describing the transfer of the kinetic energy of diffusion

$$\mathbf{Q} = \rho\mathbf{u}\left(\frac{u^2}{2} + h^{class} + \psi\right) + \sum_a \left[\frac{\rho_a \mathbf{u}_a^2}{2} + \frac{\theta_a \rho_{sa} \mathbf{u}_{sa}^2}{2}\right] + \Pi.\mathbf{u} + \mathbf{q}^{class} \qquad \text{(A2-20)}$$

The above analysis also shows that, in the classical theory, the entropy behaves formally like a weightless fluid, whereas in the extended approaches (Grad's theory, extended thermodynamics, de Broglie macro-thermodynamics and the thermal mass hypothesis analyzed here) it has an apparent inertia.

ACKNOWLEDGEMENTS

Helpful discussion with Profs. J. Nulton, B. Kupershmidt, R.E. Nettleton and E. Moszczak is gratefully acknowledged. The author acknowledges a partial support of the U.S. Department of Energy and the hospitality of Prof. R. S. Berry, The University of Chicago, and Prof. P. Salamon, San Diego State University, where a part of this work was done.

REFERENCES

Abrikosov, A.A., L.P. Gorkov, and I.E. Dzialoshinski. 1975. Methods of Quantum Field Theory in Statistical Physics. New York: Dover.

Baranowski, B. 1974. Nonequilibrium Thermodynamics in Physical Chemistry. Warsaw: PWN.

de Broglie, L.V. 1962. New presentation of the thermodynamics of an isolated particle. Compt. Rend. 255: 807-810.

de Broglie, L.V. 1964. La thermodynamique "cache des particules. Ann. Inst. Henri Poincare1: 1-19.
de Broglie, L.V. 1967. A new interpretation concerning the coexistence of waves and particles. In Perspectives in Quantum Theory, eds. W. Yourgrau and A. van der Merwe. New York: Dover.
Courant, R., and D. Hilbert. 1962. Methods of Mathematical Physics, vol. II: Partial Differential Equations. New York: Interscience Publishers.
Fitts, D.D.1962. Nonequilibrium Thermodynamics. New York: Mc Graw-Hill.
Gelfand, J.M., and S.W. Fomin. 1970. Variational Calculus. Warsaw: PWN.
Hirschfelder, J.O., C.F. Curtiss, and R.B. Bird. 1963. Molecular Theory of Gases and Liquids. New York: J. Wiley.
Huang, K. 1967. Statistical Mechanics. New York: J. Wiley.
Jou, D., J. Casas-Vazquez, and G. Lebon. 1989. Extended Irreversible Thermodynamics. Rep. Progr. Phys. 51: 1105-1115.
Kupershmidt B.A. 1990. Hamiltonian Formalism for Reversible Non-Equilibrium Fluids with Heat Flow. J. Phys. A: Math. Gen. 23: L529-L532.
Landau, L., and E. Lifshitz. 1974. Field Theory. London: Pergamon.
Landau, L., and E. Lifshitz. 1974. Mechanics. London: Pergamon.
Landau, L., and E. Lifshitz. 1974. Mechanics of Continua. London: Pergamon.
Lavenda, B.H. 1985. Nonequilibrium Statistical Thermodynamics. New York: Wiley Interscience.
Nettleton, R.E. 1986. Lagrangian Formulation of Non-linear Extended Irreversible Thermodynamics. J. Phys. A: Math. Gen. 19: L295-L297.
Nonnenmacher, T.F. 1986. Functional Poisson brackets for nonlinear fluid mechanics equations. In Recent Developments in Nonequilibrium Thermodynamics: Fluids and Related Topics, eds. J. Casas-Vazquez, D. Jou and J. M. Rubi, 149 -174. Berlin: Springer.
Sieniutycz, S. 1986. On the relativistic temperature transformations and the related energy transport problem. Intern. J. Heat Mass Transfer 29: 651-654.
Sieniutycz, S. 1987. From a least action principle to mass action law and extended affinity. Chem. Engng Sci. 42: 2697-2711.
Sieniutycz, S. 1988. Hamiltonian energy-momentum tensor in extended thermodynamics of one-component fluid. Inz. Chem. Proc. 4: 839-861.
Sieniutycz, S., and R.S. Berry. 1989. Conservation laws from Hamilton's principle for nonlocal thermodynamic equilibrium fluids with heat flow. Phys. Rev. A. 40: 348-361.
Sieniutycz, S. 1989a. Energy-momentum tensor and conservation laws for extended multicomponent fluids with the transport of heat, mass and electricity. Unpublished research report of Inst. Chem. Engng. Warsaw: WTU.
Sieniutycz, S., and R.S. Berry. 1991. Field thermodynamic potentials and geometric thermodynamics with heat transfer. Phys. Rev. A, in print.
Sieniutycz, S. 1990a. Wave equations of heat and mass transfer. In Advances in Thermodynamics 6, eds. S. Sieniutycz and P. Salamon. New York: Taylor and Francis.
Sieniutycz, S. 1990b. Thermal momentum, heat inertia and a macroscopic extension of de Broglie microthermodynamics I. The multicomponent fluid with sourceless continuity constraints. In Advances in Thermodynamics 3, eds. S. Sieniutycz and P. Salamon. New York: Taylor and Francis.

Wei, Ch.C. 1958. Relativistic hydrodynamics for a charged nonviscous fluid. Phys. Rev. 113: 1414.

Yourgrau, W., and S. Mandelstam. 1968. Variational Principles in Dynamics and Quantum Mechanics, 3rd ed. Philadelphia: W.B. Saunders.

Glossary of Symbols

$\mathbf{A}$ vector of thermodynamic conjugates of internal variables
A action, availability, affinity, extremized objective
A^{AB} energy-momentum tensor
a_i state variables
$\mathbf{B}$ magnetic induction
B_{ij} strain rate
b_i state variables, coefficients
$\mathbf{C}$ conformation tensor, inelastic strain rate
$\mathbf{C}_i$ peculiar molecular speed
C heat capacity, configuration , correlatioin function
c, c_0 light speed and propagation speed, respectively
c_i molar concentration of i-th species
$\mathbf{D}$ strain-rate tensor
D diffusion coefficient, dissipation function
d thickness, derivative
$\mathbf{E}$ electric field vector, infinitesimal strain tensor
E energy, energy density
$\mathbf{e}$ unit vector
e unit energy
$\mathbf{F}, \mathbf{f}$ external forces
F, f free energy, unit free energy
f distribution function, magnetic loading function
f_i body forces, noise
F_i conjugates of state variables, α_i
F, F_{ij}, F_{ijk} fields of extended theory
$\mathbf{G}$ relaxation function, elastic modulus
$\mathbf{G}'$ relaxation kernel
G Gibbs free energy, shear modulus
G_{jk} energy-momentum tensor
$\mathbf{g}$ gradient vector
g_{jk} non-convective part of energy tensor
$\mathbf{H}$ magnetic field
H hamiltonian function, enthalpy, Boltzmann's H function, Hubble constant
h magnetic hardening coefficient, entropy density, heat supply, reduced Hubble parameter
h^A relativistic four-vector of entropy
h Planck constant
I invariant, integral
I_v specific intensity
$\mathbf{J}$ density of total mass flux

$\mathbf{J}_a, \mathbf{j}_a$ total and diffusional mass flux of a-th species, respectively
$\mathbf{J}_e, \mathbf{J}_s$ electric flux and entropy flux, respectively
$\mathbf{J}_{sa}, \mathbf{j}_{sa}$ total and diffusional entropy flux of a-th species, respectively
J chemical flux
K momentum density, extra term of entropy flux
$\mathbf{K}_a$, $\mathbf{K}_{sa}$ wave (propagation) vectors, momentum densities
K chemical equilibrium constant, memory kernel
k Boltzmann constant
k rate constant, thermal conductivity
L conductivity matrix
L_{ik} Onsager's coefficients
L lagrangian, kinetic potential, length, phenomenological coefficient
l characteristic length, phenomenological coefficient of reaction rate expression
M molar mass, mass property, magnetization
m unit direction vector
m mass of microobject
N mole number, power output
Nu Nusselt number
n dimensionality of state vector, number density
P pressure tensor, polarization vector
P pressure, power, projection operator, chemical rate function
P_{ijk} production terms of extended theory
Pr Probability
p_i probabilities
p_v viscous pressure
p_{ij} pressure tensor, non-conservative part of momentum tensor
Q energy flux density
Q generalized coordinate, heat, electric charge
q heat flux density
q_i electric charges
R resistance matrix, directed distance, vector of extra variables in Gibbs equation
R universal gas constant, resistance, Rayleigh dissipation function
r radius, reaction rate
S entropy flux, Piola-Kirchoff stress tensor
S, s entropy, unit entropy
S_{ijk} source terms of extended theory
s_{ij} rate of deformation
T Cauchy stress tensor
T absolute temperature
t time, coordinate time

t_{ab} Cauchy stress tensor
U, u internal energy, unit internal energy
u velocity vector
$\mathbf{u}_a$, $\mathbf{u}_{sa}$ transport velocities for mass and entropy of a-th component, respectively
u_i, u_{ij} deformation vector and deformation tensor, respectively
V volume, scalar potential
V_{ij} symmetric tensor of velocity gradient
v speed
v velocity vector
W probability density
w work
X thermodynamic force
X vector of material coordinates
X_i macroscopic extensities, state variables, extra variables
x state vector, radius vector
x mole fraction
Y_i thermodynamic conjugates of variables X_i
y thermodynamic conjugate of state x
Z impedance matrix
Z partition function
Z_i state variables
z_α charge per unit mass of α-th component

α absorption coefficient, thermal diffusivity, fluid particle label, fugacity
α_i state variables
β_i nonequilibrium variable vector, vector of Lagrangian multipliers, internal variable vector
γ relativistic factor, coefficient of Cattaneo equation, shear rate
δ thickness, effective diameter
ε electric field, internal energy function
ε_{ij} strain tensor
Φ dissipation function, phase vector, Gibbs free energy, thermodynamic conjugate of flux J
ϕ distribution function correction, material action variable, dissipation function, electrostatic potential, extra variable
ϕ_i i-th component of entropy flux
η thermal action variable, efficiency, extended entropy density, viscosity
Θ specific energy source, conjugate variable

θ inertial coefficient, phase function, nonequilibrium temperature

κ diffusional momentum variable, bulk relaxation function, canonical part of conformation, conductivities

κ_{ij} conductivity tensor

Λ Lagrangian density of fluid

Λ_i Lagrangian multipliers

λ Lagrangian multiplier, thermal conductivity, wave length, eigenvalue

μ chemical potential

ν frequency, stoichiometric coefficient, dynamic viscosity

ν_α charge supply

Π specific polarization, Peltier coefficient

Π_{ij} nonequilibrium pressure tensor

ρ mass density

ρ_{ea} energy density of a-th component

ρ_a, ρ_{sa} density of mass and entropy of a-th component, respectively

ρ_s total entropy density

σ entropy source, Stefan-Boltzmann constant, viscous pressure

σ_a, σ_{sa} sources of mass and entropy of a-th component, respectively

σ_e electric conductivity

σ_{ij} stress tensor components

τ characteristic time, relaxation time, proper time, transition time

τ_{ij} stress components

Ω grand potential, statistical weight, solid angle, vorticity

$d\Omega$ element of solid angle

ω angular frequency, vorticity tensor

ψ free energy density, external scalar field

ξ_i internal variables, progress variables

Index